CHEMICAL ECOLOGY

Chemical Ecology of Insects 2

Ring T. Cardé
Department of Entomology, University of Massachusetts, Amherst, MA

William J. Bell
Department of Entomology, University of Kansas, Amherst, MA

New York • Albany • Bonn • Boston • Cincinnati • Detriot • London • Madrid • Melbourne
Mexico City • Pacific Grove • Paris • San Francisco • Singapore • Tokyo • Toronto • Washington

Cover design: Trudi Gershenov
Cover illustration: Joseph Diefenbach

Printed in the United States of America

For more information, contact:

Chapman & Hall
115 Fifth Avenue
New York, NY 10003

Thomas Nelson Australia
102 Dodds Street
South Melbourne, 3205
Victoria, Austrailia

Nelson Canada
1120 Birchmount Road
Scarborough, Ontario
Canada M1K 5G4

International Thomson Editores
Campos Eliseos 385, Piso 7
Col. Polanco
11560 Mexico D. F. Mexico

Chapman & Hall
2-6 Boundary Row
London SE1 8HN
England

Chapman & Hall GmbH
Postfach 100 263
D-69442 Weinheim
Germany

International Thomson Publishing Asia
221 Henderson Road #05-10
Henderson Building
Singapore 0315

International Thomson Publishing - Japan
Hirakawacho-cho Kyowa Building, 3F
1-2-1 Hirakawacho-cho
Chiyoda-ku, 102 Tokyo
Japan

1 2 3 4 5 6 7 8 9 10 XXX 01 00 99 98 97 96 95

Library of Congress Cataloging-in-Publication Data
(Revised for vol. 2)

Chemical ecology of insects 2.

Vol. [2] published by : New York : Chapman & Hall.
Includes bibliographies references and indexes.
1. Insects--Ecophysiology. 2. Chemical ecology. 3. Chemical senses. I. Bell, William J. 2. Cardé, Ring T.
QL495.C47 1994 595.7 ' 05 83-20212
ISBN 0-412-03951-6 (v. 2 : HB)
ISBN 0-412-03961-3 (v. 2 : PB)

British Library Cataloguing in Publication Data available

Please send your order for this or any other Chapman & Hall book to
Chapman & Hall, 29 West 35th Street, New York, NY 10001, Attn: Customer Service Department.
You many also call our Order Department at 1-212-244-3336 or fax you purchase order to 1-800-248-4724.

For a complete listing of Chapman & Hall's titles, send your request to
Chapman & Hall, Dept. BC, 115 Fifth Avenue, New York, NY 10003.

Contents

Preface vii
Contributors ix

Chemoreception and Integration

1. Behavior and Integration 3
   Marion O. Harris and Stephen P. Foster
2. Effects of Experience on Host-Plant Selection 47
   Elizabeth A. Bernays
3. Parasitoid Foraging and Learning 65
   Louise E. M. Vet, W. Joe Lewis, and Ring T. Cardé

Orientation Mechanisms

4. The Role of Chemo-Orientation in Search Behavior 105
   William J. Bell, Larry R. Kipp and Robert D. Collins

Plant-Insect Interactions

5. Host-Tree Chemistry Affecting Colonization of Bark Beetles 154
   John A. Byers
6. Host-Plant Choice in *Pieris* Butterflies 214
   F. S. Chew and J. A. A. Renwick

Insect-Insect Interactions

7. Trail and Territorial Communication in Social Insects 241
   James F. A. Traniello and Simon K. Robson
8. The Chemical Basis for Nest-Mate Recognition and Mate Discrimination in Social Insects 287
   Brian H. Smith and Michael D. Breed

9. Chemical Communication in the True Bugs and Parasitoid Exploitation 318
Jeffrey R. Aldrich
10. Propaganda, Crypsis, and Slave-making 364
Ralph W. Howard and Roger D. Akre

Preface

The first volume of *Chemical Ecology of Insects* was published ten years ago. Then this field was characterized by rapidly expanding knowledge of the chemical structures used by insects in a variety of tasks: communication, defense, and finding and identifying resources. In parallel, the plant and animal resources exploited by insects were documented to invest in chemicals to manipulate insects' behavior and physiology. The ecological complexities of these interactions, however, were often unapparent. For example, although odor plumes were known to have considerable fine-scale structure, the significance of such wispiness to the orientation maneuvers of flying insects was not suspected. Similarly, although a few early studies implicated learning in the recognition by wasp parasitoids of chemical cues from their prospective hosts, the widespread nature of this phenomenon remained to be uncovered. These examples illustrate the point that our understanding of the ecology of interactions mediated by these chemicals often lagged behind our ability to characterize the chemicals themselves.

The current volume brings together a narrower selection of reviews than did its predecessor. Our intent has not been to span the entire field of insect chemical ecology—given the growth of knowledge in this area, such a task now would of necessity cover topics superficially. Instead, we have selected topics that offer a perspective on some of the most interesting advances in insect chemical ecology. We encouraged the authors to provide proximate examinations of the ways in which chemical cues modify ecological interactions and to speculate when feasible on the ultimate selective forces that maintain and may have molded these relationships.

Chapter 1 offers a compelling case for abandoning a "chemocentric" approach to understanding how insects react to chemical cues, and instead for developing assays that account for the integration of sensory modalities in behavior output. Chemoreception and integration also are considered in two chapters of the widespread role of prior experience in host finding and acceptance. Chapter 2 reviews the effects of epxerience in host acceptance by herbivores, and Chapter 3 examines how experience modifies searching for and accepting of a host by parasitoids. Chapter 4 considers the topic of detecting and then locating chemical resources, offering new perspectives on the diversity of strategies available.

The complexities of plant-insect interactions are reviewed in the cases of bark beetles (Chapter 5) and *Pieris* butterflies (Chapter 6). The last section focuses on insect-insect interactions. The behavioral ecology of trails and territories in social insects is examined in Chapter 7, and in Chapter 8 the role of chemical signals in nest-mate and mate recognition is analyzed. Chapter 9 details a specialized example of chemical communication and its exploitation: pheromones of true bugs and their use by parasitoids as kairomones. In Chapter 10 cases of chemical espionage and deceit among inquilines and slavemakers offer the most complex examples of the exploitation of chemical signals.

One value of such reviews lies in their identification of the boundaries of our current knowledge and the prospect of identifying the most profitable areas in which we should expect these topics to develop. A consensus in these reviews may be that assays of insect behavior conducted in the field or incorporating a realistic simulation of field conditions will prove instructive to establishing natural patterns of response and variation in behavior. In turn, these observations can help identify the most relevant cues to incorporate into laboratory tests that seek to partition the importance of chemical and other sensory inputs.

We thank John Byers for his suggestions on the organization of topics and Greg Payne, our editor at Chapman and Hall, for his patience throughout this project's somewhat long gestation.

Ring T. Cardé, William J. Bell

Chemical Ecology of Insects 2

Chemoreception and Integration

1

Behavior and Integration[1]

Marion O. Harris
Department of Plant Science, Massey University, Private Bag, Palmerston North, New Zealand

Stephen P. Foster
The Horticulture and Food Research Institute of New Zealand Ltd, Private Bag 11 030, Palmerston North, New Zealand

> "All behaviour is the product of central nervous integration of different stimuli from both outside and inside the insect, and likewise of different responses."
>
> —J.S. Kennedy (1978).

1. Introduction

Chemicals are involved in mediating a wide range of insect behaviors, from communication between conspecific individuals to the recognition of specific features of the environment, such as a food source. Anyone who has observed the effect of a chemical on the behavior of an insect cannot fail to be impressed by the apparent power of the chemical. Metcalf and Metcalf (1992) related how their interest in chemical ecology was stimulated upon observing the extraordinary response of the oriental fruit fly to the plant kairomone methyl eugenol. "We vividly recall walking into our bedroom where a handkerchief with a trace of methyl eugenol was left, to find a screened window several meters away literally black with bemused *Dacus dorsalis* males."

The powerful effect that chemicals can have on the behavior of insects may give the impression that such behaviors are controlled by chemicals, that is, that the behavior is merely a stereotyped response to a chemical stimulus (or stimuli). Despite the apparently powerful behavioral effects of chemicals, it is generally accepted that the behavior of insects is not controlled by any one external stimulus (chemical or otherwise), but is mediated by a large number of external and internal stimuli (Kennedy, 1978). As Prokopy commented "There may in fact exist few, if any, cases wherein an insect uses only a single sensory modality to find a resource" (Prokopy, 1986).

While there can be little question that any behavior of an insect is mediated

[1]Dedicated to the memory of Professor J.S. Kennedy

by a large number of internal and external stimuli, it is probably not unfair to say that most studies of insect behavior (involving chemical or other stimuli) are primarily concerned with one stimulus (or one class of stimulus) and its effect on the behavior. Although such studies are necessary in order to elucidate the stimuli that are involved in a behavior, they ultimately fail to tell us how an insect uses and processes multiple stimuli, and hence how the behavior is organized in the central nervous system (CNS) and what causes the behavior (i.e., its mechanism).

In this chapter, we consider the chemically-mediated behavior of insects and how it results from the processing of a variety of inputs from chemical, other external, and internal stimuli. The processing of inputs is a physiological phenomenon, and the discussion of the processing of chemical and other inputs in insect behavior has largely been restricted to a neurophysiological perspective (Horn, 1985; Light, 1986; however, see Prokopy, 1986). A major aim of this chapter is to discuss this physiological phenomenon from a behavioral perspective. In particular, we consider what the behavioral manifestations of the processing of multiple inputs may be, and how these may be studied. By so doing, we hope to encourage more study on multiple inputs in behavior and a greater meld of physiological and behavioral studies investigating this phenomenon.

We used the quote by J.S. Kennedy at the start of this chapter because of its pertinence. While we were writing this chapter, J.S. Kennedy passed away. His contributions over the last 50 or so years to the study of insect behavior have been immense. Our ideas in this chapter have been strongly influenced by his work, and we sincerely hope that this chapter, in some way, reflects this.

2. Behavior and its Causes

Before discussing chemically-mediated behaviors, it is necessary, first, to define what is meant by the term *behavior*. Dawkins (1983) described it as: "In one sense, behaviour is nothing but movement—the movement of whole animals in space . . . In another sense behaviour is much more. The movements we see are simply the outward and visible signs of highly complex 'programs' or sequences of instructions to the muscles." Behavior is therefore, movement, be it of part of an organism (e.g., the antennae), of a whole organism, or of a group of organisms (e.g., the migratory flight of a swarm of locusts), and the physiological basis of that movement.

In behavior, movement is rarely studied for its own intrinsic interest. Rather, movement is studied, generally, in order to find out what causes the movement. Tinbergen (1951, 1963) formulated four questions concerning behavior and its causes: (1) What is the mechanism of the behavior (i.e., its immediate cause)? (2) How does the behavior develop over the lifetime of an individual? (3) How did the behavior evolve? (4) What is the function of the behavior? In this chapter we are principally concerned with the first of these questions, a question, as pointed out by several authors (Dawkins, 1989; Kennedy, 1992), that has been less fashionable in recent years than the questions concerned with function and evolution.

To understand a behavior and its immediate cause, ideally one would characterize the stimuli (we use the term *stimulus* for the actual factor or event that is sensed by the nervous system), the various inputs (we use the term *input* for the signal generated from the stimulus that is sent to the CNS) and how they are processed by the CNS, the motor programs that control muscle movements, and the actual movements that result. Although physiologists are attempting to understand behavior in this way (Camhi, 1984), the complexity of the CNS, and the way it processes inputs and regulates outputs, will defy such an understanding for many years. Thus, most researchers interested in behavior do not attempt to understand it at a level that is purely physiological, but rather at a higher, behavioral level. That is, they attempt to understand a behavior in terms of the organization of the movements, and how these movements result from particular combinations of stimuli (or inputs).

3. Studying Behavior

3.1. Classifying Behavior

To study a behavior, the behavior must first be defined, in order that it may be separated from other movements (or behaviors) of the insect. The most usual way to define a particular behavior is to classify it according to an obvious function, such as finding a mate or a host, and limit it to the events that appear most relevant to the perceived function.

This classification of behaviors along perceived function is the one we shall use in this chapter. However, we use it with caution, as a functional classification has several problems. First, it may give the impression that the life of an insect can be partitioned into discrete phases, each of which can be studied separately from all others. In many insects, the end of one functional class of behavior and the beginning of the next may be unclear (e.g., egg-laying and nectaring bouts in female butterflies). A functional classification of a behavior also tends to focus on the endpoints of the behavior (e.g., copulation, feeding, oviposition), as though these somehow encapsulate all the relevant movements that have preceded the endpoint. In this way, studies on the endpoint of the behavior are sometimes substituted for studies on all the movements that lead to (and often follow) the so-called endpoint. As we shall demonstrate, an understanding of a behavior and its immediate cause can only be achieved through studying the whole behavior, not just its endpoint. That is not to say that studying the endpoint of a behavior is not useful. The endpoint is often, somewhat paradoxically, a useful starting place to work backward from and identify the movements which are associated with it (Martin and Bateson, 1986). A final problem worth mentioning is that a functional classification of behavior implies that the insect carries out the behavior with some preconceived goal. Kennedy has discussed the hazards of such teleological thinking in mechanistic behavioral studies on numerous occasions [see, for example, Kennedy (1992)].

Despite the problems with a functional classification of behavior, it is probably the one most commonly used, especially in the area of chemical ecology, and, if somewhat problematical, it is nevertheless convenient, brief, and easily understood.

3.2. Behavioral Inputs

The quote by Kennedy at the start of this chapter states that there are internal and external stimuli that generate inputs for a given behavior. The inputs are generally classified according to their action on three classes of senses: enteroceptive, proprioceptive, and exteroceptive (Schmidt, 1978). Enteroceptors monitor the internal physiological state of an animal, and include receptors that sense temperature, and concentrations of critical physiological chemicals. Proprioceptors provide information on relative positions, and changes in these positions, of parts of the body, including internal organs. Exteroceptors sense events external to the insect, and include the commonly known visual, chemical, and mechanical senses (amongst others).

While these distinctions between different classes of senses are made, it should be noted that often there is no such clear distinction; so-called exteroceptive senses can be used as proprioceptive senses as, for example, vision in monitoring self-movement. It is also worth noting that a given stimulus does not generate the same input in all contexts. The input generated by a particular stimulus is, for example, dependent upon the physiological state of the insect (see Blaney et al., 1986, Davis, 1986), and upon the behavior of the insect. Thus a flying insect will perceive the stimuli from a plant quite differently than will a stationary insect. As most behavioral studies characterize the stimulus rather than the input for a given behavior, it is critical, therefore, that the behavioral and physiological context in which the behavior is carried out is characterized as completely as possible.

3.3. Behavioral Outputs

The first step of any behavioral study is informal observation, the purpose being to define discrete and quantifiable units of the behavior [see Martin and Bateson (1986), for a discussion of this process]. The units of a behavior can be based on many things such as posture (e.g., arching of the abdomen), movements (e.g., rapid fanning of wings), spatial relationships to other individuals (e.g., the distance of a male from a calling female) or specific features in the environment (e.g., a plant), or the consequences of the behavior (e.g., find a mate). It is important that the units of a behavior be defined as precisely and objectively as possible and are easily recognizable, so that other researchers studying the same behavior can observe and measure the same units.

Consider the egg-laying behavior of mated female Hessian fly, *Mayetiola destructor*, a herbivore that specializes on wheat and related grasses. The movements that lead to the laying of eggs by mated Hessian fly females may be divided into movements that occur before, during, and after landing (Harris et al., 1993; Withers and Harris, unpublished). This effectively divides the time spent off and on the plant. Within each of these three broad divisions, several distinct units can be recognized. There are several types of flight (i.e., movements before landing): relatively straight and rapid flights from one plant to another; circular flights, which take the female back to the same group of plants; slower zigzagging flights between and around groups of plants; and vertical flights which take the female out of the plant canopy. There are also different types of landing movements: some females fly directly to the plant and land without hesitation, while others hover within 1 cm of a plant, during which their antennae, legs, and tip of their abdomen may contact the plant surface, and then land. Once on a plant, several other units of behavior can be distinguished: extension of the ovipositor, arching of the abdomen, movement of the abdomen from side to side across the leaf surface, antennating the leaf surface, laying an egg, walking up and down the leaf, and remaining motionless, in one of several postures, on the leaf.

The choice of units to be studied for a particular behavior is made usually on the basis of what is obvious and considered important to the behavior or to the hypothesis being tested. Because the amount of data increases with the number of units studied, it is usually necessary to limit the number of units recorded for a given behavior, especially if the number of experimental treatments or the number of individuals being tested for each treatment is large. Thus, the definition of units for a given behavior is a compromise between describing precisely the behavior and the amount of data able to be observed or analyzed.

Even the most comprehensive behavioral study is, however, a simplification of what the insect is actually doing. It generally assumes that the behavior or each of its units is independent of other behaviors or units (see Section 4.2 for a more thorough discussion of this point). Furthermore, by defining a stereotypical series of movements or postures as a single unit of behavior, there is a tendency to assume that the unit is immutable, when in fact the unit may vary in subtle ways over time or in slightly different contexts. For example, locusts will stop walking when subjected to certain intensities of pulses of 3-kHz sound. However, locusts that have walked for a long period will stop when subjected to pulses of lower intensities of this sound than locusts that have walked for a relatively short period (Moorehouse et al., 1987). Thus, as discussed for stimuli, the context in which behavioral units are studied is very important.

3.4. Studying Input-Output Relationships

Once measurable units of a behavior have been identified, the causes of the behavior can be determined experimentally. Halliday and Slater (1983) described the method by which this is done: "To study behaviour at its own level is to treat the animal as a black box, modifying the stimuli that impinge upon it and seeing how these affect its behavioural output without being concerned too deeply with the exact nature of the intervening mechanisms . . . This enables behaviour to be predicted, to be changed and its causes understood without any direct knowledge of the mechanisms within the animal for translating the various inputs into the output."

Using this black box method (Fig. 1.1), it is evident that understanding the behavior and its immediate causes is dependent on (1) the detail of the behavioral analysis (i.e., the number of behavioral units composing the behavior that are measured); and (2) the researcher's ability to describe and quantify the various inputs that influence these outputs. In most cases, the inputs for a given behavior are not studied directly (i.e., by electrophysiological methods), but rather are characterized in terms of their respective stimuli (e.g., an amount of a specific chemical) or by simply stating a gross change in the external or internal environment of the test insect (e.g., the addition of a virgin female; the precise stimuli associated with that female and perceived by the male are not characterized).

To prove that a particular stimulus is involved in (i.e., generates an input for) a particular behavior, it is necessary first to isolate the stimulus and to demonstrate that it elicits a behavioral response within the normal context of the behavior. However, the isolation of a particular stimulus (and therefore its effects) is not always facile. For example, the predatory beetle, *Rhizophagus grandis*, flies upwind in response to odor from the frass of its prey, the bark beetle, *Dendroctonus micans*. In a series of experiments designed to elucidate the effect of visual and mechanical effects on the odor-mediated flight and landing behaviors of *R. grandis*, Wyatt et al. (1993) found that the beetles responded equally well to a hidden (by a mesh screen) Mylar object as to a black silhouette drawn on the

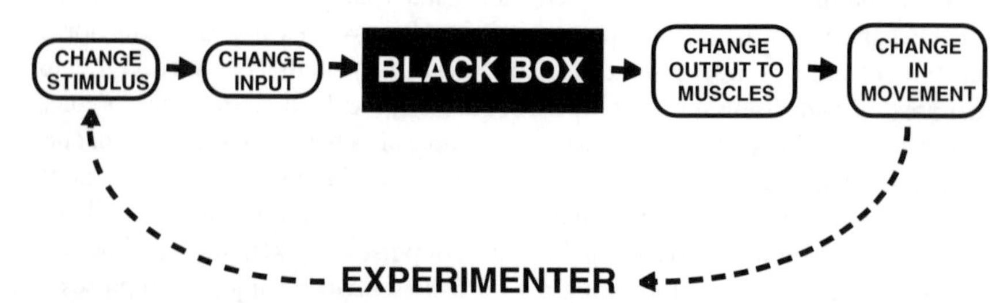

Figure 1.1. The "black box" approach to studying behavior, illustrated here for a single-input-output approach. The experimenter changes the stimulus and quantifies the change in movement of the insect.

screen plus the hidden Mylar object, suggesting that the turbulence induced by placing the object in the tunnel had a more significant effect on the behavior of the beetles than did the visual silhouette. However, as pointed out by the authors, Mylar reflects some ultraviolet light and hence may have been visible through the screen. Furthermore, while the turbulence created by the object may have affected the behavior of the insect (rather than, or in addition to, visual inputs associated with the object), precisely how this turbulence affected the sensory input of the insect is unknown. The turbulence, for example, may be sensed directly, through mechanoreceptors, as abrupt changes in air movement. Alternatively, the turbulence could affect the structure of the chemical plume, which would change the inputs from chemoreceptors and result in different flight maneuvers.

The principal aim of any study of the immediate cause of a behavior, is, however, not to isolate a particular stimulus and demonstrate whether it elicits a change in the behavioral output, but to understand and characterize the range of stimuli or inputs that elicit the complete behavioral output. In most studies, the stimuli to be studied are usually narrowed initially to those that are more easily manipulated or to those that are of interest to the researcher. Not surprisingly, in the field of chemical ecology the stimuli most commonly studied first are chemicals.

The approach used to elucidate whether a particular chemical is involved in an insect behavior is to study the relationship between the chemical and the behavioral output. The simplest way to do this is to change a single, highly defined, chemical stimulus within a context of otherwise unchanged stimuli. We will refer to this as the single input-output approach. This single input-output approach can be carried out in one of two ways: (1) In a laboratory assay in which all stimuli except the chemical one are simplified and standardized; and (2) in field or near-field assays in which the chemical stimulus is tested in a context of natural stimuli. In the study of chemically-mediated behaviors, the former way has probably been used more commonly, because it allows, among other things, experimentation all year round and at convenient times of the day, greater control over the stimuli involved in (and hence the inputs to) the behavior, and more detailed observation on the changes in the behavior. Several laboratory assays, such as *y*-tube olfactometers, four-arm olfactometers, and wind tunnels [see Baker and Cardé, (1984)], have been developed and used prolifically for studying input-output relationships between chemicals and particular behaviors. The widespread use of these standardized assays has resulted in a consistency in approach, as well as in the behavioral units used, in the study of chemically-mediated behaviors.

While such laboratory studies have the above advantages, they also have the drawback that they can yield false, negative results. For example, a chemical (or any other) stimulus can appear to have no effect because the response to that stimulus only occurs in the presence of some other input(s) which may not be present, as is evident in a study of the host-finding behavior of alfalfa seed chalcids (Kamm, 1990). In this study, adult females flew to sources of host-plant odors in the greenhouse, but not in artificially illuminated laboratory assays. By manipulating the light source, it was found that oriented flights to plant odors only occurred when the test arena was illuminated with polarized light.

Studying the chemical input(s) to a behavior within a natural or near-natural context makes it more likely that most, if not all, of the important stimuli for a behavior will be included in the assay. However, this approach by itself does not allow the identification of other inputs involved in or necessary to the behavior, such as the effects of polarized light in the seed chalcid example. Therefore, when studying the single input-output relationship between a chemical (or any) stimulus and a behavioral output, it is better to utilize both approaches (i.e., laboratory and more natural assays), and compare the respective behavioral outputs. If the behavioral output elicited by a stimulus in a more natural context cannot be repeated in a laboratory assay, it usually implies that inputs present in the natural assay are lacking in the laboratory assay; conditions in the natural assay can then be varied systematically until other inputs influencing the behavior are identified. Similarly, if a behavioral output is elicited by a stimulus in a laboratory assay but not in a more natural assay, the laboratory assay can be made progressively more complex to determine what other factors, in nature, influence or mask the behavioral output.

The single input-output approach for studying behavior is useful in that it usually requires a small number of experimental treatments (generally just the treatment or various levels of a single treatment and a control) and the statistical analysis of the resulting data is generally straightforward. Perhaps, most importantly, this approach provides a useful starting place for studying, individually, each of the stimuli (and inputs) for a given behavior. In chemically-mediated behavior, this usually directs the researcher to examine chemical stimuli first. However, once the behavioral effects of the chemical stimulus (or stimuli) have been established, the same single input-output approach can be used to explore other external and internal stimuli influencing the behavior.

Although useful, the single input-output approach is restricted in what it is actually able to tell us about the causes of behaviors. Often this approach tends to limit the researcher to focus solely on the stimulus being tested, and the actual behavior of the insect is used as a means of documenting the effect of the stimulus. This approach can "imply that each response has one self-sufficient kind of stimulus, as if each kind of stimulus acted on its own. This is to lose sight of the rich variety of interactions that are known." (Kennedy, 1965).

To think of behavior as a response to an input is to ignore the complexity of behavior and the cardinal role of the CNS to process or *integrate* sensory inputs and organize the behavior. In the remainder of this chapter, we will discuss the "variety of interactions that are known" within the context of the integration of chemical and other inputs by the CNS.

4. Behavior and Integration of Inputs

4.1. The Physiological Basis of Integration

Integration is the combination of parts into a whole. Thus, the integration of sensory inputs in a behavior must involve a convergence and physiological processing of two or more inputs (external or internal). In theory, the integration of different inputs can occur anywhere within the nervous system. However, in practice, sites of integration have proven more likely to be found in higher levels than in lower levels of the nervous system (Horn, 1985).

Much of the physiological evidence for the integration of a chemical input with other external (including chemical) inputs comes from the discovery of neurons within the CNS that respond to multiple inputs (usually from peripheral stimulation). Most commonly, neurons of this type that respond to a particular chemical input also respond to other chemical, mechanical, or visual inputs. Multimodal neurons, coding information about chemical and other external stimuli, have been found in a range of insects including crickets (Schildberger, 1984), bees (Homberg, 1984), cockroaches (Waldow, 1975), and moths (Olberg and Willis, 1990). The existence of such neurons implies that these different inputs converge and are therefore integrated within some structure of the CNS (not necessarily inside the neuron for which the electrophysiological recordings have been carried out).

The integration of olfactory inputs with other inputs has proven particularly amenable to investigation. Information generated from the perception of chemicals by an insect antenna goes to the deutocerebrum and then onto the protocerebrum. It appears that much of the processing and integration of chemical inputs from the antennae and other inputs occurs within these structures of the CNS. The distinct antennal receptor neurons sensitive to the different components of the sex pheromone of a moth, such as *Manduca sexta*, terminate within structures called glomeruli in the deutocerebrum (Hildebrand and Montague, 1986). In both of the moths, *M. sexta* and *Agrotis segetum*, the arborizations of different populations of receptor neurons terminate primarily in distinct glomeruli, although some neurons appear to have arborizations in more than one glomerulus. This suggests that the integration of the inputs from the different pheromone components largely does not occur in the glomeruli (Hansson et al., 1991, 1992), and therefore probably occurs in higher centers of the brain.

In the honey bee, *Apis mellifera*, interneurons that extend to the mushroom bodies (in the protocerebrum) show a range of inhibitory and excitatory responses when the insect is stimulated with combinations of chemical and mechanical stimuli, but not light (Homberg, 1984). The mushroom bodies also have some visual inputs, and thus it is likely antennal and some visual information converge and are integrated within these structures (Kaulen et al., 1984).

Because of its complexity, the understanding of the physiological processing and integration of behavioral inputs in the CNS is still rather limited (Horn, 1985; Light, 1986). However, the available evidence strongly suggests that the integration of chemical and other inputs occurs largely, if not exclusively, within higher centers of the CNS.

4.2. Integration and Behavioral Studies

The integration of chemical and other inputs can conceivably occur in three forms: the simultaneous integration of the inputs (Fig. 1.2a), where the two or more inputs (x and y) are present and integrated simultaneously and elicit the behavioral output (C); the successive integration of inputs (Fig. 1.2b), where one input (x) in effect triggers the responsiveness to a successive input (y); and a combination of these two (Fig. 1.2c), in which one input (or an integrated combination of inputs) triggers responsiveness to an integrated combination of inputs (or a single input).

Given the complexity of behavior and the variety of internal and external stimuli that are known to influence many insect behaviors, there can be little doubt that the integration of different inputs plays a part in all insect behaviors. However, while all behaviors probably involve integration, the dilemma is that the study of behavior cannot establish, unequivocally, that the process of integration actually occurs. The integration of inputs is a physiological process and therefore can be proven only by studies at the level of the neuron. Nevertheless, it is possible to infer from behavioral studies some information concerning the integration of inputs, and such information should be an important basis for physiological studies investigating the role of integration in behavior.

To infer from a behavioral study that integration occurs, one must assume that the process has, in many cases, some recognizable behavioral manifestation. When the presence of two or more distinct inputs results in a behavior that is different, either qualitatively (i.e., a completely different behavioral unit) or quantitatively (i.e., a change in the frequency of a behavioral unit), from the behavior when the inputs are presented separately, the behavior can be assumed to result from the integration of inputs. This is not to say that the integration of inputs always involves some discernible behavioral manifestation. The physiological integration of inputs may occur, but there may be no apparent behavioral manifestation. Rather, the process of integration can only be inferred from behavioral studies (within the constraints discussed later) when there is a recognizable behavioral manifestation.

It follows from our definition of the behavioral manifestation of integration, that a more complex experimental design is necessary to elucidate whether the integration of inputs may be occurring in a behavior. Thus, where previously in the single input-output approach, only one stimulus (for example, a chemical) was studied in a minimal design (a single dose of the chemical and a solvent

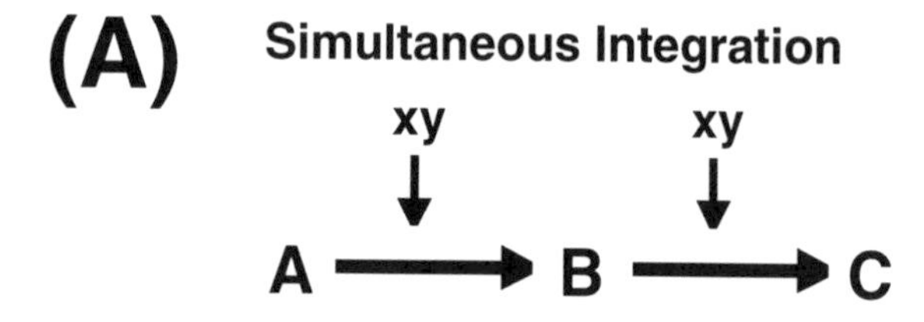

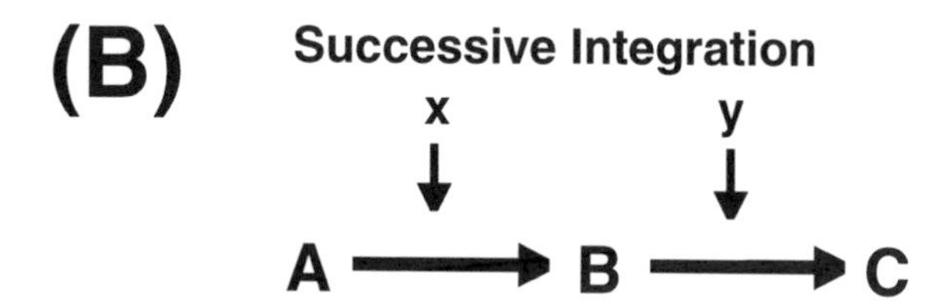

(C) **Combinations of Simultaneous and Successive Integration**

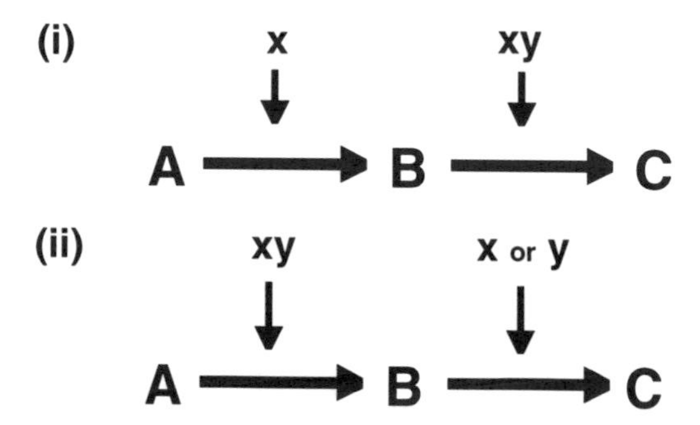

Figure 1.2. Three possible types of integration of the inputs x and y that involve the behavioral outputs A, B, and C (the endpoint of the behavior): (a) Simultaneous integration, where the inputs x and y must be present simultaneously for the behavioral state A to proceed to the behavioral endpoint C; (b) successive integration, where input x elicits output B , followed by input y which elicits the endpoint C (from output B); and (c) two combinations of simultaneous and successive integration.

control), the design must now be expanded to include other stimuli. Thus, a minimum of four experimental treatments is presented to the test insect in a factorial design (Fig. 1.3): a treatment in which neither stimulus is present (−/−), a treatment in which the chemical stimulus is present but the other stimulus (a tactile one in Fig. 1.3) is not (−/+); a treatment in which the other stimulus is present but the chemical is not (+/−); and a treatment in which both the chemical stimulus and the other stimulus are present (+/+).

Although the above example contains only two quantitative levels (i.e., present or absent) for each of the two stimuli, in theory each stimulus can be studied at any number of levels, by testing different intensities of the stimulus. For example, two or more different dosages (levels) of a plant chemical may be tested, as well as two or more densities of a tactile stimulus, in various combinations in a factorial design (Fig. 1.3). Multilevel factorial designs add further resolution to

(a)

	Chemical Stimuli	
Tactile Stimuli	absent	present
absent	-/-	-/+
present	+/-	+/+

(b)

	Chemical Stimuli		
Tactile stimuli	absent	1 plant equivalent	5 plant equivalents
absent	- / -	- / 1 PE	- / 5 PE
6 hairs/cm^2	6 h / -	6 h / 1 PE	6 h / 5 PE
12 hairs/cm^2	12 h / -	12 h / 1 PE	12 h / 5 PE

Figure 1.3. Experimental treatments involving combinations of chemical and tactile stimuli presented to a test insect in (a) a simple 2 × 2 factorial design with the factors being either absent (−) or present (+); and (b) in a multi-level factorial design with the factors present at different dosages (plant equivalents, PE), or at different densities (hairs/cm^2).

the study of input-output relationships and may be useful in the elucidation of how various internal inputs (including those resulting from experience) are involved in the integration of externally-sensed inputs (see sections 4.4 and 4.5).

The problem with such multi-input, multilevel factorial experiments lies in the large number of treatments required. For example, if one wishes to test simultaneously the effects of three stimuli (I) and four levels (L) of each input, the experiment would consist of $4 \times 4 \times 4 = 64$ (L^I) experimental treatments. The difficulties of such large experimental designs have been pointed out by Sokal and Rohlf (1981): "It may not be possible to run all the tests in one day or to hold all of the material in a single controlled environment chamber. Thus, treatments may be confounded with undesired effects if treatments are applied under not quite the same experimental conditions." Experiments with a large number of treatments are also usually harder to interpret, as establishing statistical differences between means becomes increasingly difficult with a greater number of treatments (Sokal and Rohlf, 1981).

In spite of these added difficulties, factorial, multi-input designs are preferable to ones where the change in each stimulus is tested separately, as in single input-output designs. First, all the stimuli being tested are present and changed in the same experiment under the same conditions, rather than in different experiments and possibly under slightly different and unaccounted for conditions. Second, a stimulus may only have an effect on the behavior in combination with another stimulus. Thus, when tested in a single input-output design, such stimuli may not appear to be an input for the behavior if the second input is not present. This can apply to inputs that have a positive effect (i.e., they elicit an increase in the behavioral output when present with another input) or a negative effect (i.e., they elicit a decrease in the behavioral output when present with another input) on the behavior.

The statistical analysis of a factorial experimental design provides information on the effects of each stimulus singly (so-called 'main effects') and the interactions that occur between the multiple stimuli (so-called 'higher order interactions'). The main effects of stimuli tested in a factorial design may be [but not always, see Harris and Rose (1990)] similar to the effects observed in a single input-output design; however, because the stimuli are tested simultaneously in the factorial design, more can be said about the relative effects of each on the behavior. In a factorial design, the experimental finding that two or more stimuli have significant main effects indicates that they are inputs to the behavior and furthermore suggests that these inputs are integrated at the level of the behavior being measured.

The interaction term, generated from the statistical analysis of a factorial design, addresses the question of whether the presence or amount of one stimulus affects the behavioral output to another stimulus. As such, if a significant interaction between two stimuli is found, it suggests that their resultant inputs are integrated. Such a result can be especially important when one or both of the main effects involved in the interaction is not statistically significant; i.e., if only the main effects were considered, one would falsely conclude that one or both of the stimuli has no effect on the behavior and that there can be no integration of inputs because the stimulus (or stimuli) have no resultant input into the behavior. It is tempting to think that a statistical interaction must tell us something additional (to the main effects, assuming they are significant) about the integration of inputs. However, it is important to recognize that this term is a statistical one which simply indicates that, at the level of the population, the behavioral output when both (or more) stimuli are presented together is not explained by the main effects of the stimuli presented alone. Nothing further concerning physiological integration, other than that it appears to have occurred with the resultant inputs, can be gleaned from such a result. This again emphasizes the limitations of purely behavioral data for understanding the physiological process of integration.

In the following sections, we consider specific ways that behavioral studies may be used to explore how chemical inputs may be integrated, or interact in other ways, with other inputs for a particular behavioral unit.

4.3. The Integration of Inputs from Chemoreceptors and Inputs from Other Exteroceptors

In theory, both simultaneous and successive integration of chemical and other (including chemical) inputs in a behavior are possible. In practice, however, it is difficult, if not impossible, to distinguish between these two types of integration in behavioral experiments. To demonstrate whether successive integration takes place would require the rapid removal of the former input after it has triggered responsiveness to the latter input (see Fig. 1.2), so that they are not present concurrently (and hence could not be integrated simultaneously). Whether this phenomenon can actually be addressed experimentally depends upon how feasible it is to remove quickly the former stimulus, after it has triggered responsiveness to the latter stimulus. The problem may be further exacerbated by the processing of the inputs. For example, even when the stimulus that elicits the former input has been removed, neurons that convey the input may still be conveying this information to the CNS when the second stimulus is presented and the latter input processed. Thus, what appears to be a successive presentation of two stimuli from the experimenter's point of view may not necessarily be translated into a successive presentation of inputs to the CNS.

Another limitation concerning the inference from a behavioral study of whether integration of chemical and other inputs occurs, is that any conclusion regarding integration is highly qualified by the behavioral units that have been studied. This is because the larger (as a proportion of the whole behavioral sequence) and less precisely defined the units of the behavior are, the greater the chance that inputs involved in the behavior will appear to be integrated. This phenomenon, which is a relation between successive behavioral outputs, rather than an actual physiological event, is illustrated in simplistic form in Figure 1.4. With the insect in an initial behavioral state of *A*, input *x* elicits the behavioral output *B*, and another input (*y*) elicits the behavioral endpoint *C*. If outputs *B* and *C* are studied, then significant main effects for each of the stimuli that generate the respective inputs would be observed only for their respective behavioral outputs, and no interaction between the stimuli would be observed; one would therefore conclude that the two inputs are probably not integrated. However, if only the endpoint *C* is studied, the stimuli that generate the inputs *x* and *y* would both have significant effects, apparently concurrently, and a statistical interaction between these stimuli may be observed, suggesting that the two inputs are integrated. Thus, the smaller and more precisely defined the units of a behavior that are studied, the greater confidence with which it can be inferred that the inputs are integrated or not within each individual behavioral unit and therefore within the behavioral sequence as a whole.

It is worth noting in the above example that the behavioral sequence resulting in endpoint *C* appears to result from the successive integration of the inputs *x* and *y*. However, the situation in the example is distinct from that of successive integration of inputs in that the inputs are related through successive behavioral outputs. This distinction should be evident in certain cases where the sequence of behaviors being studied corresponds to a specific sequence of encounters with different resources (e.g., as in parasitoids foraging for herbivorous hosts, where the host's habitat, the host's host plant, the host's damage to that plant, and the host itself might be encountered sequentially). In such a situation, the insect could be manipulated so as to start at different points in the behavioral sequence. Thus, in the example in Figure 1.4, if the outputs were related through successive behavioral outputs, then starting the insect at state *B* in the behavioral sequence (i.e., omitting the stimulus that generates input *x*) and presenting it with the

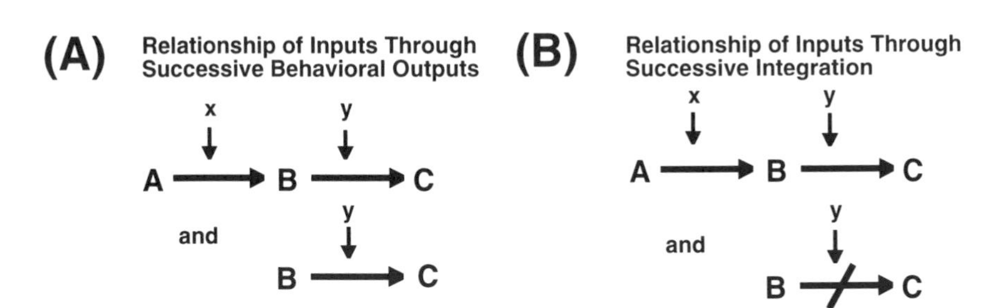

Figure 1.4. The difference between the relationship of inputs *x* and *y* through (a) successive behaviors *A*, *B*, and *C*; and (b) successive integration. In both situations, input *x* elicits output *B* (from initial state *A*), and input *y* elicits endpoint *C* (from output *B*). For the relationship of inputs through successive behaviors, the behavioral transition from *A* to *B* (triggered by input *x*) does not have to precede the transition of *B* to *C*, whereas it is necessary when the inputs are related through successive integration.

stimulus that generates input *y* would result in the insect proceeding to the endpoint *C*. However, if the inputs *x* and *y* were integrated successively, then starting the insect at *B* along with the respective stimulus for *y* (without the insect having experienced input *x*), would not result in the endpoint *C*.

The importance of studying smaller behavioral units when examining the effects of multiple inputs is illustrated by the behaviors involved in female Hessian flies ovipositing on a plant. When only the endpoint of the behavior (i.e., the number of eggs laid) was considered, chemical, tactile, and visual stimuli were all found to influence significantly the behavior (Fig. 1.5). Using a three-factor, factorial design, all possible first-order interactions (chemicals x color, chemicals x tactile, tactile x color) as well as the second-order interaction (chemicals x color x tactile) were found to be significant statistically (Fig. 1.5) (Harris and Rose, 1990).

A two-factor, factorial design was used to examine the roles of visual and chemical stimuli in Hessian fly females finding oviposition sites (Harris et al., 1993). The following behavioral units were measured: approach (the number of females entering a 1-cm zone around the model plants), landing (the percentage of females that entered the 1-cm zone that landed on the model plant), and egg-laying (the number of eggs laid per landing). Both chemical and visual stimuli influenced the number of females approaching each model (Fig. 1.6). Similarly, both stimuli influenced the landing of females (Fig. 1.6). Once females had landed on the plant, visual stimuli, such as color, had no effect on the number of eggs laid, but chemical stimuli did (Fig. 1.6).

Given the first set of experiments (i.e., studying only the endpoint of the behavior), it would have been tempting to conclude that both chemical and visual stimuli influence female Hessian flies when they are on the plant and laying eggs, and that therefore the resulting inputs are integrated during this endpoint. The second series of behavioral experiments, however, demonstrated that visual stimuli do not influence females when they are laying eggs, and hence it is unlikely that the integration of chemical and visual inputs influences this behavioral output. Instead, the integration of chemical and visual stimuli occurs earlier in the behavioral sequence.

4.4. The Integration of Inputs from Chemoreceptors and Enteroceptors

Internal stimuli mediating the behavior of insects are not as well understood as external stimuli and are often described in broad terms (e.g., a mated or virgin female). The collective set of internal stimuli corresponding to such a broad, physiological description, is often referred to as the *internal state* of the insect (Halliday and Slater, 1983).

One reason that internal stimuli are not as well understood is that it is generally difficult to manipulate these stimuli other than at a gross level. Unless some substance can be physically injected into the relevant part of the body of the

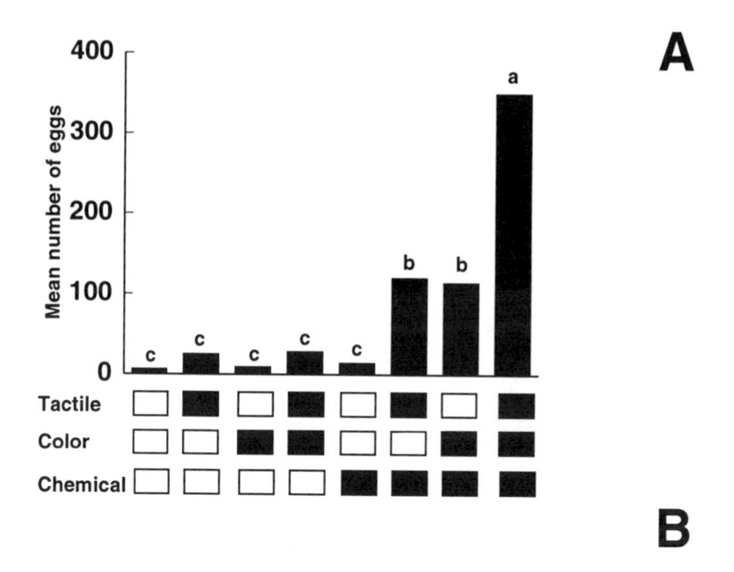

Tests	Treatment	F value	Significance
Main effect	chemicals	89.4	<0.0001
Main effect	color	40.1	<0.0001
Main effect	tactile	47.6	<0.0001
First order interaction	chemicals x color	37.7	<0.0001
First order interaction	chemicals x tactile	34.3	<0.0001
First order interaction	color x tactile	8.6	<0.006
Second order interaction	chemicals x color x tactile	8.6	<0.006

Figure 1.5. (a) The number of eggs laid by Hessian flies on plant mimics comprising different combinations of tactile, color, and chemical stimuli. The tactile stimulus consisted of parallel, vertically oriented grooves to mimic leaf venation (the absence of the tactile stimulus consisted of a smooth, waxed surface). Green filter paper constituted the color stimulus (in the absence of green, the filter paper was white), and the chemical stimulus consisted of one plant equivalent of a chloroform extract of epicuticular waxes of wheat foliage (in the absence of the chemical stimulus, the leaf-mimic was treated with chloroform). The absence or presence of a particular stimulus type is indicated by a white or black box, respectively. (b) The main effects of each stimulus type and the higher-order interactions between these effects are shown, along with their significance [from Harris and Rose (1990)].

insect to produce a specific change in inputs to the central nervous system [e.g., as in a study of stretch reception in the bursa copulatrix of *Pieris rapae*, Sugawara (1979)], internal stimuli often can only be manipulated indirectly by either testing the insect at a particular time or age, or by allowing or preventing the insect access to some resource, e.g., access to food to manipulate internal stimuli associated with the nutritional status of the insect, access to ovipositional sites

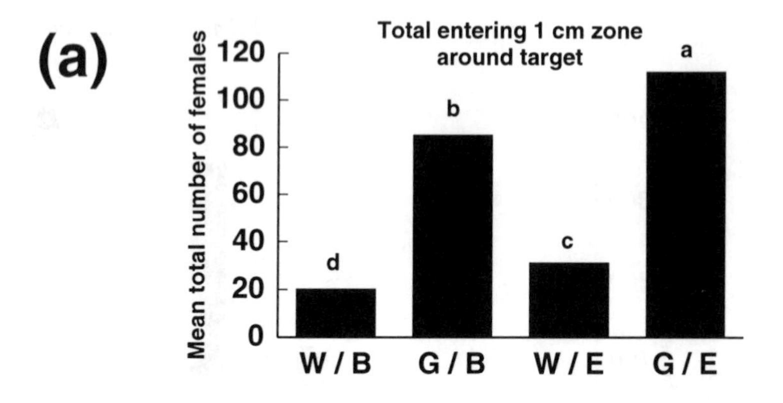

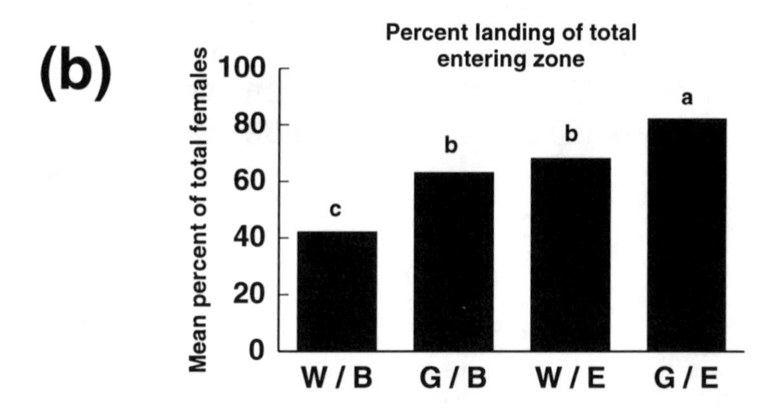

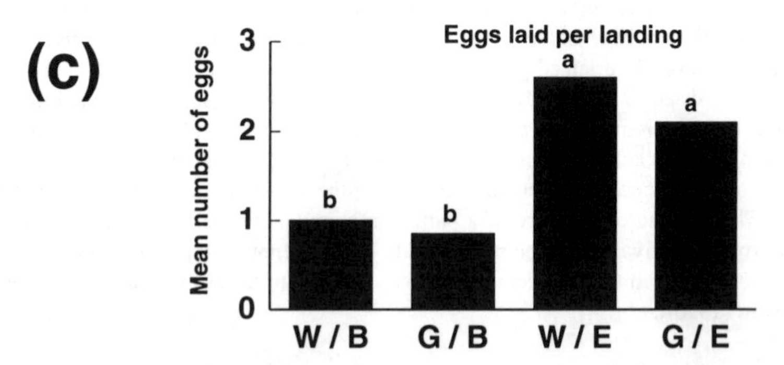

Figure 1.6. (a) Approach; (b) landing; and (c) egg-laying responses of mated female Hessian flies to differently colored objects and objects that were treated with a chloroform extract of epicuticular waxes of wheat foliage. Color and chemical stimuli were presented in various combinations using a factorial design as indicated by the letters beneath each column. W= white, G= green, B= blank with no wheat extract, E= one plant equivalent of wheat extract [from Harris et al. (1993)].

to manipulate internal stimuli associated with the egg-laying status of the insect, access to mates to manipulate internal stimuli associated with the mating status of the insect.

When such indirect and gross manipulations of internal stimuli are made, it is usually difficult to know precisely what inputs to the behavior are involved and which ones have actually changed. For example, a female *Euphydryas editha* butterfly deprived of an oviposition site behaves differently than a female that has recently oviposited (Singer, 1986). Plants that would typically be rejected by females that have recently oviposited are accepted as oviposition sites by deprived females. It is reasonable to conclude that the deprivation has changed various internal stimuli and that these putative changes result in a change in the behavior. Exactly what changes the deprivation has resulted in is, however, less clear. For example, the deprivation could result in changes in inputs from stretch receptors in the ovaries or oviducts, or in the titers of various hormones.

These difficulties make the study of the integration of inputs from chemoreceptors (or other exteroceptors) with inputs from enteroceptors more challenging than the study of the integration of inputs from exteroceptors, where the stimuli that generate the inputs can be usually manipulated and measured more precisely. Questions such as whether inputs are integrated simultaneously or successively cannot be addressed when studying chemical and internal stimuli, because internal stimuli cannot be removed or manipulated rapidly enough to switch on or switch off their inputs. Indeed, because of the way that chemically-mediated behavior is usually studied (and indeed most behavior of insects), as a response mediated by external stimuli, it must be assumed that any appropriate inputs resulting from internal stimuli are integrated simultaneously by the CNS with the inputs arising from external stimuli.

Because of the problems with characterizing internal stimuli, it is usually more appropriate to consider how internal states modulate the behavioral responses of an insect to externally-sensed chemical stimuli. There are essentially two ways that the internal state of an insect can modify the behavioral response of an insect to a chemical stimulus. First, it can change the response threshold to a particular amount of the stimulus. Thus, depending upon the internal state of the insect, the insect may respond to smaller or larger amounts of the chemical stimulus, or in extreme cases not respond to any amount of the chemical. Such effects are commonly observed with changes in internal states corresponding to changes in age or circadian rhythm. For example, males of various species of *Dacus* fruit flies do not respond to methyl eugenol or raspberry ketone in traps until some two to four days after eclosion (Metcalf, 1990).

The second way that the internal state of the insect may modify the behavioral responses of an insect to a chemical (or other) stimulus is to change the set of stimuli required to trigger the behavior. For example, in the feeding behavior of locusts (see also section 5.4), individuals that have recently fed will not palpate or bite a host plant. As the time of deprivation increases, locusts will palpate and bite a host plant and, if deprived for longer periods, will eventually palpate and bite anything with suitable visual (Williams, 1954) or mechanical features (Sinoir, 1969). In this example, the internal state of the insect changes such that after feeding the locust does not respond to a set of stimuli from a plant, then with deprivation the locust responds to the same set of stimuli (probably including chemical, visual, and mechanical), and with even longer periods of deprivation, a reduced set of more general stimuli (visual and mechanical) triggers palpation and biting.

It is worth noting that these two changes in response to external stimuli elicited by changes in internal state, may not be different; i.e., changes to different sets of external stimuli may merely involve changes in the level at which a particular stimulus (or stimuli) elicits a behavioral threshold.

4.5. Experience and the Integration of Inputs from Exteroceptors

The effect of experience is a special case in which the input (or inputs) from an exteroceptor (or exteroceptors) modifies the nervous system for a sufficiently long time so that, upon subsequent exposure to the input, the behavioral output is different. The effects can be categorized according to whether they affect the peripheral or central nervous system. Those that result from the initial effect on peripheral receptors [e.g., adaptation, see Vet et al. (1990)] can be thought of as modifying the input from the exteroceptor. In this case, the only integration that occurs is that which would normally occur with the input (whether modified or not) and other inputs (from both external and internal stimuli).

In the case where the initial input affects the central nervous system (e.g., as for learning), this may result in a significant change in the way that inputs from externally-sensed stimuli are integrated with inputs from internal or other external stimuli. The situation is similar to that for changes in the internal state of an insect; however, presumably the effects of learning on the insect are more specific, in that learned inputs and their integration will be more affected than unlearned inputs.

As discussed for changes in the internal state (section 4.4), the effect of experience on the CNS of the insect may be either to modify its behavioral response threshold to the experienced stimulus or to change its response to a qualitatively different set of external stimuli. Using the responses of a female parasitoid, *Cotesia rubecula*, to host odors, Kaiser and Cardé (1992) found that the effect of experience was to increase the probability that an individual female would orient to, and land on, weaker odor sources. However, in the presence of stronger odor sources, both inexperienced and experienced females were equally successful in detecting and finding hosts. Thus, the effect of experience in this case was to make females more able to find lower-concentration odor sources, perhaps through lowering the behavioral response threshold to the odor stimulus.

4.6. Integration and Behavioral Units

The final part of the quote by Kennedy at the start of this chapter states that a given behavior is dependent upon the integration of different responses. To illustrate this, Kennedy discussed how male moths maintain contact with a pheromone plume and eventually find a pheromone source by integrating kinetic and tactic responses (Kennedy, 1978), through what is now considered to be the integration of optomotor anemotaxis with an internal program of self-steered counterturns (Baker, 1989a) (see section 5.2). In other words, what may appear to be a single behavioral unit, in this case upwind flight, may in fact be the product of two or more behavioral outputs. The coordinated behavioral unit results from the integration of the two behavioral outputs by the CNS.

In a more general sense, individual behavioral units are also dependent upon the behaviors that precede them as well as the behaviors that are expressed concomitantly. For example, the landing behavior of an insect is usually distinct from its normal flight behavior. It not only succeeds its flight behavior, in the sense of being possible only after flight has occurred, but it must also be integrated with flight behavior, because the insect is still flying when landing behaviors are initiated. The pheromone-mediated landing behavior of males of the tortricid moth, *Epiphyas postvittana*, is influenced by chemical (the pheromone) and visual inputs, as is its flight behavior (Foster and Harris, 1992, Foster and Muggleston, 1993; Foster, unpublished). The landing behavior is apparently triggered only under certain conditions of flight, most noticeably when the insect's flight maneuvers allow it to approach and contact some surface. The insect's flight maneuvers that affect altitude, track, and velocity must also affect the inputs that eventually trigger its landing behavior. Therefore, in addition to the integration of the olfactory and visual inputs, it is likely that the two behaviors are coordinated and integrated by the CNS.

One final point that we wish to reiterate is that the separation of one unit of behavior from others is usually artificial and subjective. A better understanding of any behavior will result from an understanding of the relationships between the various behavioral units that constitute it, as well as an understanding of its relationship to other behaviors that precede, succeed, or are concurrent, regardless of whether they are integrated or not.

5. Examples of Multiple Inputs and Integration

5.1. Introduction

In this section, specific examples of three different functional classes of chemically-mediated behaviors, mating, feeding, and egg-laying, are discussed. In choosing these examples, we selected insects with a range of different habits (insects feeding on blood, plants, other insects) which had been studied in some depth with regard to a particular behavior. In selecting these examples, we have tried to avoid studies which have been restricted largely to the endpoints of behaviors. Rather, we have tended to use examples in which a sequence of units, or behaviors, has been studied, so as to illustrate the complexity of each behavior and the wide range of inputs that mediate a behavior. In certain examples, we have tended to generalize between species, so as to present a more complete study, although we are aware that the generalizations may not be applicable to all species within a particular group of insects.

Where possible, we have commented on whether inputs are probably integrated or interact in some way. However, because behavioral studies whose aim is to investigate multiple inputs and their integration are rare (Dusenbery, 1992), we have often had to infer the integration of inputs from experimental designs with a single input-output approach.

5.2. Mate Location Behavior of Male Moths

Over the last 20 years or so, the study of the sex pheromone-mediated behavior of male moths has pioneered an understanding of the mechanisms used by insects to find a distant odor source. The flight behavior of male moths is dependent upon a number of inputs from the internal state, including a circadian rhythm and age [male moths usually exhibit a peak of responsiveness at a particular time of day and at a particular age, Shorey (1973)]. Inputs from pheromone receptors are also known to vary with factors such as temperature and previous exposure to sex pheromone. Changes in these factors can have a pronounced effect on the flight behavior of male moths to the same pheromone blend (Baker et al., 1988; Linn et al., 1988).

The sex pheromones of most species of moths consist of a specific blend of chemicals (Arn et al., 1992). Typically, each component of the blend is perceived by a single type of neuron on the antenna (Kaissling, 1986) [although components are perceived by a single type of neuron, recent evidence has shown that neurons may, but perhaps only in some cases, respond to more than one component found in the blend, Todd et al. (1992)]. Despite some claims that individual chemicals elicit specific events in the behavior, it is now accepted, in the cases that have been extensively studied, that the complete blend of chemicals is processed as an integrated whole and thus mediates most if not all of the behavior from activation of flight through to landing and copulation (Linn et al., 1987; Baker, 1989b).

From the few species that have been studied in depth, it appears that male moths use an integrated combination of behavioral mechanisms to maintain contact with, and thereby locate the source of, the sex pheromone (see section 4.6 and Chapter 2). Two behavioral mechanisms, optomotor anemotaxis and a program of self-steered counterturns, appear to explain most of the maneuvers exhibited by male moths during this sex pheromone-mediated flight (Baker,

1989a). Optomotor anemotaxis in male moths requires both chemical (from the sex pheromone) and visual inputs (see Fig. 1.7) in order that the male moth may steer an upwind track, while the program of self-steered counterturns (i.e., an internal input) is apparently modulated by pheromonal input (such as pheromone concentration or blend quality) (Baker, 1989a). Flight altitude is also controlled by an optomotor response, requiring the integration of both olfactory and visual inputs (Preiss and Kramer, 1986).

When nearing a pheromone source, pheromonal and visual inputs continue to be integrated and influence where the male moth lands in relation to the pheromone source. Even after landing, male moths probably need continued pheromonal stimulation, when visually orienting and moving toward the pheromone source

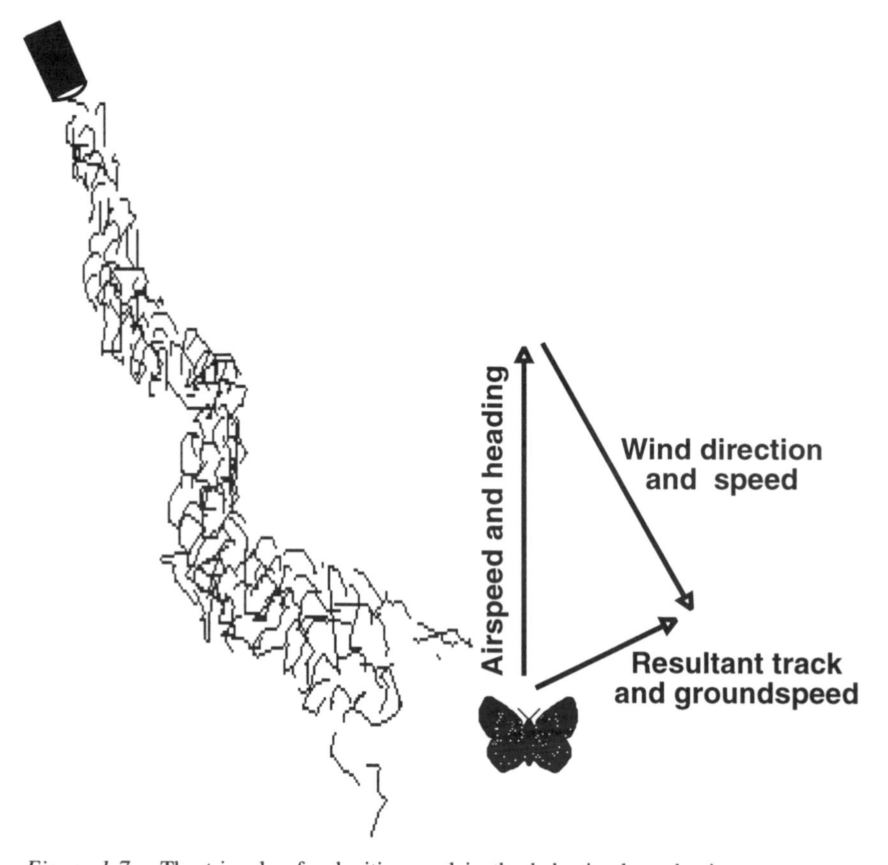

Figure 1.7. The triangle of velocities used in the behavioral mechanism, optomotor anemotaxis, of male moths during sex pheromone-mediated flight. Pheromonal stimulation (the pheromone plume is shown to the left of the insect) elicits the mechanism which allows males to use visual feedback in order to progress upwind [after Marsh et al. (1978)].

(Charlton and Cardé, 1990; Foster and Harris, 1992). In the oriental fruit moth, *Grapholita molesta*, mechanical (tactile) inputs also appear to be integrated with chemical and visual inputs during close-range interactions between the male and the source of the pheromone (Baker and Cardé, 1979).

5.3. Courtship Behavior of *Drosophila melanogaster*

Courtship behavior is a term that encompasses aspects of mating behavior when the male and female are at a somewhat arbitrarily-decided distance close to each other. In general, it is applied to situations where behavioral interactions between the two sexes are obvious. Extensive work on the courtship behavior of *Drosophila* species, especially *D. melanogaster*, has, for the most part, been carried out using genetic mutants, comparing their behavior with that of wild-type insects, as a way to probe the genetic basis of this behavior [see reviews by Ewing (1983), Tompkins (1984), and Jallon (1984)].

The courtship behavior of *D. melanogaster* has been well described (Markow and Hanson, 1981). Essentially, the behavior consists of the male orienting to face the female, tapping her abdomen with one of his foretarsi, and then extending and vibrating one of his wings. If the female moves away from the male, he follows her, continuing his wing vibration and attempting to tap her abdomen. The male extends his proboscis to contact the female's genitalia, and then curls his abdomen toward the female's genitalia. The female may then open her vaginal plates and copulation will generally ensue.

The focus in the courtship behavior of *D. melanogaster* has been on the responses of normal and mutant males to various stimuli associated with females and other males (male homosexual courtship behavior has been extensively studied). Although females clearly respond to various external stimuli, including auditory, visual, and possibly tactile, during courtship, there is no evidence to suggest that females also perceive chemical input during this behavior (Tompkins, 1984). Hence the following discussion is principally concerned with the integration of chemical and other inputs by the male.

Internal stimuli influence the courtship behavior of *D. melanogaster*. For example, young males do not exhibit courtship behaviors (Tompkins, 1984). The male's wing vibratory response (song) is believed to be affected by internal factors that also affect the internal clock controlling the circadian rhythm (Kyriacou and Hall, 1980; 1989).

Various long-chain hydrocarbons mediate the courtship behavior of male *D. melanogaster*. The compound, 7, 11-heptacosadiene, found on the cuticle of virgin females, elicits courtship responses by males (Antony et al., 1985). However, mated females generally elicit less courtship from males than do virgin females. The decreased response (of males to females) is thought to be due, in part, to 7-tricosene, which is transferred from the male to the female during mating (Antony et al., 1985; Scott and Jackson, 1988). The response of males to mated females is therefore mediated by an integration of chemical signals, one that elicits (7, 11-heptacosadiene), and the other that effectively inhibits (7-tricosene) the expression of the overt behavior.

The response of males to face the female during courtship is almost certainly mediated by the integration of chemical and visual inputs. The orientation of the male toward the female suggests that it is a direct visual orientation that is mediated by chemical input. This is supported by experiments with male visual mutants, which orient less to females than do wild-type males (Connolly et al., 1969; Willmund and Ewing, 1982). After orienting, the male contacts the female, during which he probably perceives additional chemical, as well as possibly tactile inputs. Mutant males that are both blind and olfactory-deficient (i.e., do not respond to the volatile chemicals extracted from the female) will initiate undirected courtship following contact with a female, and may eventually (after a much longer time than wild-type males) copulate with a female (Tompkins et al., 1980; Tompkins, 1984). In addition to the integration of visual, chemical, and perhaps mechanical inputs that mediate courtship at this point, there is also a possibility that males integrate an auditory input as well; the sound generated by male wing vibrations, in addition to its effect on female *D. melanogaster*, has been shown to increase the locomotor activity of males (von Schilcher, 1976).

The female's response of moving away from the male also provides an additional visual stimulus to the male. Optomotor-blind males, which show little or no response to horizontally moving objects, do not pursue a moving female as do wild-type males (Tompkins et al., 1982). Throughout the phase of courtship in which the (wild-type) male is following the female, the male taps the female's abdomen and repeatedly contacts her genitalia, suggesting that the courtship behavior continues to involve the integration of chemical, visual, and mechanical inputs.

Although the courtship behavior of *D. melanogaster* illustrates the integration of a variety of different sensory inputs during close-range behavioral interactions of mates, it should be pointed out that much of this work has been conducted in highly constrained and artificial conditions (Ewing, 1983). The natural behavior probably involves more complex behavioral interactions between mates, and hence the integration of even more inputs.

5.4. Feeding Behavior of Locusts

Studies on locusts and other acridids illustrate the variety of inputs that can be important in mediating even apparently simple behaviors, such as feeding. The following units have been distinguished for the feeding behavior of locusts (Bernays and Chapman, 1974): walking after a period of sitting, the cessation of walking after the tarsi contact a leaf, touching the leaf repeatedly with the palps (this behavior continues throughout feeding), lowering of the head so that the labrum contacts the leaf, biting the leaf, sustained feeding, and termination of feeding. Each of these behavioral units has been shown to be influenced by both internal and external stimuli.

Many different inputs appear to be integrated to produce the first of these behavioral units, movement after a period of sitting. Internal stimuli [reviewed by Simpson and Ludlow (1986), and Simpson (1992)] play an important role: they appear to trigger walking and also play a role in sustaining movement, even in what may be a fairly featureless environment. The internal stimuli producing these behavioral effects are related to processes in the digestive tract, such as the rate of gut emptying and the time of defecation, as well as the nutritional composition of the hemolymph, levels of hormones, and circadian and short-term rhythms.

The internal state of the insect is important for the responses of locusts to stimuli from plants. Starved *Schistocerca* nymphs walk upwind when exposed to grass odors in a wind tunnel (Kennedy and Moorhouse, 1969; Moorhouse, 1971). However, satiated nymphs do not respond when exposed to grass odors, suggesting that internal stimuli (associated with starvation) affect the responses of nymphs to odor through modulating response thresholds.

When a locust contacts a plant with its tarsi or antennae, it probably integrates chemical, visual, and mechanical inputs arising from the plant, as well as mechanical inputs generated during examining and feeding behaviors. Chemoreceptors on the tarsi code information about plant chemicals present in leaf surface waxes (Kendall, 1971). This information, if appropriate (e.g., stimulatory and/or not inhibitory), contributes to the triggering of palpation. During palpation, both chemical and mechanical inputs are probably integrated (Simpson, 1992). Integration of these inputs is probably involved in determining whether palpation will continue and whether the head will be lowered so that the labrum is brought into contact with the plant. Contact of the labrum with the plant stimulates mechanoreceptors on the lateral parts of the labrum (Sinoir, 1969). Inputs from mechanoreceptors, perhaps in addition to various visual inputs (Bernays and Chapman, 1974), direct the locust to the leaf edge, where the first bite is taken (Blaney and Chapman, 1970). At this point, chemoreceptors on the inside of the clypeo-labrum are brought into contact with the food and influence the initiation of feeding (Sinoir, 1970; Chapman, 1990). Throughout the course of a meal, the frequency of chewing and the power output of the muscles associated with the mandibles are modulated by feedback from various mechanical inputs (Seath, 1977a, Seath, 1977b). Inputs from chemoreceptors on the palps (palpation continues throughout the meal) and the labrum, as well as feedback from the crop and hindgut (Simpson, 1983), also affect the length of the meal (Sinoir, 1970; Blaney and Duckett, 1975).

With regard to the integration of inputs during feeding in locusts, it is interesting that the elimination of one set of chemosensory organs (and presumably the loss or change in the inputs from those sense organs) may be compensated for, to some degree, by inputs from another set of chemosensory organs. When the palps of *Schistocerca* were amputated [Mordue (Luntz), 1979], nymphs spent less time feeding during the first five minutes of the meal. However, when plant odors were present, amputee nymphs increased the time they spent feeding. This odor-mediated increase in feeding did not occur for nymphs with intact palps.

The influence of the internal state of the locust on several of the aforementioned input-output relationships is illustrated in a study by Blaney and Chapman (1970). Third instar nymphs of *Locusta* were allowed to feed to repletion, then starved for periods from 30 minutes to 3 hours. After being starved, a nymph was given a blade of grass to establish whether it would feed. Once the grass was bitten, it was removed and the nymph given either a new blade of grass or a leaf of *Bellis perennis*. Nymphs starved for less than an hour palpated both plants, but were much more likely to bite and feed on the grass than the *Bellis*. As the time of starvation increased, a larger number of nymphs bit the *Bellis* after palpating it. Thus, the internal stimuli arising from starvation modified the behavioral output of locusts to external stimuli, in this case the stimuli associated with what are usually nonpreferred plants. Starved nymphs will also palpate and bite an object with suitable visual (Williams, 1954) and mechanical (Sinoir, 1969; Dadd, 1960) features, even when the object contains no chemical stimuli.

Experience with a plant also appears to change the behavioral responses of *Locusta* nymphs. Nymphs given a nonhost plant (Blaney and Simmonds, 1985), initially palpated, bit, and then rejected the plant (Fig. 1.8). However, by the fourth encounter with the plant, rejection occurred earlier, usually after palpation.

Precisely how starvation and experience alter the behavioral input-output relationships of locusts is not known. Possibly, they change the response thresholds to particular external stimuli and thereby change the integration of the various stimuli (see section 4.5).

5.5. Host-Finding Behavior of Hemotophagous Flies

Work on the host-finding behavior of hemotophagous flies illustrates the integration of chemical and other inputs in a behavior, and also illustrates the practical benefits of investigating multiple stimuli involved in insect behavior. The two most studied groups of these flies, because of their great significance to the health of humans and domestic animals, are various species of mosquitoes and tsetses (*Glossina* spp). Research on the host-finding behaviors of these species has been driven by a desire to find ways to reduce or eliminate the health risk posed by these insects, through exploiting this behavior. The most common approach used to exploit this behavior has been the trapping of the flies, to reduce fly populations, using various stimuli that elicit the host-finding behavior. This approach has proven particularly successful for tsetse flies because of their very low reproductive potential (most females probably produce less than 10 offspring) (Colvin and Gibson, 1992).

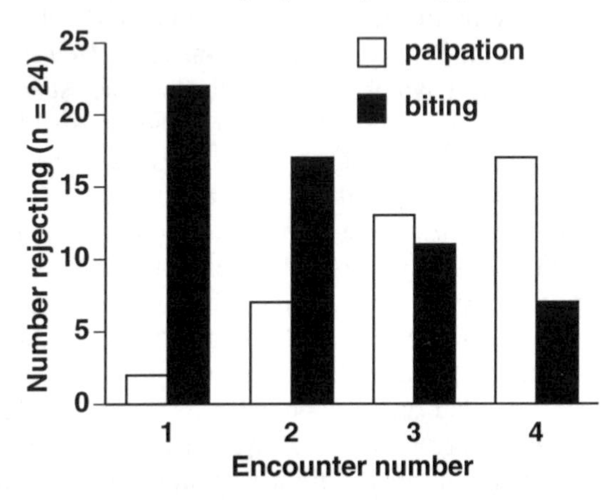

Figure 1.8. The effects of experience on the rejection behavior of fifth instar nymphs of *Locusta migratoria* to the plant *Senecio vulgaris*. Individual nymphs were placed in arenas with a plant and observed during the first four encounters with the plant for whether they rejected the plant after palpating it (open bars) or after biting it (solid bars) [from Blaney and Simmonds (1985)].

In a general sense, the host-finding behavior of hemotophagous flies is considered as comprising the range of events from activation of the fly at rest to landing on the host. Several chemicals, emanating from hosts and their excrement, are known to be involved in the host-finding behavior of both mosquitoes and tsetse flies, including CO_2, lactic acid, and 1-octen-3-ol for mosquitoes (Bowen, 1991), and CO_2, acetone, butanone, 1-octen-3-ol, and various phenols for tsetse flies (Colvin and Gibson, 1992). For both groups of insects, there is some evidence that different chemicals involved in the behavior are perceived and integrated in the CNS. Greater proportions of females of the mosquito *Aedes aegyptii* took flight and landed to specific blends of CO_2, lactic acid, and other chemicals in human sweat, than to the chemicals presented singly (Eiras and Jepson, 1991). Distinct receptor neurons for CO_2 and lactic acid have been found on the antennae and palps, respectively, of mosquitoes (Bowen, 1991), suggesting that the inputs resulting from perception of these chemicals are probably integrated in the CNS. Tsetses respond to a blend of chemicals in cattle odor, but the concomitant presence of human odor can reduce the response of tsetses to cattle (Colvin and Gibson, 1992).

In a landmark paper, Kennedy reported that the mosquito *A. aegyptii* used visual information to orient and fly upwind. He suggested that the "flying orientation to a wind-borne scent such as carbon dioxide from hosts is not, in a direct sense, easily conceivable. But the *activating* effect of scent, combined with the visual orientation upwind, would serve as an effective host-finding mechanism" (Kennedy, 1940). In effect, this laid the foundation for the elucidation of the optomotor anemotaxis mechanism used by moths and other insects for odor-mediated upwind flight.

While mosquitoes appear to integrate visual and chemical inputs for optomotor anemotaxis, the odor-mediated flight of tsetse flies appears to be slightly different. In response to host odor or CO_2, tsetses generally fly upwind, often in shallow curves. Wind tunnel studies have shown that during this odor-mediated flight, tsetses use visual feedback (i.e., optomotor anemotaxis) to control their resultant track and ground speed (Colvin et al., 1989; Paynter and Brady, 1993). In the field, tsetse flies commonly fly in short bursts, consisting of frequent landings followed by takeoffs into the wind. It has been proposed that in this type of behavior, tsetses use a mechanical stimulus, rather than the visual stimulus used during flight, to determine wind direction (Bursell, 1984; 1987; Colvin et al., 1989). Whatever, during host-finding flight it is probable that tsetse flies integrate chemical with visual, and possibly, mechanical inputs.

The role of visual stimuli in the host-finding behavior of tsetse flies has received considerable attention. It is known that the behavior of tsetses is influenced by visual stimuli, particularly, movement, shape, color, brightness, contrast, and pattern (Colvin and Gibson, 1992). The attraction of tsetse flies to visual stimuli during the host-finding behavior provides another interesting aspect of the interaction of chemical with visual inputs. The addition of CO_2 and/or host odors to an apparent (visually) object has been shown to increase trap catches of tsetse flies (Hargrove, 1980). In particular, the combination of visual and chemical stimuli appears to focus the flight and landing of flies at or near the chemical source (Torr, 1989; Bursell, 1990). However, if the apparent object is separated (by at least 3 m) from the odor source, then flies exhibiting odor-mediated flight will divert their track out of the odor plume towards the object. Although in this situation, landing on or near the object occurs strictly in the absence of the chemical stimulus, there is evidence to suggest that the visual attraction to the object is mediated by chemical input (Warnes, 1990; Bursell, 1990). Visual stimuli in combination with odor also appear to attract and stimulate landing of female mosquitoes (Hocking, 1971).

Internal stimuli influence the host-finding behavior of both mosquitoes and tsetse flies. The feeding (and therefore probably the host-finding) behavior of tsetse flies occurs mostly in the morning and late afternoon, being regulated by a circadian rhythm, which is modified by temperature (Brady and Crump, 1978). The visual responses of tsetse flies to a moving object are affected by starvation, with more starved flies exhibiting greater activity, and being more likely to land on a target. These responses are also influenced by age, sex, and pregnancy status, with, for example, pregnant females being approximately half as responsive to a moving target as mature males (Brady, 1972). The odor-mediated, host-finding behavior of female mosquitoes is also similarly influenced by circadian rhythm (Reisen and Aslamkhan, 1978; Taylor et al., 1979), age (Bowen, 1991), and level of starvation (Klowden, 1988).

5.6. Host-Finding Behavior of Onion Flies

The host-finding behavior of female onion flies, *Delia antiqua* has been studied for many years, especially within the context of chemical inputs influencing this behavior. Field studies (Martinson et al., 1988; Martinson et al., 1989) indicate that onion flies are quite mobile, moving distances of up to 1.5 km from overwintering sites. Many individuals probably find onion fields simply by chance encounter (Martinson et al., 1989). However, some proportion of the population exhibit anemotaxis from up to 100 m downwind of a source of *n*-dipropyl disulfide (Judd and Borden, 1988; Judd and Borden, 1989), a volatile chemical emitted by onions. Female onion flies typically fly upwind to the host odor source in a series of short flights, each flight punctuated by the female landing on the ground, orienting in an upwind direction and again taking flight. There are probably two distinct types of odor-mediated flight behavior: anemotaxis, involving mechanical stimuli while the fly is in contact with the ground, and optomotor anemotaxis involving visual stimuli while the fly is in flight (as described for the cabbage root fly, Nottingham, 1988).

The responses of onion flies to dipropyl disulfide appear to be influenced strongly by internal stimuli. Recapture rates of males and females and the frequency of approaches toward dipropyl disulfide-baited traps are both dependent upon age and mating status (Judd and Borden, 1992). For example, virgin females move longer distances and show directed responses to wind in the presence of host odors, while mated females apparently do not direct their movements in relation to the wind but exhibit increased frequencies of turning or slower movements in the presence of more complex blends of host odors (such as aged onions).

Visual stimuli from plants also influence this behavior. Both virgin and mated females landed more frequently on cylinders baited with dipropyl disulfide than on spheres baited with the same chemical (Judd and Borden, 1991). However, the color responses of mated and virgin females differed somewhat; when traps were baited with dipropyl disulfide, mated females landed more frequently on green rather than white cylinders, whereas virgin females did not discriminate between cylinders of the two colors.

After a female lands on an onion plant, inputs from chemoreceptors are also probably integrated with various other inputs from external stimuli. For example, onion foliage influences responses to dipropyl disulfide and other odors. In assays (Harris and Miller, 1982) comparing onion foliage positioned vertically in a dish of sand, with an onion bulb placed below the surface in a dish of sand, females laid virtually all their eggs around the foliage, even though the onion bulb contained greater amounts of stimulatory odors. When given a choice between plant mimics representing various components of the foliage (Fig. 1.9), female onion flies laid few eggs in dishes with only odor added, or in dishes with only the foliar form, or form and color added. Significantly more eggs were laid in dishes with both host odor and foliar form added, but this increase was small compared to the increase that occurred when all three components were combined (Harris and Miller, 1982). In no-choice assays, foliar form, color and odors all influenced the numbers of eggs laid (Harris and Miller, 1988).

As discussed previously (section 4.3), measuring only the endpoint of a behavior (here, the number of eggs laid) can be misleading about how and when various stimuli influence the series of behaviors or units leading to the endpoint. For the onion fly, the behaviors that occur on the plant and which may be influenced by plant chemicals, form, and color (Harris and Miller, 1991) are as follows. After landing, the onion fly female initially sits motionless on the plant or grooms herself, but within approximately 10 s begins to run. If she is on the foliage, the run is usually oriented toward the base of the plant; if she is on the soil, the run is oriented up the vertical surfaces of the foliage. During these vertical runs, females generally move no more than 4–6 cm, then turn 180° and run in the opposite direction. These runs are usually repeated several times, and sometimes are followed by short, circling runs around the base of the plant. After approximately 10 s of running, the female extends her proboscis so that her labellum contacts the foliage or soil. These proboscis extensions are repeated until the female arches her abdomen and extends her terminal abdominal segments

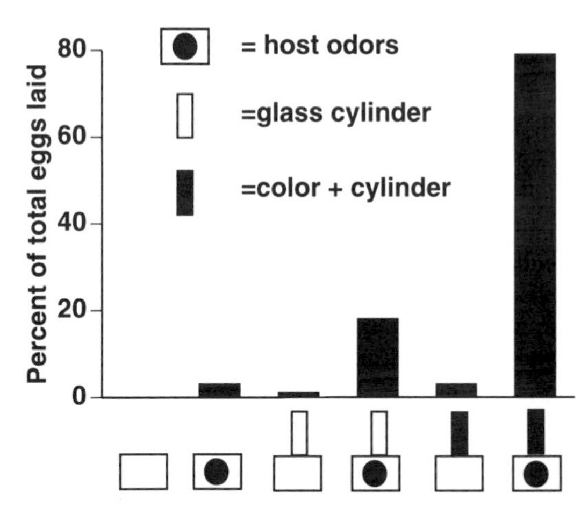

Figure 1.9. The influence of chemical, visual, and form stimuli from onion foliage on the percentage of eggs laid by female onion flies, *Delia antiqua*. Stimuli were tested in choice assays. Host odors emanated from chopped onion placed beneath sand, a glass cylinder was used to mimic the form of onion foliage, and a yellow piece of paper was inserted in the glass cylinder to introduce color [from Harris and Miller (1982)].

so that the tip of her abdomen contacts the foliage. This probing of plant (and soil) surfaces with the tip of the abdomen (surface probing) continues while the female moves over the plant. Eventually, the female places her ovipositor into the soil (subsurface probing) and probes the sides and bottom of crevices (Mowry et al., 1989). An egg is propelled into the bursa copulatrix, fertilized, and expelled through the gonopore. After depositing one such egg clutch, females frequently resume examining behaviors, such as surface probing and running, and lay more eggs.

Multiple inputs appear to be important throughout the sequence of behaviors leading to the laying of eggs by female onion flies (Harris and Miller, 1983; 1984; 1988; 1991; Harris et al., 1987). For example, runs over the plant surface appear to be triggered by foliar color and form, but are only sustained if host odors (e.g., dipropyl disulfide) are present and accompanied by specific forms which seem to serve as tracks for running females. The detection of the shape, size, and orientation of onion foliage seems to occur during this running phase. For example, when females, in the presence of odor, were on cone-shaped rather than cylindrical models, they started running, but soon slowed their forward movement and deviated from their normal vertical running tracks, wandering randomly over the surface of the model. Eventually they stopped moving and usually flew away without finishing or repeating runs.

These on-plant examining behaviors of onion flies illustrate the difficulty that can be experienced with identifying the nature of inputs for a behavior. The precise inputs coding the information concerning the shape, size, and orientation of onion foliage are not known. For females to construct an image of the plant form (in the CNS) based on inputs from mechanoreceptors [in a way similar to that proposed for *Trichogramma minutum* females examining host eggs, Schmidt and Smith (1986)] during these runs would require keeping track of inputs from many different mechanosensory systems over time. A simpler way to detect the overall form of a plant might be via the triggering of a motor pattern after landing, with this motor pattern only being: (1) Possible, if the object landed on allowed the stereotyped movements (short straight runs up and down vertical surfaces); and (2) sustained, if feedback regarding these movements are integrated with visual (color of the plant), chemical (odor), and mechanical (vertical alignment) inputs.

5.7. Host-Finding Behavior of Parasitoids

Many parasitoids respond to chemicals from intact or damaged plants, and from various larval products such as frass, during the series of movements that leads to the laying of eggs (van Alphen and Vet, 1986). These behavioral responses to chemicals are particularly interesting because they are frequently altered by previous experience (Vet and Groenewold, 1990) and because they may be useful tools for improving the efficacy of natural enemies for pest control (Tumlinson et al., 1993). However, given the large amount of research on the chemically-mediated host-finding behavior of parasitoids, it is surprising that relatively little is known about other external and internal stimuli that contribute to these behaviors, and how the integration of the resultant inputs is altered by experience (but see Wardle and Borden, 1985, Wardle, 1990, and Drost and Cardé, 1992).

Recently, McAuslane et al. (1991) have shown that visual stimuli from plants can have a significant effect on the host-finding behavior of parasitoids. In these experiments, concentrations of host-plant odors were held constant in a wind tunnel with odor sources placed on the upwind side of a screen to obscure attendant visual cues. Downwind of the screen, the size of visual stimuli was changed by presenting either a single damaged leaf or an entire plant with a single damaged leaf. *Campoletis sonorensis* were released downwind of either the leaf or the plant. Both naive and experienced females completed more flights to an odor source when a whole plant rather than a single leaf was presented (Fig. 1.10). This suggests that visual stimuli associated with either the color or the form of the plant are important at some stage during orientation. Visual stimuli had a greater effect on the host-finding behavior of naive rather than experienced parasitoids.

For parasitoids, visual stimuli from a plant could influence long-range orientation ($\geq$50 cm from the plant) or just the final stages of flight and landing (a few cm from the plant). There is some evidence for the former possibility (McAuslane et al., 1991). After taking flight 50 cm downwind of the leaf or plant, female *C. sonorensis* (both naive and experienced) were more likely to exhibit oriented flight when the plant (i.e., the larger visual stimulus) was presented (Fig. 1.10). In this example, whether females detected plant-based visual stimuli before or after takeoff from the release platform could not be distinguished. However, in a study on a different parasitoid, *Dacnusa sibirica* (Dicke and Minkenberg, 1991), responses of host-experienced parasitoids to infested and noninfested tomato leaves (plant-host odors plus visual stimulus) were compared to responses to the plant-host odors alone. Although the experiments were not strictly designed to make these comparisons, the latencies to flight were shorter (Fig. 1.11) when visual plus odor stimuli were presented than when only the odor stimulus was presented alone. Additionally, a higher proportion of these flights occurred in an upwind direction (Fig. 1.11). Thus, in the case of *D. sibirica*, it appears that the presence of odor elicited the parasitoid to initiate upwind flight sooner when the visual stimulus of a plant was also present. This suggests that the integration of inputs from both chemical and visual receptors plays an important role in the orientation behavior of this parasitoid and perhaps other species.

6. Why Study Multiple Inputs?

The examples show that even the most apparently simple and straightforward of chemically-mediated behaviors is complex and probably involves a multitude of

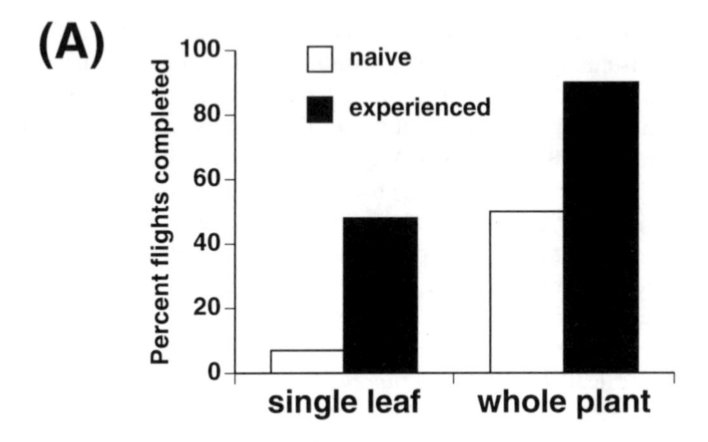

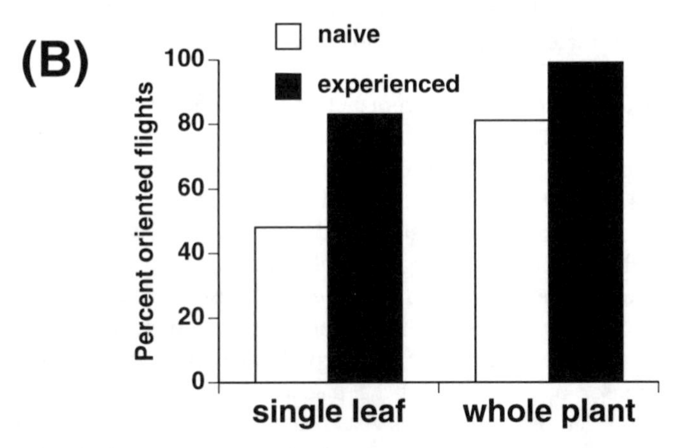

Figure 1.10. The influence of odor and visual stimuli on the host-finding behavior of female *Campoletis sonorensis* parasitoids: (a) Percentage of successful upwind flights and (b) percentage of oriented flights, exhibited by naive (open bars) and experienced (solid bars) females in the presence of a single cotton leaf (smaller visual stimulus) or cotton plant (larger visual stimulus) infested with *Heliothis virescens* larvae [from McAuslane et al. (1991)].

inputs, arising from external and internal stimuli, that are processed by the CNS. Thus, an understanding of the immediate cause of a behavior is not provided by characterizing one stimulus and ascertaining whether it elicits a response from the insect. Rather, an understanding of a behavior and its immediate cause results from studying a range of stimuli, how they interact, how their inputs are processed, and how the behavior is organized by the CNS.

The substitution of a simple approach (i.e., studying the response of the insect

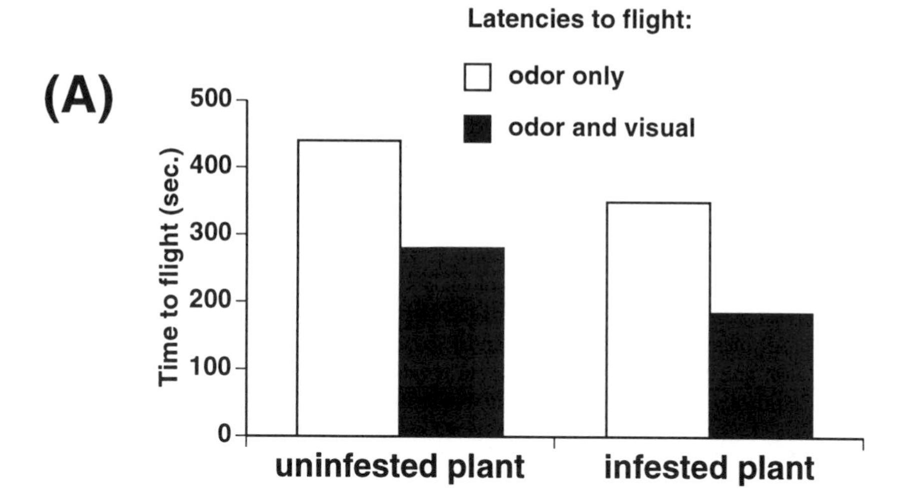

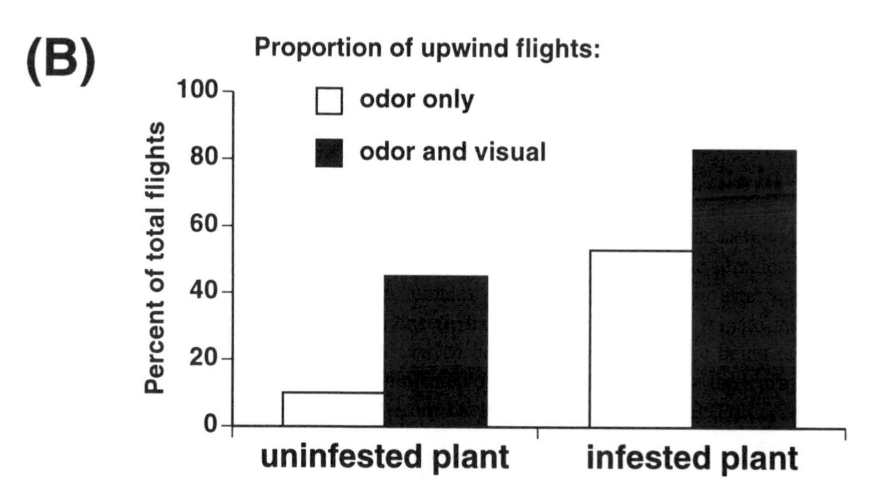

Figure 1.11. The influence of odor and visual stimuli on the host-finding behaviors of female *Dacnusa sibirica* parasitoids: (a) Latencies to flight from the time females were released and (b) the percentages of these females exhibiting upwind flight after leaving the release site. All females were experienced, having been exposed to tomato leaflets infested with larvae of the leafminer, *Liriomyza bryoniae,* for 4 hr before being tested. Odor sources were tomato plants either uninfested or infested with larvae of the leafminer. The visual stimulus was a tomato plant [from Dicke and Minkenberg (1991)].

to a single external stimulus) by a much more complex approach, begs the question of, what is the value of such a change, particularly if one is only interested in the behavior as a character of the insect so as to address economic or evolutionary questions concerning the role of the chemical? Indeed, many studies are not interested in a complex understanding of an insect behavior, and therefore it may seem superfluous to study the behavior in such depth. There are a number of examples where the primary aim of control of a pest insect using a chemical that influences a behavior has been achieved by such an approach without resorting to detailed behavioral studies involving other (than the chemical) stimuli (see Ridgway et al., 1990). However, there can be little doubt that any behavior of a pest insect would be better exploited for control by an increased understanding of the behavior (see Dent, 1991), what stimuli influence it, and how inputs to the behavior are processed and organized; the study of the host-finding behavior of tsetse flies is evidence of this (see section 5.5). Similarly, an understanding of the evolution or function of a behavior cannot be achieved without a proper understanding of the immediate causes of the behavior.

In 1972, in his address to the International Congress of Entomology entitled "The Emergence of Behaviour," J. S. Kennedy pointed out some of the problems with the study of insect behavior and its causes, and described how the science was progressing (Kennedy, 1972). Some 20 years or so later, in large part due to the contributions and influence of Kennedy, the study of the immediate causes of insect behavior has progressed to the point where certain chemically-mediated behaviors and some of their neurophysiological inputs are quite well understood. The challenge ahead for behaviorists and physiologists is to link research on these two extremes by understanding the processing and integration of the many inputs involved in these behaviors.

Acknowledgments

We are grateful to Drs. C.E. Linn, F. Marion-Poll, D.R. Papaj, and W.L. Roelofs, and to Ms. T. Withers and the editors of this book, for their helpful comments on an earlier version of this manuscript.

References

Antony, C., Davis, T., Carlson, D., and Pechiné, J. (1985) Compared behavioral responses of male *Drosophila melanogaster* (Canton-S) to natural and synthetic aphrodisiacs. *J. Chem. Ecol.* 11: 1617–1629.

Arn, H., Toth, M., and Priesner, E. (1992) *List of sex pheromones of Lepidoptera and related attractants*. 2nd Ed., OILB/IOBC-WPRS, Wädenswil, Switzerland.

Baker, T.C. (1989a) Pheromones and flight behavior. In: *In Insect Flight*, (Goldsworthy, G.J. and Wheeler, C.H., eds.), pp. 231–255. CRC Press, Boca Raton, Fla.

Baker, T.C. (1989b) Sex pheromone communication in the Lepidoptera: New research progress. *Experientia* 45: 248–262.

Baker, T.C. and Cardé, R.T. (1979) Courtship behavior of the Oriental fruit moth *(Grapholitha molesta)*: Experimental analysis and consideration of the role of sexual selection in the evolution of courtship pheromones in the Lepidoptera. *Ann. Entomol. Soc. Am.* 72: 173–188.

Baker, T.C. and Cardé, R.T. (1984) Techniques for behavioral bioassays. In: *Techniques in Pheromone Research*, (Hummel, H.E. and Miller, T.A., eds.), pp. 45–73. Springer, New York.

Baker, T., Hansson, B., Löfstedt, C., and Löfqvist, J. (1988) Adaptation of antennal neurons in moths associated with cessation of upwind flight. *Proc. Natl. Acad. Sci. USA* 85: 9826–9830.

Bernays, E.A. and Chapman, R.F. (1974) The regulation of food intake by acridids. In: *Experimental Analysis of Insect Behaviour*, (Barton Browne, L., ed.), pp. 48–59. Springer Verlag, Berlin.

Blaney, W.M. and Chapman, R.F. (1970) The functions of the maxillary palps of Acrididae (Orthoptera). *Entomol. exp. appl.* 13: 363–376.

Blaney, W.M. and Duckett, A.M. (1975) The significance of palpation by the maxillary palps of *Locusta migratoria* (L.): An electrophysiological and behavioural study. *J. Exp. Biol.* 63: 701–712.

Blaney, W.M., Schoonhoven, L.M., and Simmonds, M.S.J. (1986) Sensitivity variations in insect chemoreceptors: a review. *Experientia* 42: 13–19.

Blaney, W.M. and Simmonds, M.S.J. (1985) Food selection by locusts: An analysis of rejection behaviour. *Entomol. exp. appl.* 38: 35–40.

Bowen, M.F. (1991) The sensory physiology of host-seeking behaviour in mosquitoes. *Annu. Rev. Entomol.* 36: 139–158.

Brady, J. (1972) The visual responsiveness of the tsetse fly *Glossina morsitans* West. (Glossinidae) to moving objects: The effects of hunger, sex, host odour and stimulus characteristics. *Bull. Entomol. Res.* 62: 257–279.

Brady, J. and Crump, J. (1978) The control of circadian rhythms in tsetse flies: Environmental or physiological clock? *Physiol. Entomol.* 3: 177–190.

Bursell, E. (1984) Observations on the orientation of tsetse flies to wind-borne odours. *Physiol. Entomol.* 9: 133–137.

Bursell, E. (1987) The effect of wind-borne odours on the direction of flight of in tsetse flies, *Glossina* spp.. *Physiol. Entomol.* 12: 149–156.

Bursell, E. (1990) The effect of host odour on the landing responses of tsetse flies (*Glossina morsitans morsitans*) in a wind tunnel with and without visual targets. *Physiol. Entomol.* 15: 369–376.

Camhi, J. (1984) *Neuroethology*. Sinauer Associates Inc., Sunderland, Mass.

Chapman, R.F. (1990) Food selection. In: *Biology of Grasshoppers*, (Chapman, R.F. and Joern, A., eds.), pp. 39–72. John Wiley & Sons, New York.

Charlton, R.E. and Cardé, R.T. (1990) Orientation of male gypsy moths, *Lymantria*

(L.) to pheromone sources: The role of olfactory and visual cues. *J. Insect Behav.* 3: 443–469.

Colvin, J. and Gibson, G. (1992) Host-seeking behavior and management of tsetse. *Annu. Rev. Entomol.* 37: 21–40.

Colvin, J., Brady, J., and Gibson, G. (1989) Visually-guided, upwind turning behaviour of free-flying tsetse flies in odour-laden wind: a wind-tunnel study. *Physiol. Entomol.* 14: 31–39.

Connolly, K., Burnet, B., and Sewell, D. (1969) Selective mating and eye pigmentation: An analysis of the visual component in the courtship behavior of *Drosophila melanogaster*. *Evolution* 23: 548–559.

Dadd, R. H. (1960) Observations on the palatability and utilisation of food by locusts, with particular reference to the interpretation of performances in growth trials using synthetic diets. *Entomol. exp. appl.* 3: 283–304.

Davis, E. (1986) Peripheral chemoreceptors and regulation of insect behaviour. In: *Mechanisms in Insect Olfaction*, (Payne, T.L., Birch, M.C. and Kennedy, C.E.J., eds.), pp. Clarendon Press, Oxford, U.K.

Dawkins, M.S. (1983) The organisation of motor patterns. In: *Animal Behaviour: Causes and Effects*, (Halliday, T.R. and Slater, P.J.B., eds.), pp. 75–99. Blackwell Scientific Publications, Oxford, U.K.

Dawkins, M.S. (1989) The future of ethology: How many legs are we standing on? *Persp. Ethol.* 8: 47–54.

Dent, D. (1991) *Insect Pest Management*. C.A.B. International, Wallingford, U.K.

Dicke, M. and Minkenberg, O.P.J.M. (1991) Role of volatile infochemicals in foraging behavior of the leafminer parasitoid *Dacnusa sibirica* (Diptera: Agromyzidae). *J. Insect Behav.* 4: 489–500.

Drost, Y.C. and Cardé, R.T. (1992) Use of learned visual cues during habitat location by *Brachymeria intermedia*. *Entomol. exp. appl.* 64: 217–224.

Dusenbery, D.B. (1992) *Sensory Ecology*. W.H. Freeman and Co., New York.

Eiras, E. and Jepson, P. (1991) Host location by *Aedes aegypti* (Diptera: Culicidae): A wind tunnel study of chemical cues. *Bull. Entomol. Res.* 81: 151–160.

Ewing, A. (1983) Functional aspects of *Drosophila* courtship. *Biol. Rev. of the Cambridge Phil. Soc.* 58: 275–292.

Foster, S.P. and Harris, M.O. (1992) Factors influencing the landing of male *Epiphyas postvittana* (Walker) exhibiting pheromone-mediated flight. *J. Insect Behav.* 5: 699–720.

Foster, S.P. and Muggleston, S.J. (1993) Effect of design of a sex pheromone-baited delta trap on behavior and catch of male *Epiphyas postvittana* (Walker). *J. Chem. Ecol.* 19: 2617–2633.

Halliday, T.R. and Slater, P.J.B. (1983) Introduction. In: *Animal Behaviour: Causes and Effects*, (Halliday, T.R. and Slater, P.J.B., eds.). Blackwell Scientific Publications, Oxford, U.K.

Hansson, B.S., Christensen, T.A., and Hildebrand, J.G. (1991) Functionally distinct subdivisions of the macroglomerular complex in the antennal lobe of the male sphinx moth *Manduca sexta*. *J. Comp. Neurol.* 312: 264–278.

Hansson, B., Ljungberg, H., Hallberg, E., and Löfstedt, C. (1992) Functional specialization of olfactory glomeruli in a moth. *Science* 256: 1313–1315.

Hargrove, J. (1980) Improved estimates of the efficiency of traps for *Glossina morsitans morsitans* Westwood and *G. pallidipes* Austen (Diptera: Glossinidae), with a note on the effect of the concentration of accompanying host odour. *Bull. Entomol. Res.* 70: 579–587.

Harris, M.O. and Miller, J.R. (1982) Synergism of visual and chemical stimuli in the oviposition behaviour of *Delia antiqua*. In: *Proceedings of the 5th Insect-Plant Relationships*, pp. 117–122. Pudoc, Wageningen.

Harris, M.O. and Miller, J.R. (1983) Color stimuli and oviposition behavior of the onion fly, *Delia antiqua* (Meigen) (Diptera: Anthomyiidae). *Ann. Entomol. Soc. Am.* 76: 766–771.

Harris, M.O. and Miller, J.R. (1984) Foliar form influences ovipositional behaviour of the onion fly. *Physiol. Entomol.* 9: 145–155.

Harris, M.O. and Miller, J.R. (1988) Host-acceptance behaviour in an herbivorous fly, *Delia antiqua*. *J. Insect Physiol.* 34: 179–190.

Harris, M.O. and Miller, J.R. (1991) Quantitative analysis of ovipositional behavior: Effects of a host-plant chemical on the onion fly (Diptera: Anthomyiidae). *J. Insect Behav.* 4: 773–792.

Harris, M.O. and Rose, S. (1990) Chemical, color, and tactile cues influencing the oviposition behavior of the Hessian fly (Diptera: Cecidomyiidae). *Environ. Entomol.* 19: 303–308.

Harris, M.O., Keller, J.E., and Miller, J.R. (1987) Responses to n-dipropyl disulfide by ovipositing onion flies: Effects of concentration and site of release. *J. Chem. Ecol.* 13: 1261–1277.

Harris, M.O., Rose, S., and Malsch, P. (1993) The role of vision in the host plant-finding behaviour of the Hessian fly. *Physiol. Entomol.* 18: 31–42.

Hildebrand, J.G. and Montague, R.A. (1986) Functional organization of olfactory pathways in the central nervous system of *Manduca sexta*. In: *Mechanisms in Insect Olfaction*, (Payne, T.L., Birch, M.C. and Kennedy, C.E.J., eds.), pp. 279–285. Clarendon Press, Oxford, U.K.

Hocking, B. (1971) Blood-sucking behaviour of terrestrial arthropods. *Annu. Rev. Entomol.* 16: 1–26.

Homberg, U. (1984) Processing of antennal information in extrinsic mushroom body neurons of the bee brain. *J. Comp. Physiol. A* 154: 825–836.

Horn, E. (1985) Multimodal convergences. In: *Comprehensive Insect Physiology, Biochemistry and Pharmacology*, (Kerkut, G.A. and Gilbert, L.I., eds.), pp. 653–671. Vol. 6. Pergamon Press, New York.

Jallon, J. (1984) A few chemical words exchanged by *Drosophila* during courtship and mating. *Behav. Genet.* 14: 441–478.

Judd, G.J.R. and Borden, J.H. (1988) Long-range host-finding behaviour of the onion fly *Delia antiqua* (Diptera: Anthomyiidae): Ecological and physiological constraints. *J. Appl. Ecol.* 25: 829–845.

Judd, G.J.R. and Borden, J.H. (1989) Distant olfactory response of the onion fly, *Delia antiqua*, to host-plant odour in the field. *Physiol. Entomol.* 14: 429–441.

Judd, G.J.R. and Borden, J.H. (1991) Sensory interaction during trap-finding by female onion flies: implications for ovipositional host-plant finding. *Entomol. exp. appl.* 58: 239–249.

Judd, G.J.R. and Borden, J.H. (1992) Influence of different habitats and mating on olfactory behavior of onion flies seeking ovipositional hosts. *J. Chem. Ecol.* 18: 605–620.

Kaiser, L. and Cardé, R.T. (1992) In-flight orientation to volatiles from the plant-host complex in *Cotesia rubecula* (Hym.: Braconidae): Increased sensitivity through olfactory experience. *Physiol. Entomol.* 17: 62–67.

Kaissling, K.-E. (1986) Chemo-electrical transduction in insect olfactory receptors. *Annu. Rev. Neur.* 9: 121–145.

Kamm, J.A. (1990) Control of olfactory-induced behavior in alfalfa seed chalcid (Hymenoptera: Eurytomidae) by celestial light. *J. Chem. Ecol.* 16: 291–300.

Kaulen, P., Erber, J., and Mobbs, P. (1984) Current source-density analysis in the mushroom bodies of the honey bee (*Apis mellifera carnica*). *J. Comp. Physiol. A* 154: 569–582.

Kendall, M.D. (1971) Studies on the tarsi of *Schistocerca gregaria* Forskal. Ph.D. Thesis University of London, London.

Kennedy, J.S. (1940) The visual responses of flying mosquitoes. *Proc. Zool. Soc.London (A)* 109: 221–242.

Kennedy, J.S. (1965) Mechanisms of host plant selection. *Ann. Appl. Biol.* 56: 317–322.

Kennedy, J.S (1972) The emergence of behaviour. *J. Aust. Entomol. Soc.* 11: 168–176.

Kennedy, J.S. (1978) The concepts of olfactory "arrestment" and "attraction". *Physiol. Entomol.* 3: 91–98.

Kennedy, J.S. (1992) *The New Anthropomorphism*. Cambridge University Press, Cambridge, U.K.

Kennedy, J.S. and Moorhouse, J.E. (1969) Laboratory observations on locust responses to windborne grass odour. *Entomol. exp. appl.* 12: 489–503.

Klowden, M. (1988) Factors influencing multiple host contacts by mosquitoes during a single gonotrophic cycle. *Entomol. Soc. Am. Misc. Publ.* 68: 29–36.

Kyraicou, C. and Hall, J. (1980) Circadian rhythms mutations in *Drosophila melanogaster* affect short-term fluctuations in the male's courtship song. *Proc. Natl. Acad. Sci. USA* 77: 6729–6733.

Kyraicou, C. and Hall, J. (1989) Spectral analysis of *Drosophila* courtship song rhythms. *Anim. Behav.* 37: 5: 850–859.

Light, D.M. (1986) Central integration of sensory signals: an exploration of processing of pheromonal and multimodal information in lepidopteran brains. In: *Mechanisms in Insect Olfaction*, (Payne, T.L., Birch, M.C. and Kennedy, C.E.J., eds.), pp. 287–301. Clarendon Press, Oxford, U.K.

Linn, C.E., Jr., Campbell, M.G., and Roelofs, W.L. (1987) Pheromone components and active spaces: What do moths smell and where do they smell it? *Science* 237: 650–652.

Linn, C.E., Jr., Campbell, M.G., and Roelofs, W.L. (1988) Temperature modulation of behavioral thresholds controlling male moth sex pheromone response specificity. *Physiol. Entomol.* 13: 59–67.

Markow, T. and Hanson, S. (1981) Multivariate analysis of *Drosophila* courtship. *Proc. Natl. Acad. Sci. USA* 78: 430–434.

Marsh, D., Kennedy, J.S., and Ludlow, A.R. (1978) An analysis of anemotactic zigzagging flight in male moths stimulated by pheromone. *Physiol. Entomol.* 3: 221–240.

Martin, P. and Bateson, P. (1986) *Measuring Behaviour*. Cambridge University Press, Cambridge, U.K.

Martinson, T.E., Nyrop, J.P., and Eckenrode, C.J. (1988) Dispersal of the onion fly (Diptera: Anthomyiidae) and larval damage in rotated onion fields. *J. Econ. Entomol.* 81: 508–514.

Martinson, T.E., Nyrop, J.P., and Eckenrode, C.J. (1989) Long-range host-finding behavior and colonization of onion fields by *Delia antiqua* (Diptera: Anthomyiidae). *J. Econ. Entomol.* 82: 1111–1120.

McAuslane, H.J., Vinson, S.B., and Williams, H.J. (1991) Stimuli influencing host microhabitat location in the parasitoid *Campoletis sonorensis*. *Entomol. exp. appl.* 58: 267–277.

Metcalf, R. (1990) Chemical ecology of Dacinae fruit flies (Diptera: Tephritidae). *Ann. Entomol. Soc. Am.* 83: 1017–1030.

Metcalf, R. and Metcalf, E. (1992) *Plant Kairomones in Insect Ecology and Control*. Vol. 1. Contemporary Topics in Entomology, Miller, T. and van Emden, H., eds. Chapman and Hall, New York.

Moorehouse, J.E., Fosbrooke, I.H.M., and Ludlow, A.R. (1987) Stopping a walking locust with sound: An analysis of variation in behavioural threshold. *J. Exp. Biol.* 46:193–201.

Moorhouse, J. E. (1971) Experimental analysis of the locomotor behaviour of *Schistocerca gregaria* induced by odour. *J. Insect Physiol.* 17: 913–920.

Mordue (Luntz), A.J. (1979) The role of the maxillary and labial palps in the feeding behaviour of *Schistocerca gregaria*. *Entomol. exp. appl.* 25: 279–288.

Mowry, T.M., Spencer, J.L., Keller, J.E., and Miller, J.R. (1989) Onion fly (*Delia antiqua*) egg depositional behaviour: Pinpointing host acceptance by an insect herbivore. *J. Insect Physiol.* 35: 331–339.

Nottingham, S.F. (1988) Host-plant finding for oviposition by adult cabbage root fly, *Delia radicum*. *J. Insect Physiol.* 34: 227–234.

Olberg, R.M. and Willis, M.A. (1990) Pheromone-modulated optomotor response in male gypsy moths, *Lymantria dispar* L.: Directionally selective visual interneurons in the ventral nerve cord. *J. Comp. Physiol. A* 167: 707–714.

Paynter, Q. and Brady, J. (1993) Flight responses of tsetse flies (*Glossina*) to octenol and acetone vapour in a wind-tunnel. *Physiol. Entomol.* 18: 102–108.

Preiss, R. and Kramer, E. (1986) Pheromone-induced anemotaxis in simulated free flight. In: *Mechanisms in Insect Olfaction*, (Payne, T.L., Birch, M.C. and Kennedy, C.E.J., eds.), pp. 69–79. Clarendon Press, Oxford, U.K.

Prokopy, R.J. (1986) Visual and olfactory stimulus interaction in resource finding by insects. In: *Mechanisms in Insect Olfaction*, (Payne, T.L., Birch, M.C. and Kennedy, C.E.J., eds.), pp. 81–89. Clarendon Press, Oxford, U.K.

Reisen, W. and Aslamkhan, M. (1978) Biting rythms of some Pakistan mosquitoes (Diptera: Culicidae). *Bull. Entomol. Res.* 68: 313–330.

Ridgway, R.L., Silverstein, R.M., and Inscoe, M.N. (1990), *Behavior-Modifying Chemicals for Insect Management: Applications of Pheromones and Other Attractants*. Marcell Dekker Inc., New York.

Schildberger, K. (1984) Multimodal interneurons in the cricket brain: properties of identified extrinsic mushroom body cells. *J. Comp. Physiol. A* 154: 71–79.

Schmidt, J.M. and Smith, J.J.B. (1986) Correlations between body angles and substrate curvature in the parasitoid *Trichogramma minutum*: A possible mechanism of host radius curvature. *J. Exp. Biol.* 125: 271–285.

Schmidt, R. (1978) *Fundamentals of Sensory Physiology*. Springer-Verlag, Berlin and New York.

Scott, D. and Jackson, L. (1988) Interstrain comparison of male-predominant antiaphrodisiacs in *Drosophila melanogaster*. *J. Insect Physiol.*. 34: 863–871.

Seath, I. (1977a) The effects of increasing mandibular load on electrical activity in the mandibular closer muscles during feeding in the desert locust, *Schistocerca gregaria*. *Physiol. Entomol.* 2: 237–240.

Seath, I. (1977b) Sensory feedback in the control of mouthpart movements in the desert locust *Schistocerca gregaria*. *Physiol. Entomol.* 2: 147–156.

Shorey, H.H. (1973) Behavioral responses to insect pheromones. *Annu. Rev. Entomol.* 18: 349–380.

Simpson, S.J. (1983) The role of volumetric feedback from the hindgut in the regulation of meal size in fifth-instar *Locusta migratoria* nymphs. *Physiol. Entomol.* 8: 451–467.

Simpson, S.J. (1992) Mechanoresponsive neurones in the subesophageal ganglion of the locust. *Physiol. Entomol.* 17: 351–369.

Simpson, S.J. and Ludlow, A.R. (1986) Why locusts start to feed: A comparison of causal factors. *Anim. Behav.* 34: 480–496.

Singer, M.C. (1986) The definition and measurement of oviposition preference in plant-feeding insects. In: *Insect-Plant Interactions*, (Miller, J.R. and Miller, T.A., eds.), pp. 65–94. Springer-Verlag, New York.

Sinoir, Y. (1969) Le role de palpes et du labre dans le comportement de prise de nourriture chez la larve du criquet migrateur. *Ann. Nutr. Aliment.* 23: 167–194.

Sinoir, Y. (1970) Quelques aspects du comportement de prise de nourriture chez la larve de *Locusta migratoria migratorioides* R. & F. *Ann. Soc. Entomol. France* 6: 391–405.

Sokal, R.R. and Rohlf, F.J. (1981) *Biometry*. W.H. Freeman and Co., New York.

Sugawara, T. (1979) Stretch reception in the bursa copulatrix of the butterfly, *Pieris rapae crucivora*, and its role in behavior. *J. Comp. Physiol. A* 130: 191–199.

Taylor, D., Bennett, G., and Lewis, D. (1979) Observations on the host-seeking activity of some Culicidae in the Tantramar marshes, New Brunswick. *J. Med. Entomol.* 15: 134–137.

Tinbergen, N. (1951) *The Study of Instinct*. Clarendon Press, Oxford, U.K.

Tinbergen, N. (1963) On aims and methods in ethology. *Z. Tierpsych.* 20: 410–433.

Todd, J.L., Haynes, K.F., and Baker, T.C. (1992) Antennal neurones specific for redundant pheromone components in normal and mutant *Trichoplusia ni* males. *Physiol. Entomol.* 17: 183–192.

Tompkins, L. (1984) Genetic analysis of sex appeal in *Drosophila*. *Behav. Genet.* 14: 411–440.

Tompkins, L., Gross, A., Hall, J., Gailey, D., and Siegel, R. (1982) The role of female movement in the sexual behavior of *Drosophila melanogaster*. *Behav. Genet.* 12: 295–307.

Tompkins, L., Hall, J., and Hall, L. (1980) Courtship-stimulating volatile compounds from normal and mutant *Drosophila*. *J. Insect Physiol.* 26: 689–697.

Torr, S. (1989) The host-orientated behaviour of tsetse flies (*Glossina*): The interaction of visual and olfactory stimuli. *Physiol. Entomol.* 14: 325–340.

Tumlinson, J.H., Turlings, T.C.J., and Lewis, W.J. (1993) Semiochemically mediated foraging behavior in beneficial parasitic insects. *Arch. Insect Biochem. Physiol.* 22: 385–391.

van Alphen, J.J.M. and Vet, L.E.M. (1986) An evolutionary approach to host finding and selection. In: *Insect Parasitoids*, (Waage, J. and Greathead, R., eds.), pp. 23–61. Academic Press, London.

Vet, L.E.M., De Jong, R., Giessen, W.A.v., and Visser, J.H. (1990) A learning-related variation in electroantennogram responses of a parasitic wasp. *Physiol. Entomol.* 15: 243–247.

Vet, L.E.M. and Groenewold, A.W. (1990) Semiochemicals and learning in parasitoids. *J. Chem. Ecol.* 16: 3119–3137.

von Schilcher, F. (1976) The role of auditory stimuli in the courtship of *Drosophila melanogaster*. *Anim. Behav.* 24: 18–26.

Waldow, U. (1975) Multimodale neurone im deutocerebrum von *Periplaneta americana*. *J. Comp. Physiol. A* 116: 329–341.

Wardle, A.R. (1990) Learning of host microhabitat colour by *Exeristes roborator* (F.) (Hymenoptera: Ichneumonidae). *Anim. Behav.* 39: 914–923.

Wardle, A.R. and Borden, J.H. (1985) Age-dependent associative learning by *Exeristes roborator* (F.) (Hymenoptera: Ichneumonidae). *Can. Entomol.* 117: 605–616.

Warnes, M. (1990) The effect of host odour and carbon dioxide on the flight of tsetse flies (*Glossina* spp.) in the laboratory. *J. Insect Physiol.* 36: 607–611.

Williams, L.H. (1954) The feeding habits and food preferences of Acrididae and the factors which determine them. *Trans. Roy. Entomol. Soc. Lond.* 105: 423–454.

Willmund, A. and Ewing, A. (1982) Visual signals in the courtship of *Drosophila melanogaster*. *Anim. Behav.* 30: 209–215.

Wyatt, T.D., Phillips, A.D.G., and Grégoire, J.-C. (1993) Turbulence, trees and semiochemicals: Wind-tunnel orientation of the predator, *Rhizophagus grandis*, to its bark beetle prey, *Dendroctonus micans*. *Physiol. Entomol.* 18: 204–210.

2

Effects of Experience on Host-Plant Selection

E.A. Bernays
Department of Entomology and Center for Insect Science, University of Arizona, Tucson, Arizona

1. Introduction

Anyone who has watched insects selecting their host plants in nature is aware of the variability of responses shown by the insects with respect to attraction, landing, acceptance, and oviposition or feeding. Some of this variation results from genetic differences among individuals, and some of it is due to variation among plants that we usually cannot detect (Singer and Parmesan, 1993). However, a third variable may be even more important, and that is the effect of past experience on the herbivore. The experiential effects may be short-term and readily reversible or they may be long-term and less readily reversed. In almost all cases, plant chemicals are involved, and the experiences can profoundly influence which of the various available host plants will be selected at any given time.

2. Simple Neural Changes

2.1. Habituation

This is defined behaviorally as the waning of a response to a stimulus with repeated exposure to that stimulus. The classical example is that of an animal hearing a loud noise, which initially elicits an escape response, but with repetition is ignored. Habituation has been shown to involve synaptic changes in specific neural pathways to the central nervous system. In short-term habituation there is a decrease in neurotransmitter release at particular synapses; in longer-term habituation there may be a decrease in productivity of neurotransmitters at synapses. It is possible that in some cases of habituation to chemicals there is a reduction of acceptor sites on the taste neurons.

Habituation to chemical deterrents (as opposed to just eating more of a deterrent food as a result of prior deprivation) is potentially a means of altering host-selection behavior, enabling an insect to eat a previously unacceptable food. It has been demonstrated in a few cases. In a study with locusts, the deterrent, nicotine hydrogen tartrate, was painted onto leaves of sorghum and presented to individual insects on consecutive days for 18 hours each day. In the remaining hours each day they received uncontaminated food so that they could retain the status of well-fed insects, and grow at a normal rate. On the first day of the experiment the treated leaves were deterrent to all individuals. By the third day, the nicotine-treated leaves were relatively acceptable to individuals of the desert locust, *Schistocerca gregaria*, although naive insects of similar age and similarly fed, were still deterred by them. The nicotine-experienced insects had obviously habituated to this deterrent. Interestingly, individuals of the oligophagous migratory locust, *Locusta migratoria*, showed little habituation (Jermy et al., 1982). Further studies with *S. gregaria* showed that the changes were in the central nervous system and that postingestive factors also played a role (Szentesi and Bernays, 1984).

Experiments with caterpillars similarly showed habituation to deterrents, and the polyphagous species habituated more than the oligophagous ones. A more detailed study of habituation to deterrents using a different experimental design was carried out with the polyphagous noctuid *Pseudaletia unipuncta*. Individual caterpillars were exposed to deterrent caffeine-treated maize leaves or untreated control leaves overnight. Then each caterpillar was allowed to take two meals on untreated corn to equalize their feeding state. Then, individuals of both groups were presented with caffeine-treated leaves: those that had previously experienced caffeine ate nearly twice as much as those which were naive (Usher et al., 1988). Clearly, they had habituated to this deterrent.

Considerable variation in habituation may occur between different individuals. An example of the variation was demonstrated in experiments on the polyphagous noctuid *Mamestra brassicae*. Some individuals did not alter in their behavior over time, some habituated, and others apparently became sensitized (i.e., they showed increased deterrence) (Jermy et al., 1987).

Habituation may be very common especially since many plant secondary compounds are deterrent but not toxic (Bernays and Chapman, 1987; Bernays, 1990). Thus, there may be biological reasons to select a particular subset of plants in an environment, but this bias is overcome with no immediate costs, if the preferred plants are absent at any time. Field observations are required to determine how frequently these changes might occur in nature.

2.2. Sensitization

This is the opposite of habituation, and involves an increased response to a stimulus on repeated exposure, without any learned associations. Several different physiological mechanisms have been proposed for sensitization, including presynaptic facilitation in interneurons and a heightened general state of excitation.

In contrast to habituation, where the stimuli tend to be relatively weak, sensitization is characteristically associated with intense or highly significant stimuli such as chemicals that are extremely deterrent or extremely phagostimulatory. A deterrent that causes a food to be ingested for only a few seconds before it is rejected, may, on subsequent encounter prevent feeding at all. In some situations the taste of a very deterrent compound not only increases the responsiveness to that deterrent, but, in addition leads temporarily to a reduced responsiveness to any food. One of the functional values of sensitization is that there may be an increase in foraging efficiency due to sensitization; the decision not to eat may be made immediately, and the inspection and selection of potential foods can then be relatively more rapid.

Sensitization to stimuli producing positive effects always seems to be associated with increased activity. If contact is lost with a highly acceptable food, an insect may move around extremely rapidly, giving the appearance of being "excited" (Bernays and Chapman, 1974). An increasing propensity to search and find food after a very positive experience, such as tasting a highly acceptable food, may increase the chances of obtaining more of the same without any associative learning. This is especially likely if the food resource occurs in discrete patches, and is probably the basis of the phenomenon of "area restricted search" (see Bell, 1991, Chapter 1).

The level of overall excitation or arousal may have a role in selectivity also. Butterflies that have been aroused by contact with a strong oviposition stimulant and then lose contact will often lay an egg on a nearby nonhost. Grasshoppers artificially aroused by presentation with a highly stimulating sugar, will ingest water which they otherwise would have rejected. However, these types of arousal are probably very short-lived.

3. Associative Learning

Here, an animal learns to associate a stimulus having no specific meaning (i.e., it is neutral), with some meaningful stimulus producing either positive or negative effects. As a result, on subsequent encounter the response elicited previously only by the meaningful stimulus is then elicited by the neutral stimulus. In the literature on learning the meaningful stimulus is termed the *unconditioned stimulus* (US), while the neutral one that comes to be associated with it is termed the *conditioned stimulus* (CS). Associative learning is a process that usually occurs in the brain and a number of different models for potentiation of pathways as a result of experience have been developed. No studies with definitive conclusions on neural mechanisms are available for phytophagous insects.

3.1. Positive Associative Learning

Positive associative learning has been demonstrated with many species of phytophagous insects. In the case of oviposition, where the taste of a specific chemical is the unconditioned stimulus, leaf color or shape may be the conditioned stimulus. For example in the cabbage butterfly, *Pieris rapae,* females tended to prefer sites of the same appearance as those accepted previously for oviposition. Thus, following oviposition on a cabbage leaf, they increased their tendency to oviposit on substrates with a similar reflectance pattern. Sinigrin stimulates oviposition in this species, and individuals can be trained with sinigrin to oviposit on discs of different colored paper. It was found that, after laying an egg on such a paper, females will preferentially land on discs of that color, even if sinigrin is absent (Traynier, 1986). The memory lasted for at least a day, and the butterflies could be retrained to prefer different colors or shades of green. The neutral stimulus (color) became associated with the meaningful stimulus (the chemical, sinigrin).

A variety of field observations with butterflies demonstrate that this kind of apparent learning is widespread. After an experience with a host plant, individual females tend to select plants with one or more of the characteristics of the first rewarding host plant. The conditioned stimuli appear generally to be visual and the unconditioned stimuli are chemical (Papaj and Prokopy, 1989; Szentesi and Jermy, 1990). Although this enhanced focusing on a particular plant species has been referred to as associative learning, or as search image formation, the processes involved are controversial. It is possible, for example, that the phenomenon is closer to sensitization and that the insect is being specifically attentive to a subset of possible cues. However, there is no doubt that the responsiveness to different plant species changes.

The apple maggot fly, *Rhagoletis pomonella*, shows abilities to learn a variety of different aspects of its host, including visual and chemical factors. Experiments with different host fruit of similar appearance indicated that the host chemicals can provide the conditioned stimulus. Females were given experience of apple or hawthorn fruit and then tested with models of apple or hawthorn size. These models were coated with parafilm wax which had been earlier placed around apples or hawthorn fruit to adsorb wax chemicals. The females were offered these models and their acceptance or rejection of them was recorded; apple-experienced flies more often accepted models with apple-contaminated parafilm, while hawthorn-experienced flies more often accepted models with hawthorn-contaminated parafilm (Papaj and Prokopy, 1986). The unconditioned stimulus is presumably related to the appropriate chemicals from within the fruit, while the conditioned stimulus consists of the superficial chemistry.

Where the unconditioned stimulus is the taste of ingested food, both visual and chemical stimuli have been shown to be relevant conditioned stimuli. The grasshopper, *Melanoplus sanguinipes*, learned to forage preferentially in a place associated with a particular color/light intensity if that visual stimulus was associated with good-quality food. In one experiment individuals were observed in an arena where two differently colored boxes were provided (dark green or yellow to mimic foliage and flowers of sunflower, which are both fed upon by this species in nature). The ambient temperature was suboptimal, and a light provided warmth on the roost or resting place, so that between foraging bouts insects returned to the resting place. Naive individuals left the roost to forage and would eventually stumble on the box with the food. After feeding they would return to the roost. The next feed was preceded by a very short search time, even if the positions of the colored boxes were reversed, indicating an ability to learn features of the environment related to color or light intensity, and return to the food source (Bernays and Wrubel, 1985). The conditioned stimulus in this case was spectral reflectance, and the unconditioned stimulus, the food.

In a lab study with the desert locust, *Schistocerca gregaria*, individuals were trained to respond to specific odors associated with major essential nutrients. The locusts were trained for two days in boxes with free access to two artificial foods that were similar in all respects except that one lacked digestible carbohydrate and the other lacked protein. Each was paired with one of two distinctive odors, carvone or citral. After the training period, insects were put in clean boxes having no added odors and were given only one of the two diets so that they became deficient in either carbohydrate or protein. On retesting, insects deprived of carbohydrate did not respond differentially to the odors paired with carbohydrate, but those deprived of protein responded very differently to the two odors, approaching the source of the odor previously paired with the protein diet significantly more often than expected (Simpson and White, 1990). The experiments demonstrated that the insects had learned an association between the odors and the availability of a key nutrient (i.e., protein) that they were lacking. The conditioned stimulus was the odor of the chemical, and the unconditioned stimulus, while unidentified, was presumably some nutrient feedback associated with ingestion of protein. In a field situation, it may be possible for individual grasshoppers to learn to feed on a particular food plant that best satisfies a regular requirement for such a major nutrient.

It is usual to consider the sensory input to the brain of an animal to be consistent. That is, one expects an odor or taste that is perceived by a chemoreceptor to be relayed to the central nervous system in a manner that is similar on different occasions. However, there are several indications that, among insects, chemoreceptor thresholds vary with experience. Among parasitic Hymenoptera, changes in sensitivity of antennal receptors have been shown using electroantennograms, and the changes are correlated with changes in behavior. When an individual wasp experiences one host and its attendant odor, it later preferentially selects that odor, apparently learning that it is associated with the presence of a rewarding host. When electroantennograms were performed on wasps which had been trained to different odors, it was found that the relative sensitivity of individual wasp antennae was greatest for the odors that had been experienced (Vet et al.,1990). This could suggest that the learning process is, in part at least, due to an increase in sensitivity of odor receptors to volatiles associated with a reward. Although not yet demonstrated, it is likely that the phenomenon would occur also in herbivorous species.

3.2. Food Aversion Learning

Although the phenomenon of learning to reject food as a result of negative experience is obviously associative learning, it has been given special treatment in earlier literature because of the characteristic delay that occurs between ingestion of a food and the consequent deleterious effects. However, it has now become obvious that there is a continuum between avoidance learning unrelated to food (such as to an unacceptable physical stimulus), avoidance learning of bad taste, and avoidance learning of food that causes deleterious consequences upon ingestion. The latter two are of most relevance to food selection in phytophagous insects [see Bernays (1992b) for a review].

Clearly, an ability to learn to avoid a plant due to a noxious effect following ingestion would be of considerable potential value. Field behavioral studies would not allow a distinction between aversion learning, sensitization, or changes in behavior brought about by other, unknown, variables. Its potential importance in food selection must initially be demonstrated in laboratory experiments, where relevant controls can ensure that the results are unambiguous.

In order to demonstrate an ability to associate a taste with a noxious effect following ingestion, nymphs of the grasshopper *Schistocerca americana*, were injected with nicotine after feeding on a novel food, and then tested on the same or a different plant. It was found that individuals given novel plants known to be acceptable, after this experience ate normal-sized meals. On the other hand, individuals given the plants that had been eaten immediately before the injection ate little or none (Bernays and Lee, 1988). This demonstrates that the insects were not too sick to eat, but rejected the type of plant eaten just before they suffered the symptoms. Additional controls showed that the chemical injected was critical, and not the injection process itself. The conditioned stimulus, the taste of the first plant, and the unconditioned stimulus, a factor associated with the chemically-induced sickness, became associated. The results were clear-cut when test plants were species of moderate or low acceptability, but plants that were very highly acceptable in the first instance never became unacceptable. In a series of experiments, a variety of different secondary metabolites were shown to cause food aversion learning (Lee and Bernays, 1990).

The probable occurrence of aversion learning with a natural food has been shown with spinach leaves. Successive meals by grasshoppers on this plant became smaller, until after about four meals when it was rejected (Lee and Bernays, 1988). Spinach contains phytosterols which are totally unsuitable for this grasshopper, and, since dietary sterols are essential nutrients, it appeared that rejection might have been associated with the absence of usable sterols. Later experiments demonstrated that, when appropriate sterols were added to the spinach leaves, the decline in acceptability did not occur (Champagne and Bernays, 1991). The sterols themselves appear not to be tasted, so that the results indicate that the flavor of spinach, which was initially acceptable, became unacceptable as the insect obtained feedback concerning its nutritional unsuitability.

Aversions may also be induced by an inadequate protein concentration in the food. In a series of experiments with *S. americana*, artificial diets were prepared that were either low in protein (2% wet weight) or higher in protein (4% wet weight). Either tomatine or rutin was added at concentrations that could be detected by the grasshoppers but were not deterrent. Individuals were fed on one of the diets for four hours and then offered the low-protein diet with the familiar or a novel flavor (tomatine, rutin or nontoxic levels of nicotine). The insects that had experienced the lower-protein foods fed relatively longer on the diets with novel flavors rather than those with the same flavor, or those which had experienced the higher protein (Bernays and Raubenheimer, 1991). The conclusion was that insects fed protein-deficient foods subsequently showed an aversion for the flavor associated with the poor food as well as a preference for novel foods (neophilia). Similar kinds of results have often been demonstrated in studies on vertebrates, whereby learned aversions of one food go hand in hand with an increased acceptability of a novel food. In fact, in choice tests, the two cannot be separated. In any case the learned aversion for flavor coupled with low protein indicates that the neutral flavor had acquired significance (become the conditioned stimulus) related to the poor protein content of the food.

These data indicate that aversion learning, in grasshoppers at least, may have an important role in dietary mixing to obtain a suitable nutrient balance. Field studies have shown considerable individual polyphagy in grasshoppers and the dietary mixing may have some basis in improving nutrient mixes ingested. It is possible for example, that successive aversion learning experiences on a series of plants that are each imperfect would lead to a better diet than remaining on one plant only. A laboratory experiment to test this idea was carried out also with *S. americana*. Two unbalanced but complementary artificial diets were offered to grasshoppers, with or without distinctive flavors, and their behavior monitored over three days. In all cases the grasshoppers were mobile enough to encounter and eat both diets but the added flavors enhanced the amount of mixing (Bernays and Bright, 1991). By analogy, it may be that the distinctive secondary chemistry of plants is of considerable value to phytophagous insects that are individually polyphagous, because they would provide distinctive signatures and aid in the process of learning (Bernays and Bright, 1993).

The evidence so far indicates that aversion learning relating to nutrient profiles causes altered preferences among different foods, and that relative acceptability of a food will vary according to its nutrient content and, in addition, the insect's nutrient status/experience. *S. americana* is highly polyphagous and it is perhaps to be expected that an ability to learn from experience in this manner would be highly valuable, just as has been proposed in vertebrates. In the grasshopper experiments showing learned aversion to poisons the memory of the experience was extinguished after four days. This period would probably allow plenty of time to move into different areas with different foods to choose from. With nutrient deficiency, the length of the memory has not been tested, but short-term effects would be adequate to improve the efficiency of foraging.

There is one example with polyphagous arctiid caterpillars, which appears also to show aversion learning. *Diacrisia virginica* and *Estigmene congrua* initially respond positively to petunia leaves. After ingestion of this plant the caterpillars regurgitated the leaf material and, when subsequently given a choice of petunia and another plant, they selected the alternative (Dethier, 1980). The effect was not produced in the oligophagous caterpillar *Manduca sexta*, even though it too, regurgitated, and there is the suggestion that polyphagous species are better able to learn such negative associations (Dethier and Yost, 1979). Gelperin and Forsythe (1975) also suggest that polyphagous species would learn more readily than oligophagous or monophagous ones: species which are hardwired to accept a narrow range of plants and reject all others, are less likely to be able to learn to avoid plants through experience. However, while extreme specialists may never ingest food that is toxic to them, some at least move from low- to high-quality plants. For example, Wang (1990) showed that the creosote bush grasshopper, *Ligurotettix coquilletti,* moved away from bushes they started to develop on and accumulated on bushes known to be of higher quality for development. Similarly, Parker (1984) found that the grasshopper, *Hesperotettix viridis,* had a shorter tenure time on damaged host plants than on undamaged ones.

Because there is often a time delay between sensory patterns associated with food intake, and the (negative) consequences of ingestion, it is to be expected that certain patterns of feeding would enhance the likelihood of aversion learning. For example, discrete meals on single food items will allow associations to be made more readily than grazing on a mixture of foods within a meal (Zahorik and Houpt, 1981). Another feeding habit that may allow learned aversions to form is that of short-term fidelity to a particular resource such that a learned aversion can develop over a series of meals on a single food type, whereupon rejection and movement away may follow if the food is unsuitable. Species that tend to rest on or near their food, as do most plant-feeding insects, are in this category. It may be that many such insects have extensive capabilities to learn associations between food characteristics and unsuitability, but establishing this with certainty requires long-term continuous observations.

Inability to move readily from one food resource to another is a constraint in many insect species, and in others, distances between potential alternative foods may be prohibitive for any learning to be really useful. For example many holometabolous larvae have no alternative but to remain on or in their food source and aversion learning has no relevance for them. This is clearly the case with many fly larvae, like leaf miners or carcass dwellers; and for many beetle larvae, such as wood-boring species. Certain homopterans such as scale insects show the most extreme restriction; they are totally immobile and must feed in the position first selected.

Blaney and others [e.g., Blaney and Simmonds (1985, 1987), Blaney et al. (1985), and Chapman and Sword (1993)] have demonstrated with both grasshoppers and caterpillars that individuals presented with a relatively unfavorable plant will bite and reject the plant, but that on subsequent contacts, rejection tends to occur after palpation only. On successive contacts a greater proportion of individuals reject at palpation (i.e., before biting). They presumed that insects learn to associate the superficial taste or smell with the internal constituents of the plant. Avoidance learning appeared to be occurring, even though there had been no ingestion.

4. Induction of Preference

The term *induction of preference* is used almost exclusively in studies with phytophagous insects in which individuals tend to prefer the plant they have already experienced over one they have not experienced, whether or not this plant is most appropriate for development. It has been demonstrated in different insect groups. Some authors have drawn parallels between induction and imprinting (Szentesi and Jermy, 1990), though the classical cases of imprinting described in young birds is different in many ways. The physiological basis of induction is unknown, but probably involves a mixture of processes. In some cases changes in thresholds of chemoreceptors to chemicals that have been experienced are correlated with induction; there may be elements of sensitization, habituation, and/or associative learning.

Induction has been extensively studied among larvae of Lepidoptera where individuals of over 24 species have been shown to develop an altered preference in favor of the plant already experienced. Induction has also been demonstrated in Phasmatodea, Heteroptera, Homoptera, and Coleoptera (De Boer and Hanson, 1984; Szentesi and Jermy, 1990). In some extreme cases, such as the saturniid *Callosamia promethea*, experience on one host plant apparently precluded acceptance of an alternative one (Hanson, 1976). Similarly, it was shown that the cabbage butterfly, *Pieris brassicae*, could be reared on *Brassica* or on *Tropaeolum*, yet if larvae were first induced on *Brassica*, they refused *Tropaeolum* and died of starvation (Ma, 1972). In most published cases, induction of preference is much less extreme. Often, the normal host plant will remain favored, but if larvae are forced onto an alternative plant that is accepted although it is not a normal host the new plant may become relatively more acceptable thereafter, although not as acceptable as the normal host.

In induction studies, the oligophagous larvae of *Manduca sexta* have had considerable attention by various authors, and recently it has been shown that the induction shown on foods not normally eaten in nature can be explained by an increased acceptability of the plant, probably due mainly to decreases in deterrence of certain secondary metabolites (De Boer, 1992).

There is evidence that there can be increased responses to stimulating extracts and odors of plants upon which larvae were induced. In earlier experiments with *M.sexta* fed on an artificial diet with or without added citral, it appeared that induction on citral-containing diet led to an ability to orient toward citral. Insects having had a plain diet on the other hand turned significantly more frequently toward the plain diet and away from citral (Saxena and Schoonhoven, 1982). More realistic experiments were carried out using larvae of *Trichoplusia ni*. They were fed on either mint or basil for one week and then tested in arenas with discs of both leaf types. Not only did the larvae feed more on the plant they had experienced, as expected in an induction experiment, they also oriented to, and arrived first at, the discs of plants they had experienced previously (Lee,Bernays and Buttolph, unpublished). Clearly the particular plant odors became attractants with experience. The process of induction of preference thus appears to be a mixed-process phenomenon.

In examining the physiological basis of induction, some studies have shown modifications in firing rates of chemoreceptors to various chemicals. Behavioral and electrophysiological tests with caterpillars fed on artificial diets containing deterrents indicated that the larvae became less sensitive to the deterrents after experiencing them in the food for a period. A few studies have shown that experience of a plant secondary metabolite in artificial diet actually leads to a reduced sensitivity of chemoreceptor cells to that chemical. This has been shown with several species of caterpillars including *Manduca sexta*. For example, individuals that were force-fed a diet containing salicin, which is normally a deterrent, were found to have a reduced sensitivity to salicin at the chemoreceptor level. Furthermore, the individuals that were most tolerant of salicin behaviorally showed the lowest firing rates to salicin in their maxillary sensilla (Schoonhoven, 1976). Similarly, with *Spodoptera littoralis,* given experience of nicotine and subsequently tested with nicotine, there was a reduction in firing rates of the chemoreceptors when stimulated with it (Blaney and Simmonds, 1987). These data strongly suggest that experience-induced changes in the sensitivity of chemoreceptors to particular compounds could influence the choice of foods thereafter. No investigation of how such changes might influence selection of natural foods has been undertaken, but it seems very likely that they would play a role in induction of preference. They do correlate with behavioral changes [see Blaney et al. (1986)], but at present it is not really known if this is cause or effect.

In comparing results of induction in different lepidopterous larvae, considerable variation between species has been found. It occurs equally readily in species with different host ranges, although the most extreme cases seem to be polyphagous. It seems that the more taxonomically different the plants in any comparison, the greater the likelihood of induction. That is, the experience of very different plants leads to a greater difference in their relative acceptability in a choice test thereafter.

In addition to interspecific variation there is almost invariably a great deal of variation among individuals within a species, and often there are trends but not significant effects overall in experiments. Also, most experiments are performed with a choice test, where relative amounts eaten are measured, so it is not easy to distinguish increased acceptability of the plant experienced from decreased acceptability of the alternative plant, and such experiments tend to enhance very minor differences in acceptability that may not always have ecological relevance. In recent extensive studies (M.Weiss, unpublished) it has become clear that even among species in which induction has been repeatedly demonstrated the effects are extremely variable and induction is always absent in a proportion of individuals.

A different type of induction or imprinting may occur even before contact with the food plant. In adults, emergence from the pupal case may involve experience of remnants of the food of the larval stage. If such food remnants then cause a preference for those chemicals, this may be the simplest type of induction of preference. This process has been demonstrated in *Drosophila* where it has been shown that larvae reared on a flavored medium give rise to adults with a preference for this flavor over alternatives. However, if the pupae are thoroughly washed before the adult emerges this preference is eliminated (Jaenike, 1988). Such experiments have not been carried out with phytophagous insects but the phenomenon may have been responsible for the earlier belief, now disfavored, that larval food plant preferences were transmitted to adults.

Why do so many phytophagous insects show an induction of preference? In the few cases where starvation rather than acceptance of a novel food has been shown, it appears to be maladaptive. However, in nature, rejection behavior might lead to movement away from a novel plant if it were encountered, but not necessarily to starvation. The process needs to be examined under natural conditions in the field to determine its prevalence and possible significance, but in any case there has been considerable speculation as to its biological significance. In some field settings one can envisage a benefit of induction, as for example when a larva falls off its host and must refind it, a heightened sensitivity to host odors may be useful. There may be some benefit of induction during normal feeding activities on the host plant if it heightens arousal and minimizes interruptions to feeding (Bernays and Wcislo, 1994).

It would be an advantage to have an induced preference if larvae became particularly adept at digesting or detoxifying the preferred plant, but evidence in this direction is contradictory so far. However in one case there is evidence for this (Karowe, 1989) and it may be that sufficiently detailed studies have not yet been done. This is mainly because it is difficult to separate the effects of simply not feeding from reduced digestive performance.

Since ecological or physiological benefits are not very clear it may be that the phenomenon has its origin in terms of a more general requirement to narrow or simplify neural pathways, allowing the individual to make decisions more rapidly. Speed of decisions can, at least in theory, improve efficiency and decrease danger, allowing a relatively polyphagous animal to attain the benefits of a genetically determined narrow host range. It is also possible that many alterations in food acceptance behavior have no particular significance. Experiments with grasshoppers showed that prior experience could totally alter the acceptability hierarchy of novel plants, although there were no obvious patterns and no explicable benefits to be derived from the change (Howard and Bernays, 1991).

The fact that there is almost invariably a proportion of noninducers in "induction experiments" would at least suggest that there may be polymorphism for the trait, and that there may be benefits associated with a lack of ability to induce.

5. Postingestive Feedback

Several studies have demonstrated that the threshold response of insect taste receptor neurons to sugars and amino acids varies in relation to the current requirement for that nutrient, which is in turn a function of levels of nutrients in the hemolymph. For example, in locusts, Simpson and his coworkers have shown that when individuals were fed on food rich in carbohydrate and low in protein, they developed a higher threshold response to sugars in the chemoreceptors on the tips of the maxillary palps, while threshold responses to amino acids remained low. Conversely, insects fed on food rich in protein and low in carbohydrate showed a high threshold response to amino acids and a low threshold response to sugars. These authors have proposed that the changes could be responsible for the sophisticated ability of these insects to select mixtures of foods with complementary nutrients (Simpson et al., 1991).

Similar abilities and similar changes in chemoreceptor thresholds have been demonstrated in caterpillars. In terms of natural foods, the use of these altered thresholds in determining host selection would require that levels of sugars and amino acids in plants are correlated with available carbohydrate and protein levels. In addition, the palps are used primarily on the surface of the leaf, so that, to be useful there should be levels of sugars and amino acids on the leaf surface that somehow correlate with the available nutrients within the leaf. Although we have no evidence that this is so, and the effects could simply be an effect rather than a cause of the behavior, the changing inputs from the chemoreceptors are worth further study. It may be that the changes described are biologically most significant in relation to nutrient status in general. For example, when insects are deprived of food they tend to accept nonhosts more readily. If all chemoreceptors responding to nutrients are at their lowest thresholds, then positive input could override input from deterrents, allowing the insects to accept a food that would be unacceptable to a well-fed insect.

As discussed in the section on induction, chemoreceptor changes may be important in the changes in relative acceptability of different host plants. It is not yet known whether a secondary metabolite absorbed from the gut during and after a meal could influence the sensitivity of taste receptors from the hemolymph side. If the chemical were able to reach the site of the relevant receptors, it is conceivable that a certain amount of long-term adaptation could occur, reducing the sensitivity of the cells.

Some insects have been shown to select wet or dry foods according to previous experience/current needs, and the changes in preference may be so rapid that individuals may switch among foods with different water content within a meal. The locust, *Schistocerca gregaria*, for example, was offered in experiments fresh wheat seedlings and lyophilized wheat seedlings from the same batch. These were assumed to all be similar in their levels of basic nutrients. All individuals showed clear-cut alternating preference for the wet and dry foods indicating short-term postingestive feedbacks driving preference first for the food that had excess water and then for the food with inadequate water (Lewis and Bernays, 1985).

In individuals that have been relatively deprived of an important nutrient and then allowed access to it, rapid positive feedback from the gut can influence the amounts of the food then eaten. For example, experiments with the grasshopper, *Schistocerca americana*, showed within-meal feedbacks in relation to satisfying a sterol deficit. Individuals that were short of utilizable sterols were allowed access to sugar-impregnated filter paper with either a utilizable sterol or a nonutilizable sterol. In the latter case, meal lengths were approximately ten minutes, whereas with the satisfactory sterol, meal lengths were significantly longer, often twenty minutes (Champagne and Bernays, 1991). Since the insects did not appear to taste the sterols, it appears that feedback enhancing food intake occurred in approximately ten minutes. Although this could be a learned response to the sensory features of the available disk, it could also be a direct effect of sterol satiation causing continuation of feeding.

Some known poisons may be ingested by insects if the chemical is not deterrent. Sometimes, there is an effect on feeding within a minute or two, presumably as soon as the chemical reaches the midgut epithelium. Either directly, or through aversion learning (see above), feeding stops, and unless the observations have been continuous the chemical may be considered deterrent.

6. Compulsive Requirement for Novelty

In at least one species, an absolute requirement for novel flavors seems to exist. The highly polyphagous grasshopper *Taeniopoda eques* refuses to eat the same food plant in the laboratory over a long period, and cannot be reared without having a mixture of plants available. In the field it switches frequently between food items (Raubenheimer and Bernays, 1993). Laboratory tests showed that every one of ten foods tested on isolated individuals showed a declining acceptability over time. Whatever the initial level of acceptability, the food was rejected by about the fifth encounter. Further study was undertaken with good-quality artificial diets that were all identical but had different added harmless flavors. After two meals on food with one flavor, individuals were more likely to eat large meals on food with a novel flavor and small meals on the food with the flavor already experienced (Bernays et al., 1992).

After experience with one of two different odors in the absence of food, *T.eques* ate sooner and for longer on an artificial diet laced with the novel odor than on the diet laced with the one already experienced (Bernays, 1992a). This may be an exaggerated example of the rather well-known general phenomenon of an increased state of arousal when environmental stimuli change. It is perhaps also related to the process of sensitization. It is of course unlikely to occur in any but polyphagous species, and of those, only species that are mobile and individually polyphagous. It is the precise opposite of an induction of preference.

7. Conclusions

Experience, especially of different chemicals, has a variety of effects on food selection behavior and is one of the factors leading to the variability of response seen in nature. Several different neural phenomena are involved, and in any one behavioral change there may be a mixture of processes. An apparent difference from vertebrates is the plasticity of the chemoreceptors, which may then govern some of the changes found.

There has been discussion in the literature of behavioral plasticity and its possible fitness benefits. In one sense, it is self-evident that learning would be beneficial, yet experiments are lacking that actually measure it and there are some situations in which learning might be a disadvantage. The fact that there is variation in learning ability suggests that it is possible to select for improvements and that there may be some benefit to not learning too much. For example, a butterfly may find a good resource and remember it by its characteristic taste, and preferentially search thereafter for this taste. If the resource is readily available an advantage may accrue in some currency relating to efficiency. However, if the resource is very rare or unpredictable the learning process may hinder acceptance of alternatives, and thus be a disadvantage. If within-generation predictability of particular potential food plants is high and between-generation predictability is low, learning will be most favored.

Recently, there has been some exploration of the possibility that learning may, in itself, influence the direction of evolutionary change [see Jaenike and Papaj (1992)]. An example may be given in the case of a female butterfly which is encountering plants other than its normal host plants for oviposition. A large egg load may result in her decision to oviposit on nonhosts. If oviposition is an unconditioned stimulus for learning, she may learn cues representing this new oviposition substrate and continue to oviposit on the same plant species. Such persistent oviposition on a novel or nonpreferred host will increase the likelihood of there being individual offspring that accept the new plant. Among those that accept it there may be some that grow well on it. It could be the first stage of a host shift. The phenomenon involves a physiological state variable that alters acceptance levels of a plant, combined with learning, and a resultant increase in selection for larvae with improved ability to utilize a novel host.

References

Bell, W.J. (1991) *Searching Behaviour*. Chapman and Hall, New York.

Bernays, E.A. (1990) Plant secondary compounds deterrent but not toxic to the grass specialist *Locusta migratoria*: Implications for the evolution of graminivory. *Entomol. exp. appl.* 54: 53–56.

Bernays, E.A. (1992a) Dietary mixing in generalist grasshoppers. In: *Proc.8th Symp.Insect-Plant Interactions*, (Menken, S.B.J., Visser, J.H., and Harrewijn, P., eds). pp. 146–148. Kluwer, Dordrecht.

Bernays, E.A. (1992b) Food aversion learning. In: *Insect Learning* (Lewis, A.C. and Papaj, D., eds). pp. 1–17. Chapman & Hall, New York.

Bernays, E.A. and Bright, K.L. (1991) Dietary mixing in grasshoppers: Switching induced by nutritional imbalances in foods. *Entomol. exp. eppl.* 61: 247–254.

Bernays, E.A. and Bright, K.L. (1993) Dietary mixing in grasshoppers: A review. *Comp.Biochem.Physiol.* 104A: 125–131.

Bernays, E.A. and Chapman, R.F. (1974) The regulation of food intake by acridids. In: *Experimental Analysis of Insect Behaviour* (Barton Browne, L., ed). pp. 48–59. Springer-Verlag, Berlin.

Bernays, E.A. and Chapman, R.F. (1987) Evolution of deterrent responses in plant-feeding insects. In: *Perspectives in Chemoreception and Behavior* (Chapman, R.F., Bernays, E.A., and Stoffolano, J.G., eds). pp 1–16. Springer-Verlag, New York.

Bernays, E.A. and Lee, J.C. (1988) Food aversion learning in the polyphagous grasshopper *Schistocerca americana*. *Physiol. Entomol.* 13: 131–137.

Bernays, E.A. and Raubenheimer, D. (1991) Dietary mixing in grasshoppers: Changes in acceptability of different plant secondary compounds associated with low levels of dietary protein. *J. Insect Behav.* 4: 545–556.

Bernays, E.A. and Wcislo, W. (1994) Sensory capabililties, information processing and resource specialization. *Q. Rev. Biol.* 69: 187–204.

Bernays, E.A. and Wrubel, R.P. (1985) Learning by grasshoppers: Association of colour/light intensity with food. *Physiol. Entomol.* 10: 359–369.

Bernays, E.A., Bright, K.L., Howard, J.J., and Champagne, D. (1992) Variety is the spice of life: Compulsive switching between foods in the generalist grasshopper *Taeniopoda eques*. *Anim. Behav.* 44: 721–731.

Blaney, W.M. and Simmonds, M.S.J. (1985) Food selection by locusts: The role of learning in rejection behaviour. *Entomol. exp. appl.* 39: 273–278.

Blaney, W.M. and Simmonds, M.S.J. (1987) Experience: A modifier of neural and behavioral sensitivity. In: *Insects-Plants. Proc 6th Int Symp Insect-Plant Relat.* (Labeyrie, V., Fabres, G., and Lachaise, D., eds). pp 237–241. Junk, Dordrecht.

Blaney, W.M., Schoonhoven, L.M., and Simmonds, M.S.J. (1986) Sensitivity variations in insect chemoreceptors: A review. *Experientia* 42: 13–19.

Blaney, W.M., Winstanley, C. and Simmonds, M.S.J. (1985) Food selection by locusts: An analysis of rejection behaviour. *Entomol. exp. appl.* 38: 35–40.

Champagne, D.E. and Bernays, E.A. (1991) Phytosterol unsuitability as a factor mediating food aversion learning in the grasshopper *Schistocerca americana*. *Physiol. Entomol.* 16: 391–400.

Chapman, R.F. and Sword, G. (1993) The importance of palpation in food selection by a polyphagous grasshopper (Orthoptera:Acrididae). *J. Insect Behav.* 6: 79–92.

Dethier, V.G. (1980) Food aversion learning in two polyphagous caterpillars, *Diacrisia virginica* and *Estigmene congrua*. *Physiol. Entomol.* 5: 321–325.

Dethier, V.G. and Yost, M.T. (1979) Oligophagy and the absence of food-aversion learning in tobacco hornworms, *Manduca sexta*. *Physiol. Entomol.* 4: 125–130.

De Boer, G. (1992) Diet-induced food preference by *Manduca sexta* larvae: Acceptable non-host plants elicit a stronger induction than host plants. *Entomol. exp. appl.* 63, 3–12.

De Boer, G. and Hanson, F. (1984) Food plant selection and induction of feeding preferences among host and non-host plants in larvae of the tobacco hornworm *Manduca sexta*. *Entomol. exp. appl.* 35, 177–194.

Gelperin, A. and Forsythe, D. (1975) Neuroethological studies of learning of mollusks. In: *Simpler Networks and Behavior* (Fentress, J.C., ed). pp. 239–250. Sinauer, New York.

Hanson, F.E. (1976) Comparative studies on induction of food choice preferences in lepidopterous larvae. *Symp. Biol. Hung.* 16: 71–77.

Howard, J.J. and Bernays, E.A. (1991) Effects of experience on palatability hierarchies of novel plants in the polyphagous grasshopper *Schistocerca americana*. *Oecologia* 87: 424–428.

Jaenike, J. (1988) Effects of early adult experience on host selection in insects: Some experimental and theoretical results. *J. Insect Behav.* 1: 3–15.

Jaenike, J. and Papaj, D. (1992) Behavioral plasticity and patterns of host use by insects. In: *Insect Chemical Ecology: An Evolutionary Approach*. (Roitberg, B.D. and Isman, M.B., eds). pp. 245–264. Chapman and Hall, New York.

Jermy, T., Bernays, E.A., and Szentesi, A. (1982) The effect of repeated exposure to feeding deterrents on their acceptability to phytophagous insects. In: *Insect-plant Relationships*. (Visser, J.H. and Minks, A.K., eds). pp. 25–30. Pudoc, Wageningen.

Jermy, T., Horvath, J., and Szentesi, A. (1987) The role of habituation in food selection of lepidopterous larvae: The example of *Mamestra brassicae*. In: *Insects-Plants*. (Labeyrie, V., Fabres, G., and Lachaise, D., eds). pp. 231–236. Junk, Dordrecht.

Karowe, D. (1989) Facultative monophagy as a consequence of prior feeding experience: Behavioral and physiological specialization in *Colias philodice* larvae. *Oecologia* 78: 106–111.

Lee, J.C. and Bernays, E.A. (1988) Declining acceptability of a food plant for the polyphagous grasshopper, *Schistocerca americana:* The role of food aversion learning. *Physiol. Entomol.* 13: 291–301.

Lee, J.C. and Bernays, E.A. (1990) Food tastes and toxic effects: Associative learning by the polyphagous grasshopper *Schistocerca americana*. *Anim. Behav*. 39: 163–173.

Lewis, A.C. and Bernays, E.A. (1985) Feeding behaviour: Selection of both wet and dry food for increased growth in *Schistocerca gregaria* nymphs. *Entomol. exp. appl.* 37: 105–112.

Ma, W.C. (1972) Dynamics of feeding responses in *Pieris brassicae* Linn. as a function of chemosensory input: A behavioural, ultrastructural and electrophysiological study. *Meded. Landbouwhogesch*. Wageningen 11: 1–162.

Papaj, D.R. and Prokopy, R.J. (1986) Phytochemical basis of learning in *Rhagoletis pomonella* and other herbivorous insects. *J. Chem. Ecol.* 12: 1125–1143.

Papaj, D.R. and Prokopy, R.J. (1989) Ecological and evolutionary aspects of learning in phytophagous insects. *Annu. Rev. Entomol.* 34: 315–350.

Parker, M.A. (1984) Local food depletion and the foraging behavior of a specialist grasshopper, *Hesperotettix viridis*. *Ecology* 65: 824–835.

Raubenheimer, D. and Bernays, E.A. (1993) Patterns of feeding in the polyphagous grasshopper, *Taeniopoda eques*: A field study. *Anim. Behav*. 45: 153–167.

Saxena, K.N. and Schoonhoven, L.M. (1982) Induction of orientational and feeding preferences in *Manduca sexta* larvae for different food sources. *Entomol. exp. appl.* 32: 172–180.

Schoonhoven, L.M. (1976) On the variability of chemosensory information. *Symp. Biol. Hung*. 16: 261–266.

Simpson, S.J and White, P.R. (1990) Associative learning and locust feeding: Evidence for a "learned hunger" for protein. *Anim. Behav*. 40: 506–513.

Simpson, S.J., James, S., Simmonds, M.S.J., and Blaney, W.M. (1991) Variation in chemosensitivity and the control of dietary selection behaviour in the locust. *Appetite* 17: 141–154.

Singer, M.C. and Parmesan, L. (1993) Sources of variation in patterns of plant-insect association. *Nature* 361: 251–253.

Szentesi, A. and Bernays, E.A. (1984) A study of behavioural habituation to a feeding deterrent in nymphs of *Schistocerca gregaria*. *Physiol. Entomol.* 9: 329–340.

Szentesi, A. and Jermy, T. (1990) The role of experience in host plant choice by phytophagous insects. In: *Insect-plant Interactions* Vol. 2. (Bernays, E.A., ed). pp. 39–74. CRC Press, Boca Raton, Fla.

Traynier, R.M.M. (1986) Visual learning in assays of sinigrin solution as an oviposition releaser for the cabbage butterfly, *Pieris rapae*. *Entomol. exp. appl.* 40: 25–33.

Usher, B.F., Bernays, E.A., and Barbehenn, R.V. (1988) Antifeedant tests with larvae of *Pseudaletia unipuncta*: Variability of behavioral response. *Entomol. exp. appl.* 48: 203–212.

Vet, L.E.M., De Jong, R., Giessen, W.A.van, and Visser, J.H. (1990) A learning-related variation in electroantennogram responses of a parasitic wasp. *Physiol. Entomol.* 15: 243–247.

Wang, G.Y. (1990) Dominance in territorial grasshoppers: Studies of causation and development. PhD Thesis, University of California, Los Angeles.

Zahorik, D.M. and Houpt, K.A. (1981) Species differences in feeding strategies, food hazards, and the ability to learn aversions. In: *Foraging Behavior: Ecological, Ethological and Psychological Approaches*. (Kamil, A. and Sargent, T., eds). pp 289–310. Garland Press, New York.

3

Parasitoid Foraging and Learning

Louise E.M. Vet
Department of Entomology, Wageningen Agricultural University

W. Joe Lewis
Agricultural Research Service, United States Department of Agriculture

Ring T. Cardé
Department of Entomology, University of Massachusetts

1. Introduction

The diminutive size of most parasitoids undoubtedly has limited their choice as subjects for behavioral study, despite their great diversity in lifestyles and reproductive strategies. The present chapter addresses their foraging behavior as influenced by learning. Most of their adult life female parasitoids search for host insects which, in turn, are under selection to avoid being found and devoured. This scenario sets the stage for the evolution of diverse hide-and-seek games played by parasitoids and their victims, most often herbivores. That parasitoids are successful in their quest for hosts is evidenced by the vast number of parasitoid species and their importance in insect management.

1.1. The Fixed Response Perspective

The study of parasitoid foraging behavior entered a new era around the time of publication of Vinson's review (Vinson, 1984) on parasitoid-host relationships in the first edition of *Chemical Ecology of Insects* (Bell and Cardé, 1984). Prior to that time, students of parasitoid foraging behavior mainly were involved in identifying and describing the steps of how an insect parasitoid finds a potential host in which to lay its eggs (e.g., Lewis et al. 1976). Although a major emphasis was placed on the importance of chemical stimuli in guiding this process, an impressive array of stimuli was revealed. In spite of this diversity in stimuli and behaviors involved, some generalizations were possible. For example, different searching phases, such as habitat location, host location, host examination, were distinguished. The importance of host-derived cues including frass, webbing, mandibular secretions, and scales of adult moths, in attracting and arresting parasitoids at a short distance also seemed a general phenomenon (Weseloh,

1981). Each parasitoid species was envisaged to respond to a stimulus or combinations of stimuli in a hierarchy of stimulus-response sequences. As the sequence of search events was known for only a limited number of species, much remained to be uncovered. Furthermore, the origin and exact chemical identity of the major cues needed to be elucidated. The approach was largely descriptive and mechanistic, and lacked a theoretical basis and hypothesis-driven experiments. Furthermore, most chemical ecologists appreciated the presumed consistency of the behavioral outcome.

1.2. The current view

In the 1980's studies on model species intensified, research questions arose from field observations, and novel bioassays expanded the initially limited focus on short-range searching to include orientation at long range. These approaches revealed the variability and previously hidden complexity of the response patterns. The significance of parasitoid experience intruiged behavioral ecologists who attempted to unravel the function of such behavioral plasticity (e.g., van Alphen and Vet, 1986; Papaj and Lewis, 1993).

The profound effects of learning on parasitoid foraging were not recognized until the early 80's, although a few anecdotal examples of learning in parasitoids were known (e.g., Thorpe and Jones, 1937; Arthur, 1971; Vinson et al., 1977). Encouraged by extensive research experience on behavioral plasticity in other Hymenoptera, especially learning in social bees, students of parasitoid foraging felt challenged to tackle this unexplored realm. Now it is generally accepted that the patterns of parasitoid foraging are determined by the interactions of genetic, physiological, environmental and experiential factors (Lewis et al. 1990; Vet et al., 1990; Kester and Barbosa, 1991; Drost and Cardé, 1992b; Vet and Dicke, 1992; Turlings et al., 1993).

The first task was to document the effects of experience on the response to specific stimuli in a selection of species; such empirical studies are still accumulating (reviewed in Wardle and Borden, 1990; Vet and Groenewold, 1990; Kester and Barbosa, 1991; Turlings et al., 1993). The questions asked are becoming increasingly ecologically relevant. What is the value of learning for a particular species (Vet and Dicke, 1992; Turlings et al., 1993; Tumlinson et al., 1993)? What is the function of behavioral variability (Papaj and Vet, 1990; Vet and Papaj, 1992; Papaj, 1993a)? Recently the nutritional state of the parasitoid has been recognized as important in learning studies. Most parasitoids also need to forage for food, such as nectar, and this foraging activity, wherein learning also can play an essential role, can interact with host foraging (Lewis and Takasu, 1990). Both the informational state of parasitoids based on cues encountered in association with food and the current nutritional state can substantially affect host foraging behavior (Takasu and Lewis, 1993, 1995).

Comprehensive studies have been limited to a few species, but yet there is increasing interest in recognizing general patterns of underlying learning mechanisms and generic types of problems parasitoids solve. Many studies emphasize both proximate and ultimate factors influencing learning in parasitoid foraging (Vet et al., 1990; Lewis et al., 1990; Vet and Dicke, 1992; Turlings et al., 1993).

The approaches that influence the way learning-induced variability in searching behavior is studied and interpreted include:

—*tritrophic approach*. Parasitoids are recognized to have evolved and function within a multitrophic context (Price et al., 1980). Consequently their behavior is influenced by several trophic levels (e.g., Turlings et al., 1990; Kester and Barbosa, 1991; Vet and Dicke, 1992).

—*optimality approach*. Successful foraging in parasitoids is linked with Darwinian fitness, which makes parasitoids highly suitable for studying the adaptiveness of variation in foraging decisions and for testing hypotheses on optimization, often with the use of theoretical models (e.g., Stephens and Krebs, 1986; Roitberg et al., 1993).

—*comparative approach*. Comparative studies can reveal correlations between species characteristics and ecological factors indicative of adaptation. A clear example is the comparison of the learning capabilities of specialist and generalist species (e.g., Poolman Simons et al., 1992; Rosenheim, 1993).

In the present chapter we do not review extensively the empirical data on learning in parasitoids. Instead we focus on hypotheses on learning and foraging. Are there general patterns and underlying mechanisms? What are the major hypotheses?

After a brief review of the definitions of learning, we examine the general value of learning in parasitoids and the complexities of using host cues in foraging. Next we consider how learning is expressed and speculate on the mechanisms that underly learning. We then address how experience can affect foraging efficiency and in which stages of the foraging this occurs. We attempt to identify general patterns and generate hypotheses on the value of learning in species with specific ecological characteristics, such as the host stage they attack and their diet breadth at two trophic levels. A closing glimpse of unanswered questions and future directions may encourage future research to be guided by testable hypotheses.

2. Definitions and criteria of learning

It is difficult, if not impossible, to give a satisfying and general definition of learning (Papaj and Prokopy, 1989; Papaj, 1993b; Stephens, 1993). Divergent views of learning are hardly surprising, as scientists from the disparate fields of physiology, neurobiology, psychology, ethology and behavioral ecology study learning. Consequently, all inclusive definitions such as: "modification of behavior by experience" are not satisfying to researchers interested in the underlying mechanisms of learning, but such explanations can be 'sufficient' to someone that is studying the function of phenotypic plasticity in behavior (e.g., Stephens, 1993). A recognized problem with such a broad definition is, however, that the behavioral change observed may be due to physiological changes such as in egg load, which is generally not considered learning (Jaenike and Papaj, 1992; Rosenheim, 1993). Narrow definitions are mechanistic ones in that learning is classified according to the identified underlying mechanism such as associative vs. nonassociative, classical conditioning, instrumental learning, sensitization and habituation, each with its own definition (Rescorla, 1988; Smith, 1993) although these are not always unambiguous (see e.g., Turlings et al., 1993 for sensitization). These subdivisions of learning arise from a behaviorist-psychological approach. Using exemplar animal species, behaviorists try to search for general mechanisms underlying learning in all animal species. However, currently some animal psychologists are trying to integrate this mechanistic approach with an evolutionary one, recognizing the existence of learning specializations in specific species. The ecological context in which an animal functions can determine which learning abilities are favored (Roper, 1983; Shettleworth, 1984; Papaj, 1993b). Gould (1993) argues that terminology used by behaviorists is in fact comparable to that used by ethologists who traditionally are interested in both the mechanism and the evolution of behavior, and consequently study learning in a species-specific context.

Thus the way in which learning is defined and studied is in large measure a matter of disciplinary orientation and intended use. We employ an ecological approach to learning, although we consider the effect of learning on behavioral expression as well as the underlying mechanisms. Compared to honey bees, the basis of learning is practically unstudied in parasitoids. We follow Papaj and Prokopy (1989) and Jaenike and Papaj (1992) in applying the following criteria to learning, thereby avoiding the broad definition of "behavioral change with experience." 1. Behavior can change in a repeatable way through experience; 2. behavioral change is gradual with continued experience up to an asymptote; and 3. learned responses can be forgotten (wane) or disappear as a consequence of another experience. Criteria 1 and 3 are more easily applied to parasitoids than 2, because learning can be essentially instantaneous (Vet and Dicke, 1992; Kerguelen and Cardé, 1995).

3. Adaptive value for learning in parasitoids

3.1. The reliability-detectability dilemma

Parasitoids search for hosts that are presumed to have evolved strategies to avoid detection. Continuous selection on herbivores to be "inconspicuous" should be a major constraint on the evolution of long-distance host searching and may drive the evolution of 'indirect' search strategies, i.e. the use of information which is not derived directly from the potential host but from other sources that predict host presence (Vet and Dicke, 1992). The informational value of stimuli ultimately depends on two factors: 1) their reliability in indicating available and suitable hosts and 2) the detectability of the stimulus, i.e. the ease of stimulus discovery. Both stimulus characteristics are assumed to enhance searching efficiency and hence Darwinian fitness (Vet et al., 1991; Vet and Dicke, 1992; Wäckers and Lewis, 1994). The reliability and detectability of stimuli in general are expected to be inversely correlated (Figure 3.1).

Information from the host itself is highly reliable, but hosts are only one component of a complex environment and if they release information at all, such as odors, it will be in minute quantities, which hampers detection especially at a long distance. Furthermore, selection against detectability is expected. Plant cues, or other cues from the host's environment, are more detectable, but generally less reliable both over evolutionary time and over the foraging life of an individual. After all, the presence of the host's food plant does not guarantee the presence of a suitable host (Zanen and Cardé, 1991). Furthermore, plant habitus and volatile production may be highly variable, for example due to differences in growing conditions (Visser, 1986). These factors may hamper the evolution of invariable responses to such cues. Selection will favor animals that

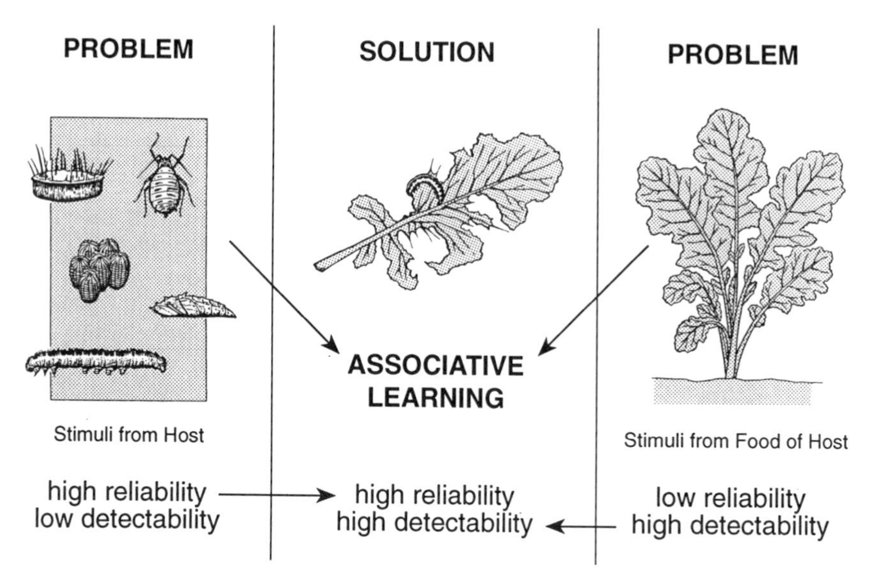

Figure 3.1. Associative learning is a major solution to the reliability-detectability problem in foraging parasitoids. By linking highly detectable cues to highly reliable cues through associative learning, parasitoids can temporarily increase the reliability of detectable indirect cues (adapted from Vet et al., 1991).

solve this dilemma as each suitable host found adds directly to their reproductive fitness.

Vet et al. (1991) and Vet and Dicke (1992) discuss several solutions to this reliability-detectability problem. One is the *infochemical detour*: parasitoids resort to information from other, more conspicuous, host stages than the one they attack. The sex pheromone communication system of moths, for example, is exploited by *Trichogramma* egg parasitoids (Lewis et al., 1982; Noldus, 1989). A tritrophic solution is the ability to respond to so called *herbivore-induced synomones*: plant volatiles that are only released upon damage by certain herbivores (reviews: Vet and Dicke, 1992; Turlings et al., 1993; Dicke, 1994). These stimuli are highly detectable as the synomone is released systemically by the whole plant under attack, (also by undamaged leaves) (Dicke et al., 1990, 1993; Turlings and Tumlinson, 1992). The information can also be quite reliable as the synomone can be plant and herbivore specific (Dicke and Takabayashi, 1991; Turlings et al., 1993; Dicke, 1994). However, due to their plant origin, herbivore-induced synomones are expected to be more variable than direct host cues. A third solution is *associative learning*. By linking highly detectable cues to highly reliable cues through associative learning, parasitoids can temporarily increase the reliability of detectable indirect cues (Fig. 3.1). Host-food cues are likely candidates to be learned. Associative learning of host-food odors has been shown for several species of parasitoids (e.g., Lewis and Tumlinson, 1988; Turlings et al., 1989; Vet and Groenewold, 1990; Lewis and Takasu, 1990; de Jong and Kaiser, 1991, 1992) and this mechanism is frequently suggested in other learning studies.

Foraging is thus constrained by the low detectability of direct host cues and the unreliability of indirect cues such as plant odors. These conditions favor learning of host-associated cues.

3.2. *Response potential: innate versus learned responses*

Naive emerging parasitoids are not condemned to search for their first victims in a random fashion. Naive parasitoids are expected to have a potential to respond innately to an array of stimuli derived from their host (these stimuli mainly for use at shorter distances), from the food and environment of the host, from organisms associated with the host. For present purposes we speak of 'innate,' although we are aware that the use of this term has created confusion when it lacked clear definition (Bateson, 1984; Tierney, 1986; Papaj, 1993a). We use the term innate in the sense of 'unlearned,' without suggesting an instinct-learning dichotomy. The 'innate part' of a response is shown by naive individuals, without the animal having had apparent experience with the stimuli concerned. However, later on in life such innate behavioral responses can be modified by experience. However, we do expect 'strong' innate responses to stimuli to be less likely modified by learning than 'weak' innate responses. In that sense the use of the word innate is comparable to 'canalized' as used by Tierney (1986), who contrasts 'strongly canalized' behaviors to 'weakly canalized' behaviors, the latter being relatively amenable to modification by learning.

A problem remains of what is meant by a naive parasitoid. The earlier addition of 'apparent' in 'no apparent experience' reflects the uncertainty of excluding experiental effects on parasitoid behavior. Due to their parasitic life style, parasitoids develop in and emerge from a host. Hence, they are inevitably exposed to many stimuli, some of which may play a role in their foraging as an adult. Still we refer to naive animals as 'inexperienced' unless experiments specifically address development as an experience factor.

Vet et al. (1990) developed a variable-response model based on observations of parasitoid responses to cues involved in foraging. The model specifies how the intrinsic variability of a response and the possibility for learning depends on the strength of the response. Each cue perceived by a parasitoid, as represented by S_1 through S_j of Figure 3.2a, has some potential of evoking an innate response in a naive parasitoid. The strength of this so called *response potential* is set by natural selection and can vary greatly between stimuli.

The way the strength of the response potential is expressed depends on the behavior under consideration. If, for example, an odor can stimulate upwind anemotaxis in a parasitoid, the probability that an individual parasitoid in fact starts to walk upwind upon encountering this odor is a measure of strength of the response. In behavioral experiments this probability is often expressed as a population measurement, i.e., the number of animals responding in a bioassay, with a maximum of 100% response. Another measure could be the walking speed of the parasitoid that responds to the odor. In the latter case the maximum response potential is set by constraints on the motor patterns elicited by stimuli, e.g., a maximal walking speed.

Stimuli perceived by the parasitoid can be ranked according to the strength of their response potential in the naive insect. The response potential is a way of assigning all incoming stimuli a relative value in common units, regardless of whether those stimuli evoke fundamentally different responses. When actually measured, the overt behavioral response will always show some variability. The model predicts that the magnitude of this variability depends on the strength of the response potential. When response potentials are high, the individual has low variability in its response (Fig. 3.2b). The inverse correlation between strength of the response and variability has been verified empirically (Vet et al., 1990; Vet and Papaj, 1992; Papaj, 1993a; see further on).

Due to their reliability, host-derived stimuli are likely to evoke strong innate responses that are little subject to learning. Examples are 'indispensible' cues used in the location and acceptance of highly suitable hosts. In fact, the stimuli on the left of the curve (Fig. 3.2a) that invariably elicit a maximal response, may act as the reinforcer in associative learning of other stimuli by the naive parasitoid or, using Pavlov's (1941) terminology, they can function as uncondi-

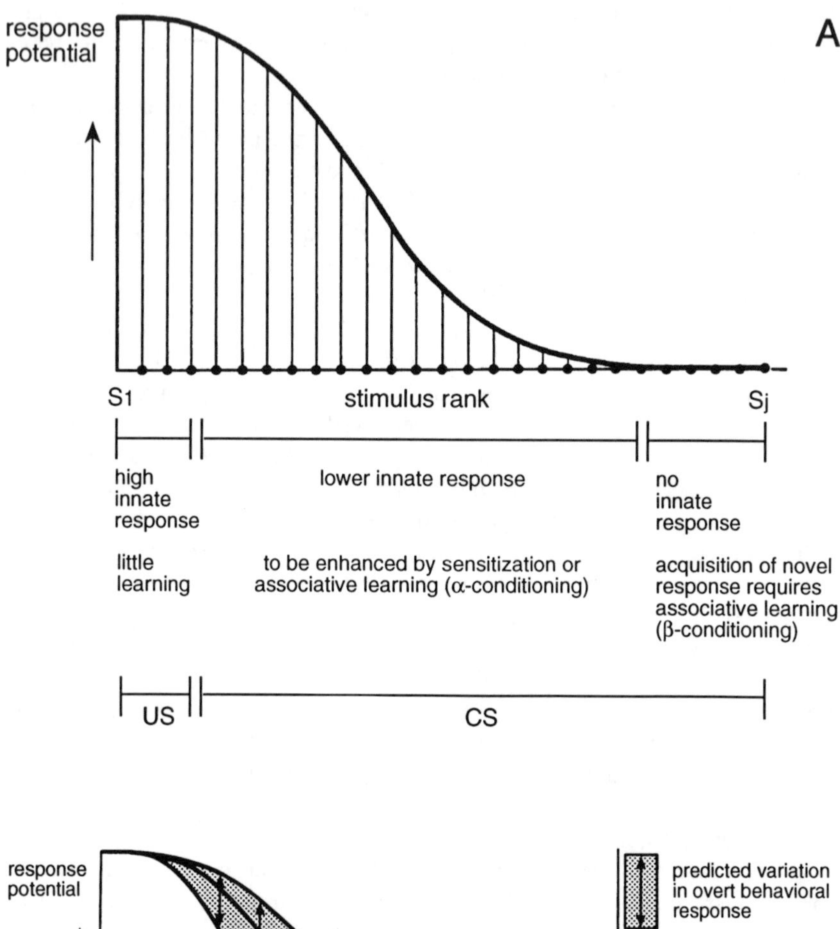

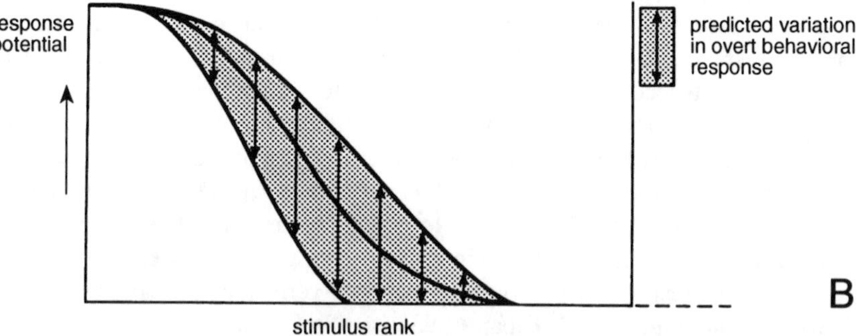

Figure 3.2. Diagram of a female parasitoid's potential behavioral response to a variety of environmental stimuli and the relationship with learning. A. All stimuli perceived by the insect are ranked according to their response potential in the naive insect. Stimuli beyond S_j are outside the range of sensory perception of the animal and can not be learned; B. Relationship between response potential level and variation in *overt* behavioral response (adapted from from Vet et al., 1990).

tioned stimuli (US). Plant and other host-environmental cues also may elicit some innate responses, the strength of which generally depends on their predictability as indicators of suitable hosts over evolutionary time. Due to their more variable nature and lower reliability (see above), innate responses to these cues will be lower than to host-derived cues. However, plant and other host-environmental cues are likely to become the most important and useful cues in foraging once learning has taken place. They are the conditioned stimuli (CS) (*sensu* Pavlov, 1941). This learning mechanism has been called alpha-conditioning and is defined as the enhancement of weak existing innate responses through associative learning (Carew et al., 1984; Menzel et al., 1993). We expect that parasitoids are likely to be able to process and are neurologically prepared to learn these biologically meaningful cues, in combination with specific reinforcing stimuli. This is the basis of so-called learning biases that are very pronounced in bees and vertebrates (Gould and Marler, 1984; Garcia and Garcia Y Robertson, 1985; Menzel et al. 1993), but practically unstudied in parasitoids (but see Lewis and Tumlinson, 1988). Another possibility is that responses to such biologically relevant cues are enhanced through sensitization, whereby the response is increased, without reinforcement, through mere exposure of the animal to the stimulus (e.g., Kaiser and Cardé, 1992). Such learning without reinforcement will be less likely for 'novel' stimuli that have no obvious biological meaning and elicit no initial response in an inexperienced animal (stimuli in the tail of the sigmoid). Associative learning will be required to acquire a new response to these novel cues (called beta-conditioning Carew et al. 1984; Menzel et al., 1993). Such cues may be the principal source of flexible foraging in parasitoids!

So when associative learning takes place during host foraging, the reward is expected to be one of the stimuli on the left of the curve (Fig. 3.2a), generally a host-derived stimulus, as indeed has been found for several parasitoid species (e.g., Drost et al., 1986; Lewis and Tumlinson, 1988; Turlings et al., 1989, 1990; Wardle and Borden, 1989; Vet and Groenewold, 1990).

When a parasitoid population is generally associated with a particular host species and its food plants, the parasitoid is expected to show high innate responses to those cues that, over evolutionary time, have shown to be reliable indicators of host presence and suitability. The innate stimulus-response profile is expected to be adapted to the most prevailing and suitable plant-host complexes, with the highest responses to direct host cues and more plastic responses to more variable indirect cues such as those derived from the host plant. Consequently, parasitoid populations are expected to differ in, for example, their innate responses to volatiles from a particular plant-host complex (Kester and Barbosa, 1991; Powell and Wright, 1992). As argued by Vet et al. (1990) and Lewis et al. (1990), this presumably genetic predisposition in response potential is flexible enough to be modified during adult life through changes in physiological state and through experience. Learning can move stimuli along the continuum and change their rank order. The individual's initial response profile is 'refined' to match the prevailing foraging circumstances, with learning as a major optimization mechanism.

As we will argue later on, the adaptive value of learning is expected to vary significantly between species and is expected to depend on factors such as predictability of host presence and availability, host stage attacked and foraging constraints related to it, life history, and dietary specialization in a tritrophic context.

4. Variation in how learning is expressed

4.1. The effect of experience on general and specific responsiveness: the learning mechanisms

Some parasitoid females can be trained to the odor of perfume, because in a laboratory experiment they once found and parasitized a suitable host while smelling a commercial perfume (De Jong and Kaiser, 1991). In nature, the effects of experience are expected to be more adaptive than enhancing the location of perfume-contaminated hosts!

In food learning by bees, exposure to an US-sugar solution is thought to have a threefold effect (Menzel et al., 1993): 1. releaser function: several responses, such as the extension of the proboscis, are released; 2. modulator function: the strength of responses to other stimuli are enhanced; and 3. reinforcer function: conditioned stimuli are reinforced. All three functions act together in classical conditioning and each initiates a different memory (see Menzel et al., 1993 for details). In spite of the gap in knowledge on neuroethology of learning in parasitoids compared to bees, it may be instructive to search for analogies. For a food-deprived parasitoid that finds food, the situation is likely to be very similar to bees. A sucrose stimulus given to a food-deprived *Microplitis croceipes* female was shown to: 1. release feeding behavior, 2. enhance an upwind flight response to odors they have not previously experienced (Takasu, unpublished data) and 3. reinforce responses to conditioned stimuli such as vanilla or chocolate odor to which they previously did not respond (Lewis and Takasu, 1990).

But what about host foraging? Are the processes involved in food-associated learning in honeybees and parasitoids similar to those governing learning in a parasitoid searching for a host? What is the exact nature of the US in learning during host foraging? What behaviors compare to the proboscis extension? To start with the last question, the responses released by an US in host foraging obviously are not as simple as the proboscis extension reaction. The responses are likely to depend on how hosts are located, and so may vary between species: e.g., antennation, release of ovipositor, ovipositor probing, and change in walking behavior. With regard to the nature of the US, the US in parasitoid learning during foraging is a stimulus that is coupled with oviposition. Parasitoids associate odors, colors, shapes, mechanosensory stimuli with the presence of hosts (Turlings et al., 1993). However, physical contact with a host is not always required for learning to occur. Contact kairomones also can act as an US (Lewis and Tumlinson, 1988; Vet and Groenewold, 1990). An inexperienced parasitoid *M. croceipes* recognizes a host-derived water-soluble component in the frass of its host *Helicoverpa zea*. Upon contacting this component, a female learns odors that are present during the contact, such as the volatile plant-derived components of the frass (Lewis and Tumlinson, 1988). Similarly, *Leptopilina heterotoma,* a parasitoid of *Drosophila* larvae, can also be conditioned to a novel odor, a green leaf volatile (Z)-3-hexen-1-ol, if this odor is presented to the wasp during search on a yeast patch containing a host kairomone solution. The learning effect was enhanced if it was reinforced with an oviposition (Vet and Groenewold, 1990).

Vet et al. (1990) hypothesize that *any* stimulus with a high response potential could *potentially* act as an US in associative learning of stimuli with a lower response potential, and that the higher the response potential of the US, the greater its effect as a reinforcer, i.e. the greater the behavioral change it induces or the longer it is remembered. Returning to the three functions of the US in food learning by bees (releaser, modulator and reinforcer), we suggest that the higher the response potential of a stimulus, the more likely the stimulus is to perform all of the three functions. The establishment of a reinforcement is not seen as a 'yes' or 'no' but occurs gradually. For example, a strong US may perform all three functions and associative learning of a CS may take place rapidly. Another stimulus with a somewhat lower response potential may have a more limited function of releaser and modulator only. For example it may release antennal responses or ovipositor probing, and enhance the general responsiveness to other stimuli, also called priming (Turlings et al., 1993) but it may not reinforce learning of a CS such as a novel odor, or it may only do so to a lesser extent. The latter is shown by an increase in the number of learning trials necessary to induce the behavioral change or by a limited memory of the information learned. In studies with *M. croceipes*, Martin and Lewis (1995) showed that memory for a CS encountered at a host site and reinforced by oviposition was more persistent than frass contact alone. When female parasitoids were provided experience involving only contact with frass, immediate subsequent responses were as strong as when contact with a host and oviposition were included in the experience. However, decay in the level of response to the CS was much more rapid when the oviposition was absent.

In summary, some stimuli may induce a general increase in responsiveness, i.e. an increase in response that is not restricted to the stimulus experienced (termed 'priming' by Turlings et al., 1993; e.g., Eller et al., 1992; McAuslane et al., 1991ab) whereas other, 'stronger' stimuli may do the same, but may additionally reinforce learning of specific responses to CS (called 'preference learning' by Turlings et al., 1993). In the following section we will mainly discuss the latter and illustrate how a change in specific responsiveness, induced by learning, can modify foraging behavior and potentially improve foraging success in the parasitoid.

4.2. How experience affects foraging behavior

We can identify several mechanisms by which experience can improve foraging success. Firstly, foraging parasitoids can *expand their stimulus set* by acquiring responses to novel stimuli that initially have little meaning (beta-conditioning). Learning to respond to a perfume may be quite adaptive if perfume is the odor of a temporary microhabitat with abundant hosts. There are quite a few examples of parasitoids learning novel cues. All proof for learning of novel stimuli, however, is limited so far to laboratory and 'semi-field' assays. Vinson et al., (1977), for example, showed that *Bracon mellitor,* a parasitoid of the boll weevil, learned to respond (probing with the ovipositor) to an antimicrobial additive in its artificial host diet. Lewis and Tumlinson (1988) showed that *M. croceipes* learn to fly to the arbitrary odor vanilla. The stimuli learned can be from numerous features of the host's environment and can involve different sensory modes. Wäckers and Lewis (1994) showed the importance of visual cues in foraging and learning for *M. croceipes.* These parasitoids learned to use 'targets' that either differ in color, pattern or shape after these were reinforced with a host encounter.

Secondly, foraging parasitoids can *alter their response to the stimulus set* (alpha-conditioning). Innate responses to biologically meaningful stimuli can be modified through rewarding and unrewarding experiences. Little is known of how an unrewarding experience affects innate responses (but see McAuslane et al., 1991b; Vinson and Williams, 1991; Papaj et al., 1994). Many examples of learning in parasitoid host foraging fall into the category 'increase after reward.' All are obviously system-specific examples showing the predisposition of species to learn biologically relevant cues. For example, in larval parasitoids of *Drosophila,* learning increases weak innate attraction to odors from the food of their host (Vet, 1988). Parasitoids of plant-feeding caterpillars learn to increase the response to volatiles from the plant, i.e. experienced females are more likely to make an upwind flight towards the odor source compared to inexperienced females (e.g., Vinson and Williams, 1991). Although both the acquisition of responses to novel stimuli and a change in responses to a fixed set of stimuli are generally assessed in laboratory and semi-field bioassays, their role in enhancing foraging success in nature can be deduced. In the case of the expansion of the stimulus set, it is the addition of a cue proven to be reliable in indicating host presence. Additionally it is also the increase in the number of sensory modes that can be beneficial if we assume that the more modes a parasitoid can use, the more efficiently she can forage. This was shown by Wäckers and Lewis (1993) who found that *M. croceipes* was more efficient in distinguishing host from non-host sites if olfactory and visual cues were combined. An increase in an existing response to a certain stimulus may help the animal to focus on this particular stimulus. When a given stimulus evokes a strong response the insect is more liable to ignore sensory inputs from other stimuli that might evoke different motor patterns (Vet et al., 1990).

A third and ecologically important effect of experience is that *preferences are altered* and temporary specializations to preferred stimuli may arise. Induction of preference is a likely mechanism to optimize essential foraging decisions such as where to go, what to neglect, where to remain and for how long. Field experiments by Papaj and Vet (1990) revealed that female *L. heterotoma*, experienced with host-infested food substrates such as mushrooms or fermenting apples, were more likely to locate a host-food substrate and do so more quickly than naive females. In addition, learning greatly influenced the choice of substrate. Mushroom-experienced females arrived more often at mushroom baits and apple-experienced at apple baits. These field experiments supported previous findings of such preference inductions in olfactometer experiments (Vet 1988).

A fourth effect of experience deals with *the variability of the response*. Strength and variability of responses have been shown to be inversily correlated (Vet et al., 1990) and this correlation is an essential element in the variable response model (Fig. 3.2b). Since learning usually increases the response to a stimulus, it should also reduce the varability of the response, and consequently we expect responses of naive animals to be more variable than those of experienced ones. Probably many students of parasitoid learning have observed that behavior appears more predictable following learning. Vet and Papaj (1992) and Papaj (1993a) quantified this effect in detail for the response of *L. heterotoma* to odors from food substrates of its drosophilid hosts. Oviposition experience in host larvae feeding in either fermenting apple-yeast or decaying mushroom substrate affected movement in odor plumes of each substrate. Using a locomotion compensator, it was shown that females walked faster and straighter, made narrower turns and walked more upwind toward the source, in a plume of odor from the substrate they were experienced with when compared to movement in odor plumes from the alternative substrate. Experience not only affected the mean values of the movement parameters, but also affected variability around these means. When walking speed or path straightness was increased following experience, variability was correspondingly decreased among individuals (Vet and Papaj, 1992) and within individuals (Papaj, 1993a). Using a stochastic simulation model of upwind movement by parasitoids in odor plumes, the effect of within-individual variability in movement patterns on the mean distance moved toward the odor source was simulated (Papaj, 1993a). While the mean walking speed and turning angle remained more or less constant over simulation runs, an increase in the variability of these parameters drastically decreased the mean net distance moved toward the source, which can be interpreted as a measure of foraging success. If we extrapolate this to field conditions, we expect that a reduction in variability through learning is likely to enhance the chance an animal finds the source once the odor plume is detected. In addition, consistent behavior will reduce the time required to find the source. As mentioned above, both effects were actually measured in the earlier mentioned field release experiments with trained and naive female *L. heterotoma* (Papaj and Vet, 1990). However, Kaiser et al., (1994) showed that flight track characteristics of inexperienced female *Cotesia rubecula* and those experienced on the plant-host complex were not separable. Perhaps flight toward odor is not as subject as walking to modification through experience, although a number of cases will need to be investigated to verify this possibility.

4.3. Foraging stages affected by learning?

Parasitoids that emerge outside a potential host population must go through a process of long-range and short-range orientation to find hosts and learning is expected to affect these stages to different extents. These stages can be delineated in the following generalized sequence: 1. host community location: behavior involved in the detection of an area with host microhabitats (e.g., a plant population that consists of host-infested and uninfested plants); 2. microhabitat location: behavior involved in the detection of a host-infested site (e.g., a host-infested plant, an infested leaf); 3. habitat acceptance: the 'attention' given to a microhabitat after arrival, often quantified as time spent searching, or inclination to antennate or probe with ovipositor; 4. host detection (*senso stricto)*: behavior involved in the detection of an individual host; 5. host acceptance: the examination behavior leading to the decision to reject or accept the host individual for oviposition.

4.3.1 Host-community location

As host-derived stimuli are not likely to be used in long-range search due to their low detectability, plants are an essential source of information for host community location (but see section 5 for exceptions). Following our previous reasoning on the general value of learning to deal with the variability and low reliability of plant information, we can expect learning to play an essential role in shaping host community location. However, the difficulty of studying long-range behaviors under field conditions so far has relegated this scanning behavior to the realm of theoretical discussion. Most studies on 'long-range' search are conducted at the level of the host's microhabitat where, as mentioned above, learning of host environmental (often plant) cues, is common.

4.3.2. Microhabitat location

The location and selection of a microhabitat, i.e. the host's plant or food substrate are little guided by direct host-derived cues (Vet and Dicke, 1992), and subject to learning. The expected degree and function of plasticity in these responses differs greatly between species and depends on the expected variability in these cues both over the foraging life of the parasitoid and over evolutionary time (see dietary specialization and learning, section 5). The response to reliable herbivore-induced (but plant-produced) synomones, indicating damage inflicted by a specific herbivore, may be innate but learning is still expected to play a major role in tracking the expected variability in these plant cues. Improving responses to these herbivore-induced synomones may not need associative learning but may be achieved simply through sensitization (Dicke et al., 1990; Kaiser and Cardé, 1992). Different behaviors involved in microhabitat location appear subject to learning. Learning may alter movement, aiding in finding and following of odor plumes leading to the microhabitat (Vet and Papaj, 1992; Papaj, 1993a, but see Kaiser et al., 1994). A form of location learning was shown for the parasitoid *M. croceipes* which attacks solitarily feeding *H. zea* caterpillars (Van Giessen et al., 1993; Sheehan et al., 1993). Van Giessen et al. found that learning plays a role in host-site discrimination: females learn to avoid previously visited host sites where oviposition has taken place. Drost and Cardé (1992a) found that *Brachymeria intermedia*, an endoparasitoid of gypsy moth pupae, learns whether to forage along tree trunks or in the duff, dependent on patterns of previous host encounters.

4.3.3. Habitat acceptance

At the level of the habitat acceptance, learning plays a significant role. Having encountered hosts in a certain microhabitat type, a parasitoid generally increases the time that is spent subsequently in such a microhabitat (e.g., Vet and Schoonman, 1988; Papaj et al, 1994), wherein initially strong innate responses seem to be less enhanced by learning than initially weak responses (Sheehan and Shelton, 1989; Poolman Simons et al., 1992). Throughout foraging, by sampling the environment and thus by gaining experience, the animal builds up an 'expectation' of the resource value of a microhabitat and acts according to this learned expectation (Stephens and Krebs, 1986; Krebs and Kacelnik, 1991). From a proximate point of view, such changes in behavior due to learning may be very difficult to separate from changes in behavior due to other factors, such as a decreasing egg load (Rosenheim, 1993; Henneman, et al., 1995).

4.3.4. Host detection (senso stricto)

Host detection is guided primarily by host-derived stimuli and, as previously argued, responses to these reliable stimuli are expected to be primarily innate. This expectation seems supported by empirical evidence as responses to kairomones are not or only slightly modifiable (see Vet and Dicke, 1992). If we expect learning it is in cases where kairomones are variable. This may be in polyphagous parasitoids that have the potential to attack a large range of host species. In this case it may be impossible to have innate responses to all possible kairomones and mothers and daughters may be faced with different situations

(Zanen and Cardé, 1991). Learning then may be used to track the specific kairomones of those host populations the parasitoid encounters during her foraging life (see also later, dietary specialization and learning, section 5). Another situation is when kairomone specifics are strongly influenced by non-host factors such as the diet of the host. In studies with *M. croceipes*, Nordlund and Sauls (1981) found that antennation and probing responses to frass of host caterpillars varied with the diet of the host. However, frass can not be considered as fully host-derived cues. Lewis et al. (1991) showed that responses to frass (from hosts feeding on cowpeas) by *M. croceipes* involve a combination of host-derived and plant-derived chemicals. The host-derived compounds were a host-specific, nonvolatile, water-soluble compound together with 13-methylhentriacontane (earlier described by Jones et al. (1971)). The hexane-soluble volatiles were plant-derived. Samples containing all three of these factors elicited the strongest antennation response upon contact and induced strongest subsequent flights to the plant volatiles in the hexane extract. Antennation of the host-derived components alone was less intense and produced a smaller effect on subsequent responses to the plant volatiles. Experience with only the plant-derived components had little impact on subsequent flights. Apparently associative reinforcement of the plant volatiles (CS) with the host-derived components (US) during an experience with frass resulted in the learning of the plant volatiles (Lewis and Tumlinson, 1988; Lewis et al., 1991). In choice tests in a wind tunnel Zanen and Cardé (1991) found *M. croceipes* females experienced on a given *H. zea*-plant complex were most likely in subsequent assays to fly toward odor sources with the same combination of stimuli; a novel plant complex was less favored.

Both at the level of habitat acceptance and of host detection, habituation, a non-associative form of learning plays an essential role. Habituation, being the waning of the response to a stimulus with repeated exposure to that stimulus, to host kairomones is expected to be the underlying mechanism of departure of the parasitoid from sites that are not or no longer profitable (Waage, 1978). Similarly, waning of previously learned responses is likely to play a role in 'giving up' and dispersal at all foraging levels, such as abandoning a leaf, the plant, the plant population. Waning occurs in the absence of the stimulus and of repeated reinforcement and is reported for parasitoids (e.g., Poolman Simons et al., 1992; Lewis and Martin, 1990).

4.3.5. Host acceptance

Arguments similar to those given for host detection may be used for the final stage of host selection, the examination of the host leading to rejection or acceptance for oviposition. Again we expect mainly innate responses to the major cues (being primarily host-derived) that elicit examination behavior but learning may be used to track variability in these cues. Associative learning of variable host details, such as chemistry, size or shape, may occur being reinforced by some major and strong host-derived cue. In *Brachymeria intermedia*, experience increases acceptance of *Lymantria dispar* pupae (Drost and Cardé, 1992c). In addition females learn to accept an unnatural host *Holomelina lamae* when these hosts are contaminated with kairomone of *L. dispar*; this kairomone most likely serves as the US in a conditioning process. Experience with clean *H. lamae* pupae does not increase subsequent response to other *H. lamae* pupae, suggesting that only the kairomone of its favored host is a strong enough stimulus to reinforce learning of other characteristics. The nature of the learned characteristics remains unknown. After a single oviposition in their favored host, a gypsy moth pupa, female *B. intermedia* become more apt to accept this host in subsequent ovipositions. There does not seem to be any definitive 'point' in the sequence of host acceptance and oviposition behaviors at which a female acquires such 'experience,' that is becomes more apt to accept its favored host. Females interrupted in a normal oviposition sequence at the stage of antennal drumming or just as they plunge their ovipositor into a pupa but before oviposition, nonetheless acquire sufficient experience to elevate their future host acceptance (Kerguelen and Cardé, 1995). Precisely when and how these cues are processed remains an open question.

Another effect of experience we expect at this level is operant conditioning, also called instrumental or trial-and-error learning, i.e. 'learning to do' rather than 'learning to recognize' as during associative learning (Gould, 1993). Experience was indeed shown to alter the sequential and temporal organization of host-acceptance behavior in *B. intermedia* (Drost and Cardé, 1990). It was demonstrated that the behavioral sequence leading to host acceptance in *B. intermedia* is highly canalized through experience, increasing the probability that oviposition actually takes place. Both associative and instrumental learning may be mechanisms behind changes in host species preference through experience (Cornell and Pimentel, 1978; Mandeville and Mullens, 1990). Ovipositional experience by the parasitoid of fly pupae, *Muscidifurax zaraptor* on *Musca domestica* or on *Fannia canicularis* significantly biased subsequent host species preference in favor of the host on which experience occurred (Mandeville and Mullens, 1990). Similar to the 'acceptance' of the habitat, both mechanistic and functional arguments should be considered when looking at whether a host is accepted or not. The foraging history of the parasitoid, and so both their remaining egg load and their informational state based on memory of previous experience, will determine whether it is more adaptive to accept a host than to reject it (Visser et al., 1992; Drost and Cardé, 1992bc; Henneman, et al., 1995).

5. Ultimate factors promoting learning

5.1. Host stage attacked and the value of learning

Some parasitoids search for and develop in small and sessile insect host eggs. Others hunt for larvae that may be actively feeding and moving about on plants.

Pupal parasitoids try to locate hosts that may be removed from any usable indirect cues, such as plant damage, as larvae often disperse from their feeding site. How different are search strategies for parasitoids that attack these different host stages? Can we recognize a pattern in specific foraging constraints and to what extent does learning help solve specific foraging problems?

In Table 3.1 we focus on broad generalizations, recognizing that there are undoubtedly many exceptions to the proposed characteristics, due to the diversity in life histories and degrees of specificity of species, even when attacking the same host stage.

Because of mortality of the host stages, eggs are more abundant than pupae, and so broadly generalizing we could say that, per host species, availability of hosts decreases with the stage attacked: host densities for egg parasitoids are higher than for parasitoids that search for larvae and larval hosts are again more available than pupal hosts. Restricted host availability may be partly solved by increasing the range of acceptable host species but let us ignore this solution at first. Besides there are also pupal parasitoids that are very host specific.

5.1.1. Egg parasitoids

Eggs are small in size and of course so are the parasitoids that emerge from them. Small size imposes major constraints on the foraging potential of these parasitoids. It will restrict upwind movement, necessary to actively track an odor plume (Noldus, 1989). Egg parasitoids are more likely to disperse by drifting passively downwind, and so long-distance 'search' will be difficult. They will likely have a low chance of surviving aversive climatic conditions and generally are short-lived. Hence, locating a fruitful host habitat is seemingly difficult and hazardous for egg parasitoids. What kind of cues guide them in long-distance search? There are limited indirect cues, such as plant damage. Pheromones from the adult stage of their host such as sex pheromones have been shown to be important cues in locating eggs of certain noctuid moths (Lewis et al., 1982; Noldus, 1989). One underlying behavioral mechanism for these responses to pheromones seems to be arrestment (Noldus, 1989 (review); Noldus et al., 1991). Information from plants does not seem to be used in attraction, but may also play a role in arrestment, important for habitat acceptance (reviewed by Noldus, 1989). At short distances host finding is readily accomplished. Egg parasitoids are strongly arrested by host kairomones (such as scales from the host adults wings, egg-adhesive materials (e.g. Lewis et al., 1972; Jones et al., 1973; Strand and Vinson, 1982; Nordlund et al., 1983; Noldus and van Lenteren, 1985; Pak and de Jong, 1987), responses that seem to be unaffected by experience (Gardner and van Lenteren, 1986; Zaborski et al., 1987). Visual cues such as color and shape of the host egg may be used to locate the host egg at a very close distance, although they are most likely to function as host-examination cues (e.g., Strand and Vinson, 1983; Pak and de Jong, 1987). Although the availability of host

Table 3.1. The effect of host stage attacked on parasitoid foraging characteristics, foraging constraints and learning

Host stage attacked	Egg	Larva	Pupa
Host density (per host species)	high	intermediate	low
Expected number of long-distance foraging decisions	low	high	low
Major long-distance cues (innate)	host adult pheromones	plant-derived: feeding damage (also learned) if possible, host-derived: host adult pheromones	host-derived: volatiles pupae if possible, plant-derived: larval feeding damage??? (also learned)
Within-generation predictability in major foraging cues	high	high	low?
Between-generation predictability in major foraging cues	high	low	low
Learning	unlikely	any predictable environmental cue, especially plant-derived	any predictable environmental cue, but possibilities limited due to absent association with host plant
Major constraints on host finding	small size, short longevity	time limited: host encounter rate egg-limited: encounters with suitable hosts	lacks help of plant, lack of predictable indirect cues, use of adult pheromones unlikely
Solutions	utilize cues from the adult stage (pheromones) and phoresis to increase chance of successful habitat location	abundant learning of plant and environmental cues	learn any predictable environmental cue, learn to refine host acceptance, live long, invest in quality offspring or broaden host range

eggs is high, the short longevity of egg parasitoids in combination with their limited ability to search may limit the number of life-time foraging decisions that lead to host encounters. However, once a host community is located an abundance of hosts may be encountered, especially if the host eggs are laid in clusters. At that point they may become egg-limited and may invest in quality by maximizing fitness gain per host encountered. The fact that eggs are sessile and can not actively defend themselves against the act of oviposition allows meticulous examination to optimize the assessment of host suitability (e.g., Schmidt and Smith, 1987).

Stephens (1993) postulates that learning is most valuable when the environment changes unpredictably between generations but is predictable within generations. How predictable is the environment (i.e., with respect to essential foraging cues) for egg parasitoids? Again due to their typically short life they are likely to experience limited within-generation variability. If the findings for *Trichogramma* egg parasitoids can be generalized, pheromones of major host species may be important foraging cues for egg parasitoids and, as argued earlier, parasitoids are expected to have innate responses to such reliable host-derived cues. As well, the pheromone of a certain host species constitutes a relatively invariable source of information (Vet and Dicke, 1992), which also favors the development of prewired responses. So although mothers and daughters may be searching for different host species, this variability is likely to be tracked by innate responses. Another argument against learning in egg parasitoids can be derived from a dynamic model of Roitberg et al. (1993). They conclude that the frequency of decisions can strongly affect the value of learning about environmental variability. They argue that animals making few major decisions (as in choosing mates) in general will be less likely to learn about their local environment than animals making many small decisions (as in choosing hosts). Although the model deals with adjustment of preference for host or mate quality according to local availability, the same arguments may apply to preference for and learning of host location cues. We suggest that the more foraging decisions, the more likely it is that learning of foraging cues is adaptive. Hence, for egg parasitoids learning of foraging cues is not expected because they make possibly one or very limited long-distance foraging decisions.

If learning is of little use to egg parasitoids, how do they solve the major foraging constraints that are set by their life history? Egg parasitoids seem to rely heavily on the adult stage of their host to solve their problems. A few species have been demonstrated to respond to their host's pheromone. Some other species are even physically transported (phoresy) by the adults, thereby securing the discovery of a host habitat and encounters with suitable host eggs that are about to be laid (Clausen, 1976). The latter obviously requires the development of great host specificity. Quite an opposite solution may be the development of polyphagy as in *Trichogramma* species. Polyphagy increases the potential number of hosts for attack whereby a gregarious reproductive mode can help to optimize clutch size and sex ratio in hosts of various quality (size) to maximize parental fitness (Waage and Ng, 1984; Godfray 1987). In summary, we expect little learning in egg parasitoids. If learning were to occur it would be most likely in the more polyphagous species that attack hosts that lay their eggs in a solitary fashion, not in clutches (lowest between-generation predictability, highest number of foraging decisions) (see Kaiser et al., 1989).

5.1.2. Larval parasitoids:

The story is quite different for parasitoids attacking the larval stage. The expected number of foraging decisions is much higher which by itself favors learning if the arguments of Roitberg et al. (1993) are applicable. Selection will favor maximization of the rate of host encounter, especially for proovigenic species (*sensu* Flanders, 1950: oogenesis has ceased when oviposition starts, i.e. females have all their eggs available at once) that are limited in the time available for search and oviposition. The non-sessile and often agressive and defensive host larvae may be more difficult to examine thoroughly for suitability, and certainly compared to eggs and pupae, oviposition may be a hazardous experience (e.g., Stamp, 1982).

Host-derived components of larval feeding damage and frass will be important cues to which innate responses may exist. When possible, an infochemical detour may be expected whereby the larval parasitoid, like the egg parasitoid, spies on the intraspecific communication system of the adult stage of the host, such as its pheromone. *L. heterotoma*, a parasitoid that attacks larvae of *Drosophila* species, shows a strong attraction and arrestment to (Z)-11-octadecenyl acetate, also known as *cis*-vaccenyl acetate (cVA), the major component of the aggregation pheromone of two of its hosts, *Drosophila melanogaster* and *D. simulans*. As expected responses (attraction and arrestment) to these reliable host-derived infochemicals are strong and innate (Wiskerke et al., 1993). This strategy needs to be investigated for other parasitoids that attack larval stages.

As argued earlier, plants can provide important information for foraging parasitoids, although innate responses to plant cues are probably restricted. The plant cues are expected to vary greatly between generations, as mothers and daughters may be foraging for different host species feeding from different host plants. Although the nature of plant information is by itself quite variable due to, for example, differences in growing conditions, the information is likely to be quite predictable over a particular period of a parasitoid's foraging life. This strongly favors learning (Stephens, 1993).

5.1.2.1. Time versus egg limitation.

Whether an individual is likely to run out of eggs or time may have interesting implications for the value of learning. Under time-limited conditions, selection will favor increasing the host encounter rate rather than selecting the better hosts.

In time-limited situations we therefore expect any predictable environmental cue, mostly plant-derived, to be learned as long as its use enhances the host encounter rate. These cues can be rather general, indicating merely host presence, not suitability. Smith (1993) discussed the adaptive value of stimulus generalization, a concept used in psychology learning studies, being defined as the tendency to respond to stimuli that were not experienced during conditioning trials, but which vary from a learned CS along a defined perceptual dimension (e.g., wavelength, brightness, size for visual stimuli; for odors this dimension may be more difficult to define as odors may vary in many respects, e.g., carbon-chain length, structure, functional groups). Smith argues that generalization of learned stimuli is not 'a mistake' by the animal in that it responds to something else than what was reinforced, but generalization may be highly adaptive when 'error' costs are low. For parasitoids under time-limited conditions error costs are low for finding less suitable hosts and generalization of learned information can be expected. For individuals that are egg-limited there is stronger selection on fitness gain per host and therefore also on investment in quality, especially when the life-time number of eggs is small (as already mentioned for egg parasitoids). This means that the premium will be placed on obtaining specific and therefore reliable information about the presence of *suitable* hosts. In egg-limited situations learning may therefore be of less importance, as they need to rely more on innate responses to reliable host-derived cues to avoid making mistakes. As the error cost of getting unreliable information is high, generalization of learned stimuli is not expected under egg-limitation circumstances.

Time and egg limitations are probably a characteristic of a foraging individual rather than of a species (Driessen and Hemerik, 1992; Visser et al., 1992), although at the species level we may expect a correlation between being time limited and being proovigenic. Little is known of egg vs. time limitation in parasitoids under field conditions, and the differences in learning between egg- and time-limited individuals and species remain to be tested.

5.1.3. Pupal parasitoids:

Considering pupal parasitoids we expect host densities to be comparatively low and consequently, compared to larval parasitoids, the expected number of foraging decisions leading to successful host encounters should be less frequent. The availability of long-distance cues may be a major constraint on host finding in pupal parasitoids. As for the other parasitoids, the use of direct host-derived cues is limited. However, pupae can have an additional constraint: pupae typically are not associated with a predictable food substrate and this feature of their biology also restricts the use of predictable indirect cues. Pupal parasitoids thus generally have to forage without the plant's 'help.' After all, food substrates may be ephemeral and so gone by the time the larvae have pupated (e.g., *Drosophila* in decaying mushrooms). Additionally, larvae often disperse from their food to pupate, enhancing their inconspicuousness. Besides, the process of dispersion itself induces random distribution which hinders foraging efficiency in the searching parasitoid. Thus, the solution of using larval feeding damage is limited to situations where the pupae stay in or on the plant and larval and pupal stages more or less co-occur. Another difficulty may be that some of the volatile cues emanating from the pupae are expected to be quite variable if dependent on the food of the larvae (as compared to, e.g., sex pheromones of adult hosts). This would limit the possibility for innate responses to specific kairomones and select for the use of a general kairomone (US) and for learning of specific details of the kairomones (CS) during foraging.

So how do pupal parasitoids forage and does learning help? Our expectation is that they do learn predictable environmental cues when possible, even very general ones as long as they predictably indicate host presence. It was shown that *B. intermedia*, an endoparasitoid of pupae of the gypsy moth, learns such general cues: whether to forage along tree trunk or on the ground (Drost and Cardé, 1992a). Schmidt et al. (1993) showed that a parasitoid of the genus *Pimpla* learns to find *Pieris* pupae in certain colored 'microhabitats' (paper cylinders). The extended life span of this large parasitoid could increase the fitness value of such learned responses.

What other 'solutions' are there? Increasing host range and reproducing gregariously (with a consequence of being small) may be one of the solutions to increase host encounter rate (investment in quantity). Being host specific but large and long lived may be another (investment in quality), and both strategies are encountered. Many other factors of course can drive the evolution of these alternative reproductive strategies. Being limited as a pupal parasitoid in the possibility of increasing host encounter rate through learning, learning instead may help to increase the investment in quality and fitness gain per host. Being sessile, like eggs, pupae can also be examined meticulously to optimize oviposition decisions on clutch size and sex allocation. During oviposition in a first suitable host, other host recognition characteristics such as odor, shape, color, may be learned thereby increasing the reliability of future host acceptance (see above, 4.3) (Drost and Cardé, 1992c.)

5.2. Dietary specialization and learning

Many parasitoid species, but especially those attacking larvae, function within a tritrophic context and so information from both the first and second trophic level can shape their foraging strategy. Parasitoids attacking a single host species that feeds on a single plant species requires a different foraging strategy than a species attacking a wide variety of host insect species, each with a variable diet. The degree of specificity at two trophic levels can influence the variability that parasitoids encounter both over generations and within their foraging life. Hence we expect dietary specialization to be an ecological factor setting specific values for learning (Vet and Dicke, 1992).

Figure 3.3 shows the expected response patterns by parasitoids that differ in their dietary specialization at two trophic levels. Learning is expected to play an insignificant role in species that experience little variation in host and plant species (A) and strong innate responses to kairomones, and stimuli from the host's food are expected in this situation. In situation B, information from the host's food is expected to be relatively important in foraging, and hence strong innate responses to plant volatiles and other plant characteristics can be expected. However, these plant specialists still have to deal with the intrinsic variabilty in these first trophic level cues which may dictate a need for some learning. With an increase in host range attacked, we expect innate responses to specific kairomones of each possible host species to become less likely. First of all, there may be little value in such specific information and it may also be less feasible

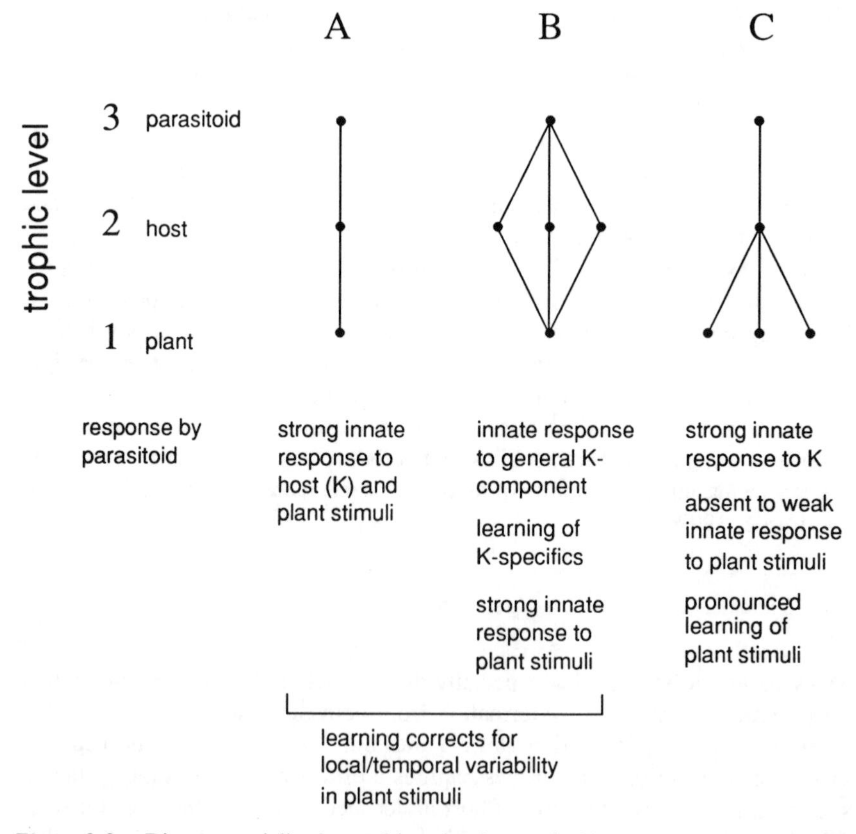

Figure 3.3. Dietary specialization and learning in parasitoids in a tritrophic context. K = kairomone (adjusted from Vet and Dicke, 1992).

at the sensory and processing level to encode all this information. An innate response to a general kairomone cue may be sufficient to overcome between-generation variability in kairomone cues and, if necessary, learning of details of a kairomonal mixture can help when a parasitoid is confronted with a temporal or spatial clustering of a certain host species. In situation C, we expect strong innate responses to any cue that is host-derived. Weak innate responses are expected to important host food cues (i.e., in an evolutionary sense), but learning is expected to easily modify these responses and adjust preferences. An example of situation C is *M. croceipes,* a specialist parasitoid of *H. zea*, that itself feeds on a wide diversity of plant species. Thus, we expect most learning to occur in situation C due to two reasons. First, the reliability-detectability dilemma already argues for learning of first trophic level stimuli and, second, the wide range at this trophic level adds to the variability of these cues.

The idea that diet breadth and learning may be correlated is not new and has been applied to many different species of animals ranging from insects to mammals (Gould and Marler, 1984; Shettleworth, 1984). Generally the discussion deals with the relation between diet breadth and learning *ability* and here the evidence that generalists would be more *able* to learn is conflicting (e.g., for insect herbivores, Papaj and Prokopy, 1989). Also in parasitoids learning has been reported for several specialist species (e.g., Arthur, 1971; Sheehan and Shelton, 1989; Kaiser and Cardé, 1992). However, to answer the evolutionary question of whether generalists and specialists have different learning abilities, sound comparative studies with closely-related species are needed and these kinds of studies are still rare (Rosenheim, 1993). After all, as argued by Tierney (1986) and Papaj (1993b), learning may be an old and general optimization mechanism. In other words, the ability to learn may have evolved in ancestral species for quite other reasons, but may still be used in present day species to solve a diversity of problems, including those dealing with foraging. Poolman Simons et al.(1992) postulate that it may be equally interesting, if not more valid, to investigate if generalists and specialists use their learning ability in a different way, instead of having evolved different abilities. By comparing learning in closely related generalist and specialist parasitoids of *Drosophila* larvae, they showed that the two specialist species *Leptopilina boulardi* and *L. fimbriata* learn patch characteristics, although their behavior was less flexible compared to the generalist *L. heterotoma*. While patch times are strongly influenced by experience in the generalist, the specialists have fixed search times on their natural host food substrates and learning only affects their search time on less preferred substrates. It thus seems that similar learning abilities may be used in a different way, for different purposes: the generalist can achieve a great flexibility in substrate selection, whereas the specialist may employ learning to divert temporarily to a less preferred substrate when their preferred substrate is temporarily absent.

Comparative studies, wherein phylogenetically close species are used to search for correlations between species characteristics and ecological factors, are valuable in delineating the adaptive value of learning traits and answering questions such as: How does learning enhance foraging efficiency in different species? How much learning is involved and do the findings support the hypotheses posed above? The function of stimulus generalization—to what extent should an animal generalize from a learned to a novel stimulus (Smith, 1993)—could be explored in such comparative studies. We could test the hypothesis that the adaptive value of generalizing learned stimuli differs among specialists and generalists. However, to design ecologically relevant bioassays with highly comparable stimuli we would need to use species with highly comparable or even overlapping niches. This sort of comparison enhances the chance of drawing unambiguous conclusions that differences in learning observed are indeed correlated to diet breadth and not to some other ecological or phylogenetic factor.

6. Physiological state and learning

6.1. Learning of food- and host-related cues

The behavioral output of a parasitoid in part will be dictated by internal factors such as its nutritional and egg-load states (Bell, 1990; Minkenberg et al., 1992; Drost and Cardé, 1992b; Barton Browne, 1993). Most parasitoids require food during the adult stage (Jervis and Kidd, 1986). Hence, foraging for food is yet another task to be fulfilled. Feeding activities can interfere with host foraging for several reasons. Food sources are likely to be separated temporally or spatially from hosts. Moreover, cues used in food and host foraging may differ and compete as the parasitoid is not likely to attend to them simultaneously. Lewis and Takasu (1990) showed that *M. croceipes* can learn to associate novel odors with a sucrose reward. Such learning may play a role in optimizing the task of foraging for food just as it does in foraging for hosts. Associative learning during feeding is not surprising. More intriguing is the way parasitoids deal with the combination of food- and host-related learning. Lewis and Takasu (1990) and Takasu and Lewis (1993) showed that *M. croceipes* could learn two different odors, one associated with hosts, another with food, and that their preference for these odors depends on whether they are food- or host-deprived (Figure 3.4).

Apparently learned information acquired during feeding and during foraging is used according to the parasitoid's physiological state. A remarkably sophisticated system of olfactory information processing and adaptive use that may rival that described for bees and even rats. These findings raise interesting mechanistic questions such as: are parasitoids more prepared to link food-related cues, such as flower odors, with food and host-related cues, such as plant odors, with hosts? What information is processed through which receptors on the antennae and mouth? Where is the olfactory and gustatory information integrated and what is stored peripherally versus centrally?

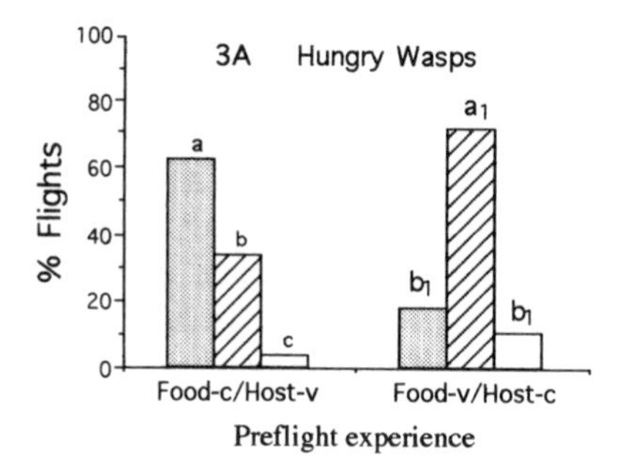

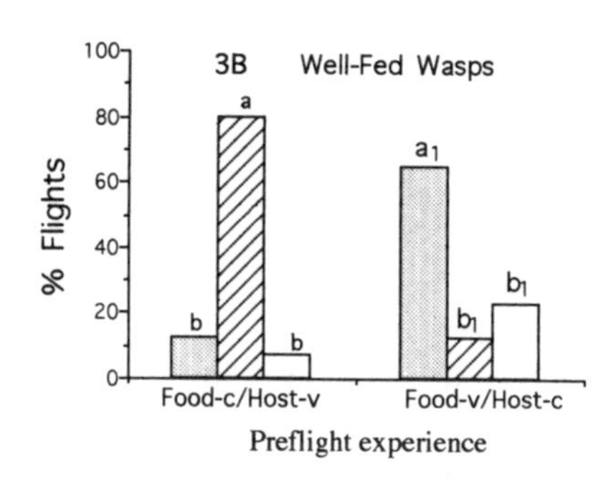

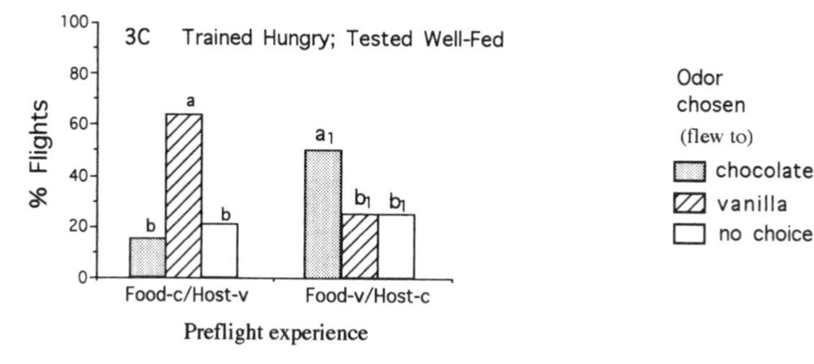

Figure 3.4. [from Lewis and Takasu, (1990)]. Flight responses in a flight tunnel to vanilla or chocolate extract by (a) hungry, (b) well fed, or (c) trained hungry then well fed females with preflight training indicated. Food-C/Host-V: females given training experience of chocolate-to-vanilla and vanilla-to hosts; Food /Host C; females given training experience to-vanilla-to-food and chocolate-to hosts (the order of food-odor and host-odor training was alternated and the results combined). Bars within same treatment group capped by different letters are significantly different, $P<0.01$ Waller-Duncan K-ratio t-test; minimum significant difference, 10.5% for a and b, and 16.3% for c; $n=62$ wasps for each test group (six replications, six wasps per replication per preflight treatment). The training procedure involved exposure permitting feeding on sucrose or stinging of a host while smelling the respective odor of vanilla or chocolate.

6.2. Egg load and learning

In addition to nutritional state, a parasitoid's egg load can have a proximate and ultimate effect on host foraging decisions. For example, a female parasitoid that still has all her eggs at the end of her life is expected to have a high 'motivation' to lay eggs and therefore accept any host for oviposition. Theoretical optimization models predict a fitness pay-off for lower selectivity under such sub-optimal conditions and so females should indeed accept less suitable hosts for oviposition (e.g., MacArthur and Pianka, 1966; Janssen, 1989; Mangel, 1992). Nutritional state may be important in that the intake of honeydew or nectar increases longevity which consequently releases some of the pressure of time-limitation and host-deprivation.

Let us assume that high egg load indeed decreases selectivity of responses to foraging stimuli. Does this have any consequence for the value of learning? After all, learning often *increases* selectivity, by enhancing the response to and preference for specific foraging stimuli after a rewarding experience. Is 'learning to prefer' less functional under high egg-load conditions? If so, is the propensity to learn or the rate of learning different under high egg-load compared to low egg-load conditions? Are parasitoids likely to learn different things under these different conditions? So far little is known about egg-load effects on learning processes in foraging parasitoids. Here we restrict ourselves to some ideas that may be open to experimental testing. Firstly, under high egg-load conditions we are dealing with time-limited search and we can repeat the arguments put forth in the previous section on learning in time-limited parasitoids: learning of general cues is to be expected to increase general host detection and generalization of learned stimuli (*sensu* Smith, 1993) is likely to occur. So if animals are not host-deprived we may indeed expect them to learn different things and utilize learned information in foraging compared to when they are loaded with eggs and running out of time. This could be tested empirically in comparative studies where egg load is manipulated independent of experience [which may however be very difficult! See Rosenheim (1993) for a discussion on confounding variables in experimental analyses of learning]. Secondly, host deprivation may lower stimulus thresholds for important host foraging stimuli. Vet et al. (1990) hypothesize that in associative learning the strength of the response to the US determines the effect of the reinforcement. If the response to the US itself is affected by egg load and higher under high egg-load conditions, a stronger learning effect under such conditions is expected. For example, we could compare learning rates under high and low egg-load conditions. Thirdly, when host deprivation takes place after some hosts have been found and therefore after an initial learning experience, strong learning-induced preferences will wane as a consequence of a lack of reinforcement and parasitoids then are likely to express their innate preferences (Vet et al., 1988; Poolman Simons et al., 1992). Whether forgetting, or memory processes in general, are affected by egg load is unknown.

Hence, the interaction of physiological state, learning and foraging is yet another research area that is largely uncharted. Stochastic dynamic programming, whereby behavioral modification due to learning (change in informational state) and due to changes in egg load are incorporated, can be a powerful tool for theoretical development, especially when used in combination with experimental work (Mangel, 1993).

7. Future directions and unanswered question

In this chapter we attempted a synthesis of present knowledge on learning by parasitoids. In our search for generalizations it has become clear that the evidence for our generalizations is limited. The field of learning and parasitoid foraging is still fragmented and in need of contributions from disparate disciplines, ranging from chemical ecology to neuroethology. In this section we identify the gaps in our knowledge and unanswered questions.

First, the range of species studied needs to be expanded. The hypotheses given in the present chapter may prove useful in guiding the choice of new experimental species, although in practice the choice of species to be studied is dictated in most cases by the presumed value of the parasitoid in pest management. There is considerable knowledge on learning in larval parasitoids and a comparative lacuna on species with other life styles. And parasitoids offer such a diversity from which to chose.

Secondly, insight into the adaptive value of learning-induced behavior requires field studies or at least questions which arise from the field. What are the exact problems parasitoids solve with learning? Answers to this question are essential if we want to use parasitoids as objects for the study of the evolution of learning. Are learning abilities indeed species specific or is learning a general optimization mechanism that is applied in a species specific context [see Papaj (1993b) for a general discussion on the evolution of learning]. Comparative studies, using species with a span of phylogenetic relatedness and ecological requirements is the most useful approach for answering these questions.

To answer evolutionary questions on learning, insight into mechanisms that manifest in behavioral plasticity is crucial. How generic in a physiological sense are the learning mechanisms? How physiological state and learning interact in modifying behavior is barely known. As well, how variation in sensory perception of stimuli influences learning and foraging in general needs definition. There is a clear need to know the chemical identity of the stimuli involved, but the other sensory modalities also deserve more attention than they have accorded to date. Sensory modalities have been shown to interact in determining foraging behavior and learning. A research focus on a single sensory modality may provide a misleading picture

Last but not least, little effort has been put into extrapolating effects of variation due to learning from the level of the individual to that of the population. This may turn out to be one of the great challenges of the future. How do learning-induced changes in foraging strategy and host-location efficiency of parasitoids affect the dynamics of both parasitoid and host populations? The answers may be provided by population ecologists who include observations of searching behavior into their models of dynamics of parasitoid-host systems. Population consequences of learning are of particular interest to applied entomologists who intend to manipulate the behavior of natural enemies used in biological control of insect pests.

There is an obvious need for an interdisciplinary approach to study learning and foraging in parasitoids and a fertile ground for future collaborations among chemical and population ecologists, ethologists and physiologists.

Acknowledgments

The senior author thanks Marcel Dicke, Felix Wäckers and Ted Turlings for many fruitful discussions. Comments on the manuscript were given by Marcel Dicke, Ted Turlings, Felix Wäckers and Olivier Zanen. Piet Kostense is acknowledged for drawing some of the figures.

References

Alphen, J.J.M. van and Vet, L.E.M. (1986) An evolutionary approach to host finding and selection, in *Insect Parasitoids*, (eds J.K. Waage and D.J. Greathead), Acad. Press, London, pp. 23–61.

Arthur, A.P. (1971) Associative learning by *Nemeritis canescens* (Hymenoptera: Ichneumonidae). *Can. Entom.* 103: 1137–1141.

Barton Browne, L. (1993) Physiologically induced changes in resource-oriented behavior. *Annu. Rev. Entomol.* 38: 1–25.

Bateson, P.P.G (1984) Genes, evolution and learning, in *The Biology of Learning* (eds P. Marler and H.S Terrace), Springer-Verlag, Berlin pp. 75–88.

Bell, W.J. (1990) Searching behavior patterns in insects. *Annu. Rev. Entomol.* 35: 447–467.

Bell, W.J. and Cardé, R.T. (eds) (1984) *Chemical Ecology of Insects*, Chapman & Hall, London.

Carew, T.J., Abrams, T.W., Hawkins, R.D., and Kandel, E.R. (1984) in *Primary Neural Substrates of Learning and Behavioural Change*, (eds D.L. Alkon and J. Farley), Cambridge University Press, Cambridge, pp. 169–184.

Clausen, C.P. (1976) Phoresy among entomophagous insects. *Annu. Rev. Entomol.* 21: 343–368.

Cornell, H. and Pimentel, D. (1978) Switching in the parasitoid *Nasonia vitripennis* and its effects on host competition. *Ecology* 59: 297–308.

De Jong, R. and Kaiser, L. (1991) Odor learning by *Leptopilina boulardi*, a specialist parasitoid (Hymenoptera: Eucoilidae). *J. Insect Behav.* 4: 743–750.

De Jong, R. and Kaiser, L. (1992) Odour preference of a parasitic wasp depends on order of learning. *Experientia* 48: 902–904.

Dicke, M. (1994) Local and systemic production of volatile herbivore-induced terpenoids: their role in plant-carnivore mutualism. *J. Plant Physiol.* 143: 465–472.

Dicke, M., Sabelis, M.W., Takabayashi, J., Bruin, J., Posthumus. M.A. (1990) Plant strategies of manipulating predator-prey interactions through allelochemicals: prospects for application in pest control. *J. Chem. Ecol.* 16: 3091–3118.

Dicke, M., Baarlen, P. van, Wessels, R., Dijkman. H. (1993) Herbivory induces systemic production of plant volatiles that attract herbivore predators: extraction of endogenous elicitor. *J. Chem. Ecol.* 19: 581–599.

Dicke, M. and Takabayashi, J. (1991) Specificity of induced indirect defence of plants against herbivores. *Redia* 74 (3) Append: 105–113.

Driessen, G. and Hemerik, L. (1992) The time and egg budget of *Leptopilina clavipes* (Hartig), a larval parasitoid of *Drosophila*. *Ecol. Entomol.* 17: 17–27.

Drost, Y.C. and Cardé, R.T. (1990) Influence of experience on the sequential and temporal organization of host-acceptance behavior in *Brachymeria intermedia* (Chalcididae), an endoparasitoid of gypsy moth. *J. Insect Behav.* 3: 647–661.

Drost, Y.C. and Cardé, R.T. (1993a) Use of learned visual cues during habitat location by *Brachymeria intermedia*. *Entomol. exp. appl.* 64: 217–224.

Drost, Y.C. and Cardé, R.T. (1993b) Influence of host deprivation on egg load and oviposition behaviour of *Brachymeria intermedia*, a parasitoid of gypsy moth. *Physiol. Entomol.* 17: 230–234.

Drost, Y.C. and Cardé, R.T. (1993c) Host switching in *Brachymeria intermedia* (Hymenoptera: Chalcididae), a pupal endoparasitoid of *Lymantria dispar* (Lepidoptera: Lymantriidae). *Environ. Entomol.* 21: 760–766.

Drost, Y.C., Lewis, W.J., Zanen, P.O., Keller, M.A. (1986) Beneficial arthropod behavior mediated by airborne semiochemicals. I. Flight behavior and influence of preflight handling of *Microplitis croceipes* (Cresson). *J. Chem. Ecol.* 12: 1247–1262.

Eller, F.J., Tumlinson, J.H., Lewis, W.J. (1992) Effect of host diet and preflight experience on the flight responses of *Microplitis croceipes* (Cresson). *Physiol. Entomol.* 17: 234–240.

Flanders, S.E. (1950) Regulation of ovulation and egg disposal in the parasitic Hymenoptera. *Can. Entomol.* 82: 134–140.

Garcia, J. and Garcia Y Robertson, R. (1985) The evolution of learning mechanisms. *Amer. Psych. Assoc.: Master Lecture Series*, 4: 191–243.

Gardner, S.M. and Lenteren, J.C. van (1986) Characterisation of the arrestment responses of *Trichogramma evanescens*. *Oecologia* 68: 265–270.

Giessen van, W.A., Lewis, J., Vet, L.E.M., and Wäckers, F.L. (1993) The influence of host site experience on subsequent flight behaviour in *Microplitis croceipes* (Cresson) (Hymenoptera: Braconidae). *Biol. Control* 3: 75–79.

Godfray, H.C.J. (1987) The evolution of clutch size in invertebrates, in *Oxford Surveys in Evolutionary Biology*, 4, (eds. P. Harvey & L. Partridge), Oxford University Press, Oxford. pp.117–154.

Gould, J.L. (1993) Ethological and comparative perspectives on honey bee learning, in *Insect Learning. Ecological and Evolutionary Aspects*, (eds D.R. Papaj and A.C. Lewis), Chapman & Hall, New York, pp.18–50.

Gould J.L. and Marler, P. (1984) Ethology and the natural history of learning, in *The Biology of Learning*, (eds P.Marler and H.S. Terrace), Springer-Verlag, Berlin, pp. 47–74.

Henneman, M.L., Papaj, D.R., Figueredo, A.J., Vet, L.E.M. (1995) Egg-laying experience and acceptance of parasitized hosts by the parasitoid *Leptopilina heterotoma*. *J. Insect Behav.* in press.

Jaenike, J. and Papaj, D.R. (1992). Behavioral plasticity and patterns of host use by insects, in *Insect Chemical Ecology, An Evolutionary Approach,* (eds. B.D. Roitberg and M.B. Isman), Chapman & Hall, New York, pp. 245–264.

Janssen, A. (1989) Optimal host selection by *Drosophila* parasitoids in the field. *Func. Ecol.* 3: 469–479.

Jervis, M.A., Kidd, N.A.C. (1986) Host-feeding strategies in Hymenopteran parasitoids. *Biol. Rev. Camb. Phil. Soc.* 61: 395–434.

Jones, R.L., Lewis, W.J., Beroza, M., Bierl, B.A., Sparks, A.N. (1973) Host seeking stimulants (kairomones) for the egg-parasite *Trichogramma evanescens*. *Environ. Entomol.* 2: 593–596.

Jones, R.L., Lewis, W.J., Bowman, M.C., Beroza, M., Bierl, B.A. (1971) Host seeking stimulants for parasite of corn earworm: isolation, identification and synthesis. *Science* 173: 842–843.

Kaiser, L. and Cardé, R.T. (1992). In-flight orientation to volatiles from the plant host complex in *Cotesia rubecula* (Hym.: Braconidae): increased sensitivity through olfactory conditioning. *Physiol. Entomol.* 17: 62–67.

Kaiser, L. Pham-Delegue, M.H., Bakchine, E., Masson, C. (1989) Olfactory responses of *Trichogramma maidis* Pint. et Voeg.: effects of chemical cues and behavioral plasticity. *J. Insect Behav.* 2: 701–712.

Kaiser, L., Willis, M.A. and Cardé, R.T. (1994) Flight manoeuvers used by a parasitic wasp to locate host-infested plant. *Entomol. exp. appl.* 10: 285–294.

Kerguelen, V. and Cardé, R.T. (1995) Increased host acceptance in experienced females of the parasitoid *Brachymeria intermedia:* which oviposition behaviours contribute to experience. (unpublished study).

Kester, K.M. and Barbosa, P. (1991) Behavioral and ecological constraints imposed by plants on insect parasitoids: implications for biological control. *Biol. Control* 1: 94–106.

Krebs, J.R. and Kacelnic, A. (1991) Decision-making, in *Behavioural Ecology. An Evolutionary Approach,* (eds J.R. Krebs and N.B. Davies), Blackwell Scientific Publications, Oxford, pp. 105–137.

Lewis, W.J., Jones, R.L., Gross, H.R., Nordlund, D.A. (1976) The role of kairomones and other behavioral chemicals in host finding by parasitic insects. *Behav. Biol.* 16: 267–289.

Lewis, W.J., Jones, R.L., Sparks, A.N. (1972) A host seeking stimulant for the egg-parasite *Trichogramma evanescens:* its source and a demonstration of its laboratory and field activity. *Ann. Entomol. Soc. Am.* 65: 1087–1089.

Lewis, W.J. and Martin, W.R.jr. (1990) Semiochemicals for use with parasitoids: status and future. *J. Chem. Ecol.* 16: 3067–3089.

Lewis, W.J., Nordlund, D.A., Gueldner, R.C., Teal, P.E.A., Tumlinson, J.H. (1982) Kairomones and their use for management of entomophagous insects. XIII. Kairomonal activity for *Trichogramma* spp. of abdominal tips, excretion, and a synthetic sex pheromone blend of *Heliothis zea* (Boddie) moths. *J. Chem. Ecol.* 8: 1323–1331.

Lewis, W.J. and Takasu, K. (1990) Use of learned odours by a parasitic wasp in accordance with host and food needs. *Nature* 348: 635–636.

Lewis, W.J. and Tumlinson, J.H. (1988) Host detection by chemically mediated associative learning in a parasitic wasp. *Nature* 331: 257–259.

Lewis, W.J., Tumlinson, J.H., Krasnoff, S. (1991) Chemically mediated associative learning: an important function in the foraging behavior of *Microplitis croceipes* (Cresson). *J. Chem. Ecol.* 17: 1309–1325.

Lewis, W.J., Vet, L.E.M., Tumlinson, J.H., Lenteren, J.C. van, Papaj, D.R. (1990) Variations in parasitoid foraging behavior: essential element of a sound biological control theory. *Environ. Entomol.* 19: 1183–1193.

MacArthur, R.H. and Pianka, E.R. (1966) On optimal use of a patchy environment. *Am. Nat.* 916: 603–609.

Mandeville, J.D. and Mullens, B.A. (1990) Host species and size as factors in parasitism by *Muscidifurax* spp and *Spalangia* spp. (Hymenoptera: Pteromalidae) in the field. *Ann. Entomol. Soc. Am.* 83: 1074–1083.

Mangel, M. (1992) Descriptions of superparasitism by optimal foraging theory, evolutionarily stable strategies and quantitative genetics. *Evol. Ecol.* 6: 152–169.

Mangel, M. (1993) Motivation, learning, and motivated learning, in *Insect Learning. Ecological and Evolutionary Aspects*, (eds D.R. Papaj and A.C. Lewis), Chapman & Hall, New York, pp. 158–173.

Martin, W.R. and Lewis, W.J. (1995) Timing of host and host feces encounters: influence on learning and memory in the parasitoid *Microplitis croceipes*. (unpublished study).

McAuslane, H.J., Vinson, S.B., Williams, H.J. (1991a). Influence of adult experience on host microhabitat location by the generalist parasitoid, *Campoletis sonorensis* (Hymenoptera: Ichneumonidae). *J. Insect Behav.*, 4: 101–113.

McAuslane, H.J., Vinson, S.B., Williams, H.J. (1991b) Stimuli influencing host microhabitat location in the parasitoid *Campoletis sonorensis*. *Entomol. exp. appl.* 58: 267–277.

Menzel, R., Greggers, U., Hammer, M. (1993) Functional organization of appetitive learning and memory in a generalist pollinator, the honey bee, in *Insect Learning. Ecological and Evolutionary Aspects*, (eds D.R. Papaj and A.C. Lewis), Chapman & Hall, New York, pp.79–125.

Minkenberg, O., Tatar, M., Rosenheim, J.A. (1992) Egg load as a major source of variability in insect foraging and oviposition behavior. *Oikos* 65: 134–142.

Noldus, L.P.J.J. (1989) Semiochemicals, foraging behaviour and quality of entomophagous insects for biological control. *J. Appl. Entomol.* 108: 425–451.

Noldus, L.P.J.J. and Lenteren, J.C. van (1985) Kairomones for the egg parasite *Trichogramma evanescens* Westwood. II. Effect of contactchemicals produced by two of its hosts, *Pieris brassicae* L. and *Pieris rapae* L.. *J. Chem. Ecol.* 11: 793–800.

Noldus, L.P.J.J., Lenteren, J.C. van, Lewis, W.J. (1991) How *Trichogramma* parasitoids use moth sex pheromone as kairomone: orientation behaviour in a wind tunnel. *Physiol. Entomol.* 16: 313–327.

Nordlund, D.A., Lewis, W.J., Gueldner, R.C. (1983) Kairomones and their use for management of entomophagous insects. XIV. Response of *Telenomus remus* to abdominal tips of *Spodoptera frugiperda,* (Z)-9-tetradecene-1-ol acetate and (Z)-9-dodecene-1-ol acetate. *J. Chem. Ecol.* 9: 695–701.

Nordlund, D.A. and Sauls, C.E. (1981) Kairomones and their use for management of entomophagous insects. XI. Effect of host plants on kairomonal activity of frass from *Heliothis zea* larvae for the parasitoid *Microplitis croceipes. J. Chem. Ecol.* 7: 1057–1061.

Pak, G.A. and Jong, E.J. de (1987) Behavioural variations among strains of *Trichogramma* spp.: host recognition. *Neth. J. Zool.* 37: 137–166.

Papaj, D.R.(1993a) Automatic behavior and the evolution of instinct: lessons from learning in parasitoids, in *Insect Learning. Ecological and Evolutionary Aspects*, (eds D.R. Papaj and A.C. Lewis), Chapman & Hall, New York, pp.243–272.

Papaj, D.R. (1993b) Afterword: learning, adaptation, and the lessons of O, in *Insect Learning. Ecological and Evolutionary Aspects*, (eds D.R. Papaj and A.C. Lewis), Chapman & Hall, New York, pp.374–386.

Papaj, D.R. and Lewis, A.C. (eds) (1993) *Insect Learning. Ecological and Evolutionary Perspectives,* Chapman & Hall, New York.

Papaj, D.R. and Prokopy, R.J. (1989) Ecological and evolutionary aspects of learning in phytophagous insects. *Annu. Rev. Entomol.* 34: 315–350.

Papaj, D.R., Snellen, H. Swaans, K. and Vet, L.E.M. (1994) Unrewarding experiences and their effect on foraging in parasitic wasps. *J. Insect Behav.* 7: 465–481.

Papaj, R.D. and Vet, L.E.M. (1990) Odor learning and foraging success in the parasitoid, *Leptopilina heterotoma. J. Chem. Ecol.* 16: 3137–3150.

Pavlov, I.P. (1941) *Lectures on Conditioned Reflexes,* 2 vols. International Publishers, New York.

Poolman Simons, M.T.T., Suverkropp, B.P., Vet, L.E.M., Moed, G. de (1992) Comparison of learning in related generalist and specialist eucoilid parasitoids. *Entomol. exp. appl.* 64: 117–124.

Powell, W. and Wright, A.F. (1992) The influence of host food plants on host recognition by four aphidiine parasitoids (Hymenoptera: Braconidae). *Bull. Entomol. Res.* 81: 449–453.

Price, P.W., Bouton, C.E., Gross, P., McPheron, B.A., Thompson, J.N., Weis, A.E. (1980) Interactions among three trophic levels: influence of plant on interactions between insect herbivores and natural enemies. *Annu. Rev. Ecol. Syst.* 11: 41–65.

Rescorla, R.A. (1988) Behavioral studies of Pavlovian conditioning. *Annu. Rev. Neurosci.* 11: 329–352.

Roitberg B.D., Reid, M.L., Li, C. (1993) Choosing hosts and mates: the value of learning, in *Insect Learning. Ecological and Evolutionary Aspects,* (eds D.R. Papaj and A.C. Lewis), Chapman & Hall, New York, pp.174–194.

Roper T.J. (1983) Learning as a biological phenomenon, in *Genes, Development and Learning* (eds T.R. Halliday and P.J.B. Slater), W.H. Freeman, New York, pp. 178–212.

Smith, B.H. (1993) Merging mechanism and adaptation: an ethological approach to learning and generalization, in *Insect Learning. Ecological and Evolutionary Aspects*, (eds D.R. Papaj and A.C. Lewis), Chapman & Hall, New York, pp.126–157.

Schmidt, J.M. and Smith, J.J.B. (1987) Short interval time measurement by a parasitoid wasp. *Science* 237: 903–905.

Schmidt, J., Cardé, R.T., Vet, L.E.M. (1993) Host recognition by *Pimpla instigator* F. (Hymenoptera: Ichneumonidae): preferences and learned responses. *J. Insect Behav.* 6: 1–11.

Sheehan, W. and Shelton, A.M. (1989) The role of experience in plant foraging by the aphid parasitoid *Diaeretiella rapae* (Hymenoptera:Aphidiidae). *J. Insect Behav.* 2: 743–759.

Sheehan, W., Wäckers, F.L., Lewis W.J. (1993) Discrimination of previously searched, host free sites by *Microplitis croceipes* (Hymenoptera: Braconidae). *J. Insect Behav.* 6: 323–331.

Shettleworth, S.J. (1984) Learning and behavioural ecology, in *Behavioural Ecology. An evolutionary approach*, (eds J.R. Krebs and N.B. Davies), Blackwell Scientific Publications, Oxford. pp. 170–194.

Stephens, D.W. (1993) Learning and behavioral ecology: incomplete information and environmental predictability, in *Insect Learning. Ecological and Evolutionary Aspects*, (eds D.R. Papaj and A.C. Lewis), Chapman & Hall, New York, pp.195–218.

Stephens, D.W. and Krebs, J.R. (1986) *Foraging Theory*, Princeton University Press, Princeton.

Stamp, N.E. (1982) Behavioral interactions of parasitoids and Baltimore checkerspot caterpillars (*Euphydrias phaeton*). *Environ. Entomol.* 11: 100–104.

Strand, M.R. and Vinson, S.B. (1982) Behavioral response of the parasitoid *Cardiochiles nigriceps* to a kairomone. *Entomol. exp. appl.* 31: 308–315.

Strand, M.R. and Vinson, S.B. (1983) Factors affecting host recognition and acceptance in the egg parasitoid *Telenomus heliothidis* (Hymenoptera: Scelionidae). *Environ. Entomol.* 12: 1114–1119.

Takasu, K. and Lewis, W.J. (1993) Host- and food-foraging of the parasitoid *Microplitis croceipes*: learning and physiological state effects. *Biol. Control* 3: 70–74.

Takasu, K. and Lewis, W.J. (1995) Importance of adult food sources to host searching of the larval parasitoid *Microplitis croceipes*. *Biol. Control* 5: 25–30.

Thorpe, W.H. and Jones, F.G.W. (1937) Olfactory conditioning in a parasitic insect and its relation to the problem of host selection. *Proc. R. Soc. Lond. B* 124: 56–81.

Tierney, A.J. (1986) The evolution of learned and innate behavior: contributions from genetics and neurobiology to a theory of behavioral evolution. *Anim. Learning Behav.* 14: 339–348.

Tumlinson, J.H., Lewis W.J., Vet, L.E.M. (1993) Parasitic wasps, chemically guided intelligent foragers. *Sci. Amer.* 268: 100–106.

Turlings, T.C.J., Scheepmaker, J.W.A., Vet, L.E.M., Tumlinson, J.H., Lewis, W.J. (1990) How contact foraging experiences affect preferences for host-related odors in the larval parasitoid *Cotesia marginiventris* (Cresson) (Hymenoptera:Braconidae). *J. Chem. Ecol.* 16: 1577–1589.

Turlings, T.C.J. and Tumlinson, J.H., (1992) Systemic release of chemical signals by herbivore-injured corn. *Proc. Natl. Acad. Sci.* USA89: 8399–8402.

Turlings, T.C.J., Tumlinson, J.H., Lewis, W.J. (1990) Exploitation of herbivore-induced plant odors by host-seeking parasitic wasps. *Science* 250: 1251–1253.

Turlings, T.C.J., Tumlinson, J.H., Lewis, W.J., Vet, L.E.M. (1989) Beneficial arthropod behavior mediated by airborne semiochemicals. VIII. Learning of host-related odors induced by a brief contact experience with host by-products in *Cotesia marginiventris* (Cresson), a generalist larval parasitoid. *J. Insect Behav.* 2: 217–225.

Turlings, T.C.J., Wäckers, F.L., Vet, L.E.M., Lewis, W.J., Tumlinson, J.H. (1992) Learning of host-finding cues by hymenopterous parasitoids, in *Insect Learning. Ecological and Evolutionary Aspects*, (eds D.R. Papaj and A.C. Lewis), Chapman & Hall, New York, pp.51–78.

Vet, L.E.M. (1988) The influence of learning on habitat location and acceptance by parasitoids. *Les Colloq del'INRA* 48: 29–34.

Vet, L.E.M. and Dicke, M. (1992) Ecology of infochemical use by natural enemies in a tritrophic context. *Annu. Rev. Entomol.* 37: 141–172.

Vet, L.E.M. and Groenewold, A.W. (1990) Semiochemicals and learning in parasitoids. *J. Chem. Ecol.* 16: 3119–3135.

Vet, L.E.M. and Papaj, D.R. (1992) Effects of experience on parasitoid movement in odour plumes. *Physiol. Entomol.* 17: 90–96.

Vet, L.E.M., Lewis, W.J., Papaj, D.R., Lenteren, J.C. van (1990) A variable-response model for parasitoid foraging behavior. *J. Insect Behav.* 3: 471–490.

Vet, L.E.M. and Schoonman, G. (1988) The influence of previous foraging experience on microhabitat acceptance in *Leptopilina heterotoma*. *J. Insect Behav.* 1: 387–392.

Vet, L.E.M., Wäckers, F.L., Dicke, M. (1991) How to hunt for hiding hosts: the reliability-detectability problem in foraging parasitoids. *Neth. J. Zool.* 41: 202–213.

Vinson, S.B. (1984) Parasitoid-host relationships, in *Chemical Ecology of Insects,* (eds. W.J. Bell and R.T. Cardé), Chapman and Hall, London, pp. 205–233.

Vinson, S.B., Barfield, C.S., Henson, R.D. (1977) Oviposition behaviour of *Bracon mellitor*, a parasitoid of the boll weevil (*Anthonomus grandis)*. II. Associative learning. *Physiol. Entomol.* 2: 157–164.

Vinson, S.B. and Williams, H.J. (1991) Host selection behavior of *Campoletis sonorensis:* a model system. *Biol. Control* 1: 107–117.

Visser, J.H. (1986) Host odor perception by phytophagous insects. *Annu. Rev. Entomol.* 31: 121–144.

Visser, M.E., Alphen, J.J.M. van, Nell, H.W. (1992) Adaptive superparasitism and patch time allocation in solitary parasitoids: the influence of pre-patch experience. *Behav. Ecol. Sociobiol.* 31: 163–171.

Waage, J.K. (1978) Arrestment responses of the parasitoid, *Nemeritis canescens,* to a contact chemical produced by its host, *Plodia interpunctella*. *Physiol. Entomol.* 3: 135–146.

Waage, J.K. and Ng, S.M. (1984) The reproductive strategy of a parasitic wasp. I. Optimal progeny and sex allocation in *Trichogramma evanescens*. *J. Anim. Ecol.* 53: 401–415.

Wäckers, F.L. and Lewis, W.J. (1993) Olfactory and visual learning and their combined influence on host site location by *Microplitis croceipes*. *Biol. Control,* 4: 105–112.

Wardle, A.R. (1990) Learning of host microhabitat colour by *Exeristes roborator* (F.) (Hymenoptera: Ichneumonidae). *Anim. Behav.* 39: 914–923.

Wardle, A.R. and Borden, J.H. (1989) Learning of an olfactory stimulus associated with a host microhabitat by *Exeristes roborator*. *Entomol. exp. appl.* 52: 271–279.

Wardle, A.R. and Borden, J.H. (1990) Learning of host microhabitat form by *Exeristes roborator* (F.) (Hymenoptera: Ichneumonidae). *J. Insect Behav.* 3: 251–263.

Weseloh, R.M. (1981) Host location by parasitoids, in: *Semiochemicals: their Role in Pest Control,* (eds D.A. Nordlund, R.L. Jones and W.J. Lewis), Wiley, New York, pp. 79–95.

Wiskerke, J.S.C., Dicke, M., Vet, L.E.M. (1993) Larval parasitoid uses aggregation pheromone of adult hosts in foraging behaviour: a solution to the reliability-detectability problem. *Oecologia* 93: 145–148.

Zaborski, E., Teal, P.E.A., Laing, J.E. (1987) Kairomone-mediated host finding by the spruce budworm egg parasite, *Trichogramma minutum*. *J. Chem. Ecol.* 13: 113–122.

Zanen, P.O. and Cardé, R.T. (1991) Learning and the role of host-specific volatiles during in-flight host-finding in the specialist parasitoid *Microplitis croceipes*. *Physiol. Entomol.* 16: 381–389.

Orientation Mechanisms

4

The Role of Chemo-orientation in Search Behavior

William J. Bell, Larry R. Kipp and Robert D. Collins
Department of Entomology, University of Kansas

1. Introduction

Chemo-orientation is a diverse topic, intersecting the disciplines of fluid dynamics, chemistry, physiology, cell biology, genetics, behavior, ecology, and evolution. Chemo-orientation is important for such diverse life processes as sperm orientation in fertilization, white blood cell orientation preceding phagocytosis in the autoimmune response, and locating resources such as food, mates, and oviposition sites. In this chapter, with a few exceptions, we limit the discussion to chemo-orientation processes involved in mate finding by insects. Not all insect mate-finding strategies require chemo-orientation processes, of course, and even in those instances where chemo-orientation is involved, insects often employ other sensory modalities (sequentially or contemporaneously) when locating mates.

Chemo-orientation involves the use of an animal's sense of smell (olfaction) or taste (contact chemoreception) to locate a resource or to avoid a stress source. The process functions as a one-way communication system between a signal sender and a receiver. In order for the receiver to orient successfully toward the sender, the receiver must execute "a series of responses based on a combination of internal and external factors" (Hansell, 1985). Some behaviors, such as local search patterns, are guided entirely by internal controls; other behaviors, such as chemotaxis, are guided by external chemical gradients. Chemo-orientation requires the integration of external and internal factors to generate one of several possible searching strategies. A search strategy comprises a series of tactics, including (1) initiation, during which an insect waits for or searches for an active chemical space; (2) orientation, during which the animal moves within the active space and toward the source of the chemical; (3) arrestment, including stopping or landing; and (4) closure, a tactic for achieving the final approach to the resource after stopping or landing.

2. Locating a Pheromone Signal

Prior to mating, an insect must be motivated to search for, locate, and court a prospective mate. Insects have evolved numerous ways to "switch on" mate-finding behavior and to switch off competing drives such as finding food. Specific mechanisms vary among species. For example, short-lived adult insects that do not feed, such as the Hessian fly, *Mayetiola destructor*, or cecropia moth, may initiate mate search soon after adult eclosion, whereas long-lived adult insects first deal with other priorities, such as protein gathering, or nest building.

Ultimately, a mate-seeking male insect must encounter an active chemical space. This space can be defined as the volume within which is the threshold of a chemical (or blend of chemicals) that can initiate chemo-orientation in a particular species. The active space is influenced by many factors, including the rate of emission of the chemical signal and the perceptual abilities of the receiving organism, as well as abiotic factors such as ambient temperature, and biotic factors such as period of exposure to the odor (habituation may occur).

Two extreme tactics, with many intermediates, can be discerned with regard to how an insect may increase its chances of detecting an active space. The first is *perching*, in which an insect moves to a position in space that is optimal in height or microhabitat features and then waits. The second is *ranging* in which an insect actively searches for a stimulus (Jander, 1975). Ranging may be guided by internal information, visual, wind, or gravitational cues, or some combination of these. Perching is energy-conserving and risk-averse, especially if the animal is cryptic. Ranging can expose the searcher to greater predation risk and may require greater expenditure of energy.

Successful searchers depend on their ability to choose an optimal perch site or ranging locale, and on the reliability and predictability of the signal emitter. Such abilities include ranging or perching during the season and time of day or night when the signal emitter is active. Ultimately, selection will favor those searchers who are most efficient in locating mates.

The distribution and density of the emitting animals (usually females) could influence the tactic employed by searchers. Most likely, the perching strategy would be most efficient when the emitters are both numerous and widely distributed; when the emitters are aggregated during some phase of their lives, such as emergence sites, feeding sites, and oviposition sites, perchers would have to move from site to site in order to perceive pheromone. A major problem for both perching and ranging is unproductive sites or paths. As noted below, both perchers and rangers have evolved ways to avoid unproductive zones.

2.1. Perching

To optimize their chances of receiving olfactory signals, perchers would be expected to station themselves at the right time and at an appropriate place. An insect can generally remain nearly motionless and often hidden while scanning with chemosensory receptors, being inconspicuous to predators during this initial phase of information gathering. For example, male cockroaches perch at night on leaves where they can detect vibrations of approaching predators and at the same time extend their antennae to pick up olfactory signals (Schal et al., 1983; Seelinger, 1984). If sufficient chemosensory information is received, a perching insect may then expose itself to potential predators and initiate local search or orientation relative to wind, or gravitational or visual cues (Bell, 1990).

Silverman and Bell (1979) observed a population of *Periplaneta americana* in a cubical 1.73 m^3 chamber with a horizontal divider shelf. Cockroaches could position themselves below the shelf or above the shelf, and they could rest horizontally or vertically on walls or floors or ceilings. The results revealed a vertical stratification, with males positioned above females. Later, Schal (1982) and Schal and Bell (1986) found that males of several tropical cockroach species also occur higher than females in rain-forest vegetation. This intersexual variation in perch heights (described in Section 3.3) allows gravitational orientation in which males run downward to locate pheromone-secreting females.

Presumably, insects that perch have ways to prevent the possibility of continuously resting at a nonproductive site. One mechanism is for perch duration to be time-dependent. For example, male cockroaches perching on vegetation in the rain forest changed sites several times during a 4-hr period (Bell, unpublished).

2.2. *Ranging*

Theoretically, the ranging movements of a walking insect could be generated entirely by internal information. For example, one way to localize a resource is to move in a spiral in which the distance between paths is no greater than the perceptual abilities of the insect. This tactic only works, however, if the resource is quite close and does not move; if the resource is far away, the searcher may expire before reaching it. As far as we know, the only example of fairly regular spirals is in termites that lose an odor trail (Jander, unpublished). More often, tracks of this kind tend to be a series of loops, and spiralling is exhibited only periodically. Another possible tactic is to walk straight, perhaps using some external cue, and thereby avoid revisits to areas previously searched (Jander, 1975). In straight-line ranging there must be a mechanism to prevent an insect from becoming locked into an unproductive pattern of movement. For example, blowflies, *Phormia regina*, walk straight for some distance, and then turn and walk in another direction (Dethier, 1976). Straight search may either be continuous or discontinuous (Dusenbery, 1992). In discontinuous search, also referred to as saltatory search (O'Brien et al., 1989), an insect may move some distance and stop to search for the resource or a cue emanating from the resource (Dusenbery, 1992).

Dusenbery (1992) suggests that a terrestrial odor trail, such as that produced by an ant, could be located by moving in a fairly straight course, in which the path is made up of nonoverlapping segments. Some sort of zigzag might fall into this category. How do insects move in a relatively straight line with periodical changes in direction? Pline and Dusenbery (1987) define a *collimating stimulus* as a stimulus (such as a wind current or a visual cue) that may be followed even if the stimulus is not directly connected to the target. For insects, these cues may be visual or gravitational stimuli or wind currents, as in the examples discussed below.

2.2.1. Wind Cues

It has been suggested that insects should search for an odor plume by walking or flying across the wind if the plume length exceeds its width, and upwind or downwind if the reverse (Linsenmair, 1969; Cardé, 1981b). Generally a plume presents a larger target area when approached from crosswind than when approached parallel to the wind.

Crosswind ranging is less than optimal when the wind direction fluctuates significantly. If the range of current directions exceeds ±30° from the mean current direction, the plume will be wider than it is long, and the greater cross section across the current will make a search along the mean current direction more efficient than one across it (Sabelis and Schippers, 1984). These maneuvers would be the favored tactics for flying insects because we know that they can orient upwind in a pheromone plume, and that they can slip downwind, whereas it would be more difficult for them to orient nearly exactly crosswind (i.e., 90° to the wind) (Murlis et al., 1992). Even if they could orient crosswind, such orientation would be energetically expensive owing to numerous fluctuations in wind direction (Elkinton et al., 1987). The prediction, therefore, would be for flight somewhat between upwind and crosswind, or between downwind and crosswind.

Zanen et al. (1994) showed that *Drosophila* spp. in a wind tunnel tended to fly 90° to the wind when the wind had a steady direction, but when the wind shifted more than 60° the flies flew parallel to the wind direction. These observations suggest that *Drosophila* could shuttle between either an upwind or crosswind strategy, depending on whether the wind direction was variable or stable. Thus, *Drosophila* conforms to the models of Sabelis and Schippers (1984) on optimal search for odor plumes.

The most complete information about ranging with wind orientation derives from studies of walking arthropods. They use wind currents to orient when other cues are not available, and they tend to orient at angles between crosswind and upwind. For example, dung beetles, *Geotrupes* spp., and scorpions prefer angles between 27° and 32° relative to exact upwind (Linsenmair, 1968, 1969), and the median anemotactic course direction of the cockroach, *Blaberus craniifer* is 30° (Bell and Kramer, 1979).

2.2.2. Visual Cues

Some arthropods use topographical features in their ranging. For example, the predatory mite, *Amblyseius potentillae*, searching for prey on a rose leaf follows the leaf edges (Sabelis and Dicke, 1985), and butterflies, *Pieris rapae*, orient to "leading lines" such as telephone lines, hedgerows, and roadways (Fig. 4.1) (Baker, 1978). In the evident absence of sex pheromone, ranging male gypsy moths spend more time in tree-oriented vertical flight than in any other behavior; this result is consistent with the significantly higher trap catch at trees than away from trees at corresponding heights (Elkinton and Cardé, 1983). Tree-oriented vertical flight most commonly commences at the base of the tree. The moths

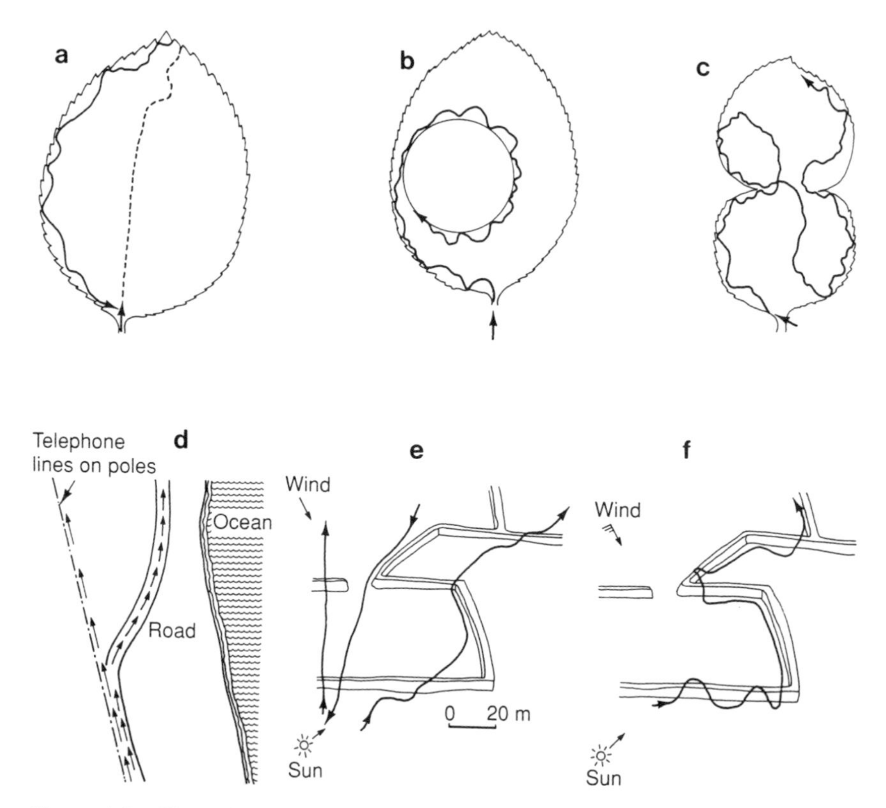

Figure 4.1. Examples of ranging. (a, b, and c) Edge-oriented walking pattern of a predatory mite on a rose leaf. Solid lines are paths on top of the leaf; dotted lines are paths under the leaf. By cutting the leaf in (b) and (c), it was shown that the cue is any edge, not just the outer edge [after Sabelis and Dicke (1985)]. (d, e, and f) Ranging of butterflies, *Pieris rapae*. (d) Orientation to leading lines such as roads, telephone poles or shoreline. Flight relative to wind (e) and leading lines (f) [after Baker (1978)].

then hover near the tree and then fly at ground level out of the observation zone or move up the tree and fly into the canopy. Many males move up and down a tree repeatedly before leaving. Similarly, the western budworm, *Choristoneura occidentalis*, and spruce budworm, *C. fumiferana*, "buzz" branches visually during the day (Greenbank, 1973; Kipp et al., 1994), presumably searching for mates.

3. Response to Pheromone Signals

Once an active chemical space has been located through ranging or while perching, the chemo-orientation tactic is activated. These tactics may initially involve local search, presumably because in some species the pheromone emitter may be close by. If local search fails to allow location of a resource, it may be followed by orientation with respect to a wind current or gravitational or visual cue. Those species that do not aggregate are more likely to immediately walk or fly when a sex pheromone is detected, as in most species of moths.

3.1. Activation

Insects often exhibit a stereotypical behavioral response when they first detect sex pheromone. For example, male green June beetles, *Cotinus nitida*, and granary beetles, *Trogoderma variabile*, exhibit the following behaviors when they detect sex pheromone: turning to face upwind, extending and waving the antennae with the lamellae spread apart, raising the body up by lifting the forelegs, and finally upwind running or flight (Domek et al., 1990; Tobin and Bell, 1986). Presumably these behaviors are designed to gather as much sensory information as possible before engaging in walking search or flight. Male gypsy moths, *Lymantria dispar*, and many other moths respond to sex pheromone by wing fanning for several minutes prior to flight (Cardé and Hagaman, 1979). These behaviors are likely related to the warming up of flight muscles prior to taking off.

3.2. Local Search

The function of local search is to find a resource that is nearby. An example of purely endogenous local search orientation is the behavior of courting male *Blattella germanica* in darkness. When the cockroach perceives the nonvolatile female sex pheromone through contact with a female, it raises its wings and turns 180°. If the female does not mount, or perhaps walks away, the male engages in local search behavior in the area where wing raising occurred. Schal et al. (1983) allowed a male to run on a locomotion compensator for a few minutes and when it stopped an experimenter "fenced" with the male using a sex pheromone impregnated antenna. Wing raising and turning was elicited, and after a few seconds the male began to search. Whereas movement prior to stimulation is quite straight, local search is characterized by fast running in overlapping loops. Return to straight walking occurs after several minutes.

In species that aggregate, such as the granary beetle, *Trogoderma variabile*, males may engage in local search, comprising looping orientation near where the odor was detected, enhancing its chances of locating a calling female (Tobin and Bell, 1986). Similarly, dermestid beetles, *Dermestes maculatus*, stop, and then turn in short loops when exposed to a puff of aggregation pheromone (Rakowski et al., 1989). Local search can occur in flying insects as well. Douwes (1968) plotted the flight paths of the geometrid moth, *Cidaria albulata*, as they approached plants. When approaching host plants they increased their turning rate, tending to remain in the area, followed by landing. This effect was not noticed when moths approached nonhost plants, suggesting that plant odor stimulated the local search. The local search tactic is adaptive in that the insect tends to remain near where the resource odor was detected, enhancing the probability of locating the resource or of detecting visual cues. The gradual manner in which looping changes to relatively straight movements, eventually taking the insect away from unproductive local search, has been found in all cases of local search (Bell, 1991).

3.3. Gravity Orientation

Orientation to gravity may explain the behavior of male cockroaches in the rain forest, where males perch higher than females (see Section 2.1) (Fig. 4.2a). This height stratification of cockroaches in the tropical rain forest may be a mechanism that enhances the efficiency of the mate-finding process. Schal (1982) found that the nighttime temperature near the ground remains higher than at 1–2 meters (Fig. 4.2b). The higher temperature near the ground causes convection currents to move upward at night, as demonstrated by titanium tetrachloride smoke plumes (Figs. 4.2d and 4.2e). The resulting wind speeds (Fig. 4.2c) range from 7 cm sec^{-1} near the ground to 16 cm sec^{-1} at 2 m. This model was tested in the laboratory in a vertical temperature gradient, and bioassays revealed that male *P. americana* stationed above a sex pheromone source responded with courtship behavior. Thus, the upward-rising convection currents at night in the rain forest are sufficient to carry pheromones upward in the foliage.

To test whether male cockroaches perching higher than females tend to run downward when they detect sex pheromone, male *P. americana* were tethered by thin wires attached to their pronota, and a syringe was used to blow a puff of air onto the back of the cockroach, perpendicular to a vertical arena. When clean air was directed at cockroaches, both males and females ran upward in the vertical plane, but when the air contained female sex pheromone, females ran upward, but males ran downward (Silverman and Bell, 1979). The upward running responses of both males and females may represent escape from preda-

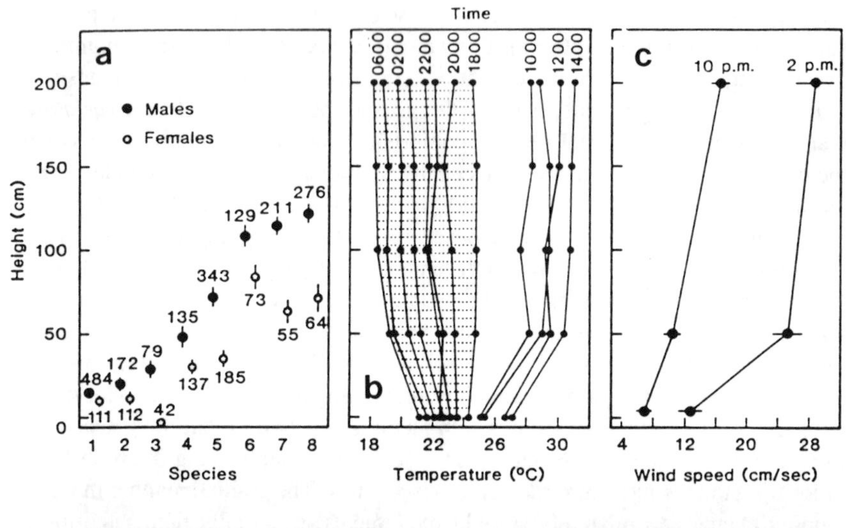

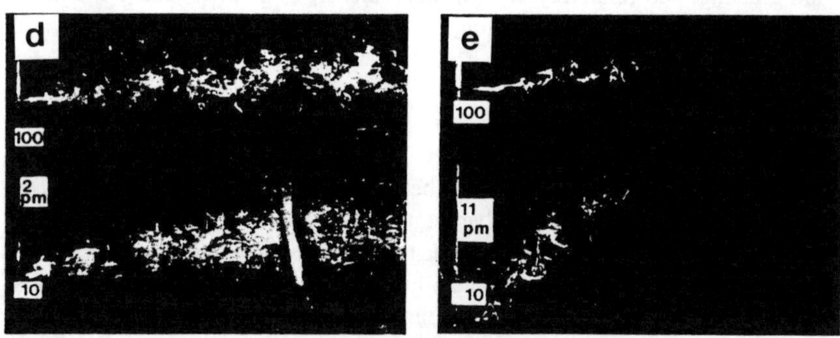

Figure 4.2. (a) Height stratification of eight species of cockroaches in the rain forest. (b) Temperature profiles for a 24- hr period in the dry season. Stippled area is nighttime. (c) Wind profile for day (1400 hr) and night (2200 hr) conditions. (d and e) Titanium tetrachloride used to illustrate wind patterns. (d) Afternoon, (e) nighttime. (Schal, 1982).

tors, such as mice, rats, and centipedes, which are normally found on the floor. For males perceiving sex pheromone, the downward running response is appropriate if females occur lower on a vertical surface than males.

As discussed in Section 5, responses to gravity are also involved in search patterns that occur after male moths land near target females.

3.4. Anemotaxis

Discussions of the mechanisms by which insects orient upwind in a pheromone plume have often been controversial [e.g., Tobin and Bell (1981), Baker (1985), Cardé and Charlton (1984)]. This is largely because flying insects seem to use somewhat different mechanisms than walking insects. The following discussion is not a complete review of the literature, but merely an attempt to compare some of the more important points distinguishing flying and walking in a pheromone plume.

The ability of a walking insect to maintain a relatively straight course in a wind current is easily achieved because walking insects, such as immature locust hoppers and cockroaches, detect wind speed and direction by using mechanoreceptors such as the Johnston's organ at the base of the antenna or fields of mechanoreceptive hairs on the head (Gewecke, 1974; Gewecke and Philippen, 1978; Bell and Kramer, 1979). In addition, walking insects can stop and move their antennae or their bodies to monitor the wind direction to guide the next series of movements. For example, as shown in Fig. 3, the larder beetles, *Dermestes ater*, slow down when they perceive a directional change in a wind

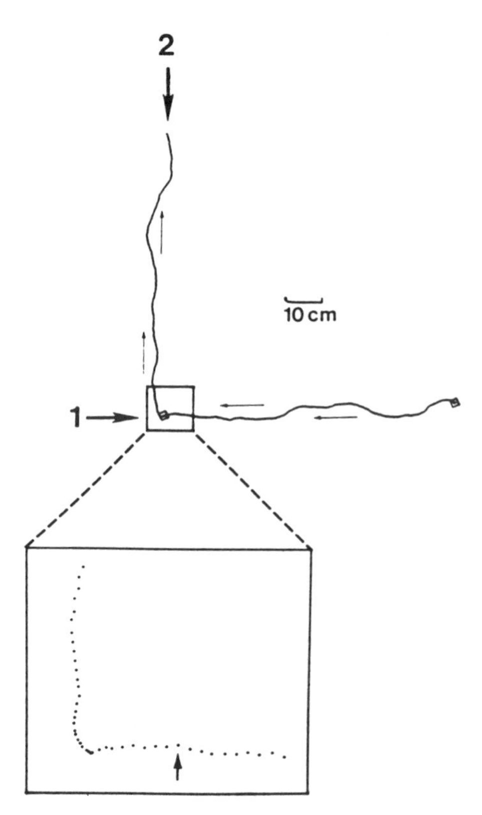

Figure 4.3. Pathway of a larder beetle, *Dermestes ater*, orienting to a wind current moving from left to right (1) and then after the current is switched 90° (2). Inset shows enlargement of track representing points observed every 0.1 s. Arrow shows point at which wind shifts. (Bell et al., 1989).

stream; they stop, turn to the new direction and proceed (Bell et al., 1989). In contrast, flying insects are not in contact with the ground, and must use the optomotor reaction, comparing the difference between the heading and the visually detected drift or side slip in order to orient upwind.

3.4.1. Flying Insects

With some exceptions, sex pheromone releases two programs in flying insects: optomotor anemotaxis (upwind orientation) and a program of internally controlled zigzag counterturns up a pheromone plume (Kennedy, 1986) (Fig. 4.4a). Whereas an odor plume was once viewed as a more or less uniform cloud of pheromone-laden air, at least on a time-averaged basis, it has become evident that pheromone plumes comprise many tortuous, discontinuous filaments (Murlis and Jones, 1981; Baker et al., 1985). This finding has greatly contributed to understanding behavioral chemo-orientation mechanisms of flying insects.

The filamentous nature of pheromone plumes appears to be critical for successful chemo-orientation by flying insects (Baker, 1989). The flying insect briefly surges upwind when a filament of pheromone is encountered and then returns to endogenously programmed zigzag turns. The rate at which pheromone filaments with detectable concentrations are encountered by male moths flying upwind may be relatively low (ca. once per second) (Baker and Haynes, 1989). Because small discontinuous filaments of pheromone are less likely to become mixed with other compounds, fine-scale structure of plumes may help to keep the target signal separated and discernible from background chemical noise, such as pheromone emitted by heterospecifics (Liu and Haynes, 1992).

Why do moths zigzag? Baker (1985) contends that the zigzag movements back and forth across the plume allow the insect more opportunities to scan repeatedly the changes in odor concentration, thereby allowing it to remain in the plume more efficiently than if it flew straight. Two major factors influence this behavior: (1) Changes in pheromone concentration, reflecting in part the distance of the moth from the point source; and (2) loss of pheromone when the moth strays out of the plume. With respect to changes in pheromone concentration, wide zigzags when the pheromone concentration is low and narrow zigzags when the pheromone concentration is high add to the efficiency of locking onto the plume. Thus, as a male moves closer to the pheromone source, the pheromone concentration increases, the plume becomes narrower, and the behavior of the male changes accordingly. For example, in *L. dispar* as pheromone concentration is increased in a wind tunnel, the width of zigzags decreases, and more moths sustain flight (Cardé and Hagaman, 1979). The response to straying out of the plume and into clean air is to decrease the frequency of zigzagging. The resulting lateral excursions lengthen until one of the forays reaches the plume and the frequency of zigzagging increases once again. For example, when the pheromone source is removed from a wind tunnel, male *Plodia interpunctella* and *Grapholita molesta*

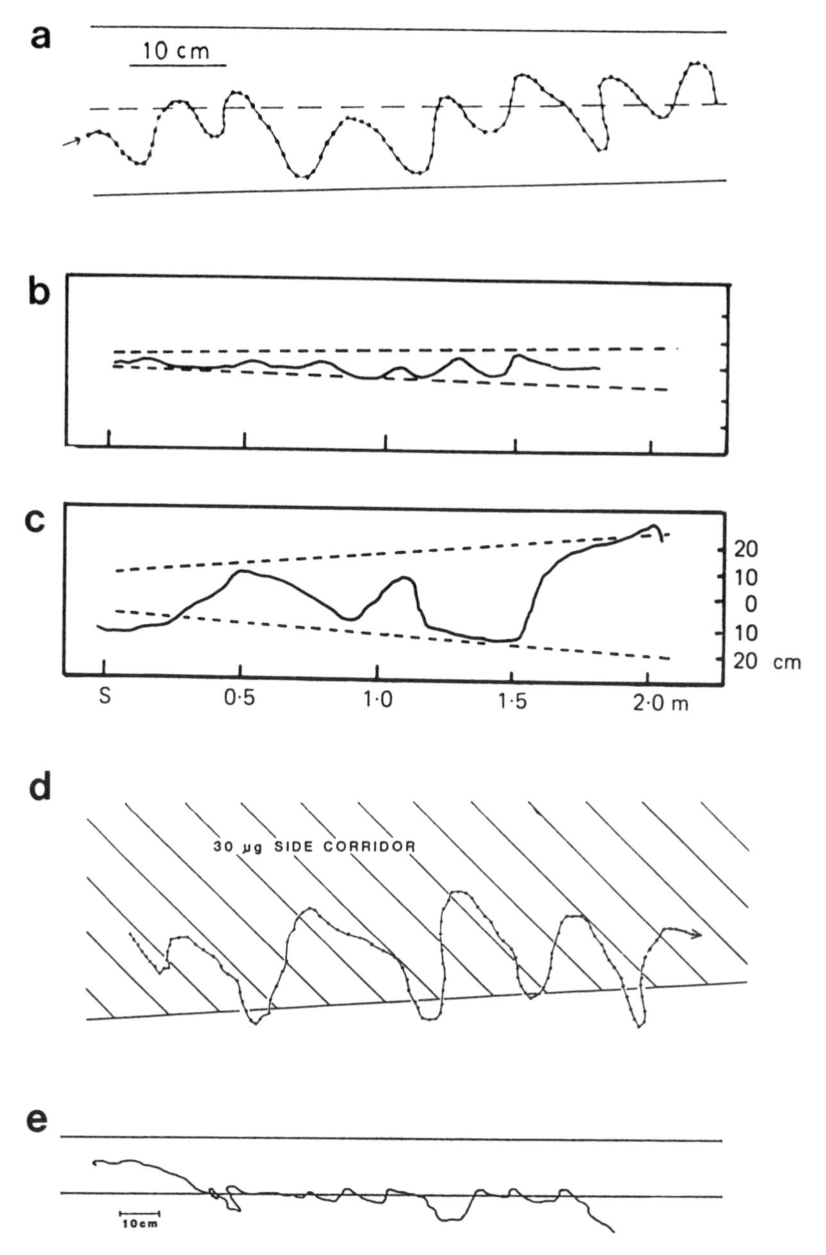

Figure 4.4. (a) Flight path of male *Grapholita molesta* in a plume of female sex pheromone; wind from right (Kuenen and Baker, 1982). (b, c) Walking path of a male cockroach, *Periplaneta americana* in a narrow plume (b) and wide plume (c) of female sex pheromone; wind from left (Tobin, 1981). (d) Flight of male *G. molesta* in a side corridor of female sex pheromone; wind from right (Willis and Baker, 1984). (e) Walking paths of a male beetle, *Trogoderma variabile*, in a "temporal" corridor of female sex pheromone; wind from right (Tobin and Bell, 1986). All tracks videotaped from above.

fly into a clean airstream. They immediately switch from upwind zigzags to casting nearly 90° across the wind (Kennedy and Marsh, 1974; Kuenen and Baker, 1983).

In 1982 David et al. described a feature of plumes in wind streams that changed the view of how moths follow a pheromone plume. They showed that in the field a plume remains fairly straight for some distance, but that wind shifts cause the plume to snake to and fro (Fig. 4.5a). Thus, flying persistently upwind would actually lead the moth out of the plume. This idea was substantiated by experiments with gypsy moths flying in a plume that could be visualized (David et al., 1983). In David's experiments, the movement of the plume in a field was indicated by soap bubbles. Individual bubbles moved away from the pheromone source in relatively straight lines, but then were displaced laterally by shifts in the wind direction. However, the relatively straight lines, though displaced, remained intact (David et al., 1982). A male *L. dispar* followed the plume by flying in shallow zigzags upwind; when contact with the plume was lost, it flew back and forth across the wind until it located the plume again (Figs. 4.5b and 4.5c). Each parcel of air carrying pheromone was nearly always moving directly away from the source. When the moth lost the plume because a wind shift displaced the plume laterally, it increased its velocity and began to cast. The casting allowed the insect not only to relocate the plume, but to advance upwind toward the source during the process. This experiment not only confirmed flight tunnel data, but it showed that moths often cast in wide lateral sweeps to locate the displaced pheromone plume. The wide sweeps could not be observed within the confines of a wind tunnel.

Elkinton et al. (1987) carried out experiments similar to those of David et al. (1983), but in the forest, which is the normal habitat of *L. dispar*. They showed that when the long axis of the plume and the wind direction are parallel the moth flies quickly and directly upwind; only during these brief periods does the male gypsy moth make substantial progress toward the pheromone source.

Can moths detect the edge of a plume? Several experiments have shown that various species of moths turn back into a corridor of sex pheromone in a wind tunnel [e.g., Willis and Baker (1984)] (Fig. 4.4d), which could be interpreted as changing tack when the pheromone concentration drops significantly. Baker (1985), however, concludes from various other studies [e.g., Kuenen and Baker (1982)] that moths do not steer according to loss of pheromone. Their evidence is that the width of the zigzags in a plume from a high-concentration pheromone source (larger active space) is actually narrower than in a plume from a lower-concentration source (smaller active space). They conclude that "This is the converse of what would be expected if each turn were steered by an excursion outside of the active space." However, if the moth in the wide plume was actually zigzagging along the plume edge, this would be consistent with changing tack according to changes in pheromone concentration.

An explanation for edge-tracking along a pheromone corridor has been offered

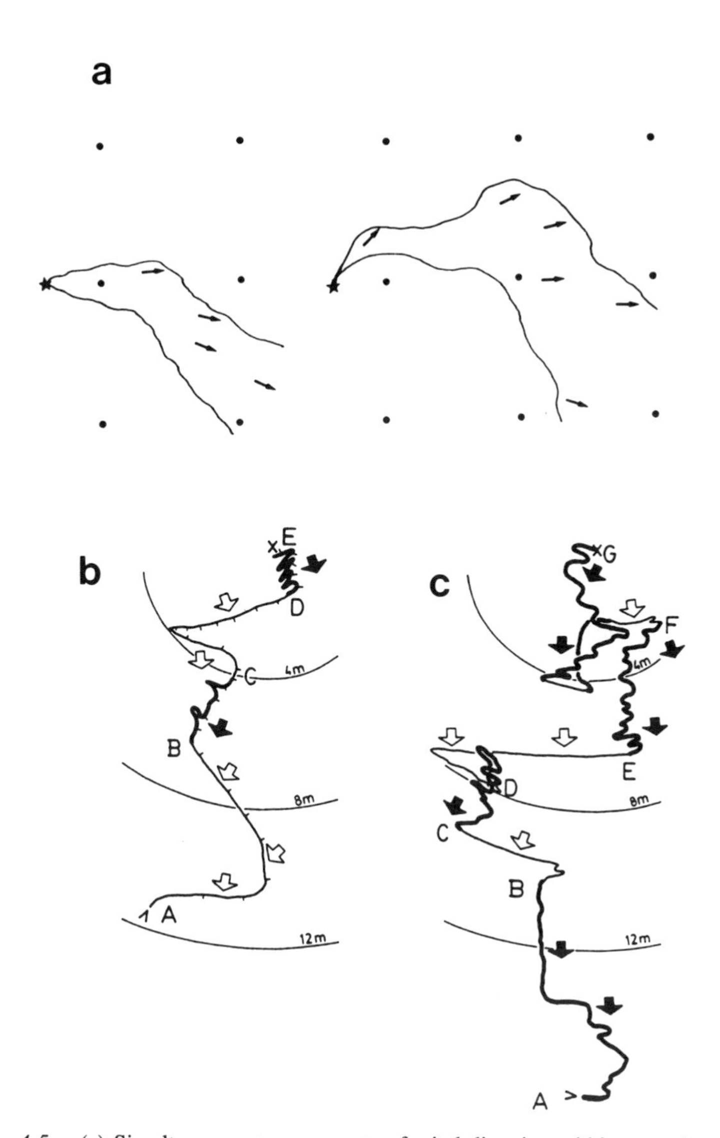

Figure 4.5. (a) Simultaneous measurements of wind direction within a smoke plume. Smoke was released at 0.5 m above ground level over a flat grassy area. Mean wind speed was 1.6+/− m sec^{-1}; distance between markers equals 5 M. (David et al., 1982). (b and c) Tracks, videotaped from above, of male gypsy moths, *Lymantria dispar*, flying toward a pheromone source (X) in the field. Thick black line is the male's track when in contact with the pheromone plume, and each solid black arrow is the wind's direction during each period of pheromone contact. Thin black line is the track when the male lost contact with the pheromone plume, and each hollow arrow denotes the wind direction during these periods of flight in clean air. Tick marks denote 1-s intervals. When the plume is lost often the wind has swung sharply (as in a-b and c-d in the left figure, and b-c and d-e in the right figure) the male's 90° crosswind flight often brings it back into contact with the plume at a point closer to the source than where the pheromone was lost. (David et al., 1983).

by Willis and Baker (1984), based on the idea that the "edge" of a pheromone cloud or corridor is probably ragged rather than sharply defined. Male moths lock onto this ragged edge because it provides a phasic stimulation, similar to that provided by a pheromone plume, that is needed to drive the internal program of upwind zigzagging. In essence a moth may lock onto and follow the edge, but not because they turn each time the pheromone concentration drops.

Cardé and Charlton (1984) pursued the idea that flying gypsy moths straying out of a plume are brought back to it by a counterturn provoked by cessation of the pheromone stimulus. Confirming Kennedy et al. (1981, their Fig. 4.9), Cardé and Charlton (1984) found that the zigzags of moths are narrower along a discrete plume than in a uniform cloud of pheromone, indicating that the moths turn back when they detect the plume border. An alternative explanation is that the moths habituate to the pheromone while in the corridor (Kennedy et al., 1981; Willis and Baker, 1984). As explained by Kennedy (1986): "For the moth, habituation of the initial narrowing-down response to entering pheromone would be like passing into clean air again, with the well-known result that the zigzags widen." Diagrams of moth flight, however, show clearly that in many cases the moths orient along the edge of the pheromone corridor, zigzagging in and out of the corridor (Fig. 4.4d); habituation would seem unlikely. An alternative explanation is that, rather than requiring chemotaxis, the turn of the moth back into the corridor is simply a truncated version of the casting normally exhibited when a moth loses the plume (Baker and Haynes, 1987). Is it not also possible that the moth's first reaction is to turn in the correct direction back toward the plume, which it seems to do, and then to cast if it cannot quickly find the plume? Is it also possible that moths utilize several different mechanisms, depending on conditions and available information, but the existence of such redundant systems remains undocumented?

Some species of flying insects exhibit alternative anemotactic behaviors. *Drosophila* fly nearly straight upwind in banana odor (David, 1986). Interestingly, *Drosophila* also walk straight upwind in a plume of banana odor (Flügge, 1934). The mosquito, *Aedes aegypti*, flying in a wind tunnel turns back on leaving a favorable airstream (high CO_2 and humidity), but does not turn back upon entering one (Daykin et al., 1965). This behavior keeps the mosquito within the favorable airstream. Thus, it is not necessary to zigzag in order for a flying insect to locate an odor source upwind. It is possible, however, that a plume containing CO_2 and humidity would be more homogeneous than pheromone in a plume, accounting for the straight flight. Some insects orient upwind by flying only in short hops. For example, the onion fly *Hylemya antiqua* and cabbage root fly *Delia brassicae* approach a source of host-plant odor in a series of short, low flights ("hops"), without the zigzagging that is typical of the sustained flight of male moths nearing a source of sex pheromone. When the fly perceives the host plant odor it turns and walks or flies in short hops upwind toward the odor source; wind direction is determined while the fly rests (Hawkes and Coaker, 1979; Dindonis and Miller,

1980). Based on studies of Brady (1970) and Bursell and Taylor (1980), which showed that tsetse flies, *Glossina* spp., alternate short flights with longer periods of rest, Bursell (1984) speculates that the flies assess wind direction when resting, and would therefore be able to fly in the upwind direction on the next flight. They may even use local visual cues during their short flight rather than an optomotor mechanism. It should be noted, however, that this mechanism would not be productive under conditions of shifting wind directions.

Finally, moths that usually zigzag upwind in a pheromone plume occasionally have been observed to fly straight [e.g., Mafra-Neto and Cardé (1994)]. To explore this behavior further, Mafra-Neto and Cardé (1994) manipulated the filamentous nature of plumes using a pheromone pulse generator in a wind tunnel. Male *Cadra cautella*, which usually zigzag slowly up a pheromone plume, also zigzagged in a continuous plume produced by the pulser, but flew straight upwind when the plume comprised discrete pulses. When low-pulse frequency plumes were used, the moth zigzagged when in clean air, and surged upwind when contacting a pheromone pulse. Males thus change their flight behavior in response to changes in the plume's overall shape, and also to its internal structure.

3.4.2. Walking Insects

When Tobin (1981) filmed male *P. americana* running in a sex pheromone plume in a wind tunnel, it was not surprising, given the data on flying moths, that cockroaches also ran upwind in a zigzag fashion. However, the zigzag path was related to the edge of the plume (dimensions visualized with titanium oxide) (Fig. 4.4c). To determine if male cockroaches direct their zigzag orientation pattern by detecting the edge of a plume, Tobin used either a narrow or a wide pheromone plume. Whereas males remained tightly within the narrow plume, turning precisely at the plume edges, in the wide plume they sometimes turned within the plume and other times ran all the way to the edge of the plume before turning (Fig. 4.4c). These observations indicate that the cockroach (1) runs upwind at some angle, and then changes tack when it leaves the pheromone plume (i.e., detects a decrease in pheromone concentration); and (2) turns stochastically within the plume even when there is no change in pheromone concentration. Thus, cockroaches combine chemotaxis (orientation driven by a change in odor concentration) and spontaneously generated turns, with anemotaxis. Turning at the edge of the plume allows the cockroach to remain within the plume, whereas the spontaneous turns ensure efficient upwind orientation when pheromone secretion from many females creates a wide swath of pheromone in an air stream. Turning frequency appears to be stochastic in the cockroach, and also in walking *Bombyx mori* moths (Kramer, 1975).

In the walking cockroach, loss of pheromone in a wind current leads to crosswind turns that are similar to casting in moths. It would appear, however, in contrast to flying moths, that the walking insect "knows" in what direction to turn to find the plume. This is an important point. The following experiments with *T. variabile* test the idea of edge following (Tobin and Bell, 1986). The experimental setup avoided the problem of a ragged corridor edge because a corridor of female sex pheromone was produced by turning the odor on or off in a wind-tunnel with the beetle on a locomotion compensator. They simulated what the beetle would perceive if it walked across a sharp boundary between clean air and air containing pheromone. The results showed that beetles walk upwind when they perceive sex pheromone, and reached the boundary as a result of net sideways displacement across the wind. When a beetle crossed the boundary, the pheromone was eliminated from the wind stream, and the beetle responded by slowing down and turning back toward the corridor. When it reentered the corridor, pheromone was again provided, and the beetle again walked upwind. The chemotactic turning response, resulting from an abrupt decrease in pheromone concentration, indicates that walking insects are capable of detecting temporal changes in odor stimuli in an air current and turning in the correct direction that takes them back toward the plume. If the insect makes an error, and turns in the wrong direction, it circles or casts, until it relocates the odor.

As with flying insects, changes in pheromone concentration also affect zigzag movements of walking insects. For example, increasing female sex pheromone concentration alters pathways of *P. americana* running on a locomotion compensator (Bell and Kramer, 1980). As the concentration increases, the male moves straighter upwind. Clean air causes escape responses, which consist of rapid downwind running. With no air current the cockroaches tend to circle about. Walking *Ips paraconfusus* orients in a more upwind direction as the concentration of pheromone is increased (Akers and Wood, 1989). The tracks of walking beetles in a wind stream containing pheromone consist of relatively straight or slightly curving sections interspersed with a few relatively abrupt course adjustments.

3.5. Chemotaxis

Insects can orient without wind or visual cues by sampling information spatially or temporally. In spatial sampling, the inputs to the two antennae are compared, and the insect moves in the direction of the antenna receiving the higher input (given that the stimulus is a favorable one). In temporal sampling the insect samples over time and determines if it is headed toward the source by comparing odor concentration at different points in time.

Male *P. americana* easily locate a source of female sex pheromone in an arena in the absence of directional wind currents (Bell and Tobin, 1981). Movements of individual males were filmed from above a 2.4-m diameter arena in which a male was introduced at the edge, and a sex pheromone source was located in the center. A male tended to remain along the edge until it detected pheromone.

After a brief stop, during which the antennae were waved about, it ran very quickly in forays toward the center of the arena, stopping at the pheromone source.

The males were manipulated to discriminate between spatial and temporal mechanisms in still air: (1) Males were unilaterally antennectomized (1-ANT) to determine if it is necessary to compare inputs from the two antennae in order to orient toward the pheromone source; and (2) the antennae were crossed, left to right and vice versa (X-ANT), and the bases were sealed with wax. In the latter case the brain should receive information reversing the apparent concentration difference between right and left antennal receptors. If a cockroach can still orient properly, it would suggest that information is not compared between antennae. The results showed that 1-ANT males located the pheromone source as frequently as control (2-ANT) males, but that X-ANT males failed to locate the source. Bell and Tobin (1981) hypothesize that male *P. americana* probably use spatial comparisons (between the two antennae) if the gradient is sufficiently steep to perceive a difference in pheromone concentration. They are also apparently capable of making temporal comparisons (1-ANT), in which pheromone concentration is tested over time; otherwise they would not be able to orient with one antenna.

It is not an easy matter to ascertain which chemo-orientation mechanism (spatial or temporal) an animal is actually using at any particular moment. Experiments with the clearest outcomes are those with honeybees (Martin, 1965) and snails (Chase and Croll, 1981) in which tubes were placed over antennae or tentacles, and different odor concentrations entered the left and right tubes. In both cases the animal moved toward the direction of the olfactory organ receiving the higher odor concentration. Martin (1965) then fixed the antennae of honeybees close together so that they were unable to sample a concentration gradient using spatial information; the bees then tended to move their heads from side to side, as though to read the gradient using temporal information. Thus, it is probable that insects have the ability to switch between temporal and spatial mechanisms, depending on the information available.

Studies have shown that in the absence of a wind current some flying insects can continue to fly in a static plume over a relatively short distance (a few meters). They apparently sample pheromone concentration from place to place (Farkas and Shorey, 1972; Kuenen and Baker, 1983; Willis and Cardé, 1990). This tactic could be useful when the wind speed is too slow for the optomotor reaction to work. Oriental fruit moths, *G. molesta*, exhibit zigzag flight, even in the absence of wind, although they do not zigzag when they are not stimulated by pheromone (Baker et al., 1984), suggesting that pheromone turns are based on an internal program of zigzag flight. Without wind, however, the zigzags have no consistent direction, since the wind polarizes the zigzag movements. Interestingly, the moths without wind were able to locate the pheromone source in some (21%) tests.

4. Landing

The strategy of a searching male would be expected to contain an "end game," such that the male would curtail locomotion, either landing or stopping, at a position in space that ensures location of the female. The problem for the male is to land at the right time and at the right place.

One tactic for a flying insect is to land whenever the individual odor pulses (filaments encountered) in an odor plume can no longer be resolved (Baker et al., 1985; Baker and Haynes, 1989). Such a condition would likely occur only when an insect is downwind of and very near to an odor source. However, given that there is variability among females in the amount of pheromone emitted [see for example, Collins et al. (1990)], there could be errors in the distance from the female at which a male alights.

Two possible visual tactics depend upon the ability of a male to visually resolve the target female. If it can, the male may land when the female comes into view, or it may land on some nearby object and engage in walking search. Obviously this task is simpler for diurnal species, but even nocturnal species can resolve silhouettes of trees and other contrasting objects.

The larch casebearer moth, *Coleophora laricella*, may orient to nonfemale objects, preferentially landing on twigs containing pheromone rather than flat surfaces containing pheromone (Witzgall and Priesner, 1984). Their tendency to orient to tree silhouettes was confirmed by placing pheromone baits in nonhost trees. Male *L. dispar* in the forest can locate an isolated sex pheromone source as readily as sources supplemented with female visual cues (Charlton and Cardé, 1990). In flight tunnel choice experiments using cylinders as surrogate trees, and with pheromone in different spatial configurations, visual attributes did not influence either a male's choice of landing site or the efficiency with which it located a female. Pheromone on the cylinder was required to elicit search orientation as well as landing.

Castrovillo and Cardé (1980) examined male codling moths, *Laspeyresia pomonella*, approaching, landing, and walking on a platform containing a sex pheromone source. The mean total orientation time and mean time in contact with the platform was unaffected by the presence or absence of a dead female. When a dead female was present, however, males spent more time near the female than in other quadrants of the platform. During these periods copulation attempts were made. Similarly, Shorey and Gaston (1970) found that when dead *Trichoplusia ni* females or black paper silhouette models were positioned 2 cm from a pheromone source, the presence of such visual cues did not influence the persistence of orientation, but it did influence both the frequency and direction of copulatory attempts.

The results of experiments with dead females or surrogates are not always easy to interpret, however. In the lightbrown apple moth, *Epiphyas postvittana*, a high proportion of males land just downwind of a pheromone source, even when visual objects such as dead moths or a rubber septa are present (Foster and Harris, 1992). But the males tend to land very near live, moving male moths (Foster et al., 1991). Thus, the behavior of the target individual may be more important than the visual pattern of a dead individual or surrogate that cannot move.

Long-range attraction in the strictly diurnal six-spotted burnet moth, *Zygaena filipendulae*, is mediated by sex pheromone, but once males are within 50 cm they switch to casting flight and utilize visual cues to locate the female (Zaggatti and Renou, 1984). In the laboratory the final approach was filmed with a video camera. At about 50 cm from the calling female the male describes a loop in the air, gently brushing the female, then climbing and descending. This loop brings him back to the female slightly underneath her. Then he either makes another loop or gets closer to the female to attempt to contact her with his antennae. Thus, the latter approach phase is visually mediated.

5. Searching after Landing or Stopping

After landing near or at the site of the pheromone emitter, the searching male may employ an internally generated search pattern and may use gravity, wind, or visual cues to localize the emitter. Charlton and Cardé (1990) found that pheromone on surrogate tree cylinders was required to elicit search orientation as well as landing in gypsy moths. If pheromone stimulation is interrupted during search, males do not proceed directly to a female that is only a few centimeters away. Rather, males initially walk downward, and then exhibit convoluted, mainly vertically directed movements (Fig. 4.6). In contrast, males receiving continual pheromone stimulation walk fairly directly toward a visually apparent female. The mechanism appears to be a combination of local search and upwind orientation. These behaviors may enhance relocating a pheromone plume and/or increase the chances of contacting the female.

6. Factors Affecting Search Tactics

Both external and internal factors may influence the search tactics used by an insect. External factors can potentially influence all members of the population, whereas internal factors influence an individual's behavior but not directly the behavior of others. Internal factors may help explain much of the observed variation of a population's daily activities.

6.1. Internal Factors

Internal factors are all of those influences and constraints originating from within the organism, such as characteristics provided by an organism's genotype, experi-

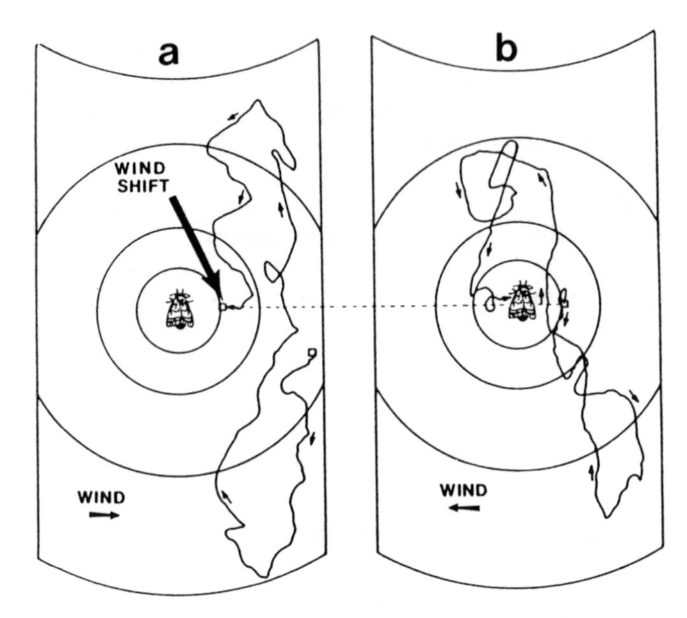

Figure 4.6. Walking path of a male gypsy moth, *Lymantria dispar*, responding to combinations of females and sex pheromone. Reactions (a) prior to and (b) following wind and pheromone plume shift induced by rotation of the cylinder. Before wind shift, male was 5 cm directly downwind of the female and in the pheromone plume. (Charlton and Cardé, 1990).

ential effects, and acquired characteristics. Internal factors include available energy stores, motivational state, age, previous experience, parasite load, size, and learning capability. At any given time, researchers will observe behavioral differences within a population that could be attributable to differences in age.

6.1.1. *Biological Rhythms*

Biological rhythms are responsible for synchronizing female calling and male responses to pheromones. For example, more male German cockroaches, *B. germanica*, respond to female sex pheromone in the scotophase than in the photophase (Bell et. al., 1978). In *G. molesta*, periodicities of both calling female and male response to sex pheromone by wing fanning are controlled in part by a circadian rhythm (Baker and Cardé, 1979). The response of *P. americana* is affected by the length of time since the male was previously exposed to female sex pheromone, the length of that exposure, as well as the time in the photocycle (Hawkins and Rust, 1977). Liang and Schal (1990) found that in the brown-banded cockroach, *Supella longipalpa*, the sex pheromone concentration and the time of testing relative to the entrainment conditions affect the number of insects responding. High pheromone concentration can alter the periodicity of the response of male *S. longipalpa*: males may respond to sex pheromone at any time if the concentration is sufficiently high.

6.1.2. Body Size

Male gypsy moths differ in pupal length from 14 to 26 mm. Kuenen and Cardé (1993) examined the effects of male size on flight parameters in a wind tunnel. Mean net ground speed along the wind line and airspeed was faster among large males than in medium and small males. It is, therefore, likely that larger males would be able to reach calling females before smaller males. However, Kuenen and Cardé point out that fast flight could reduce orientation precision by forcing them out of the plume if crosswinds cause the plume to snake (see David et al., 1983). Once the male finds a female there is no advantage to being large, since females mate with the first male that arrives (Cardé and Hagaman, 1984).

6.1.3. Parasitism

Parasites can have significant effects on an animal's behavior. Hurd and Parry (1991) observed reduced responses in male *Tenebrio molitor* beetles infected with the metacestode *Hymenolepis diminuta* to glass rods coated with the female sex pheromone. Likewise, infection of the cockroach, *P. americana*, with the acanthocephalan *Moniliformis moniliformis*, affects sexual behavior. Response to female sex pheromone, periplanone B, was significantly reduced in infected individuals. If an infected male has a decreased behavioral response to sex pheromone, it may also have decreased success in locating a mate, and thus decreased fitness (Carmichael et al., 1993).

6.1.4. Age

As insects age the tendency to maintain flight periodicity may decline. For example, young *Anax imperator* males mainly fly in the hours just after dawn; older males usually fly later, near midday (Corbet, 1957, 1960). In biting Diptera, periodicity changes in oviposition, feeding, and flight have also been observed (Nielson and Nielson, 1962). Recent evidence suggests that a change in the periodicity of flight also occurs with male spruce budworm moths, *C. fumiferana*, as they age. Young males (< 24 hr) primarily take flight in the evening when females are calling. Older males (>120 hr) primarily search during the day, especially during peak female emergence, but may search again at night (Fig. 7) (Kipp et al., 1994). The behavior of older males could significantly reduce the number of virgin females available to younger males that waited until evening to search.

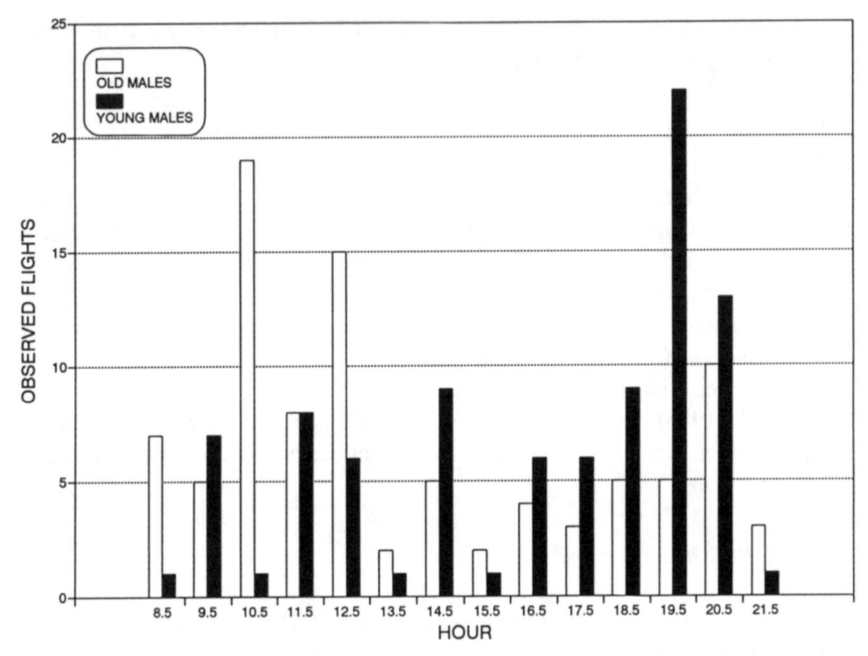

Figure 4.7. Frequency of flight activity of male *Choristoneura fumiferana* moths in a fir grove in the absence of females. Over five days in June 1989, one month prior to the natural budworm flight season, 403 young (<12 hr old) and 367 older (>120 hr old) laboratory-reared males were released and observed every other 15-min period from sunrise to sunset. A total of 88 flights by young males (21.8%) and 94 flights by older males (25.6%) were observed (Kipp et al., 1994).

6.2. *External Factors*

External factors such as day length or temperature can modulate search tactics to fit current conditions. The presence and behaviors of other individuals can also be influential factors. For example, the time for optimal search may vary with population density, which will affect the encounter rate, relative risk of predation, or the intensity of male competition.

6.2.1. *Density*

Some species of insects incur vast changes in density from year to year. Populations of the spruce budworm, *C. fumiferana*, may vary from a few to 2.6 million pupae per hectare (Seabrook et al., 1989). Under low-density conditions, males apparently locate females by upwind orientation, following a pheromone plume. Under high-density conditions, several alternative tactics may come into play. For example, a small percentage of *L. dispar* males locate females simply by walking about on the tree (Cardé and Hagaman, 1984). Thus, some females are mated before beginning pheromone emission.

Under high-density situations, older spruce budworm males may search for and mate females during the day (Kipp et al., 1990; Kipp et al., 1994). Since females usually do not call until after 1700 hr, daytime flying males should not be able to locate females by pheromone communication.

6.2.2. Ambient Temperature

Two reproductive behaviors highly influenced by shifts in ambient temperature are female calling and male responsiveness. Laboratory experiments have demonstrated that calling in *G. molesta* females is temperature dependent; the onset of calling can be advanced by as much as 4 hr when the temperature is reduced from 25° to 20° C at various times during the photophase (Baker and Cardé, 1979). In another series of laboratory experiments Linn et al. (1988) showed that the response specificity of both *Pectinophora gossypiella* and *G. molesta* male moths to sex pheromone specificity was temperature-dependent. At 20° C a narrow pheromone blend-dose specificity was observed, while at 26° C specificity decreased. Other temperature influences on female behavior are reviewed by McNeil (1991).

The timing of male moth capture at attractant- or pheromone-baited traps appears to be temperature-related. Laboratory studies of this phenomenon present the most compelling results (McNeil, 1991), whereas the interpretation of results from field studies are more problematic. Turgeon et al. (1983) found that peak trapping of armyworm, *Pseudaletia unipuncta*, males on warm nights in the field ranged from 2400 hr to 0400 hr, whereas on cool nights peak trapping occurred near 2300 hr. Comeau et al. (1976) showed that the mean hour of attraction of redbanded leafroller, *Argyrotaenia velutinana*, males to pheromone traps was significantly correlated with the evening average temperature (1800–2300 hr) ($r^2 = 0.91$). Significant correlations were observed for both the spring and summer populations. Similar correlations were observed for five other lepidopteran species [see also Cardé et al. (1975)]. These field studies are consistent with the hypothesis that male behavior is affected by temperature (McNeil, 1991).

In addition to temperature, certain biotic factors may also advance or delay the mean hour of attraction to pheromone traps (e.g., temporally nonrandom predation or migration, and population density). First, Kipp et al. (1994) observed that male moths attempting morning flight in low-density areas were attacked by birds. Removal of these males by birds in the morning would lead to a low trap catch in the morning, causing a delay in the mean time of trap catch. Such a delay would be independent of temperature. Second, daily migration of moths may affect trap catches. Kipp and Lonergan (1990), using a mark-release-recapture technique (Kipp and Lonergan, 1992) showed that the ratio of marked to unmarked males captured over the day was nonstationary: more marked males were captured earlier and more unmarked males were captured later. This resulted in significant differences in the hour of peak capture for marked and unmarked males. The increased catch of unmarked males later in the day may have resulted from immigration, which is known to occur at that time (Greenbank et al., 1980) with moths dispersing up to 600 km (Dobesberger et al., 1983). Such increased trap catch late in the evening may have had little to do with ambient temperature. Third, earlier field studies relating moth flight with ambient temperature may have also been confounded by changes in moth density. Kipp et al. (1994) have shown that with spruce budworm a single evening peak is observed when populations are relatively sparse; whereas when populations are relatively dense an earlier, morning peak also occurs. These peaks are independent of daily temperature fluctuations.

We conclude that it is difficult to infer causes for shifts in the flight behavior of moths from field-collected trap capture data. Trapping periodicity data represent the integration of many factors, possibly including age-dependent shifts on daily flight time, temporally nonrandom predation, moth densities relative to predator densities, and daily emigration and immigration rates, as well as abiotic factors such as ambient temperature.

7. Genetic Components of Variance

As information about pheromone-mediated communication systems accrued, and as techniques for identifying and synthesizing pheromone compounds became more accessible, it soon became clear that pheromone systems are quite often based on complex, multicomponent signals, and that response behaviors elicited by these signals are equally complex. It also became evident that pheromone systems are much more variable than formerly believed, and that significant variation often exists at all levels, from within individuals to among populations. Knowledge about this variation is critical to understanding pheromone-mediated search behavior.

Evolutionary responses to selective forces are only possible if there is variance in the behavioral phenotype, the phenotype is correlated with fitness, and there is a genetic component of variance (Lewontin, 1970). The primary functions of pheromone-based long-distance sexual communication systems are mate location and reproductive isolation (Cardé and Baker, 1984). Reproductive isolation may be maintained by the identity and blend of pheromone components (Roelofs and Brown, 1982), differences in concentration (Kaae et al., 1973), or both.

Because many closely related species achieve reproductive isolation behaviorally, understanding the genetic bases of pheromone-mediated communication may contribute to an understanding of speciation and the evolutionary causes of diversity in pheromone systems (Löfstedt, 1993). Information about variation in pheromone communication systems also has great practical significance for pest management tactics that utilize pheromone trapping to monitor or control pest populations, and for programs that employ synthetic pheromone to disrupt mating. Until recently, however, information on the extent of phenotypic variance in aspects of pheromone signal and response behaviors was sparse, and studies of the genetic bases of pheromone communication systems were even more uncommon.

In the preceding sections of this chapter, external and internal factors have been examined that contribute to variation in chemo-orientation behaviors. An alternative way to look at variation in these behaviors is to partition observed phenotypic variance into components that can be attributed to genetic and environmental sources. Within the past decade, information about the extent and sources of variance in these behaviors has been accumulating at an increasing rate. In this section we report briefly on some of this information. Because of the considerable research emphasis that has been placed on pheromone communication in Lepidoptera, the discussion will focus on this group. No attempt is made to provide an exhaustive review of this rapidly expanding area of research; rather, we focus on several examples of better-understood systems.

7.1. Within-Individual Variance

Pheromone systems are characterized by the emission of a narrow-variance signal. In moths it is usually females that produce long-distance pheromone signals to which males respond. Males generally exhibit relatively broad response spectra, but peak response is often elicited by the particular pheromone blend emitted by conspecific females, especially at low concentrations (Linn and Roelofs, 1985).

Total phenotypic variance observed within a population can be partitioned into within-individual variance and among-individual variance. Within-individual variance occurs when the behavior of an individual changes over time as a function of age, reproductive status, deprivation, and external environmental conditions such as climate or photoperiod (Caro and Bateson, 1986). Variability in an individual's behavior may also represent an important adaptive strategy. To directly measure within-individual variation, the behavior of individuals must be measured more than once. However, repeated measures are seldom reported for behaviors associated with chemo-orientation. Du et al. (1987) measured repeatabilities for the relative emission rates of pheromone components emitted by individual female ermine moths, *Yponomeuta padellus*. Repeatability measures among-individual variance as a fraction of total phenotypic variance, providing an estimate of the extent to which repeated measures for an individual are consistent over time (Boake 1989), and setting an upper limit to the character's heritability (Falconer, 1981). Emission rates were measured for individual females at six and eight days after adult emergence using an airborne collection technique (Baker et al., 1981). The emission rates of five of the seven known pheromone components were measured relative to the emission rate for the component with the highest emission rate. Estimated repeatabilities for these relative emission rates were high, ranging from 0.82 to 0.90. Therefore, most of the observed variance in relative component emission rates was attributable to among-individual variance; component ratios emitted by individual females were relatively constant over time. A similar result was also obtained for *Yponomeuta rorellus*.

Responding individuals appear to more variable than signalers. This was demonstrated in a field population of the pink bollworm using a mark-recapture technique (Haynes and Baker 1988). The observed broad response spectrum was attributed to variation within individuals because individual males responded to a variety of pheromone blends, rather than to multiple and invariant male response phenotypes. In a laboratory colony, the wing-fanning response of males (Collins and Cardé, 1989c) was significantly repeatable (Collins and Cardé, 1989a) but most of the observed phenotypic variance was due to within-individual variance. Variation in pheromone response by male Oriental fruit moths, *G. molesta*, was also largely due to within-individual variation (Cardé et al., 1976).

7.2. Variance Within Populations

Pheromone communication systems are under strong stabilizing selection, and are generally characterized by the emission of and response to narrow variance signals (Löfstedt 1993). Therefore, in many early studies of genetically based control in pheromone communication systems, within-population variation was either ignored or assumed to be unimportant.

The inheritance of signal production and response in chemical communication systems has been examined using interspecific crosses [e.g., Lanier (1970), Grant et al. (1975), Sanders et al. (1977), and Grula and Taylor (1979)] and intraspecific crosses between populations that differ in pheromone production or response characters [e.g., Lanier et al. (1980), Boake and Wade (1984)]. Interspecific and interpopulation hybridization studies can provide valuable information about the mode of inheritance for aspects of pheromone systems, and about possible phylogenetic relationships among closely related species. However, pheromone systems cannot evolve unless there is heritable variation within populations upon which selection can act.

Not all observable phenotypic variation is directly available to selective forces. Phenotypic variance may be partitioned into genetic and environmental components. Heritability (h^2) is a measure of the fraction of total phenotypic variance attributable to additive genetic variance (i.e., total genetic variance less nonadditive genetic variance due to dominance and interaction deviations) (Falconer, 1981). Heritability is of paramount importance to understanding behavioral evolution because it can be used to predict selection responses. Despite the presumed importance of maintaining limited variance in the pheromone communication channel, there have been few studies documenting the extent and nature of intrapopulation variation in either signal production or response.

7.2.1. Redbanded Leafroller Moth

Within-population variance in sex pheromone communication in the redbanded leafroller moth, *A. velutinana*, has been studied extensively. The pheromone of this moth consists of a blend of seven components (Bjostad et al., 1985). Miller and Roelofs (1980) found limited phenotypic variation in the relative amounts of the two major pheromone components, (*Z*)-11-tetradecenyl acetate (Z-11-14:OAc) and (*E*)-11-tetradecenyl acetate (E-11-14:OAc), based on gas-liquid-chromatographic (GLC) analysis of excised female abdominal tips. Mean ($\pm$ SD) pheromone blends in a laboratory and a field population contained 7.0 $\pm$ 1.4 and 9.1 $\pm$ 1.8% E-11-14:OAc, respectively. Low coefficients of variation (9.7% for both populations) demonstrated the narrow variance signal produced by females of this species.

The genetic basis of this variation was examined through artificial selection for altered pheromone blends in laboratory populations (Du et al., 1984; Roelofs et al., 1986). Because female moths cease or drastically reduce pheromone emission after mating, a technique was developed for removing a portion of the pheromone gland for GLC analysis without destroying the female's ability to mate and reproduce (Du et al., 1984). Experiments using several different selection protocols yielded small but significant changes in the pheromone blend, but these modest selection gains were lost after several generations. Females that produced high and low ratios were mated with males chosen randomly from a laboratory colony, yielding F_1 females with means of 9.7 and 7.8% E-11-14:OAc, respectively, compared to a mean of 8.5% in the parental generation.

Only 10% of males from the laboratory colony flew upwind to a 20% E-11-14:OAc blend. When these high-responding males were mated with females that produced high E-11-14:OAc blends, the F_1 female offspring produced blends with greater than 10% E-11-14:OAc, and more than 90% of the F_1 males flew upwind to the 20% E-11-14:OAc blend. Realized heritabilities for the ratio of E-11-14:OAc to ez-11-14:OAc were 0.42 for females mated to males chosen randomly, and 0.85 for females mated to high-responding males (Roelofs et al., 1986), perhaps reflecting a genetic correlation between loci that influence pheromone production and response.

Although heritable variance in the ratio of the two major pheromone components was limited, more extensive genetic variance may exist for minor components. Artificial selection was used to estimate the extent of genetically based variance in the amount of (*Z*)-9-dodecenyl acetate (Z9-12:OAc), a minor component, expressed as a percentage relative to the amount of E11-14:OAc (Sreng et al., 1989). Family selection (Falconer, 1981) was used to generate lines with females that produced high or low relative amounts of E9-12:OAc. After only three generations of selection the amount of the minor component in the low and high lines were 14% and 42% of E11-14:OAc, respectively. Selection was discontinued after three generations, but the lines were maintained. Heritabilities estimated from reciprocal crosses between the high and low lines were 0.64 and 0.42 measured in the fourth and seventh generations, respectively, further indicating a that a relatively large proportion of observed phenotypic variance in the production of the minor component was attributable to additive genetic variance.

7.2.2. Pink Bollworm Moth

Variation in the pheromone produced by the pink bollworm moth, *P. gossypiella*, has been studied in both field and laboratory populations (Haynes et al. 1984; Collins and Cardé, 1985). The long-distance sex pheromone of this moth is a blend of the (*Z,E*)- and (*Z,Z*)- isomers of 7,11-hexadecadienyl acetate (ZE and ZZ, respectively) (Hummel et al., 1973; Bierl et al., 1974).

Phenotypic variance in the ratio of pheromone components emitted by females is limited (Haynes et al., 1984). Consistent ratios were observed over several years and throughout the worldwide geographic range of the species, from 42.5% ZE in a Chinese population to 38.9% ZE in a Pakistani population (Haynes and Baker, 1988). Geographic variation in the amount of pheromone emitted is more extensive; emission rates ranged from 0.057 ng/min for the Chinese population to 0.119 ng/min in a California population (Haynes and Baker, 1988).

The genetic basis of variation in pheromone production and response behavior by pink bollworm moths has been studied in a laboratory colony. Based on parent-offspring mean regressions, significantly heritable variance was found for both the percent of ZE (± SE) (h^2 = 0.34 ± 0.08) and for the total amount of pheromone in excised gland extracts (h^2 = 0.41 ± 0.09) (Collins and Cardé, 1985). The genetic correlation between the amounts of the ZE and ZZ components approached unity (r_G=0.989), indicating control by identical or closely linked loci.

Intense directional selection for 12 generations increased the mean (± SD) percentage of the ZE isomer from 42.9 ± 1.0 in the parental generation to 48.2 ± 1.2 (Collins and Cardé, 1989b). Realized heritability (± SE) for the blend produced was 0.50 ± 0.04. Selection for females with a lower percentage of the ZE isomer yielded no significant change. Wing-fanning duration was measured in a still-air bioassay (Collins and Cardé, 1989c) to a 65% ZE blend for males in the female-selected line. Mean wing-fanning duration increased from 6.3 to 16.8 s after 12 generations, even though no direct selection was applied to male response behavior.

Although the selection response was statistically significant, the absolute change in the mean pheromone blend was small, and perhaps not biologically significant because of the relatively broad response spectra of males (Linn and Roelofs, 1985; Collins and Cardé, 1989c). The observed level of genetically based variance in the ratio of pheromone components produced by females may also be buffered from selection pressures in the field by environmental sources of phenotypic variation in male response behavior. For example, male response specificity has been found to vary with ambient temperature (Linn et al., 1988). Although major changes in field populations are often due to initially rare mutations with large phenotypic effects (Roush and Croft, 1986), the small size of the laboratory population makes the appearance of such mutations unlikely [but see Haynes and Hunt (1990a)].

In a separate experiment, the mean (± SD) pheromone titer increased by 91%, from 24.0 ± 8.1 ng to 45.8 ± 7.9 ng per female, in six generations of selection for increased pheromone production (Collins et al., 1990). Neither the component ratio produced by females nor the duration of wing fanning changed in this female-selected line.

Heritabilities (± SE) for mean durations of wing fanning to blends with 25, 44, and 65% ZE were 0.38 ± 0.11, 0.38 ± 0.10, and −0.14 ± 0.10, respectively (Collins and Cardé, 1989a). Response specificity was measured by comparing the duration of wing fanning for males tested first with an extreme blend (either 25% or 65% ZE) and then retested using the 44% ZE blend normally produced by females. Response specificity was significantly heritable for males exposed to the 25% ZE blends (0.12 ± 0.06), but not for those tested with the 65% ZE blend (−0.04 ± 0.07). The observed asymmetric pattern of genetic variance may reflect a correlation between male pheromone response and fitness, with relatively low heritability expected in the direction of increased fitness (Falconer 1981).

Mean (± SE) wing-fanning duration rose from 5.4 ± 1.4 s to 17.4 ± 2.7 s after six generations of selection for increased male response to a 65% ZE blend (Collins and Cardé, 1990). Realized heritability for male wing-fanning response was significant but low (0.16 ± 0.02). Neither the amount of pheromone nor the ratio of components produced by females changed significantly in the male-selected line.

In the few examples in which variance in signal production or response has been studied within populations, genetically based variance has been found, but was generally considered too limited for selection of biologically significant changes (Sreng et al., 1989). However, because of the resource-intensive nature of artificial selection experiments, and because so few such studies have been undertaken, it is notable that even minor changes in pheromone production and response behaviors have been achieved. Because the scope for selection in nature is so much greater, even modest success in laboratory populations may reflect a considerable potential for within-population selection responses in natural populations.

7.3. Variance Among Populations

Pheromone system variation among different populations of the same species have been found for many species. Examples include the red flour beetle, *Tribolium castaneum* (Boake and Wade, 1984; Boake, 1985), the western avocado leafroller, *Amorbia cuneana* (Bailey et al., 1986), and the pine engraver, *Ips pini* (Lanier et al., 1972).

7.3.1. Turnip Moth

Major differences among populations in pheromone production and response in the turnip moth, *Agrotis segetum*, have been revealed. The three main sex pheromone components in this species are: (Z)-5-decenyl acetate, (Z)-7-dodecenyl acetate, and (Z)-9-tetradecenyl acetate. Substantial variation was observed in the relative amounts of these pheromone components emitted by females among populations in Sweden, France, Armenia, and Bulgaria (47:40:13, 4:52:42, 1:52:47, and 1:42:57, respectively) (Löfstedt et al., 1986; Hansson et al., 1990). There is also considerable variation in pheromone composition among individuals within populations (Löfstedt et al., 1985).

The main pheromone components in this moth are detected by olfactory receptor cells located in physiologically distinct, component-specific sensilla types (Löfstedt et al., 1982; Van Der Pers and Löfstedt, 1986; Löfstedt et al., 1986; Hansson et al., 1992), allowing measurement of their relative abundances. Relative male receptor-type frequencies are quite variable (66:33:1, 87:12:1, 9:90:1, and 6:92:2, for the Swedish, French, Armenian, and Bulgarian populations, respectively) (Löfstedt et al., 1986; Hansson et al., 1990).

Variance observed in both pheromone production phenotypes and receptor cell types appears to be continuous among populations in Eurasia and northern Africa. However, based on extensive pheromone trapping data, populations in southern Africa have pheromone systems that are discretely different from those used by populations north of the Sahara desert (Toth et al., 1992).

Some species exhibit considerable consistency throughout their distributions. For example, both the amount and composition of pheromone produced by females varies little among field populations of the cabbage looper, *Trichoplusia ni*, (Haynes and Hunt, 1990b), and the pink bollworm moth (Haynes et al., 1984; Haynes and Baker, 1988). Gene flow among populations and the importance of behavioral coordination between the signaling and responding sexes may account for this geographic uniformity.

Differences in aspects of pheromone signal and response behaviors between laboratory and field populations have often been noted [e.g., Miller and Roelofs (1980), Haynes and Hunt (1990b)]. These differences may be due to founder effects, population size bottlenecks, inbreeding, or relaxed selection for aspects of long-range communication behaviors that have little relevance to confined laboratory populations (Haynes and Hunt, 1990b). These differences may also have limited relevance to field populations, but must be taken into account when extrapolating the results of laboratory experiments to natural populations.

7.4. Pheromone Polymorphism

Genetic control of variation in pheromone communication based on Mendelian inheritance at a small number of loci has been found in several species. These systems provide an opportunity for examining the nature of variation in pheromone systems not possible for polygenic systems where variance depends on a large number of genes, each with small and unmeasurable phenotypic effects.

7.4.1. European Corn Borer

The most thoroughly studied and probably best understood moth pheromone communication system is that of the European corn borer, *Ostrinia nubilalis*. The female-emitted sex pheromone of this moth is a mixture of two major components, Z-11-14:OAc and E-11-14:OAc.

There are two distinct pheromone morphs (Klun et al., 1973; Cardé et al., 1975; Kochansky et al., 1975). Z morph females emit a 3:97 mixture of E-11-14:OAc and Z-11-14:OAc, whereas the E morph females produce a 99:1 mixture of these components. These morphs occur sympatrically in parts of Europe and the United States (Klun and Cooperators, 1975; Roelofs et al., 1985; Peña et al., 1988). Although partial premating reproductive isolation occurs in confined situations (Cardé et al., 1978), the hybrids are fully fertile, and exhibit heterosis, developing faster and having a higher rate of survivorship (Liebherr and Roelofs, 1975).

Three independent, major genes control important aspects of pheromone signal production and response in this species. The major features of pheromone production are controlled by a simple one-locus, two-allele system (Klun and Maini, 1979). Female hybrids from crosses between E and Z morphs produce a 65:35 E:Z ratio, suggesting that the allele for the E isomer is incompletely dominant. Reciprocal crosses yield females that produce the same pheromone ratio, indicating autosomal inheritance [in Lepidoptera males are homogametic (ZZ) and females are heterogametic (ZW)]. Backcrosses produced a 1:1 ratio of hybrid and parental pheromone-production phenotypes, supporting the single-locus model.

The unsaturated E-11 and Z-11 14-carbon acyl precursors of the corresponding E-11-14:OAc and Z-11-14:OAc pheromone components occur in the same relative proportions (70:30) in E morph, Z morph, and hybrid females (Glover and Roelofs, 1988). Therefore, alleles at the autosomal pheromone production locus appear to affect an enzyme(s) involved in the final reduction and acetylation sequence in the biosynthesis of the pheromone components.

Male behavioral response is controlled independently by a sex-linked locus

(Roelofs et al., 1987). The location of the locus controlling male behavioral response on the Z-chromosome was confirmed using an allozyme marker gene (Glover et al. 1990). Postdiapause development time, an important characteristic of voltinism, is also controlled by genes on the Z-chromosome (Glover et al., 1992).

Based on recordings from single olfactory sensilla on the male antenna, a third locus was found that controlled olfactory response to sex pheromone in this species (Hansson and Löfstedt, 1987). Based on amplitude differences between E- and Z-cell action potentials, inheritance of olfactory response was found to be controlled by an autosomal locus with two alleles. Recombination experiments were used to demonstrate that the autosomal locus for pheromone production by females and the male receptor response locus are inherited independently (Löfstedt et al., 1989).

In areas of sympatry, the interaction of pheromone production and response results in assortative mating (Klun and Huettel, 1988). However, there is no evidence of sexual selection by females or males based on genotype at the pheromone production locus (Bengtsson and Löfstedt, 1990).

7.4.2. Cabbage Looper

A mutation was discovered in a laboratory population of the cabbage looper that causes a major change in the six-component pheromone blend produced by females, including reduced emission of (Z)-7-dodecenyl acetate (normally the major component) and (Z)-5-dodecenyl acetate, and a 20-fold increase in the emission of (Z)-9-tetracecenyl acetate (Haynes and Hunt, 1990a). A single autosomal locus appears to control this change. Pheromone produced by females with the mutation are ineffective in attracting conspecific males in the field (Haynes and Hunt, 1990a). The mutant gene does not appear to influence male pheromone receptors (Todd et al., 1992). The discovery of this mutation effectively illustrates the potential for major evolutionary changes in pheromone communication systems through abrupt, discontinuous shifts in the species-specific blend.

7.5. Evolution of Pheromone Systems

Considerable variation may also be found in the proposed mechanisms by which pheromone systems have evolved and the selective forces that maintain them. Competing explanations for the development of pheromone system diversity contend that pheromone systems evolve as a result of (1) selection for species recognition to impose reproductive isolation; and (2) sexual selection that is mediated by pheromone-based mate choice [reviewed by Löfstedt (1993)].

The evolutionary origin of the pheromone compounds themselves may involve the modification of chemicals already present in the environment. For example, the aggregation pheromones of some scolytid beetles are produced by altering host tree compounds (Vanderwel and Oehlschlager, 1987). Endogenous cues may also be usurped for use in chemical communication. Haynes et al. (1992) showed that adult males of the scarab beetle *Cyclocephala lurida* are attracted to a volatile sex pheromone emitted by adult females. This pheromone is also produced by larvae of both sexes but not by adult males. Because the larvae live below ground, these volatile compounds are presumed to have no communicative function until the beetles become adults.

7.5.1. Genetic Coupling

Evolutionary changes in pheromone-mediated communication systems will depend on selective forces that impinge on both pheromone production and response behaviors. Several explanations have been put forth to explain how complementary changes in signal and response behaviors can be maintained during the evolution of pheromone systems. Variance in signal and response could be controlled pleiotropically by the same genes or by closely linked genes (genetic coupling). Because coordination is so critical, co-adapted gene complexes may control both signal production and response behaviors (Alexander, 1962; O'Donald, 1962).

Alternatively, pheromone signal production and response may be genetically uncorrelated, and behavioral coordination may be maintained through complementary responses to similar selective forces (Löfstedt et al., 1989). Stabilizing selection resulting from feedback between signal production and response behaviors may limit divergence within the communication channel (Miller and Roelofs, 1980).

Apparent genetic coupling has been demonstrated in several pheromone-mediated systems, including the bark beetle, *Ips pini*, based on interpopulational crosses (Lanier et al., 1980; Mustaparta et al., 1985). Selection response and realized heritability were enhanced in a redbanded leafroller colony when females within a line selected for altered pheromone blend production were paired with males that were chosen based on their pheromone-response phenotype (Du et al., 1984). Selection for altered pheromone blends produced by female pink bollworm moths resulted in a correlated selection response in male wing-fanning behavior (Collins and Cardé, 1989b). A similar correlated selection response was found for the khapra beetle, *Trogoderma granarium* (Rahalkar et al., 1985); characteristics of pheromone produced by females were altered significantly in a line selected for reduced pheromone response by males. These results suggest possible genetic correlations between pheromone production and response behaviors, which could reinforce assortative mating based on pheromone production phenotype, and could facilitate the maintenance of communication channel exclusivity.

Evidence of a gene complex affecting both signal production and response has been provided by interspecific hybridization studies between two sympatric sulfur butterflies, *Colias eurytheme* and *C. philodice*. Reproductive isolation is based on species-specific differences in male production of and female response to visual and pheromonal cues (Grula, 1978; Silberglied and Taylor, 1978).

Male courtship pheromones comprise shared and dissimilar elements. Both species produce the same four straight-chain hydrocarbons. *C. philodice* males also produce three esters (n-hexyl myristate, n-hexyl palmitate, and n-hexyl stearate) that are nearly absent in *C. eurytheme*. A branched hydrocarbon (13-methyl heptacosane) is produced only by *C. eurytheme* males (Grula and Taylor, 1979). Females of both species respond to pheromonal signals, but only *C. eurytheme* females respond to visual cues (Silberglied and Taylor, 1973).

Although autosomal loci control production of the three esters in *C. philodice* (Grula and Taylor, 1979), most other aspects of reproductive isolation appear to be controlled by genes located on the Z-chromosome, including the production of the branched hydrocarbon in *C. philodice* (Gurla and Taylor, 1979), the ultraviolet-reflectance pattern of male *C. eurytheme* (absent in *C. philodice*) (Silberglied and Taylor, 1973), and female response behavior (Grula and Taylor 1980b). Sexed-linked loci may also influence size, developmental rate, and wing color patterns (Grula and Taylor, 1980a).

The apparent coadapted gene complex controlling aspects of courtship and mate selection in *Colias* butterflies may be exceptional. In many moth pheromone systems, genetic coupling is absent (e.g., European corn borer) or weak (e.g., redbanded leafroller moth, pink bollworm moth). Therefore, genetic coupling may not be a general phenomenon (Gould, 1982). Because convincing evidence of significant genetic coupling has not been found in most studies in which it has been sought, Löfstedt (1990) argues that the maintenance of behavioral coordination during evolutionary change may often result from complementary selection on genetically uncorrelated characters in the signalling and responding sexes. However, these explanations are not necessarily mutually exclusive; both genetic coupling and coevolutionary behavioral coordination through reciprocal selection may influence the rate of change in these systems (Kyriacou and Hall, 1986).

7.5.2. Evolution of Resistance

Rapid evolution of resistance to synthetic pheromone used in control programs has been suggested (Beroza, 1960; Roelofs and Comeau, 1969). Mechanisms for the development of resistance could include a shift in the pheromone blend, an elevation of the emission rate, or increased reliance on nonpheromonal cues in mate location (Cardé, 1981a). Knowledge of these mechanisms may allow the development of pheromone-use strategies that retard the evolution of resistance or modify the disruptant system to restore efficacy (Croft and Hoyt, 1978; Cardé, 1976, 1981a).

The pheromone-based communication system of the pink bollworm is of particular interest in this regard because of the economic importance of this major cotton pest and the commercial use of synthetically produced pheromone for mating disruption. Synthetic pheromone was first registered as a mating disruptant for the pink bollworm in 1978 (Doane and Brooks, 1981), and is now used extensively in Arizona and California (Haynes et al., 1984).

The potential for the evolution of new strains that are resistant to disruption of long-distance communication with synthetic pheromone will depend on the nature and extent of heritable variation in aspects of both signal production and response behavior. This potential has been examined by comparing blends emitted by females from fields treated for three to five years with synthetic pheromone to blends emitted by females from fields with little or no exposure to synthetic pheromone (Haynes et al., 1984). Although significant differences were not found, only southern California populations were examined. In an expanded study of populations worldwide, the emitted blend was also consistent over time and throughout the range of the species, but the amount of pheromone emitted by females was elevated in areas where synthetic pheromone has been used (Haynes and Baker, 1988). These data suggest that the use of synthetic pheromone is more likely to result in selection for increased emission rates rather than altered blends.

The development of resistance to synthetic pheromones seems unlikely. The amount of available genetic variance appears insufficient for rapid alteration of the communication channel. Even were resistance to develop based on a change in pheromone blend, the problem for control efficacy might be surmounted simply by changing the synthetic blend used.

8. Conclusions

The most primitive searching mechanism is probably local search, in which insects follow internally-stored information to carry out a locomotory program. If such a program allows an insect to perceive sensory information emanating from a resource, it may switch to orientation stimulated or directed by visual or chemical cues.

Chemotaxis, the use of gradient information to locate a chemical source, is also probably primitive. Since cockroaches can utilize both temporal and spatial processing of olfactory information, it is likely that most insects can do so. The limiting factor would seem to be the distance between the antennae, since spatial comparisons require that the antennae are far enough apart to discriminate across the gradient.

Walking insects probably employed upwind chemo-orientation to locate calling females long before any insect flew. Thus, flying insects inherited the mechanisms of the walking insects and then evolved ways to improve the efficiency of chemo-orientation in flight. This radiation included (1) The more primitive hopping mechanism, retaining the way walking insects stop and calculate the wind direction before proceeding; (2) flying straight upwind, and dropping back downwind when the pheromone stimulus is lost; and (3) the more advanced case of an endogenous zigzag program, already found in some walking insects, turned on by phasic stimulation of pheromone.

As emphasized in this chapter, visual and gravitational cues are important in mate finding. Chemical information may initiate a search strategy, probably because it can be emitted with little risk to the sender. In walking insects visual information may not come into play until the odor source is localized, or perhaps not at all. Gravitational cues, at least for nocturnal species, may be more important. Visual information is required, however, for flying insects, during flying chemo-orientation and/or for locating the odor source.

The latter part of this chapter points out the wide range of internal and external factors that can affect the response of an insect to a chemical cue. It would seem that insects should have a matching repertoire of different search strategies to efficiently locate a chemical source. Thus, we should be looking for alternative tactics in insects, dependent upon the information that is available in time and space, rather than a single tactic that might be applicable for a specific set of internal and external factors.

Successful pheromone communication systems depend on coordination between signaling and responding individuals. Therefore, these systems are generally characterized by the emission of and response to narrow variance signals. However, it has recently become evident that pheromone systems are much more variable than formerly believed, and that significant variation exists at all levels, including within and among individuals within a population, as well as among populations. Recognition of the extent and importance of variation in signal production and response behaviors is beginning to be reflected in the rapidly expanding literature. Of particular interest is the additive genetic component of this variance, because it is upon this portion of total phenotypic variance that selection can act to bring about evolutionary change.

Because behavioral coordination between signaler and responding individuals is so critical to fitness, it has long been hypothesized that features of these systems may be inherited as blocks of coadapted genes to ensure continued coordination as these systems evolve. Based on recent studies of genetic control of pheromone communication behaviors, most notably in the European corn borer, it now appears that this expectation is rarely met; genetic variance in aspects of pheromone communication appear to be genetically independent, especially in polymorphic systems. Although there is some evidence of limited genetic coupling within populations (e.g., the redbanded leafroller moth and the pink bollworm moth), the total amount of genetic variance in these systems may be insufficient to allow for evolutionary or economically significant shifts in the communication channel.

References

Akers, R.P. and Wood, D.L. (1989) Olfactory orientation responses by walking female *Ips paraconfusus* bark beetles II. In an anemotaxis assay. *J. Chem. Ecol.* 15: 1147–1159.

Alexander, R.D. (1962) Evolutionary change in cricket acoustical communication. *Evolution* 16: 443–467.

Bailey, J.B., McDonough, L.M., and Hoffmann, M.P. (1986) Western avocado leafroller, *Amorbia cuneana* (Walsingham), (Lepidoptera: Tortricidae), discovery of populations utilizing different ratios of sex pheromone components. *J. Chem. Ecol.* 12: 1239–1245.

Baker, R.R. (1978) *The Evolutionary Ecology of Animal Migration*, Hodder & Stoughton, London.

Baker, T.C. (1985) Chemical control of behaviour. In: *Comprehensive Insect Physiology, Biochemistry and Pharmacology*, Vol. 9 (Kerkut, G.A. and Gilbert, L.I., eds.) pp. 621–672. Pergamon Press, New York.

Baker, T.C. (1989) Sex pheromone communication in the Lepidoptera: New research progress. *Experientia* 45: 248–262.

Baker, T.C. and Cardé, R.T. (1979) Endogenous and exogenous factors affecting periodicities of female calling and male sex pheromone response in *Grapholitha molesta* (Busck). *J. Insect Physiol.* 25: 943–950.

Baker, T.C. and Haynes, K.F. (1987) Manoeuvres used by flying male oriental fruit moths to relocate a sex pheromone plume in an experimentally shifted wind-field. *Physiol. Entomol.* 12: 263- 279.

Baker, T.C. and Haynes, K.F. (1989) Field and laboratory electroantennographic measurements of pheromone plume structure correlate with oriental fruit moth behavior. *Physiol. Entomol.* 14: 1–12.

Baker. T.C., Willis, M.A., and Phelan, P.L. (1984) Optomotor anemotaxis polarizes self-steered zigzagging in flying moths. *Physiol. Entomol.* 9: 365–376.

Baker, T.C., Willis, M.A., Haynes, K.F., and Phelan, P.L. (1985) A pulsed cloud of sex pheromone elicits upwind flight in male moths. *Physiol. Entomol.* 10: 257–265.

Baker, T.C., Gaston, L.K., Pope, M.M., Kuenen, L.P.S., and Vetter, R.S. (1981) A high-efficiency collection device for quantifying sex pheromone volatilized from female glands and synthetic sources. *J. Chem. Ecol.* 7: 961–968.

Bell, W.J. (1990) Searching behavior patterns of insects. *Annu. Rev. Entomol.* 35: 447–467.

Bell, W.J. (1991) *Searching Behaviour, The Behavioural Ecology of Finding Resources*. Chapman & Hall, London.

Bell, W.J. and Kramer, E. (1979) Search and anemotaxis in cockroaches. *J. Insect Physiol.* 25: 631–640.

Bell, W.J. and Kramer, E. (1980) Sex pheromone stimulated orientation responses by the American cockroach on a servo-sphere apparatus. *J. Chem. Ecol.* 6: 287–295.

Bell, W.J. and Tobin, T.R. (1981) Orientation to sex pheromone in the American cockroach: analysis of orientation mechanisms. *J. Insect Physiol.* 27: 501–508.

Bell, W.J., Tobin T.R., and Sorensen, K.A. (1989) Orientation responses of individual larder beetles, *Dermestes ater*, to directional shifts in wind stimuli. *J. Insect Behav.* 2: 787–801.

Bell, W.J., Vuturo, S.B., Silverman, J.M., Burgstahler, A.W., and Weigel, L.O. (1978) Factors involved in the responses of male German cockroaches to synthetic sex pheromone. *J. Chem. Ecol.* 4: 495–501.

Bengtsson, B.O. and Löfstedt, C. (1990) No evidence for selection in a pheromonally polymorphic moth population. *Am. Nat.* 136: 722–726.

Beroza, M. (1960) Insect attractants are taking hold. *Acric. Chem.* 15: 37–40.

Bierl, B.A., Beroza, M., Staten, R.T., Sonnet, P.E., and Adler, V.E. (1974) The pink bollworm sex attractant. *J. Econ. Entomol.* 67: 211–216.

Bjostad, L.B., Linn, C.E., Jr., Roelofs, W.L., and Du, J.-W. (1985) Identification of new sex pheromones in *Trichoplusia ni* and *Argyrotaenia velutinana*, predicted from biosynthetic precursors. In: *Semiochemistry, Flavors and Pheromones* (Acree, T.E., and Soderlund, D.M., eds.) pp. 223–237. Walter de Gruyter and Co., Berlin, New York.

Boake, C.R. (1989) Repeatability: its role in evolutionary studies of mating behavior. *Evol. Ecol.* 3: 173–182.

Boake, C.R.B. (1985) Genetic consequences of mate choice: a quantitative genetic method for testing sexual selection theory. *Science* 227: 1061–1063.

Boake, C.R.B. and Wade, M.J. (1984) Populations of the red flour beetle *Tribolium castaneum* (Coleoptera: Tenebrionidae) differ in their sensitivity to aggregation pheromones. *Environ. Entomol.* 13: 1182–1183.

Brady, J. (1970) Characteristics of spontaneous activity in tsetse flies. *Nature* 228: 286–287.

Bursell, E. (1984) Observations on the orientation of tsetse flies (*Glossina pallidipes*) to wind-borne odours. *Physiol. Entomol.* 9: 133–137.

Bursell, E. and Taylor, P. (1980) An energy budget for *Glossina* (Diptera, Glossinidae). *Bull. Entomol. Res.* 70: 187–196.

Cardé, R.T. (1976) Utilization of pheromones in the population management of moth pests. *Environ. Health Persp.* 14: 133–144.

Cardé, R.T. (1981a) Disruption of long-distance pheromone communication in the Oriental fruit moth, camouflaging the natural aerial trails from females. In: *Management of Insect Pests with Semiochemicals* (Mitchell, E.R., ed.) pp. 385–401. Plenum Press, New York.

Cardé, R.T. (1981b) Precopulatory sexual behavior of the adult gypsy moth. In: *The Gypsy Moth, Research Toward Integrated Pest Management* (Doane, C.C. and McManus, M.L., eds.) pp. 572–587. U.S. Department of Agriculture, Washington, D.C.

Cardé, R.T. and Baker, T.C. (1984) Sexual communication with pheromones. In: *Chemical Ecology of Insects* (Bell, W.J and Cardé, R.T., eds.) pp. 355–377. Chapman & Hall, London.

Cardé, R.T. and Charlton, R.E. (1984) Olfactory sexual communication in Lepidoptera, strategy, sensitivity and selectivity. In: *Insect Communication* (Lewis, T., ed.) pp. 241–265. Academic Press, New York.

Cardé, R.T. and Hagaman, T.E. (1979) Behavioral responses of the gypsy moth in a wind tunnel to air-borne enantiomers of disparlure. *Environ. Entomol.* 8: 475–484.

Cardé, R.T. and Hagaman, T.E. (1984) Mate location strategies of gypsy moths in dense populations. *J. Chem. Ecol.* 10: 25–31.

Cardé, R.T, Baker, T.C., and Roelofs, W.L. (1976) Sex attractant responses of male oriental fruit moths to a range of component ratios: pheromone polymorphism? *Experientia* 32: 1406–1407.

Cardé, R.T., Comeau, A., Baker, T.C., and Roelofs, W.L. (1975) Moth mating periodicity: temperature regulates the circadian gate. *Experientia* 31: 46–48.

Cardé, R.T., Kochansky, J., Stimmel, J.F., Wheeler, A.G., Jr., and Roelofs. W.L. (1975) Sex pheromone of the European corn borer (*Ostrinia nubilalis*): *cis*- and *trans*-responding males in Pennsylvania. *Environ. Entomol.* 4: 413–414.

Cardé, R.T., Roelofs, W.L., Harrison, R.G., Vawter, A.T., Brussard, P.F., Mutuura, A., and Munroe, E. (1978) European corn borer: pheromone polymorphism or sibling species? *Science* 199: 555–556.

Carmichael, L.M., Moore, J., and Bjostad, L.B. (1993) Parasitism and decreased response to sex pheromones in male *Periplaneta americana*. *J. Insect Behav.* 6: 25–32.

Caro, T.M. and Bateson, P. (1986) Organization and ontogeny of alternative tactics. *Anim. Behav.* 34: 1483–1499.

Castrovillo, P.J. and Cardé, R.T. (1980) Male codling moth (*Laspeyresia pomonella*) orientation to visual cues in the presence of pheromone and sequences of courtship behaviors. *Ann. Entomol. Soc. Am.* 73: 100–105.

Charlton, R. E.,and Cardé, R.T. (1990) Orientation of male gypsy moths, *Lymantria dispar* (L.) to pheromone sources: the role of olfactory and visual cues. *J. Insect Behav.* 3: 443–469.

Chase, R. and Croll, R.P. (1981) Tenticular function in snail olfactory orientation. *J. Comp. Physiol. A* 143: 357–362.

Collins, R.D. and Cardé, R.T. (1985) Variation in and heritability of aspects of pheromone production in the pink bollworm moth, *Pectinophora gossypiella* (Lepidoptera: Gelechiidae). *Ann. Entomol. Soc. Amer.* 78: 229–234.

Collins, R.D. and Cardé, R.T. (1989a) Heritable variation in pheromone response of the pink bollworm, *Pectinophora gossypiella* (Lepidoptera: Gelechiidae). *J. Chem. Ecol.* 15: 2647–2659.

Collins, R.D. and Cardé, R.T. (1989b) Selection for altered pheromone-component ratios in the pink bollworm moth, *Pectinophora gossypiella* (Lepidoptera: Gelechiidae). *J. Insect Behav.* 2: 609–621.

Collins, R.D. and Cardé, R.T. (1989c) Wing fanning as a measure of pheromone response in the male pink bollworm, *Pectinophora gossypiella* (Lepidoptera: Gelechiidae). *J. Chem. Ecol.* 15: 2635–2645.

Collins, R.D. and Cardé, R.T. (1990) Selection for increased pheromone response in the male pink bollworm, *Pectinophora gossypiella* (Lepidoptera: Gelechiidae). *Behav. Genet.* 20: 325–331.

Collins, R.D., Rosenblum, S.L., and Cardé, R.T. (1990) Selection for increased pheromone titre in the pink bollworm moth, *Pectinophora gossypiella* (Lepidoptera: Gelechiidae). *Physiol. Entomol.* 15: 141–147.

Comeau, A., Cardé, R.T., and Roelofs, W.L. (1976) Relationship of ambient temperatures to diel periodicities of sex attraction in six species of Lepidoptera. *Can. Entomol.* 108: 415–418.

Corbet, P.G. (1957) The life cycle of the emperor dragonfly *Anax imperator* Leach (Odonata: Aeshnidae). *J. Anim. Ecol.* 26: 1–69.

Corbet, P.G. (1960) Patterns of circadian rhythms in insects. *Cold Spring Harbor Symp. Quant. Biol.* 25: 357–360.

Croft, B.A. and Hoyt, S.C. (1978) Considerations for the use of pyrethroid insecticides for deciduous fruit pest control in the U.S.A. *Environ. Entomol.* 7: 627–630.

David, C.T. (1986) Mechanisms of directional flight in wind. In: *Mechanisms in Insect Olfaction* (Payne, T.L., Birch, M.C. and Kennedy, C.E.J., eds.) pp. 49–58. Oxford Press, Oxford, U.K.

David, C.T., Kennedy, J.S., and Ludlow, A.R. (1983) Finding of a sex pheromone source by gypsy moths released in the field. *Nature* 303: 804–806.

David, C.T., Kennedy, J.S., Ludlow, A.R., Perry, J.N., and Wall, C. (1982) A reappraisal of insect flight towards a distant point source of wind-borne odor. *J. Chem. Ecol.* 8: 1207–1215.

Daykin, P.N., Kellogg, F.E., and Wright, R.H. (1965) Host-finding and repulsion of *Aedes aegypti*. *Can. Entomol.* 97: 239–263.

Dethier, V.G. (1976) *The Hungry Fly*. Harvard University Press, Cambridge, Mass.

Dindonis, L.L. and Miller, J.R. (1980) Host finding behavior of onion flies, *Hylemya antiqua*. *Environ. Entomol.* 9: 769–772.

Doane, C.C. and Brooks, T.W. (1981) Research and development of pheromones for insect control with emphasis on the pink bollworm, *Pectinophora gossypiella*. In: *Management of Insect Pests With Semiochemicals, Concepts and Practice* (Mitchell, E.R., ed.) pp. 285–303. Plenum, New York.

Dobesberger, E.J., Lim. K.P., and Raske, A.G. (1983) Spruce budworm moth flight from New Brunswick to Newfoundland. *Can. Entomol.* 115: 1641–1645.

Domek, K.M., Tumlinson, J.H., and Johnson, D.T. (1990) Responses of male green June beetles *Cotinis nitida* (L.) (Coleoptera: Scarabaeidae) to female volatiles in a flight tunnel. *J. Insect Behav.* 3: 271–276.

Douwes, P. 1968. Host selection and host finding in the egg laying female *Cidaria albulata* (Lep.: Geometridae). Opusc. Entomol. 33: 233–279.

Du, J.-W., Linn, C.E., Jr., and Roelofs, W.L. (1984) Artificial selection for new pheromone strains of red banded leafroller moths *Argyrotaenia velutinana*. *Contr. Shanghai Inst. Entomol.* 4: 21–30.

Du, J-W, Löfstedt, C., and Löfqvist, J. (1987) Repeatability of pheromone emissions from individual female ermine moths, *Yponomeuta padellus* and *Yponomeuta rorellus*. *J. Chem. Ecol.* 13: 1431–1441.

Dusenbery, D.B. (1992) *Sensory Ecology*. 558 pp. W.H. Freeman, Salt Lake City, Utah.

Elkinton, J.S. and Cardé, R.T. (1983) Appetitive flight behavior of male gypsy moths (Lepidoptera, Lymantriidae). *Environ. Entomol.* 12: 1702–1707.

Elkinton, J.S., Schal, C., Ono, T., and Cardé, R.T. (1987) Pheromone puff trajectory and upwind flight of male gypsy moths in a forest. *Physiol. Entomol.* 12: 399–406.

Falconer, D.S. (1981) *Introduction to Quantitative Genetics*, 2nd Ed. Longman, London.

Farkas, S.R. and Shorey, H.H. (1972) Chemical trail-following by flying insects: a mechanism for orientation to a distant odor source. *Science* 178: 67–68.

Flügge, C. (1934) Geruchliche raumorientierung von *Drosophila melanogaster*. *Z. vergl. Physiol.* 20: 463–500.

Foster, S.P. and Harris, M.O. (1992) Factors influencing the landing of male *Epiphyas postvittana* (Walker) exhibiting pheromone-mediated flight (Lepidoptera: Tortricidae). *J. Insect Behav.* 5: 699–720.

Foster, S.P., Muggleston, S.J., and Ball, R.D. (1991) Behavioral responses of male *Epiphyas postvittana* (Walker) to sex pheromone-baited delta trap in a wind tunnel. *J. Chem. Ecol.* 17: 1449–1468.

Gewecke, M. (1974) The antennae of insects as air-current sense organs and their relationship to the control of flight. In: *Experimental Analysis of Insect Behaviour* (Barton-Browne, L., ed.) pp. 100–113. Springer, Berlin.

Gewecke, M. and Philippen, J. (1978) Control of the horizontal flight-course by air-current sense organs in *Locusta migratoria*. *Physiol. Entomol.* 3: 43–52.

Glover, T.J. and Roelofs. W.L. (1988) Genetics of lepidopteran sex pheromone systems. *ISI Atlas Science, Plants & Animals*. 1: 279–282.

Glover, T., Campbell, M., Robbins, P., and Roelofs, W. (1990) Sex-linked control of sex pheromone behavioral responses in European corn borer moths (*Ostrinia nubilalis*) confirmed with TPI marker gene. *Arch. Insect Biochem. Physiol.* 15: 67–77.

Glover, T.J., Robbins, P.S., Eckenrode, C.J., and Roelofs, W.L. (1992) Genetic control of voltinism characteristics in European corn borer races assessed with a marker gene. *Arch. Insect Biochem. Physiol.* 21: 107–117.

Gould, J.L. (1982) *Ethology, The Mechanisms and Evolution of Behavior*, pp 298–307. Norton, New York.

Grant, G.G., French, D., and Grisdale, D. (1975) Tussock moths, pheromone cross stimulation, calling behavior, and effect of hybridization. *Ann. Entomol. Soc. Am.* 68: 519–524.

Greenbank, D.O. (1973) The dispersal process of spruce budworm moths. *Maritimes*

For. Res. Can. Info Rep No M-X–39. pp. 1–25. Maritimes Forestry Agency, Fredericton, N.B., Canada.

Greenbank, D.O., Schafer, T.W., and Rainey, R.C. (1980) Spruce budworm moth flight and dispersal, New understanding from canopy observations, radar, and aircraft. *Mem. Entomol. Soc. Canada*. No. 110.

Grula, J.W. (1978) The inheritance of traits maintaining ethological isolation between two species of *Colias* butterflies. Ph.D. Dissertation, University of Kansas.

Grula, J.W. and Taylor, O.R., Jr. (1979) The inheritance of pheromone production in the sulphur butterflies *Colias eurytheme* and *C. philodice*. *Heredity* 42: 359–371.

Grula, J.W. and Taylor, O.R., Jr. (1980a) The effect of X-chromosome inheritance on mate-selection behavior in the sulfur butterflies, *Colias eurytheme* and *C. philodice*. *Evolution* 34: 688–695.

Grula, J.W. and Taylor, O.R., Jr. (1980b) Some characteristics of hybrids derived from the sulfur butterflies, *Colias eurytheme* and *C. philodice*, phenotypic effects of the X-chromosome. *Evolution* 34: 673–687.

Hansell, M. (1985) Ethology. In: *Comprehensive Insect Physiology, Biochemistry and Pharmacology*, Vol. 9 (Kerkut, G.A. and Gilbert, L.I., eds.), pp. 1–94. Pergamon Press, New York.

Hansson, B.S. and Löfstedt, C. (1987) Inheritance of olfactory response to sex pheromone components in *Ostrinia nubilalis*. *Naturwissenschaften* 74: 497–499.

Hansson, B.S., Ljungberg, H., Hallberg, E. and Löfstedt, C. (1992) Functional specialization of olfactory glomeruli in a moth. *Science* 256: 1313–1315.

Hansson, B.S., Toth, M., Löfstedt, C., Szocs, G., Subchev, M. and Löfqvist, J. (1990) Pheromone variation among eastern European and a western Asian population of the turnip moth *Agrotis segetum*. *J. Chem. Ecol.* 16: 1611–1622.

Hawkes, C. and Coaker, T.H. (1979) Factors affecting the behavioural responses of the adult cabbage root fly, *Delia brassicae*. *Entomol. exp. appl.* 25: 45–58.

Hawkins, W.A. and Rust, M.K. (1977) Factors influencing male sexual response in the American cockroach *Periplaneta americana*. *J. Chem. Ecol.* 3: 85–99.

Haynes, K.F. and Baker, T.C. (1988) Potential for evolution of resistance to pheromones: worldwide and local variation in chemical communication system of pink bollworm moth, *Pectinophora gossypiella*. *J. Chem. Ecol.* 14: 1547–1560.

Haynes, K.F. and Hunt, R.E. (1990a) A mutation in pheromonal communication system of cabbage looper moth, *Trichoplusia ni*. *J. Chem. Ecol.* 16: 1249–1257.

Haynes, K.F. and Hunt, R.E. (1990b) Interpopulational variation in emitted pheromone blend of the cabbage looper moth, *Trichoplusia ni*. *J. Chem. Ecol.* 16: 509–519.

Haynes, K.F., Potter, D.A., and Collins, J.T. (1992) Attraction of male beetles to grubs, evidence for evolution of a sex pheromone from larval odor. *J. Chem. Ecol.* 18: 1117–1124.

Haynes, K.F, Gaston, L.K., Mistrot Pope, M., and Baker, T.C. (1984) Potential for evolution of resistance to pheromones: Interindividual and interpopulational variation in chemical communication system in pink bollworm moth. *J. Chem. Ecol.* 10: 1551–1565.

Hummel, H.E., Gaston, L.K., Shorey, H.H., Kaae, R.S., Byrne, K.J., and Silverstein, R.M. (1973) Clarification of the chemical status of the pink bollworm sex pheromone. *Science* 181: 873–875.

Hurd, H. and Parry, G. (1991) Metacestode-induced depression of the production of, and response to sex pheromone in the intermediate host *Tenebrio molitor*. *J. Invert. Pathol.* 58: 82–87.

Jander, R. (1975) Ecological aspects of spatial orientation. *Annu. Rev. Ecol. Sys.* 6: 171–188.

Kaae, R.S., Shorey, H.H., and Gaston. L.K. (1973) Pheromone concentration as a mechanism for reproductive isolation between two lepidopterous species. *Science* 179: 487–288.

Kennedy, J.S. (1986) Some current issues in orientation to odour sources. In: *Mechanisms in Insect Olfaction* (Payne, T.L., Birch, M.C. and Kennedy, C.E.J., eds.) pp. 11–26. Oxford Press, Oxford, U.K.

Kennedy, J.S. and Marsh, D. (1974) Pheromone-regulated anemotaxis in flying moths. *Science* 184: 999–1001.

Kennedy, J.S., Ludlow, A.R., and Sanders, C.J. (1981) Guidance of flying male moths by wind-borne sex pheromone. *Physiol. Entomol.* 6: 395–412.

Kipp, L.R. and Lonergan, G.C. (1990) Male spruce budworm moth mating periodicity. In: *University of New Brunswick 1989 Spruce Budworm-related Research, Final Report* (Kipp, L.R., Lonergan, G.C., and Seabrook, W.D., eds.). Minister of Natural Resources and Energy, Provence of New Brunswick, Canada.

Kipp, L.R. and Lonergan, G.C. (1992) Comparison of topically applied rubidium chloride and florescent dye markers on survival and recovery of field-released male spruce budworm moths. *Can. Entomol.* 124: 325–333.

Kipp, L.R., Ellison, R., and Seabrook, W.D. (1990) Copulatory mate guarding in the spruce budworm. *J. Insect Behav.* 3: 121–131.

Kipp, L.R., Lonergan, G.C., and Bell, W.J. (1995) Population density-related shifts in male trapping periodicity and the timing of mating in the spruce budworm, *Choristoneura fumiferana* (Clem.) (Lepidoptera: Tortricidae). *Environ. Entomol.* (in press).

Klun, J.A., and Cooperators. (1975) Insect sex pheromones, intraspecific pheromonal variability of *Ostrinia nubilalis* in North America and Europe. *Environ. Entomol.* 4: 891–894.

Klun, J.A. and Huettel, M.D. (1988) Genetic regulation of sex pheromone production and response: interaction of sympatric pheromonal types of European corn borer, *Ostinia nubilalis* (Lepidoptera, Pyralidae). *J. Chem. Ecol.* 14: 2047–2061.

Klun, J.A. and Maini, S. (1979) Genetic basis of an insect chemical communication system: the European corn borer. *Environ. Entomol.* 8: 423–426.

Klun, J.A., Chapman, D.L., Mattes, K.C., Wojtkowsky, P.W., Beroza, M., and Sonnet, P.E. (1973) Insect sex pheromones: minor amount of opposite geometrical isomer critical to attraction. *Science* 181: 661–663.

Kochansky, J., Cardé, R.T., Liebherr, J., and Roelofs, W.L. (1975) Sex pheromone of the European corn borer, *Ostrinia nubilalis* (Lepidoptera, Pyralidae), in New York. *J. Chem. Ecol.* 1: 225–231.

Kramer, E. (1975) Orientation of the male silkmoth to the sex attractant bombykol. In: *Olfaction and Taste*, Vol. 5 (Denton, D. and Coglan, J.D., eds.), pp. 329–335. Academic Press, New York.

Kuenen, L.P.S. and Baker, T.C. (1982) The effects of pheromone concentration on the flight behaviour of the oriental fruit moth, *Grapholitha molesta*. *Physiol. Entomol.* 7: 423–434.

Kuenen, L.P.S. and Baker, T.C. (1983) A non-anemotactic mechanism used in pheromone source location by flying moths. *Physiol. Entomol.* 8: 277–289.

Kuenen, L.P.S. and Cardé, R.T. (1993) Effects of moth size on velocity and steering during upwind flight toward a sex pheromone source by *Lymantria dispar* (Lepidoptera: Lymantriidae) J. Insect Behav. 6: 177–193.

Kyriacou, C.P. and Hall, J.C. (1986) Interspecific genetic control of courtship song production and reception in *Drosophila*. *Science* 232: 494–497.

Lanier, G.N. (1970) Sex pheromones: abolition of specificity in hybrid bark beetles. *Science* 169: 71–72.

Lanier, G.N., Birch, M.C., Schmitz, R.F., and Furniss. M.M. (1972) Pheromones of *Ips pini* (Coleoptera, Scolytidae): variation in response among three populations. *Can. Entomol.* 104: 1917–1923.

Lanier, G.N., Classon, A., Stewart, T., Piston, J.J., and Silverstein, R.M. (1980) *Ips pini:* the basis for interpopulational differences in pheromone biology. *J. Chem. Ecol.* 6: 677–687.

Lewontin, R. D. (1970) The units of selection. *Annu. Rev. Ecol. Syst.* 1: 1–18.

Liang, D. and Schal, C. (1990) Effects of pheromone concentration and photoperiod on the behavioral response sequence to sex pheromone in the male brown-banded cockroach, *Supella longipalpa*. *J. Insect Behav.* 3: 211–224.

Liebherr, J. and Roelofs, W. (1975) Laboratory hybridization and mating period studies using two pheromone strains of *Ostrinia nubilalis*. *Ann. Entomol. Soc. Am.* 68: 305–309.

Linn, C.E., Jr., and Roelofs, W.L. (1985) Response specificity of male pink bollworm moths to different blends and dosages of sex pheromone. *J. Chem. Ecol.* 11: 1583–1590.

Linn, C.E., Campbell, M.G., and Roelofs, W.L. (1988) Temperature modulation of behavioural thresholds controlling male moth sex pheromone response specificity. *Physiol. Entomol.* 13: 59–67.

Linsenmair, K.E. (1968) Anemomenotaktische orientierung bei skorpionen (Chelicerata, Scorpiones). *Z. vergl. Physiol.* 60: 445–449.

Linsenmair, K.E. (1969) Anemomenotaktische orientierung bei tenebrioniden und mistakäfern (Insecta, Coleoptera). *Z. vergl. Physiol.* 64: 154–211.

Liu, Y.-B. and Haynes, K.F. (1992) Filamentous nature of pheromone plumes protects integrity of signal from background chemical noise in cabbage looper moth, *Trichoplusia ni*. *J. Chem. Ecol.* 18: 299–307.

Löfstedt, C. (1990) Population variation and genetic control of pheromone communication systems in moths. *Entomol. exp. appl.* 54: 199–218.

Löfstedt, C. (1993) Moth pheromone genetics and evolution. *Phil. Trans. R. Soc. Lond. B* 340: 167–177.

Löfstedt, C., Hansson, B.S., Roelofs, W., and Bengtsson, B.O. (1989) No linkage between genes controlling female pheromone production and male pheromone response in the European corn borer, *Ostrinia nubilalis* Hübner (Lepidoptera: Pyralidae). *Genetics* 123: 553–556.

Löfstedt, C., Lanne, B.S., Löfqvist, J., Appelgren, M., and Bergstrom. G. (1985) Individual variation in the pheromone of the turnip moth, *Agrotis segetum*. *J. Chem. Ecol.* 11: 1181–1196.

Löfstedt, C, Löfqvist, J. Lanne, B.S., Van Der Pers, J.N.C., and Hansson, B.S. (1986) Pheromone dialects in European turnip moths *Agrotis segetum*. *Oikos* 46: 250–257.

Löfstedt, C., Van Der Pers, J.N.C., Löfqvist, J., Lanne, B.S., Appelgren, M., Gergstrom, G., and Thelin, B. (1982) Sex pheromone components of the turnip moth, *Agrotis segetum*, chemical identification, electrophysiological evaluation and behavioural activity. *J. Chem. Ecol.* 8: 1305–1321.

Mafra-Neto, A. and Cardé, R.T. (1994) Fine-scale structure of pheromone plumes modulates upwind orientation of flying moths. *Nature* 369: 142–144.

Martin, H. (1965) Osmotropotaxis in the honey bee. *Nature* 208: 59–63.

McNeil, J.N. (1991) Behavioral ecology of pheromone-mediated communication in moths and its importance in the use of pheromone traps. *Annu. Rev. Entomol.* 36: 407–430.

Miller, J.R. and Roelofs, W.L. (1980) Individual variation in sex pheromone component ratios in two populations of the redbanded leafroller moth, *Argyrotaenia velutinana*. *Environ. Entomol.* 9: 359–363.

Murlis, J. and Jones, C.D. (1981) Fine-scale structure of odour plumes in relation to insect orientation to distant pheromone and other attractant sources. *Physiol. Entomol.* 6: 71–86.

Murlis, J., Elkinton, J.S., and Cardé, R.T. (1992) Odor plumes and how insects use them. *Annu. Rev. Entomol.* 37: 505–532.

Mustaparta, H., Tommeras, B.A., and Lanier, G.N. (1985) Pheromone receptor cell specificity in interpopulational hybrids of *Ips pini* (Coleoptera, Scolytidae). *J. Chem. Ecol.* 11: 999–1007.

Nielson, H.T. and Nielson, E.T. (1962) Swarming of mosquitoes. *Entomol. exp. appl.* 5: 14–32.

O'Brien, W.J., Evans, B.I. and Howick, G.L. (1989) A new view of the predation cycle of a planktivorous fish, white crappie (*Pomoxis annularis*). *Can. J. Fish. Aquat. Sci.* 43: 1894–1899.

O'Donald, P. (1962) The theory of sexual selection. *Heredity* 17: 541–52.

Peña, A, Arn, H., Buser, H.-R., Rauscher, S., Bigler, R., Brunetti, R., Maini, S., and Toth, M. (1988) Sex pheromone of European corn borer, *Ostrinia nubilalis*, polymorphism in various laboratory and field strains. *J. Chem. Ecol.* 14: 1359–1366.

Pline, M. and Dusenbery, D.B. (1987) Responses of the plant-parasitic nematode *Meloidogyne incognita* to carbon dioxide determined by video camera-computer tracking. *J. Chem. Ecol.* 13: 873–888.

Rahalkar, G.W., Tamhankar, A.J., and Gothi K.K. (1985) Selective breeding for reduced male response to female sex pheromone in *Trogoderma granarium* Everts (Coleoptera: Dermestidae). *J. Stored Prod. Res.* 21: 123–126.

Rakowski, G., Sorensen, K.A., and Bell, W.J. (1989) Responses of dermestid beetles *Dermestes imaculatus* to puffs of aggregation pheromone extract. Entomol. Generalis 14: 211–215.

Roelofs, W.L. and Brown, R.L. (1982) Pheromones and evolutionary relationships of Tortricidae. *Annu. Rev. Ecol. Syst.* 13: 395–422.

Roelofs, W.L. and Comeau, A. (1969) Sex attractant specificity, taxonomic and evolutionary aspects in Lepidoptera. *Science* 165: 398–400.

Roelofs, W.L., Du, J.-W., Linn, C., Glover, T.J., and Bjostad, L.B. (1986) The potential for genetic manipulation of the redbanded leafroller moth sex pheromone blend. In: *Evolutionary Genetics of Invertebrate Behavior, Progress and Prospects* (Huettel, M.D., ed.), pp. 263–272. Plenum Press, New York.

Roelofs, W.L., Du, J.-W., Tang, X.-H., Robbins, P.S., and Eckenrode. C.J. (1985) Three European corn borer populations in New York based on sex pheromones and voltinism. *J. Chem. Ecol.* 11: 829–836.

Roelofs, W., Glover, T., Tang, X., Sreng, I., Robbins, P., Eckenrode, C, Löfstedt, C, Hansson, B.S., and Bengtsson, B.O. (1987) Sex pheromone production and perception in European corn borer moths is determined by both autosomal and sex-linked genes. *Proc. Natl. Acad. Sci. USA* 84: 7585–7589.

Roush, R.T. and Croft, B.A. (1986) Experimental population genetics and ecological studies of pesticide resistance in insects and mites. In: *Pesticide Resistance, Strategies and Tactics for Management* (Board of Agriculture, National Research Council). National Academy Press, Washington, D.C.

Sabelis, M.W. and Dicke, M. (1985) Long-range dispersal and searching behaviour. In: *Spider Mites. Their Biology, Natural Enemies and Control*, Vol. 1B (Helle, H. and Sabelis, M.W., eds.), pp. 141–60. Elsevier, Amsterdam.

Sabelis, M.W. and Schippers, P. (1984) Variable wind directions and anemotactic strategies of searching for an odour plume. *Oecologia* 63: 225–228.

Sanders, C.J., Daterman, G.E., and Ennis, T.J. (1977) Sex pheromone responses of *Choristoneura* spp. and their hybrids (Lepidoptera: Tortricidae). *Can. Entomol.* 109: 1203–1220.

Schal, C. (1982) Intraspecific vertical stratification as a mate-finding mechanism in tropical cockroaches. *Science* 215: 1405–1407.

Schal, C. and Bell, W.J. (1986) Interspecific and intraspecific stratification of tropical cockroaches. *Ecol. Entomol.* 11: 411–423.

Schal, C., Surber, J., Vogel, G., Tobin, T.R., Tourtellot, M.K., Leban, R., and W.J. Bell. (1983) Search strategy of sex pheromone stimulated male German cockroaches. *J. Insect Physiol.* 27: 575–579.

Seabrook, W.D., Kipp, L.R., and Lonergan, G.C. (1989) *University of New Brunswick Spruce Budworm Pheromone Project, 1988 Progress Report, Department of Natural Resources and Energy*, Provence of New Brunswick, Fredericton, N.B., Canada.

Seelinger, G. (1984) Sex-specific activity patterns in *Periplaneta americana* and their relation to mate-finding. *Z. Tierpsychol.* 65: 309–326.

Shorey, H.H. and Gaston, L.K. (1970) Sex pheromones of noctuid moths. XX. Short-range visual orientation by pheromone-stimulated males of *Trichoplusia ni*. *Ann. Entomol. Soc. Am.* 63: 829–832.

Silberglied, R.E. and Taylor, O.R. (1973) Ultraviolet differences between the sulfur butterflies, *Colias eurytheme* and *C. philodice*, and a possible isolating mechanism. *Nature* 241: 406–408.

Silberglied, R.E. and Taylor, O.R. (1978) Ultraviolet reflection and its behavioral role in the courtship of the sulfur butterflies, *Colias eurytheme* and *C. philodice* (Lepidoptera: Pieridae). *Behav. Ecol. Sociobiol.* 3: 203–242.

Silverman, J. M. and Bell, W.J. (1979) Role of strato and horizontal object orientation on mate finding and predator avoidance by the American cockroach. *Anim. Behav.* 27: 652–657.

Sreng, I., Glover, T., and Roelofs, W. (1989) Canalization of the redbanded leafroller moth sex pheromone blend. *Arch. Insect. Biochem. Physiol* 10: 73–82.

Tobin, T.R. (1981) Pheromone orientation: role of internal control mechanisms. *Science* 214: 1147–1149.

Tobin, T.R. and Bell, W.J. (1981) Guidance system for pheromone orientation in moths. *Nature* 295: 203.

Tobin, T.R. and Bell, W.J. (1986) Local search and anemotaxis in the beetle, *Trogoderma variabile*. *J. Comp. Physiol A.* 158: 729–739.

Todd, J.L., Haynes, K.F., and Baker, T.C. (1992) Antennal neurones specific for redundant pheromone components in normal and mutant *Trichoplusia ni* males. *Physiol. Entomol.* 17: 183–192.

Toth, M., Löfstedt, C., Blair, B.W., Cabello, T., Farag, A.I., Hansson, B.S., Kovalev, B.G., Maini, S., Nesterov, E.A., Pajor, I., Sazonov, A.P., Shamshev, I.V., Subchev, M., and Szocs, G. (1992) Attraction of male turnip moths *Agrotis segetum* (Lepidoptera: Noctuidae) to sex pheromone components and their mixtures at 11 sites in Europe, Asia, and Africa. *J. Chem. Ecol.* 18: 1337–1347.

Turgeon, J.J., Mcneil, J.N., and Roelofs, W.L. (1983) Field testing of various parameters for the development of a pheromone-based monitoring system for the armyworm, *Pseudaletia unipuncta* (Haworth) (Lepidoptera: Noctuidae). *Environ. Entomol.* 12: 891–894.

Van Der Pers, J.N.C. and Löfstedt, C. (1986) Signal-response relationship in sex pheromone communication, In: *Mechanisms in Insect Olfaction* (Payne, T.L., Birch, M.C., and Kennedy, C., eds.), pp. 235–241. Oxford Press, Oxford, U.K.

Vanderwel, D. and Oehlschlager, A.C. (1987) Biosynthesis of pheromones and endocrine regulation of pheromone production in Coleoptera. In:*Pheromone Biochemistry* (Preswich, G.D. and Blomquist, G.J., eds.), pp. 175–215. Academic Press, Orlando, Fla.

Willis, M.A. and Baker, T.C. (1984) Effect of intermittent and continuous pheromone stimulation on the flight behavior of the oriental fruit moth *Grapholitha molesta*. *Physiol Entomol*. 9: 341–358.

Willis, M.A. and Cardé, R.T. (1990) Pheromone mediated optomotor response in male gypsy moths *Lymantria dispar*, upwind flight in different wind velocities. *J. Comp Physiol A*. 167: 699–706.

Witzgall, P. and Priesner, E. (1984) Behavioral responses of *Coleophora laricella* male moths to synthetic sex attractant, (Z)-5-decenol, in the field. *Z. Ang. Entomol*. 98: 15–33.

Zaggatti, P. and Renou, M. (1984) Les phéromones sexualles des zygènes III. Le comportement de *Zygaena filipendulae* L. (Lepidoptera: Zyganidae). *Ann. Soc. Entomol. Fran. (N.S.)* 4: 439–454.

Zanen, P.O., Sabelis, M.W., Buonaccorsi, J.P., and Cardé, R.T. (1994) Search strategies of fruit flies in steady and shifting winds in the absence of food odours. *Physiol Entomol*. 19: 335–341.

Plant-Insect Interactions

5

Host-Tree Chemistry Affecting Colonization in Bark Beetles

John A. Byers

1. Introduction

Bark beetles (order Coleoptera: family Scolytidae) comprise a taxonomic group of species that look similar although they differ widely in their ecology and biochemical adaptations to host trees. This diversity of bark beetle biology, in which each species is adapted to only one or a few host-tree species, has probably resulted from natural selection due to the great variety of trees and their biochemicals. It also is likely that each species of tree has coevolved various chemicals to defend against the herbivorous selection pressures of bark beetles and other insects (Erlich and Raven, 1965; Feeny, 1975; Cates, 1981; Berryman et al., 1985). Host-plant chemicals can be attractive, repellent, toxic, or nutritious to bark beetles and have effects on: (1) finding and accepting the host tree (host selection and suitability); (2) feeding stimulation and deterrence; (3) host resistance; (4) pheromone/allomone biosynthesis and communication; and (5) attraction of predators, parasites, and competitors of bark beetles.

Bark and ambrosia beetles contain at least 6,000 species from 181 genera worldwide (S.L. Wood, 1982). In the United States there are 477 species, and in North and Central America a total of 1,430 species occur from 97 genera. Bark beetles may have originated as early as the Triassic period (over 200 million years ago) on conifer hosts (S.L. Wood, 1982). Some Baltic amber dating from the Oligocene period (25 million to 30 million years ago) contains entrapped insects that appear identical to bark beetles from species of present-day genera such as *Tomicus* (S.L. Wood, 1982).

Since 1970 there have been over 3,800 research papers on bark and ambrosia beetles (BIOSIS Previews computer database, Philadelphia, PA). The genus *Dendroctonus* has been the most studied with over 1,196 papers published, primarily on four pest species of North America, *D. frontalis*, *D. ponderosae*, *D. brevi-*

comis, and *D. pseudotsugae*. Other genera in order of studies were: *Ips*, *Scolytus*, *Xyleborus*, *Trypodendron*, *Tomicus*=*Blastophagus*, *Pityogenes*, *Hypothenemus*, *Pityophthorus*, *Hylastes*, and *Gnathotrichus*. The most studied *Ips* species (852 papers) also were pests, *I. typographus* of Europe, and the three North American species, *I. paraconfusus*=*confusus*, *I. pini*, and *I. grandicollis*. *Scolytus multistriatus*, vector of the Dutch elm disease, made up the majority of papers from this genus. Thus, it is clear that most biological knowledge of bark beetles derives from studies on relatively few pest species [these obligate and facultative parasites make up about 10% of scolytid species in the United States and Canada (Raffa et al., 1993)]. This focus on pests is appropriate, however, since only commonly occurring bark beetles that kill living trees or their parts would be expected to have a significant influence on evolution of the host tree and its chemistry.

1.1. Colonization and Life Cycle of Bark Beetles

Bark beetles are one of the few insect groups that as adults bore into the host plant for the purpose of laying eggs (S.L. Wood, 1982). Bark beetle adults and larvae in northern temperate climates generally feed on phloem/cambium (phloeophagy) of conifers (Gymnospermae). In the more tropical zones, the majority of species feed on wood (xylem) and on phloem of broad-leaved trees and shrubs (Dicotyledoneae). Monocotyledoneae are fed on by only a few tropical species of bark beetle. Species that feed on phloem are usually restricted to one or a few host species, whereas xylomycetophagous beetles (ambrosia beetles) that carry their own symbiotic fungi (which breaks down the xylem) may colonize a larger range of hosts (S.L. Wood, 1982).

Semiochemicals from both trees and bark beetles influence many behavioral actions of a bark beetle during its life cycle (Fig. 5.1) [for reviews see D.L. Wood (1982), Borden (1982), Lanier (1983), Birch (1984), Borden et al. (1986), Byers (1989a and 1989b), and Raffa et al. (1993)]. Most of the following presentation involves species in the genera *Dendroctonus*, *Tomicus*, *Ips*, and *Pityogenes*. In general, adults of these species overwinter in either forest litter (*Ips*, *Pityogenes*) or the brood tree (*Dendroctonus*, *Ips*, *Pityogenes*) (Fig. 5.1) (Lekander et al., 1977). In species that have several generations during the summer, emergence is from the brood tree. *Tomicus piniperda*, has a more complex life cycle in which adults overwinter in living, nonbrood trees (Salonen, 1973, Långström, 1983). After emergence the adults of all species attempt to locate a host tree (termed the dispersal flight), often by olfactory means, and determine if it is suitable for colonization and reproduction. This recognition of suitability may be in flight as well as after landing on the tree, as will be discussed later in section 2. In the monogamous *Tomicus* and *Dendroctonus* (subfamily: Hylesininae), the females select the host and a site to begin oviposition galleries that are excavated in the phloem. In contrast, males of polygynous *Ips* and *Pityogenes* (subfamily: Scolytinae) begin the entrance hole (attack) and later

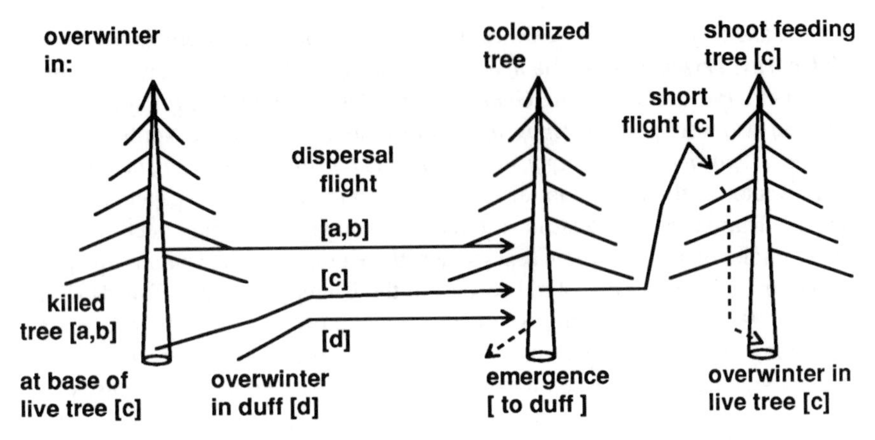

Figure 5.1. Generalized life cycles of phloem-feeding bark beetles. *Dendroctonus* (a), many *Ips* species, and *Pityogenes*, among other species (b) overwinter as larvae, pupae or callow (yellow) adults in the bark of the colonized tree. *Tomicus piniperda* (c) overwinter as adults in the outer bark of pines while some *Ips* species, e.g., *I. typographus* overwinter in forest litter as adults (d). In the spring, beetles of all species disperse and search for suitable host trees in which to reproduce. Beetles successful in colonizing a tree lay eggs, these develop to larvae, pupate, and feed as callow adults until emergence. Depending on the season, beetles either overwinter or in warmer climates complete several generations during the summer. *T. piniperda* survive the summer by flying a short distance to the crowns of nearby trees and feeding in the stems of pine shoots (c); later in the autumn they crawl down the trunk and overwinter at the base (c).

accept several females. In most cases individuals of only one sex begin the attack, and release a species-specific blend of chemicals composing an aggregation pheromone (Byers, 1989b). However, in *D. brevicomis* both the female and the joining male each produce a unique and synergistic pheromone component that when combined elicit maximal attraction response (Silverstein et al., 1968, Kinzer et al., 1969). In *T. piniperda* there is no evidence of an aggregation pheromone (Byers et al., 1985; Löyttyniemi et al., 1988); instead, host-tree chemicals induce aggregation (discussed in section 2.3).

Once an individual or pair begins to release an aggregation pheromone, the likelihood that the tree is colonized depends on (1) the population level of beetles available for recruitment to the attack and (2) the resistance (health) of the tree and its ability to produce defensive resin (discussed in section 4). Beetles of many species have specialized areas of the integument or pouches where symbiotic fungi are carried and sometimes nourished until introduced inside the entrance tunnel where they grow into the tree (Happ et al., 1976; Whitney, 1982; Bridges et al., 1985; Paine and Stephen, 1987; Levieux et al., 1991). Some of the fungal species (genera *Ceratocystis*=*Ophiostoma*, *Trichosporium*) may attack the living tissues of the tree and paralyze the tree's ability to produce and exude resin for defense against the beetle (Mathre, 1964; Horntvedt et al., 1983; Paine, 1984; Raffa and Berryman, 1987; Paine and Stephen, 1987; Paine et al., 1988). Other fungal species of the beetle's mycangium grow in the galleries after the tree has been killed and appear important to the growth of the larvae (Bridges and Perry, 1985; Paine et al., 1988; Goldhammer et al., 1991). In ambrosia beetles, which generally attack unhealthy or dead trees, the adults and larvae feed on fungi lining the galleries instead of directly on the tree's tissues (Funk, 1970; Furniss et al., 1987; Kajimura and Hijii, 1992).

Successful colonization and reproduction by a bark beetle in a living tree requires release of enough aggregation pheromone to ensure the attraction of sufficient conspecifics to overwhelm the host-tree defenses. However, after killing the tree and securing mates, pheromone should not be released any longer in order for the beetles to avoid further competition for bark areas (Berryman et al., 1985). Thus, semiochemicals play a role in cooperation among beetles when killing the tree and in their avoidance of competition (discussed in section 5.4). Pioneer beetles that attack the tree first may suffer most from the tree's defensive resin, but these beetles may have no choice but to attack due to low fat reserves (discussed in 2). The later a beetle arrives in the colonization sequence of the host, the poorer is the quality of the bark substrate due to (1) space utilization by established conspecifics (intraspecific competition); and (2) degradation by microorganisms (discussed in sections 2.4 and 4).

Under the bark, females lay eggs which hatch to larvae that feed on the phloem for several weeks. Chemicals from the plant and from microorganisms could affect larval survival at this time, but little is known about these possible interactions. However, once the tree is dead, there can be no benefits to genotypes that produce harmful chemicals after death, unless these benefits operate through kin selection, a possibility for plants with limited seed dispersal (see Tuomi and Augner, 1993). The larvae pupate in the bark and become yellow, callow adults where they feed and mature until emerging. The beetles may begin a dispersal flight during the same season, or after overwintering in either the tree (*Dendroctonus*, *Pityogenes* and many *Ips*) or in the forest litter (*I. typographus*) (Fig. 5.1) (Lekander et al., 1977; Byers and Löfqvist, 1989). *Tomicus minor* and *T. piniperda* emerge from the bark and fly relatively short distances to the tops of pine trees where they bore into a shoot during the summer (Salonen 1973; Långström and Hellqvist, 1991) (Fig. 5.1). In the autumn, beetles of *T. piniperda* crawl down the trunk and bore into its base to overwinter, whereas *T. minor* overwinters in the litter (Salonen 1973; Långström, 1983).

1.2. Sensory Organs

Bark beetle sensory organs, including visual, olfactory and gustatory receptors, are studied not only to understand the ecological interactions between host trees and bark beetles such as host finding and acceptance, but also to understand many other behavioral processes such as communication, mating, feeding, and oviposition. The eyes of bark beetles are needed for flight, and in conjunction with antennae, for orientation toward or away from semiochemical sources. The eyes of bark beetles (e.g., *Ips*, *Scolytus*, and *Pityogenes*) have about 100–240 ommatidia, relatively less than many insects (Chapman, 1972; Byers et al., 1989a, see Figure 5.2). Two color receptor types in the eyes are indicated, based on electrophysiological recordings with a maximum at 450 nm (blue) and 520 nm (green)(Groberman and Borden, 1982). Observations of *I. paraconfusus, I. typographus*, *D. brevicomis, P. chalcographus*, and *T. piniperda* in flight chambers under dim red light or in complete darkness using an electronic vibration detector indicate they will not fly after dark (Lanne et al., 1987; Byers and Löfqvist, 1989; Byers, unpublished). Bark beetles are attracted more so to traps baited with host odor or pheromone that are placed next to tree trunk silhouettes than to traps without such visual stimuli, indicating that beetles orient to the tree trunk during landing (Moser and Browne, 1978; Borden et al., 1982; Tilden et al., 1983; Lindgren et al., 1983; Bombosch et al., 1985; Ramisch, 1986; Chénier

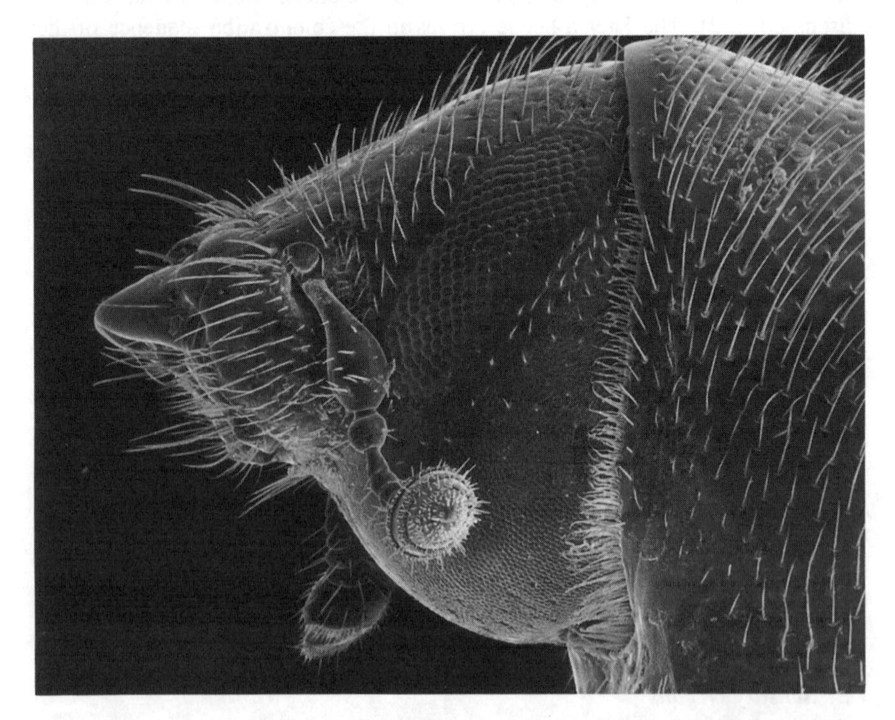

Figure 5.2. Lateral view of the head of *Tomicus piniperda* seen through the scanning electron microscope showing eye with ommatidia and antenna with club containing the olfactory and tactile sensory hairs (height of head is 800 μm, preparation by E. Hallberg and J. Byers).

and Philogène, 1989). Beetles of some species prefer to land on horizontal silhouettes rather than on vertical ones of the same size (Pitman and Vité, 1969). Bark beetles have relatively poor visual acuity; for example, my observations of *T. piniperda* indicate that males must walk within 1 cm of a female beginning her entrance hole before they can detect her and initiate guarding behavior. The same individuals as well as those of *D. brevicomis* can be induced to drop off the tree by movements of the human body about 2 m away (about the same angle of resolution and relative size).

Little is known about the sensilla on the maxillary and labial palpi as well as surrounding mouthparts in bark beetles except for morphological studies of *D. ponderosae* and *I. typographus* (Whitehead, 1981; Hallberg, 1982). In these species there is clearly a large number of chemosensilla (Fig. 5.3) and these appear important for host selection and food discrimination (as will be discussed in sections 2 and 3). The tarsi and ovipositor in other insects have chemosensilla (Städler, 1984), but these have not been studied in bark beetles; it is assumed that all important chemosensory functions involve the mouthparts and antennae.

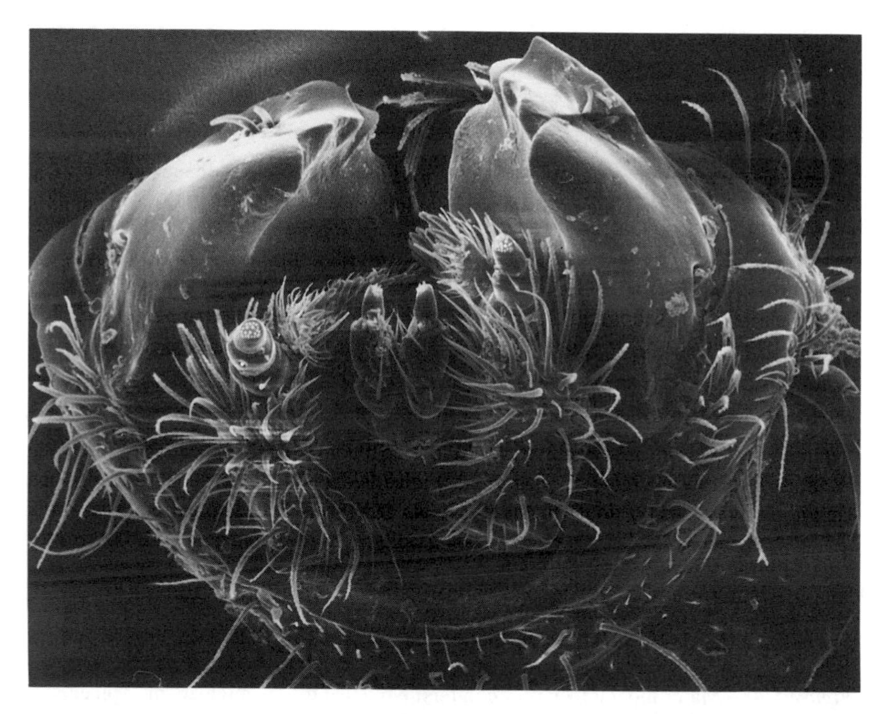

Figure 5.3. Ventral view of the mouth parts of *Ips typographus* seen through the scanning electron microscope showing the labial palpi (central pair) and maxillary palpi with their chemo- and mechanoreceptor hairs that are important in feeding behavior (maximum width is 700 μm, preparation by E. Hallberg and J. Byers).

Most work has involved the antennae (Fig. 5.2), which are known to have sensilla responsive to volatile pheromone and host components as well as other airborne chemostimulants (Borden and Wood, 1966; Payne et al., 1973; Payne, 1979; Mustaparta, 1984; Faucheux, 1989).

The electrophysiological response of an insect to semiochemicals can be studied with the electroantennogram (EAG) of the whole antenna or the single-cell technique that measures responses of specific receptor cells (Payne, 1979). Each antennal receptor cell contains multiple acceptor sites that interact with the chemicals. Bark beetle olfactory cells on the antennae have been shown to be of several functional types, which probably are found in most species: (1) A highly specific type such as the ipsdienol-sensitive cells in *I. paraconfusus* and *I. pini* that is responsive only to one of two possible enantiomers (chemicals that cannot be structurally superimposed but otherwise are identical, see α-pinene in Figure 5.4); (2) a pheromone-sensitive type that is also responsive to some other synergists or inhibitors such as the frontalin cells of *D. frontalis* (the cells have at least two acceptor types each specific for one enantiomer of frontalin, see

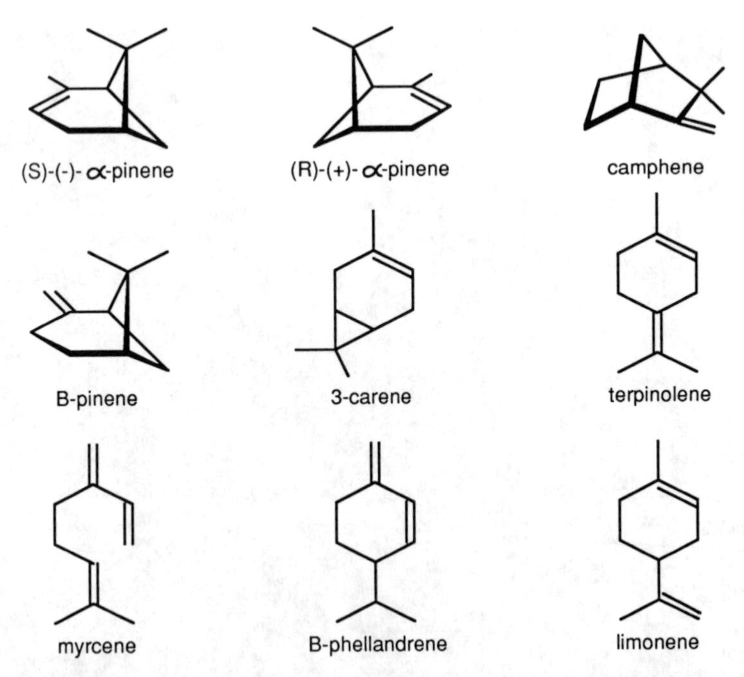

Figure 5.4. Major monoterpenes of conifers. Note that the enantiomers of α-pinene are identical except that they are non-superimposable (mirror images). Camphene, β-pinene, 3-carene, β-phellandrene, and limonene also have two enantiomers, although only (−)-β-pinene and (+)-3-carene are found in trees (Mirov, 1961). Myrcene and terpinolene are achiral.

Figure 5.5), and (3) generalist types that respond to host monoterpenes as well as pheromones to some extent (Mustaparta et al., 1980; Payne et al., 1982; Dickens et al., 1985; Dickens, 1986).

The technique of differential adaptation first attenuates an electrophysiological response by exposing a receptor cell (or antenna) to a high concentration of a compound until specific acceptor sites are saturated, then these sites are exposed to a different volatile to see if any responses are elicited (Payne and Dickens, 1976; Payne, 1979; Dickens et al., 1985). By using this technique with single-cell recordings it was shown that *D. pseudotsugae* has at least four olfactory cell types (Dickens et al., 1985). Three types are each most sensitive to either 3-methylcyclohex-2-en-1-one (3,2-MCH), seudenol, or frontalin (Fig. 5.5), although they are all stimulated somewhat by all of these pheromone components. The fourth cell type is most sensitive to host compounds released by excavating beetles and that are synergists of pheromone components. Acceptors can be specific for one enantiomer of a chiral mixture (Payne et al., 1983) and there may be either only one type of acceptor per cell [e.g. (+)- or (−)-ipsdienol in *I. paraconfusus* (Mustaparta et al., 1980)] or both types of chiral acceptors on the same cell [e.g. (+)-and (−)-frontalin in *D. frontalis* (Payne et al., 1982) and *D. pseudotsugae* (Dickens et al., 1985)].

The antennae of both sexes of *I. paraconfusus* are equally sensitive as measured with the electroantennogram (EAG) to natural pheromone and to (+)-ipsdienol (Light and Birch, 1982). However, the males have been shown to be relatively less attracted by higher concentrations of synthetic pheromone components, and they were not as likely to fly directly toward the pheromone source as were females (Byers, 1983c). Thus, the sexual differences in behavioral response of *I. paraconfusus* to aggregation pheromone appear to be the result of differences in central nervous system (CNS) integration rather than differences in peripheral receptors. *Cis*-verbenol (Fig. 5.5) elicits a similar electrophysiological dose-response curve for both sexes of *I. typographus*, but some sexual differences exist for responses to methyl butenol (Dickens, 1981). Similar to *I. paraconfusus*, males of *I. typographus* orient less directly to pheromone than females during the approach and landing on the host (Schlyter et al., 1987c); this behavioral difference could be due either to CNS differences between the sexes or to the lesser receptor sensitivity of males to methyl butenol. Mustaparta et al. (1980) have shown that eastern U.S. populations of *I. pini* have separate cells for each of the two ipsdienol enantiomers that are synergistic attractants (Lanier et al., 1980). However, the exposure of the two cells to the enantiomers together did not synergistically increase the nerve impulse rate, and so it was concluded that synergism in this case acts at the CNS level.

Of the three or four receptor types found in most species it appears that individual receptors within a type also can vary in their response spectrum to various chemicals (Dickens, 1986; Dickens et al., 1985; Mustaparta et al., 1980, 1984; Tommerås et al., 1984). In the few cases so far known, the cells responsive

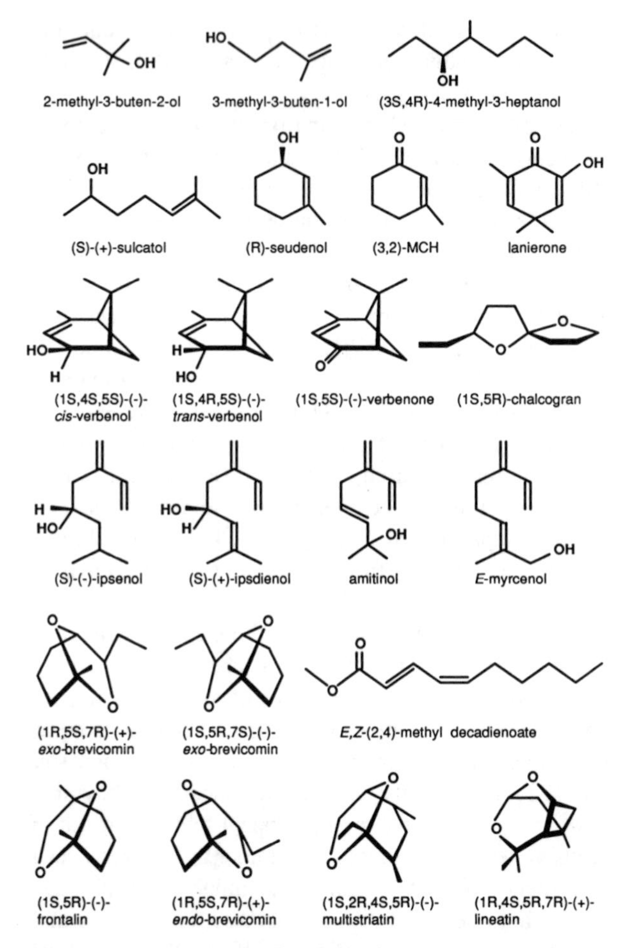

Figure 5.5. Pheromone components of bark beetles. Key: aggregation component (a), inhibitor of aggregation (i). Row 1: 2-methyl-3-buten-2-ol (a, *I. typographus*); 3-methyl-3-buten-1-ol (a, *I. cembrae*), 4-methyl-3-heptanol (a, *S. multistriatus*). Row 2: sulcatol (a, *G. sulcatus*, *G. retusus*); seudenol (*D. pseudotsugae*, *D. rufipennis*, *D. simplex*); MCH (i, *D. pseudotsugae*); lanierone (a, *I. pini*). Row 3: *cis*-verbenol (a, *I. paraconfusus*, *I. typographus*, *I. calligraphus*); *trans*-verbenol (a, *D. ponderosae*, *T. minor*; i, *D. brevicomis*); verbenone (i, *Dendroctonus*); chalcogran (a, *Pityogenes*). Row 4: ipsenol (a, *Pityokteines curvidens*, and many *Ips*: e.g. *I. paraconfusus*, *I. grandicollis*), ipsdienol (a, many *Ips*, e.g. *I. paraconfusus*, *I. duplicatus*, *I. pini*, *I. calligraphus*, *I. avulsus*); amitinol (a, *I. amitinus*); *E*-myrcenol (a, *I. duplicatus*). Row 5: (+)-*exo*-brevicomin (a, *D. brevicomis*, *Dryocoetes*); (−)-*exo*-brevicomin (a, *Dryocoetes*); methyl decadienoate (*P. chalcographus*). Row 6: frontalin (a, many *Dendroctonus*); *endo*-brevicomin (i, *D. frontalis*); multistriatin (a, *S. multistriatus*); lineatin (a, *T. lineatum*). References to above pheromones are in reviews by Borden (1982) and Byers (1989b), and the following (Bakke, 1975; Baker et al., 1977; Lanne et al., 1987; Borden et al., 1987; Byers et al., 1989b, 1990a, 1990b; Teale et al., 1991; Camacho et al., 1993).

to host-plant chemicals are present in both sexes, but the host-selecting sex (males of *Ips* or females of *Dendroctonus*) has a lower threshold to plant compounds (Dickens, 1981, 1986; Dickens et al., 1983). However, the role of plant compounds in long-range orientation of many species of *Ips* is not certain (as discussed in section 2). Probably short-range behaviors on the bark (e.g., gustatory responses) are influenced by host volatiles, but this also is poorly understood.

Some bark beetles have receptors sensitive to compounds they do not produce but are found in several other bark beetles as pheromone components (Tommerås et al., 1984). Lanne et al. (1987) showed compounds not found in *T. piniperda* (*exo*-brevicomin, ipsenol, and pheromone components of other bark beetle species) elicited an EAG response in the beetle. Some of these compounds are found in competing species of bark beetles, but other compounds are probably found only in species colonizing nonhost trees (e.g., Norway spruce). Three sympatric species of the southern United States, *Ips avulsus*, *I. calligraphus*, and *I. grandicollis*, are most responsive electrophysiologically to components of their own pheromones (ipsenol, ipsdienol, *cis*-verbenol); but they can also respond to components of their sibling species as well as frontalin and verbenone from *D. frontalis* and α-pinene from their host pine released by activities of other species (Smith et al., 1988).

The difficulties with electrophysiological methods for unraveling ecological phenomena during host finding and tree colonization are threefold: (1) The electrical measurements may not be correlated to the behavior of interest; (2) the nerve impulses vary between cell types (Mustaparta et al., 1980) of which all are not located and tested; and (3) the electrical patterns are further integrated in the complexity of the CNS.

1.3. Genetic-Based Variation of Host Plants and Bark Beetles

Evolution of plant chemicals that increase the tree's resistance to colonization by a bark beetle population requires that (1) the plant chemicals are detrimental to the beetle; (2) the host chemistry is genetically determined; (3) populational variation in genotypes of these trees exists; and (4) the bark beetle exerts selection pressure on the tree (by killing or reducing fertility). The beetle population should coevolve, if possible, by shifting their genotype frequencies to those that offer more protection against the plant chemicals. The disadvantage for the tree in this "arms race" is that the beetle population may undergo between 25 and several hundred reproductive cycles compared to one generation for the tree, thus the chances for beneficial genetic recombinations and mutations are greater for the insect. Mutant bark beetles of greater endurance would in the beginning have help in killing "resistant" trees from "normal" beetles (that would die more frequently), until gradually mutant beetles would become the most frequent genotype.

Many chemicals in the tree that affect colonization by bark beetles may not be under selection pressure from the insect. Chemicals that are required in physiological processes by the tree may not be readily dispensed with in an evolutionary response to evade their secondary use by insects. For example, various sugars are transported by the phloem and required by the tree for growth; the same sugars may be feeding stimulants for the beetle. If the tree could dispense with the sugars it would become undesirable as food. However, this is unlikely since all trees use sugars (photosynthate) in many biosynthetic pathways (e.g., cellulose).

A theory accounting for the evolution of bark beetle races of *D. ponderosae* each adapted to feeding in ponderosa, lodgepole, or limber pines has been presented by Sturgeon and Mitton (1986). The three species of trees occur together in Colorado and are colonized by *D. ponderosae*. Five enzymes, each varying in several isozymes that migrate differently in electrophoresis gels, were investigated among beetles taken from the three host-tree species. The isozyme frequencies, which represent different alleles at a polymorphic loci, were different among the beetles from the three hosts. The beetles from limber pines were less heterozygous than beetles from the other two hosts. Furthermore, heterozygous beetles were less numerous than expected, suggesting that selection had occurred against these beetles because they were not well adapted to any of these three hosts. If mating between host populations was restricted, for example by different emergence times due to differing development times in each host, then host races could develop. However, no host-related differences in isozyme variation were found for *D. frontalis* from shortleaf or loblolly pines (Namkoong et al., 1979). Langor et al. (1990) naturally reared *D. ponderosae* from limber and lodgepole pines and crossbred them in each species again. They found small reductions in egg production and hatching when pairs were mated from different pine sources compared to the same host source, although beetles from all possible crosses could reproduce—indicating the host races did not appear reproductively isolated (at least under epidemic conditions), thus precluding speciation.

Host trees also vary in monoterpenes which are undoubtedly genetically regulated. Tree monoterpenes appear to affect colonization of bark beetles in a variety of ways (discussed in sections 2 to 6). Monoterpenes (examples shown in Figure 5.4) vary little within a tree, moderately between trees of the same species within a habitat, and greatly between geographic regions; the largest differences are evident among conifer species (Mirov, 1961; Smith, 1964, 1967, 1968, 1969; Sturgeon 1979; Byers and Birgersson, 1990). Genetic differences among beetles over large geographic areas may, in part, reflect the variation in the monoterpene composition of their host. For example, bark beetle populations of *D. ponderosae*, *D. frontalis*, *D. terebrans*, and *I. calligraphus* from different regions when analyzed for certain enzymes by electrophoresis were found to vary genetically within a species (Stock et al., 1979; Namkoong et al., 1979; Stock and Amman, 1980; Anderson et al., 1979, 1983).

There is also semiochemical evidence that bark beetles vary genetically over geographic regions. *I. pini* varies geographically in their production of and response to pheromone enantiomers of ipsdienol (Lanier et al., 1972, 1980; Miller et al., 1989). Two populations of *D. pseudotsugae* from Idaho (inland) and coastal Oregon were found to differ in isozyme frequencies (Stock et al., 1979). These two populations also have a number of possible genetic-based differences in behavioral responses to semiochemicals: (1) ethanol is much more attractive to inland beetles (Pitman et al., 1975; Rudinsky et al., 1972); (2) *trans*-verbenol inhibits pheromonal response in inland beetles but not in coastal beetles (Rudinsky et al., 1972); and (3) the inhibitor 3-methylcyclohex-2-en-1-one (3,2-MCH) (Fig. 5.5) is produced by coastal females but not in inland females (Pitman and Vité, 1974; Rudinsky et al., 1976). Borden et al. (1982) found *Trypodendron lineatum* response to host-released ethanol and α-pinene differed between continents. Western North American beetles responded weakly to ethanol plus α-pinene and these compounds did not enhance a strong attraction to the aggregation pheromone lineatin; whereas beetles in England were similarly attracted to lineatin or to the two host volatiles, and their combination was synergistically active.

2. Host-Plant Selection

Bark beetles and associated beetles feeding or living in trees must locate a suitable host from among the relatively few scattered widely in the forest. The host tree is restricted usually to one or a few species and in most cases the insects seek weakened, less resistant trees, or trees that are in the initial stages of death and decay. Thus, it is expected that species have evolved behavioral responses to volatile host-plant chemicals that indicate the presence of a suitable host in which reproduction can occur.

Bark beetles emerge from the forest litter or from the brood tree and search for suitable hosts in a dispersal flight. The distances and paths of these dispersal flights are poorly known and difficult to observe. In some experiments, beetles are painted or marked with fluorescent powder and released to be recaptured by pheromone-baited traps: *I. typographus*, the major pest of Norway spruce, *Picea abies*, was found to disperse up to 8 km (Botterweg, 1982) and *Ips sexdentatus* up to 4 km (Jactel, 1991). Other mark-recapture studies have found *Ips* and *T. lineatum* disperse generally downwind (Anderbrant, 1985; Salom and McLean, 1991; Jactel, 1991) probably due to the wind-drift component, but in light winds the flight is nondirectional (Salom and McLean, 1989). Jactel and Gaillard (1991) flew *I. typographus* on rotary flight mills connected electronically to a computer. They found that 50% of *I. typographus* can fly more than 20 km and 10% more than 45 km based on about 50 interrupted flights. In another study, the longest continuous fight on a flight mill was 6 hr and 20 min (Forsse and Solbreck, 1985). This suggests that a few *I. typographus* could fly up to 46 km at a speed of 2 m/s without resting (Byers et al., 1989a), and even further when blown by winds. Of course, many individuals would be attracted to hosts or attacked trees much nearer their dispersal origin (Lindelöw and Weslien, 1986).

Several bark beetle species (*D. pseudotsugae*, *S. multistriatus*, *T. lineatum*, and *I. sexdentatus*) may require a period of flight exercise before they are fully responsive to pheromone or host attractants (Graham, 1959; Atkins, 1969; Bennett and Borden, 1971; Wollerman, 1979; Choudhury and Kennedy, 1980; Jactel, 1991). Other species such as *D. ponderosae*, *D. brevicomis*, *I. paraconfusus*, *Pityogenes chalcographus*, and *T. piniperda* are responsive to semiochemicals immediately after beginning flight (Gray et al., 1972; Byers and Wood, 1980; Byers et al., 1985, 1990a). *I. typographus* emerging from the duff responded to nearby pheromone traps without the need for an extended period of flight exercise (Lindelöw and Weslien, 1986), although experiments in the laboratory indicate that response increases with the length of flight exercise (Schlyter and Löfqvist, 1986). Bark beetles generally swallow air and inflate their ventriculus before flight, and this may function as a barometric air pressure receptor to indicate imminent stormy weather (Lanier and Burns, 1978). Artificial changes in the air pressure reduced responsiveness of *S. multistriatus* to the host compound α-cubebene and pheromone components (Lanier and Burns, 1978).

Finding and accepting host plants by insects has been reviewed by Miller and Strickler (1984). They present a model (their Fig. 6.1) by Dethier (1982) where the decision by the insect whether to accept the plant is dependent on external (olfaction, vision, mechanoreception, and gustation) stimulatory and inhibitory inputs balanced against internal excitatory and inhibitory inputs. A graphical, and simplified, model of host acceptance is shown in Figure 5.6 that is directly applicable to bark beetles. In this model, as the bark beetle flies around searching for suitable host trees (usually trees already under attack by conspecifics) they use up energy reserves of lipids (Atkins, 1969; Thompson and Bennett, 1971) and probably become increasingly "desperate" to accept a host. The beetle may by chance encounter several hosts during the dispersal flight that are more or less suitable for reproduction. The beetle will accept the host if the combination of the host suitability and fatigue level of the beetle is above the curve (Fig. 5.6); otherwise the beetle will continue searching for more suitable hosts. The curve is asymptotic to the Y-axis for those beetles that require flight before responding to semiochemicals, whereas the curve would intersect the Y-axis for species that are immediately responsive after emergence. The suitability of the host is determined by the nutritional quality as well as the density of established attacks by the same or other species of bark beetle that indicate the potential for damaging competition.

2.1. Theories of How Bark Beetles Find Suitable Host Trees

There are two theories on how bark beetles find suitable host trees (McMullen and Atkins, 1962). The first is that they locate such trees by orienting over

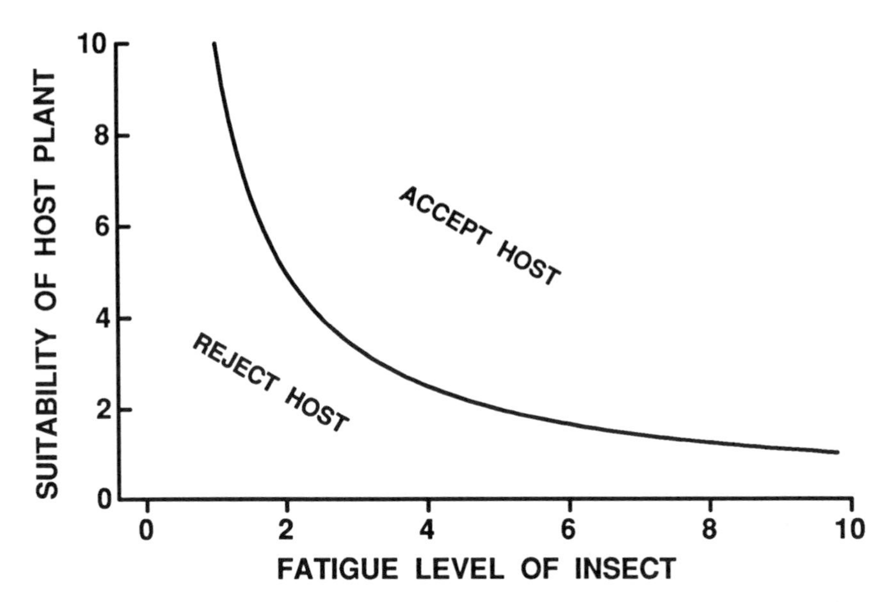

Figure 5.6. Theoretical curve for the acceptance of host trees by bark beetles depending on prerequisite flight exercise (asymptotic Y-axis), level of fatigue (amount of flight) and suitability of the host for reproduction (which depends on nutritional quality and density of colonization by competing bark beetles).

several meters to volatile chemicals usually released by damaged or diseased trees (called *primary attraction*). The second theory is that beetles fly about and encounter suitable host trees at random, whereupon they land and test them by short-range olfaction or gustation. The two theories are not mutually exclusive, and one or the other may primarily operate in a particular species. In California, host finding by the important pests *D. brevicomis* and *I. paraconfusus* is thought to be a random process. Ponderosa pines that were killed by freezing with dry ice and screened to prohibit bark beetle attack, did not have higher landing rates for the prevalent *D. brevicomis* and *I. paraconfusus* bark beetles (among other species) than did living trees. Landing rates on diseased and healthy trees also were similar; it was estimated that each tree in the forest was visited by about one *D. brevicomis* beetle each day (Moeck et al., 1981; D.L. Wood, 1982). Logs of freshly cut ponderosa pine placed in sticky screen traps did not catch beetles of these species, while at the same time high numbers were attracted to synthetic pheromone or infested logs (Moeck et al., 1981).

In addition to *I. paraconfusus* and *D. brevicomis*, many species probably visit trees at random and determine whether they are an appropriate host after landing. For example, *Scolytus quadrispinosus* was caught equally on traps placed in host shagbark hickory, *Carya ovata*, and nonhost white oak, *Quercus alba* (Goeden and Norris, 1965). Beetles also may test the resistance of a host tree when they bore into the bark during an attack. Berryman and Ashraf (1970) found attacks by *Scolytus ventralis* in the basal section of 74% of grand fir examined, while only 3.5% of these trees were killed and colonized. Most unsuccessful attacks were abandoned before beginning the gallery. The attacks on grand fir appeared random during the early part of the flight period before aggregations resulted. Hynum and Berryman (1980) caught *D. ponderosae* in traps on 96% of the lodgepole pines (*P. contorta*) sampled, but only 66% of these pines were killed. Also, they found no differences in landing rates between killed and surviving lodgepole pines or between host and nonhost trees. A direct relationship between the number of *D. ponderosae* caught on unattacked trees and the number of trees upon which beetles landed was found in a study of 123 lodgepole pines (Raffa and Berryman, 1979). *I. grandicollis* landed equally on sticky traps on trees judged resistant or susceptible based on crown area (Witanachchi and Morgan, 1981). However, Schroeder (1987) found a higher average of 35 *T. piniperda* landing on lower vigor Scots pine, *P. sylvestris* (as judged by less crown area), than on higher vigor trees (mean of 22). These differences could be due to secondary release of monoterpenes by beetles boring in the low vigor trees that were less able to resist attack.

There is some evidence that *I. typographus* is weakly attracted to host volatiles (Austarå et al., 1986; Lindelöw et al., 1992) or monoterpenes such as α-pinene (Rudinsky et al., 1971), but other studies have not observed any attraction to host volatiles or synergism of pheromone and host volatiles (Schlyter et al., 1987a). A computer model by Gries et al. (1989), in which "beetles" must take a series of flights between trees in a grid (each flight to one of eight neighboring trees) and test each tree for suitability, showed that few beetles would find the widely scattered hosts designated as susceptible. Thus, they concluded that a mechanism of long-range primary attraction would be required for maintenance of the population. However, a more recent computer model, in which beetles fly more naturally among randomly dispersed susceptible trees, can be used to show that a significant proportion of the population will find the susceptible trees (of actual trunk diameter) by chance interception (Byers, 1993b). If beetles then test the defenses of the potential host (although this rarely has been observed, see section 4) then weaker, more susceptible, trees will not exude adequate resin and allow the beetle to produce aggregation pheromone. According to the later model, this will in effect greatly increase the effective "radius" of the tree so that the remainder of the population can quickly find and colonize these trees (Byers, 1993b).

In addition to the random and host-volatile theories, some bark beetles may find weakened and susceptible host trees by orienting to volatiles produced by competing insect species during colonization. The volatiles can be host compounds that virtually any bark beetle would release upon attack (e.g., monoterpenes) or pheromone components of these other species. For example, *D. brevicomis* responds to pheromone components of *I. paraconfusus* (Byers and Wood,

1981b); *I. typographus* responds to *exo*-brevicomin (from *D. micans* and *Dryocoetes* spp.) (Borden et al., 1987) when combined with its pheromone components (Tommerås et al., 1984); and several sympatric species of *Ips* in the southeastern United States are cross-attracted to infested pine logs (Birch et al., 1980b).

2.2. Orientation Mechanisms for Attraction to Semiochemicals

Primary attraction to the host tree can be considered to occur over a long or short range. The concept of range differs between authors and depends also on the insect considered. Here I consider long-range attraction for bark beetles to be flight orientation over several meters to a semiochemical source. In reality the division is arbitrary, since bark beetles may orient over practically any distance depending on release rate of semiochemical, although at higher release rates the insect may not closely approach the source due to adaptation (Baker et al., 1988). However, the concept of range is still helpful since at natural release rates the beetle will have a range of orientation distances which can be considered long or short range in comparison to orientation distances elicited by other semiochemicals.

Attraction to pheromone is certainly long range. Three parallel lines spaced 4.6 m apart and hung with sticky screens spaced every 1.5 m intercepted *I. paraconfusus* in a V-shaped pattern narrowing to a pheromone source of 50 males boring in a pine log (Byers, 1983c). In this experiment beetles appeared to be orienting over a distance of at least 17 m. *S. quadrispinosus* beetles were intercepted by passive traps 12 m from a girdled hickory tree that was attracting these beetles (Goeden and Norris, 1964). The distance over which beetles respond anemotactically depends primarily on the release rate of the volatile (under mild wind conditions). In Denmark, I once observed *I. typographus* flying slowly upwind (0 to 0.5 m/s ground speed) in 3 m/s gusty winds to a large fallen spruce tree under massive attack. During their orientation to pheromone, beetles were flying at 3–6 m height from at least as far as 50 m downwind from the tree. Jactel (1991) estimated that the maximum attraction distance of *I. sexdentatus* to pheromone-baited traps was 80 m.

Byers et al., (1989a) proposed the "effective attraction radius" (EAR) as an index of attraction strength for a semiochemical release rate from a trap. The EAR is the radius that a trap would need to be enlarged, as a spherical passive trap, in order to intercept as many dispersing insects as were actually caught on the trap when baited with attractant (Byers et al., 1989a). For example, the EAR of *T. piniperda* to a blend of three host monoterpenes, released at rates equivalent to a cut log of Scots pine from each of 10 traps along a 12-m high pole, was largest at the lowest trap (EAR = 1.3 m). The same design found an EAR of 3.2 m for *I. typographus* response to a blend of its pheromone components (Fig. 5.5). These comparisons indicate that the effective attraction radius can be larger for a pheromone than for host volatiles. However, both these values would be greater at higher chemical release rates.

The optomotor anemotaxis mechanism for orientating to pheromone sources proposed for insects, especially moths (David et al., 1982; Baker, 1989), also appears to function in bark beetles (Choudhury and Kennedy, 1980). In this theory, a bark beetle attempts to fly directly upwind when in contact with a packet of pheromone-laden air of the plume, but casts (flies from side to side with respect to the source) when contact is lost. The beetle senses the wind direction while flying by observing the ground below: in no wind, or head-on wind, the ground moves directly underneath during flight. However, if the visual ground field also moves from right to left somewhat, for example, then wind is coming from the left, and the beetle turns to the left to minimize the transverse ground shift and keep the ground moving directly underneath so that the insect heads upwind and toward the pheromone source.

Short-range attraction could be considered to occur within one meter such as when flying along the trunk as I have observed for *T. piniperda*; however, after landing the beetle must use a different mechanism than optomotor anemotaxis. During walking the ground does not move under the beetle due to wind, but the beetle probably can sense wind direction by mechanoreceptors and use pheromone-modulated anemotaxis combined with "casting" or circling movements to locate the odor source. Beetles walking in an arena with laminar airflow respond to a point source of synthetic pheromone (or air from an attacked log) by walking directly upwind within the odor plume. If they happen to walk outside the plume as it narrows to the source, they would experience a concentration gradient decline as they walked. By turning slightly with respect to the upwind angle (as detected by tactile hairs) they would either soon recontact the odor or the concentration would further decline. In the latter case they could reverse the angle or continue turning in a circle which would bring them into odor contact, whereupon they could walk directly upwind again. This mechanism is consistent with observations of beetles responding to pheromone or host odors in a laboratory olfactometer [see Birch (1984)] for species of *Ips*, *Dendroctonus*, *Tomicus*, and *Pityogenes* (Byers et al., 1979; Byers and Wood, 1981b; Lanne et al., 1987; Byers, 1983c; Byers et al., 1990a, 1990b). Borden and Wood (1966) show tracings of tracks of *I. paraconfusus* walking upwind to pheromone.

Akers (1989) studied orientation of *I. paraconfusus* to pheromone in a laboratory olfactometer. He found that beetles increased their counterturning rate (turning left, then right, etc.) in relation to a decline in the rate of concentration increase as they approached the source. In a second study, beetles walked in all directions with respect to the wind without pheromone present, but when in a pheromone plume they decreased their angle to the source (although usually not heading directly upwind) and their turning rate increased (Akers and Wood, 1989a). These generally upwind walking angles and increased turning rates would be expected for beetles orienting to a pheromone source. An important finding was that beetles did not usually walk directly upwind but at slightly different angles. This was attributed to inaccurate anemotaxis rather than a preference for a specific angle with respect to wind and pheromone (anemomenotaxis). Bark beetles appear to have a third mechanism for finding odor sources in the absence of wind. Akers and Wood (1989b) discovered that *I. paraconfusus* can find pheromone sources in still air. The turning rate increased, but only slightly, as the beetles approached the diffusion source, while the mean heading angle to the source decreased as the beetles neared the source (but not for the last 15 cm). Thus, bark beetles are able to use any of several orientation mechanisms, depending on the environmental context, to locate hosts and mates.

2.3. Host-Plant Volatiles Attractive to Bark Beetles

Host volatiles are attractive to a number of forest scolytids including species in the genera *Scolytus*, *Dendroctonus*, *Hylurgops*, *Trypodendron* and *Tomicus* (Goeden and Norris, 1964; Rudinsky, 1966; Meyer and Norris, 1967a; Moeck, 1970; Byers et al., 1985; Lanne et al., 1987; Volz, 1988; Lindelöw et al., 1992; Hobson et al., 1993). Species of bark beetle that regularly attack and kill living trees (termed *aggressive*) have been shown invariably to possess an aggregation pheromone, usually of two or more components, but are weakly, if at all, attracted by host volatiles alone (Vité and Pitman, 1969; Byers, 1989b). However, so-called *secondary* bark beetle species (those that arrive later after the tree has already been killed by the aggressive bark beetles or that feed as saprophytes in decaying trees) may not use an aggregation pheromone, but generally are strongly attracted to either host monoterpenes, ethanol, or a combination (Kohnle, 1985; Klimetzek et al., 1986; Schroeder, 1988; Schroeder and Lindelöw, 1989).

Ethanol, probably released by microorganisms in decaying woody tissue (Graham, 1968; Moeck, 1970; Cade et al., 1970) and stressed plants (Kimmerer and Kozlowski, 1982), is attractive to a wide variety of species of bark beetles (Moeck, 1970, 1981; Magema et al., 1982; Montgomery and Wargo, 1983; Kohnle, 1985; Klimetzek et al., 1986; Schroeder 1987, 1988; Schroeder and Eidmann, 1987; Phillips et al., 1988; Volz, 1988; Chénier and Philogène, 1989; Schroeder and Lindelöw, 1989; Byers, 1992a). Primary alcohols other than ethanol have not been reported as being attractive to scolytids. However, only a few studies have tested methanol (Moeck, 1970; Montgomery and Wargo, 1983; Byers, 1992a); longer-chain alcohols up to hexanol did not attract scolytids in Sweden when they were known to be flying (Byers, 1992a). Electroantennogram (EAG) responses of *T. piniperda* to a series of straight-chain alcohols indicated that the antennae respond increasingly with longer chain length up to a maximum between pentanol and heptanol and then decrease (Lanne et al., 1987). The response spectrum could be due in part to differences in volatility. Thus, although ethanol plays a role in host selection (discussed subsequently), the EAG response for ethanol is lower than for longer-chain alcohols (which probably are not involved in behavior). Ethanol and CO_2 are the usual end products of sugar fermentation by microorganisms whereas methanol is not, which probably explains the evolution of the use of ethanol by forest insects. Moeck (1970) found methanol to be a minor constituent and ethanol a major constituent of extracts from Douglas fir sapwood attractive to *T. lineatum*.

Various tree monoterpenes (e.g. α-pinene, myrcene, terpinolene, β-pinene) (Fig. 5.4) and turpentine are also attractive to a large number of bark beetle species (Byers et al., 1985; Byers 1992a; Phillips et al., 1988; Schroeder, 1988; Chénier and Philogène, 1989; Schroeder and Lindelöw, 1989; Miller and Borden, 1990; Phillips, 1990; Hobson et al., 1993). Synergism between ethanol and various monoterpenes (or turpentine) is also of widespread occurrence (Nijholt and Schönherr, 1976; Kohnle, 1985; Vité et al., 1986; Phillips et al., 1988; Volz, 1988; Schroeder, 1988; Chénier and Philogène, 1989; Schroeder and Lindelöw, 1989; Phillips, 1990). These compounds are not only important for primary attraction to plants, but also may play a role in enhancing the bark beetles' response to aggregation pheromone (Bedard et al., 1969, 1970; Pitman et al., 1975; McLean and Borden, 1977; Borden et al., 1980, 1981; Paiva and Kiesel, 1985; Byers et al., 1988; Miller and Borden, 1990). Host-tree compounds, ethanol and monoterpenes, elicited increased entering rates of bark beetles *T. lineatum* and *P. chalcographus*, respectively, into pipe traps baited with aggregation pheromone (Vité and Bakke, 1979; Bakke, 1983; Byers et al., 1988). β-Phellandrene (Fig. 5.4) is slightly attractive alone to *I. pini* and enhances response to pheromone (Miller and Borden, 1990), and so this monoterpene might induce entering of holes.

In most of the above studies, the discovery of host compounds attractive to bark beetles has been by the comparative approach (similar species are known to be attracted) or by surmising that identified chemicals in the host would be attractive. Thus, most studies are incomplete because of the possibility that there are still undiscovered chemicals important for attraction to the host. Byers et al. (1985) used the subtractive-combination bioassay and fractionation method (Byers, 1992c) to identify rigorously the host volatiles responsible for aggregation of *T. piniperda*. A combination of (−)-(*S*)-α-pinene, (+)-(*R*)-α-pinene, (+)-3-carene, and terpinolene, or each alone, was effective in attracting both sexes (Fig. 5.4). During the isolation study, designed for detection of synergistic pheromone components, no evidence was found for beetle-produced compounds being attractive, in contrast to most bark beetles that aggregate en masse on hosts (Byers, 1989b). Byers et al. (1985) quantified the release rates of α-pinene, terpinolene and 3-carene from a freshly cut log of Scots pine (28 cm × 13 cm diameter) and found them each to be about 15 mg/day. Release of comparable amounts in the field competed favorably with a host log in attracting *T. piniperda*. They theorized that the attraction to monoterpenes functioned in the beetle's selection of both host species (other common tree species have less monoterpenes) and recognition of the host's susceptibility (storm-damaged trees are unable to resist attack and have resinous wounds that release monoterpenes).

In the isolation of host volatiles attractive to *T. piniperda*, a gas-chromatographic adsorbent (Porapak Q), widely used for trapping insect pheromones, was used to collect headspace air from the infested pine logs. Unfortunately, Porapak Q will not retain ethanol molecules due to their small size. Thus ethanol could be a constituent of the attractive host odor. Vité et al. (1986) presented evidence that ethanol enhanced the attraction of *T. piniperda* to α-pinene and terpinolene (identified above) by about eightfold, but these results are difficult to confirm since the chemical release rates were not given. They proposed that ethanol would be released from diseased trees and thus indicate their suitability to *T. piniperda*. Ethanol is attractive when released on healthy trees since *T. piniperda* were caught in ethanol-baited traps on trees, and these beetles also attacked trees baited with ethanol (Schroeder and Eidmann, 1987; Byers, 1992a). However, the attraction to ethanol in traps away from trees is weak or negligible, while monoterpenes in these traps are attractive (Schroeder, 1988; Schroeder and Lindelöw, 1989; Byers, 1992a).

Ethanol, in fact, sometimes reduces response to attractive baits. Klimetzek et al. (1986) tested different release rates of ethanol (24 to 125 mg/day) with an unreported release rate of α-pinene and terpinolene and found that the higher releases of ethanol inhibited attraction of *T. piniperda*. However, a control with ethanol alone, or terpenes alone, was not reported. Schroeder (1988) increased the release of ethanol in five dosages over an even wider range from 0 to 50 g/day in combination with a 240 mg/day release of α-pinene. In this case, the attraction of *T. piniperda* declined linearly with the logarithm of ethanol release, in conflict with the theory of Vité et al. (1986) that ethanol was synergistic with monoterpenes.

Schroeder and Lindelöw (1989) provided the first evidence that could integrate the disparate results. They found that a high release of α-pinene was most attractive to beetles and that ethanol releases alone from 0 to 3 g/day were barely attractive. At a low release rate of α-pinene (2.4 or 22 mg/day) and thus low attraction, the lower releases of ethanol from 0 to 3 g/day had a synergistic effect with α-pinene in attracting beetles (Schroeder and Lindelöw, 1989). Their results are supported by Byers (1992a); i.e., a weak enhancement of attraction by ethanol at low releases of the blend of three host monoterpenes, but no observable effect of ethanol on the greater attraction to higher releases of monoterpenes.

Ethanol released at even higher rates, 120 mg/day (Klimetzek et al., 1986) or 50 g/day (Schroeder, 1988), inhibits the response of *T. piniperda* to monoterpenes. Therefore, the beetle could find diseased, but physically uninjured, trees by a weak response to a synergism between low monoterpene release rates and moderate ethanol rates—the hypothesis of Vité et al. (1986). These trees would be tested occasionally by beetles, and if low in resistance this would permit beetles to continue feeding. Resinosis and monoterpene release would elicit increased numbers joining in a mass-attack. Injured trees with wound oleoresin, and trees under attack with pitch tubes, would have a higher monoterpene release and attract the greatest numbers of beetles, according to Byers et al. (1985). Trees with high ethanol release rates would indicate a tree in advanced decay and unsuitable for reproduction, and thus to be avoided, as theorized by Klimetzek et al. (1986). High monoterpene releases from trees (freshly wounded and not dead) would not naturally coincide with high ethanol release rates (presumably during decay after death). In addition, other compounds such as verbenone from decaying hosts would inhibit response to monoterpenes from unsuitable hosts (discussed in the next section). These studies emphasize the need for releasing semiochemicals at precise rates during tests in the field. In addition, measurements of the natural release rates of ethanol and monoterpenes from various host and nonhost substrates are necessary for further understanding of bark beetle chemical ecology.

2.4. Avoidance of Unsuitable Host Trees and Nonhosts

Host-plant suitability in insects has been reviewed by Scriber (1984). A host plant's suitability to bark beetles varies with its nutritional quality and composition of anti-insect toxins. Nonhost trees are probably less nutritional, and the beetle in most cases would not be adapted to detoxifying some of the nonhost toxins that have evolved for use against other herbivorous insects. A beetle would save much time and energy if it can discriminate the host from the nonhost and determine the suitability of a host by olfactory means without the need to land. Sometimes host and nonhost trees are in virtual contact and the beetle could land by mistake on the nonhost; however, short-range olfactory cues might indicate the inappropriateness of the bark substrate. If the beetle still could not decide, boring a short distance into the nonhost might reveal the lack of feeding stimulants or the presence of deterrents so that the beetle would leave.

According to studies in the previous section, bark beetles find their host tree by attraction to host volatiles or after random landing and probing. However, it is becoming increasingly apparent that many beetles avoid nonhost trees due to specific odors. It is inherently more difficult to isolate repellents and inhibitors used in avoidance behavior than to isolate attractants since tests of avoidance require one to first isolate the attractive host odors and then present these with and without the possibly inhibitory nonhost odors. Several studies indicate that at least some species of bark beetle avoid nonhost volatiles during their search for host trees. The attraction of both *T. piniperda* and *H. palliatus* to ethanol (1–6 g/day) was reduced by odors from cut logs of nonhost trees, birch, *Betula pendula*, and aspen, *Populus tremula* (Schroeder, 1992). In future experiments, host logs (or monoterpenes and ethanol) should be tested instead of ethanol alone to simulate the host tree. Dickens et al. (1992) reduced the attraction response of *D. frontalis*, *I. grandicollis*, and *I. avulsus* to aggregation pheromone by release of the green-leaf volatiles, 1-hexanol and hexanal. *T. domesticum* colonizes wood of deciduous trees (e.g. *Fagus sylvatica*, *Quercus* spp., *Betula* spp.) and is known to be attracted to ethanol (Magema et al., 1982; Paiva and Kiesel, 1985). Conifer monoterpenes of Scots pine and verbenone (from decaying conifers) reduced response of this bark beetle to ethanol (Byers, 1992a) and would provide a mechanism for avoiding nonhosts and unsuitable colonization areas. This also is valid for the hardwood-breeding *Anisandrus dispar* (Schroeder and Lindelöw, 1989).

In addition to their ability of discriminating the host from among nonhost trees, flying beetles are capable of determining the suitability of their host. Verbenone is found in relatively large amounts (μg) in male hindguts of several pest bark beetles of North America, *D. frontalis*, *D. brevicomis*, *D. ponderosae*, and *D. pseudotsugae* (Renwick and Vité, 1968; Rudinsky et al., 1974; Byers et al., 1984; Pierce et al., 1987), but in low amounts (ng) in *T. piniperda* (Lanne et al., 1987), or essentially absent in *I. paraconfusus*, *I. typographus*, and *P. chalcographus* (Byers, 1983b; Birgersson et al., 1984, 1990). Verbenone inhibits the attraction of these beetles to their respective aggregation pheromones (Renwick and Vité, 1969, 1970; Byers and Wood, 1980; Bakke, 1981; Byers et al., 1989c; Byers, 1993a). Some microorganisms isolated from bark beetles or their gallery walls may convert α-pinene to *cis*- and *trans*-verbenol (*Bacillus cereus*) (Brand et al., 1975), or *trans*-verbenol to verbenone (various yeasts) (Leufvén et al., 1984). A fungal culture isolated from the mycangium of *D. frontalis* was able to convert *trans*-verbenol to verbenone, and it was proposed that this process may account for termination of attack (Brand et al., 1976). Verbenone is increasingly released from aging logs of spruce and pine colonized by bark beetles, *I. typographus* (Birgersson and Bergström, 1989) and *T. piniperda* (Byers et al., 1989c), possibly due to the activity of microorganisms.

Byers (1989a, 1989b) speculated that if verbenone is a consistent signal of microbial activity in decaying hosts, then a bark beetle species may have evolved an avoidance to this compound (a kairomone) in order to avoid unsuitable hosts. This species then could have evolved to produce verbenone as a pheromone that reduced intraspecific competition, since the avoidance response was already existent. Other bark beetle species might then evolve to avoid species that produced verbenone (as an allomone), and so avoid interspecific competition. Sympatric species on the same host might coevolve response to and/or production of verbenone since the chemical could serve as a signal for several types of beneficial information (kairomone, pheromone, and allomone). Verbenone does not always inhibit bark beetles, for example, *H. palliatus* feeds in unhealthy or dying Scots pines that probably are releasing verbenone (Byers et al., 1989c), and the beetle's attraction to ethanol was not inhibited by verbenone (Byers, 1992a). Although both conifers and deciduous trees in a state of decay release ethanol that is attractive to *T. domesticum*, this beetle chooses only deciduous trees as hosts. This may occur because *T. domesticum* has evolved to avoid decadent nonhost conifers by avoiding verbenone since deciduous trees probably do not release this compound (deciduous trees lack likely monoterpene precursors of verbenone such as α-pinene) (Byers, 1992a).

In the case of conifer bark beetles, verbenone is increasingly implicated as a general sign of host unsuitability (due to microbial decay or competition with bark beetles). Therefore, it is paradoxical that conifers have not evolved the capacity of converting α-pinene, which they have in abundance, to verbenone in order to become repellent to bark beetles and thereby increase resistance.

3. Feeding Stimulants and Deterrents

Presumably, the beetle must not only determine that the bark underneath is the proper host and is suitable for reproduction, but it must also judge potential competition by whether nearby areas have bark beetles beginning their attacks. Many species of bark beetle bore their entrance holes in a spaced or uniform pattern, indicating the beetles are territorial in order to avoid competition (Byers, 1984, 1992b). In some cases the beetle will bore through the outer bark, regardless of the host, until it encounters the phloem. For example, *I. paraconfusus* will bore through the outer bark of the nonhost white fir, *Abies concolor*, as readily as through bark of the host ponderosa pine. However, the beetle only bores about 1 mm in white fir phloem and then leaves (Elkinton and Wood, 1980). At this time gustatory stimulants and deterrents (also possibly olfactory cues) are balanced in a decision whether to continue feeding and excavating the gallery. The beetle probably can determine whether the host tissue is of good quality in terms of nutritional and moisture factors (Webb and Franklin, 1978). The phloem of ponderosa pine, sugar pine (*P. lambertiana)*, Douglas fir (*Pseudotsuga menziesii*), red fir (*Abies magnifica*), and several other conifers contain about equal amounts of glucose, fructose and sucrose (Smith and Zavarin, 1960). Bark beetles have been induced to feed or lay eggs on several diets, but the most successful contain some percentage of host (usually phloem) tissue (Jones and Brindley, 1970; Richeson et al., 1970; Whitney and Spanier, 1982; Conn et al., 1984; Byers and Wood, 1981a), indicating the presence of feeding or ovipositional stimulants. Sucrose was found in preliminary experiments to increase feeding by *I. paraconfusus* in powdered cellulose diets (Byers and Wood, 1981a).

Few studies have attempted to isolate feeding stimulants in conifer-feeding bark beetles, and none have isolated specific compounds. Elkinton et al. (1981) extracted ponderosa pine phloem successively with diethyl ether (partitioned with water), water, and then methanol, and added these extracts to powdered cellulose diets. *I. paraconfusus* beetles were then given a choice between a control diet and a diet with extract. The diet with the ether extract did not cause beetles to remain longer, but did cause more feeding but no preferential boring. The water partition of the ether extract only caused beetles to remain longer. The water extract elicited more boring and feeding, while the methanol extract was inactive since feeding stimulants had already been extracted by the ether and water treatments. These results indicate that several compounds function in gustatory preferences. Solvent (methanol-water-benzene) extracts of lodgepole pine bark (*Pinus contorta*) were absorbed by tissue paper and shown to induce feeding by *D. ponderosae* (Raffa and Berryman, 1982b). The benzene fraction induced biting but not feeding while the polar fraction (water-methanol) caused continued feeding. Differences in feeding preferences for bark extracts were large between trees, but these differences could not be attributed to amounts of 13 monoterpenes as determined by gas chromatography (GC). Also, extracts of trees judged resistant, because beetles that had been forced to attack in cages either refused or discontinued attack, were as stimulatory to feeding beetles as those from susceptible trees. In contrast, Hynum and Berryman (1980) found greater feeding preferences for extracts of bark of susceptible than of resistant lodgepole pine. However, the susceptible trees had been killed by the beetles before solvent extraction, which might have allowed microorganisms to produce additional feeding stimulants. White (1981) also found differences in gustatory deterrent and stimulatory properties of bark extracts from different trees of loblolly pine, *P. taeda*.

In beetles of deciduous trees, most work on feeding stimulants and deterrents involves the elm bark beetle, *S. multistriatus*. Vanillin and syringaldehyde are short-range attractants inducing feeding in *S. multistriatus* (Meyer and Norris, 1967b). Feeding stimulants were isolated from the bark of American elm, *Ulmus americana*, of which one was partially identified as a pentacyclic triterpene (Baker and Norris, 1967). Later some lignin intermediates and phenolics were indicated (Meyer and Norris, 1974). Doskotch et al. (1973) succeeded in identifying another feeding stimulant in elm bark as a catechin xyloside. A tritiated catechol feeding stimulant was shown to penetrate the gustatory receptor of *S. multistriatus* (Borg and Norris, 1971). *Scolytus rugulosus* are stimulated to feed in fruit trees by several phenolic compounds (Chararas et al., 1982).

S. multistriatus was induced to feed on sucrose pith disks when volatiles from benzene extracts of bark of nonhost trees (Eastern cottonwood, *Populus deltoides*, and yellow buckeye, *Aesculus octandra*) were placed 7 mm away (Baker and Norris, 1968). However, these nonhost trees were not fed upon since they contained nonvolatile feeding deterrents as shown by lowered feeding on a mixture of host and nonhost extracts compared to host extracts. *S. multistriatus* beetles do not attack the nonhost hickory, *C. ovata*, due to the presence of juglone (5-hydroxy-1,4-napthoquinone), a feeding deterrent (Gilbert et al., 1967). Elm tree tissue infected by the fungus *Phomopsis oblonga* is avoided by *S. multistriatus* due to several feeding deterrents of complex structure, e.g., oblongolide (isomer of dimethylnapthofuranone), a norsesquiterpene lactone, two tiglic esters of 5,6-

dihydro-5-hydroxy-2-pyrones, nectriapyrone, 4-hydroxyphenylethanol, 5-methylmellein as well as acids of 2-furoic, orsellinic, 3-nitropropanoic, and mellein-5-carboxylic (Begley and Grove, 1985; Claydon et al., 1985).

Diterpene acids (e.g., abietic, levopimaric, neoabietic, and palustric acids) have been isolated from ponderosa pine oleoresin (Anderson et al., 1969; Himejima et al., 1992) and these are known from other conifers to be antifeedants against aphids and sawflies (Wagner et al., 1983; Schuh and Benjamin, 1984; Rose et al., 1981), but have not been tested on bark beetles. Tannins, phenolics and terpenoids that can inhibit feeding or digestion in other insects (Berenbaum and Isman, 1989) could also affect bark beetles, but again there are no reports. Ponderosa pine bark extracted first with ether yields behenic and lignoceric acids, fatty alcohols, resin acids, and flavonols (quercetin and myricetin, pinoquercetin, pinomyricetin and dihydroquercetin), a subsequent acetone extract contains tannins and phlobaphenes, while a hot-water extract has tannin (6–11% dry weight of bark) and carbohydrates (Anderson, 1962; Anderson et al., 1969). Many of these compounds are found only in the outer bark and although they may be important in deterring nonhost bark beetles, at least the host-adapted *I. paraconfusus* does not eat the outer bark (Elkinton and Wood, 1980).

4. Plant Compounds and Resin in Resistance Mechanisms

Resistance of pines to bark beetle attacks has long been attributed to the amount of resin exudation and formation of pitch tubes (Webb, 1906). It is obvious that oleoresin acts to entrap and impede the excavation efforts of bark beetles. Dead beetles can often be seen in crystallized resin of pitch tubes. However, *D. brevicomis* and other aggressive bark beetles have a great ability to survive the toxic monoterpenes and suffocating mucilage and may struggle for hours in copious resin flows (*D. frontalis*) (Hodges et al., 1979). I have observed *D. brevicomis* beetles completely covered with resin while attempting to clear the entrance tunnel (Fig. 5.7), and it appears that breaths of air are taken by slightly lifting the elytra that protect the spiracle openings on the dorsal side of the abdomen; later these beetles entered the tree and made galleries (Byers et al., 1984). Bark beetles cannot attack all hosts since in experiments where three female *D. ponderosae* were caged on each of 79 trees, only 43 trees were attacked, and of these just 15 were successful so that aggregations resulted (Raffa and Berryman, 1979).

Oleoresin may provide resistance to trees due to chemical toxicity to the beetle and associated microorganisms or to physical impedance and entrapment (Hodges et al., 1985). Oleoresin and the monoterpenes therein are repellent to bark beetles in concentrated amounts (Struble, 1957; Pitman et al., 1966; Berryman and Ashraf, 1970; Bordasch and Berryman, 1977). Drought and poor water balance lowers the resistance of conifers (Hodges and Lorio, 1975; Hodges et al., 1979)

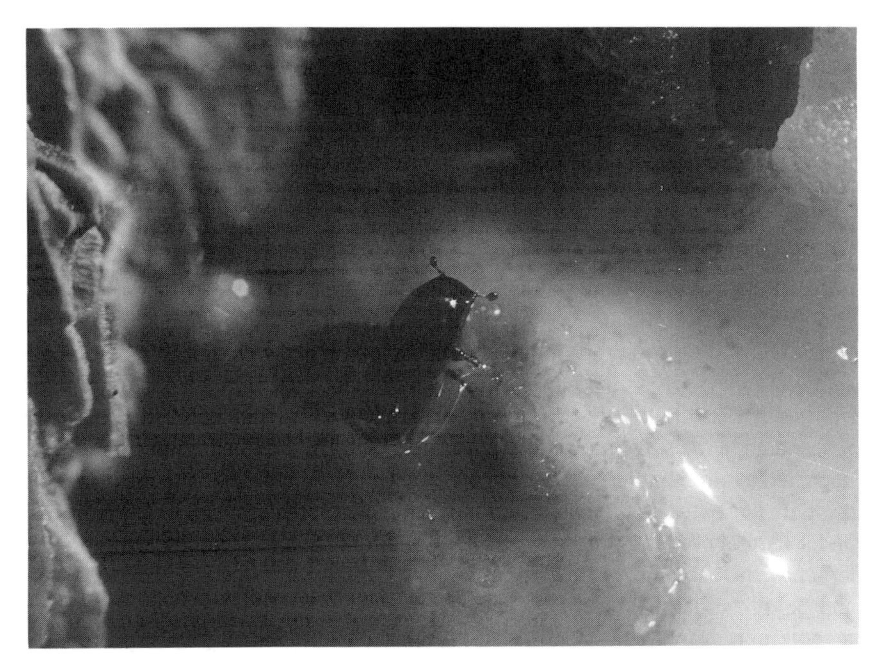

Figure 5.7. Western pine beetles, *Dendroctonus brevicomis*, "swimming" in oleoresin exuding from a pitch tube on the bark of ponderosa pine. The beetle (in center of photo) and its mate underneath were observed to burrow down through the viscous liquid into the entrance of the gallery (directly under and to right of beetles) for up to several minutes before returning and shoving out more resin. The resin is slightly toxic and may exhaust the beetles; it also may eventually crystallize to entrap them.

probably by lowering the turgidity of resin duct cells, which lowers the oleoresin exudation pressure (OEP). A correlation between higher OEP and greater resistance of ponderosa pine to attack by *D. brevicomis* and *I. paraconfusus* has been reported (Vité, 1961; Wood and Vité, 1961; Wood, 1962; Brown et al., 1987). Hodges et al. (1979) found that in more resistant pines their resin was slower to crystalize (*P. elliotii*) or had a higher resin flow (*P. palustris*) compared to more susceptible trees (*P. taeda* and *P. echinata*) colonized by *D. frontalis*. Cook and Hain (1987) also found that susceptible shortleaf pines, *P. echinata*, had a lower resin flow than resistant trees. However, Raffa and Berryman (1982c) found no relationship between resistance and the rate of resin flow or crystallization. Also, Schroeder (1990) found no difference in resin flow between resistant and susceptible Scots pine, *P. sylvestris*, fed on by *T. piniperda*. Western larch, *Larix occidentalis*, had no OEP but the trees with higher content of 3-carene in the resin were attacked less by *D. pseudotsugae* (Reed et al., 1986).

Another factor that might be more important for resistance could be the toxicity of compounds within the resin. Smith (1961, 1965a, 1965b) exposed beetles of several *Dendroctonus* species to resin vapors from host and nonhost pines and found beetles were able to tolerate vapors of their host better than nonhosts. In an attempt to determine which components of the resin vapor were toxic to bark beetles, various conifer monoterpenes were presented alone at vapor saturation to *D. brevicomis* held in a glass chamber (Smith, 1965b). The most toxic monoterpene was limonene, followed by (+)-3-carene, myrcene, (−)-β-pinene and then α-pinene. In another study, Smith (1965a) found that n-heptane, a major constituent of Jeffrey pine (*P. jeffreyi*), was quite toxic to *D. ponderosae* which does not feed in this tree, whereas *D. jeffreyi* beetles showed little mortality in saturated vapors. However, the monoterpenes when presented alone at saturation were from 40 to 80 times higher in concentration than in the natural atmosphere of a beetle's gallery, as found empirically by GC headspace analysis and explained by Raoult's law that states the vapor pressure of a compound is due to its mole percentage in the substrate/solvent mixture (Byers, 1981b). Most of these species may also avoid monoterpene vapors temporarily by breathing at ventilation holes through the outer bark. Thus, Smith's conclusion that monoterpenes are toxic under natural conditions is dubious, but his results may still indicate these monoterpenes increase mortality of bark beetles over longer periods during feeding and colonization. Raffa et al. (1985) exposed *Scolytus ventralis* to the above monoterpene vapors and also found limonene the most toxic, but myrcene, α-pinene, β-pinene, and 3-carene, in that order also caused significant mortality.

Differences in monoterpene composition within a tree are slight, even from year to year (Smith, 1964; Byers and Birgersson, 1990). Differences in monoterpene ratios can be large between trees of one area, almost as large as differences over wide geographic regions (Smith, 1964, 1966, 1967, 1969). The oleoresins of 88 trees under attack by *D. brevicomis* and subsequently killed were compared to those from 202 living trees, and the living trees had a higher content of myrcene and limonene in their resins (Smith, 1966). This correlation supports the theory that limonene is important in host resistance.

Sturgeon (1979) theorized that *D. brevicomis* is a selective force in the evolution of ponderosa pine, *P. ponderosa*, and that therefore areas of recent outbreaks might have selected trees with a higher titer of limonene (since these trees would be more resistant). Eight populations of trees, a total of 617 trees, were sampled for monoterpenes and analyzed by principal component multivariate statistics. The average proportions of myrcene and α-pinene in resins from the eight populations ranged the least (10.0–15.2% and 4.5–9.2%, respectively), while the other monoterpenes varied more (3-carene, 23.6–60.4%; β-pinene, 13–35%; limonene, 6.1–18.7%). The populations were separated by the Cascade Range in northern California and southern Oregon into two regions. The west side had higher proportions of myrcene, β-pinene, and limonene. The proportion of limonene in resin from three populations considered to have been historically under heavy bark beetle predation was higher than in populations not considered to have such a history (Sturgeon, 1979). The problem with these findings is that a correlation is made between rather accurate monoterpene results and rather poor knowledge of former predation pressure by *D. brevicomis*, or for that matter other bark beetle species in each of the areas.

Sturgeon (1979) concluded that *D. brevicomis* beetles (and others) may exert a frequency-dependent selection pressure on chemically polymorphic populations of ponderosa pine. Thus after beetles have colonized most of the low limonene and less resistant trees during an epidemic, the beetle population would either (1) Die out or disperse to areas that had more chemically susceptible trees; or (2) evolve a tolerance to limonene. The second possibility was considered less likely because of the large variation in monoterpene composition among trees that would make it improbable that selection of beetles would occur that were capable of detoxifying all of these compounds. However, bark beetles, including *I. paraconfusus* and *D. brevicomis*, already must be able to detoxify all of these compounds since they survive exposure to monoterpene vapors in part by converting them to oxygenated products that are more soluble and readily excreted (Hughes 1973, 1974; Byers, 1981a, 1981b, 1982, 1983a, 1983b; Pierce et al., 1987). Another hypothesis is that limonene might not always be the most toxic to a bark beetle population, but rather those monoterpenes that the population is not well adapted to (since they occur infrequently) are the most toxic. In this regard, lodgepole pines in some regions of California have very high titers of limonene (Byers and Birgersson, 1990), yet they are readily attacked and killed by *D. ponderosae*.

Gollob (1980) measured the monoterpene content of resin from unattacked loblolly pines, *P. taeda*. Two apparently resistant pines that had survived attack by *D. frontalis* in an epidemic area had a much higher content of myrcene compared to other trees that were killed by the beetle and had low or trace amounts of myrcene. However, no consistent differences in monoterpene composition of Douglas fir, *P. menziesii* var. *glauca*, resin were found between trees that had resisted attack by *D. pseudotsugae* and trees that had succumbed (Hanover and Furniss, 1966). Similarly, Raffa and Berryman (1982c) found no relationship between monoterpene composition and degree of resistance of lodgepole pines, *P. contorta* var. *Latifolia*, to *D. ponderosae*. Hodges et al. (1979) also did not find differences in monoterpene or resin acid composition which could account for differences in resistance among four pine species to attack by *D. frontalis*.

In addition to the wound or primary resin production, conifers have a secondary or hypersensitive response to attack (Reid et al., 1967; Berryman 1969, 1972; Berryman and Ashraf, 1970; Raffa and Berryman, 1982a, 1983; Christiansen et al., 1987). The tree responds by isolating the invading insect or fungus within a lesion of dead cells and secondary resin by autolysis of cells and formation of traumatic resin containing higher concentrations of monoterpenes and phenolics (Reid et al., 1967; Berryman, 1969, 1972; Shrimpton, 1973; Wong and Berryman, 1977; Wright et al., 1979; Raffa and Berryman, 1982a, 1982c, 1983; Hain et al., 1983; Croteau et al., 1987).

Croteau et al. (1987) identified elevated levels of monoterpenes and diterpene resin acids in stems of lodgepole pine inoculated with blue-stain fungus *Ceratocystis clavigera*, resulting in induced lesions and secondary resin production. The inoculated stems contained about three times more of the monoterpenes α-pinene, β-pinene, 3-carene, and β-phellandrene, but less limonene. *De novo* resin synthesis was indicated in the infected tissue since radiolabeled sucrose was incorporated up to 20 times faster into monoterpenes and up to 10 times faster into diterpene resin acids. Chitosan, a fungal wall fragment, induced monoterpene biosynthesis and increased levels of terpene cyclase enzyme which converted radiolabeled terpene precursors (geranyl pyrophosphate and farnesyl pyrophosphate) to labeled monoterpenes and sesquiterpene olefins, respectively (Croteau et al., 1987). Nearly all *S. ventralis* females leave their entrance holes in grand fir, *Abies grandis*, when hypersensitive reactions are evident (Berryman and Ashraf, 1970). Shorter egg galleries are made or fewer eggs are laid by females in lesions (Berryman and Ashraf, 1970; Paine and Stephen, 1988).

The hypersensitive wound reaction not only affects bark beetles directly, but inhibits symbiotic microorganisms from growing and killing the tree. Blue-stain and other symbiotic fungi are surrounded and isolated in the lesions (Wong and Berryman, 1977; Stephen and Paine, 1985; Paine and Stephen, 1987, 1988). Sometimes, phytopathogenic fungi (e.g., *Ceratocystis minor* var. *barrasii*) that are carried in the beetle's mycangium (*D. frontalis*) do not stimulate the hypersensitive response and thus spread to kill the tree (Paine and Stephen, 1987). Possibly the fungi secrete compounds, such as water-soluble glycans, that inhibit the plant's defensive hypersensitivity—as in potatoes (Doke and Tomiyama, 1980). Cobb et al. (1968) cultured four species of *Ceratocystis* fungi associated with bark beetles (species: *ips*, *minor*, *schrenkiana*, and *pilifera*) as well as the root pathogen fungus, *Fomes annosus*, during exposure to saturated atmospheres of oleoresin or monoterpenes of the host ponderosa pine. They found that all fungal species were inhibited in growth by oleoresin except *C. ips*, while all the monoterpenes (the five discussed above plus β-phellandrene and camphene) as well as undecane (present in Jeffrey pines) inhibited growth of the fungal species. Camphene and undecane were the least toxic, α-pinene was intermediate, while β-pinene, β-phellandrene, and 3-carene were more toxic, with myrcene and limonene the most toxic. However, the most toxic of all compounds tested was n-heptane (also the most volatile), a major constituent of digger pine, *P. sabiniana* (Mirov, 1961). When the monoterpenes were incorporated into the culture medium as well as in the vapor phase, myrcene and β-phellandrene appeared the most toxic to most species, but all of the monoterpenes and undecane reduced fungal growth (Cobb et al., 1968). They also indicated that (+)-α-pinene was more inhibitory than the (*S*)-enantiomer. The growth inhibition by monoterpenes was proposed to allow the tree time to synthesize phenols (Shain, 1967) such as pinosylvin (Anderson, 1962) that would kill the fungi (Cobb et al., 1968).

Bridges (1987) tested α-pinene and β-pinene on two mycangial fungi of *D.*

frontalis and the blue-stain fungi, *C. minor*. β-pinene inhibited growth of *C. minor*, both monoterpenes inhibited one mycangial fungi while the other fungus was stimulated. A phenylpropanoid, 4-allylanisole, from *P. taeda* resin inhibited growth of all fungi tested. Himejima et al. (1992) steam-distilled ponderosa pine oleoresin into a distillate of monoterpenes and sesquiterpenes and a residue of four diterpene acids. The individual monoterpenes of the distillate were not inhibitory to growth of several gram-positive bacteria (at 800 μg/ml) but did inhibit two species of fungi. Longifolene, a sesquiterpene, inhibited the gram-positive but not the gram-negative bacteria. Other species of common mold fungi were not affected by the monoterpenes, but the diterpene acid, abietic acid, was effective against three species of gram-positive bacteria. Although these species of microorganisms are not those associated with bark beetles, the results indicate that oleoresin constituents may be important as a general defense against microorganisms.

The carbon balance of a tree is the relative level of photosynthate available for growth, maintenance, and biosynthesis of defensive compounds. Christiansen et al. (1987) have reasoned that too little moisture (drought), insect defoliation, and root pathogens will reduce the amounts of carbon available for biosynthesis of primary and secondary resin. Mild drought may actually increase resistance of trees by lowering growth rates and shifting the use of photosynthate to the biosynthesis of defensive chemicals, while extended drought will increase the probability of bark beetle outbreaks due to the depleted carbon reserves (Christiansen et al., 1987). A similar theory of tree resistance and attack by southern pine beetles was developed by Lorio (1986).

5. Plant Compounds as Precursors of Semiochemicals

Bark beetle pheromones used in aggregation and for avoidance of competition consist of many varied structures (Fig. 5.5). Many of the same pheromonal compounds are used by species in the same genus, such as ipsenol, ipsdienol, and *cis*-verbenol in the genus *Ips*, or *exo*-brevicomin, frontalin, *trans*-verbenol and verbenone in the genus *Dendroctonus* [see Borden (1982), Birch (1984), and Byers (1989b)]. Some compounds such as *cis*- and *trans*-verbenol may be found in *Ips* (Scolytinae) as well as *Dendroctonus* and *Tomicus* (Hylesininae). However, *cis*-verbenol has so far only been proven as an aggregation pheromone component for species in the subfamily Scolytinae, whereas *trans*-verbenol has semiochemical activity only in the subfamily Hylesininae (Fig. 5.5). Verbenone, although inhibiting attraction response of most bark beetle species tested so far, is produced in significant amounts only by species of Hylesininae (primarily or exclusively by males).

The base structure of ipsenol, ipsdienol, and *E*-myrcenol resembles the plant monoterpene, myrcene; likewise, *cis*- and *trans*-verbenol as well as verbenone resemble α-pinene (Figs. 5.4 and 5.5). These structural similarities support the hypothesis, considered next, that in many cases bark beetles use plant compounds as precursors for their pheromone components. This hypothesis also seems logical in that a beetle would save energy by reducing the biosynthetic costs in the production of pheromones.

5.1. Precursors of Aggregation Pheromone Components

Aggregation pheromone components were first identified in bark beetles from male *Ips paraconfusus* as a synergistic blend of (*S*)-(−)-ipsenol, (*S*)-(+)-ipsdienol, and (4*S*)-*cis*-verbenol (Silverstein et al. 1966, 1967; Wood et al. 1968). Several other *Ips* species were soon discovered to produce and respond to various blends of these compounds (Vité et al., 1972). The similarity of chemical structure between a major host monoterpene, myrcene, and ipsenol and ipsdienol (Figs. 5.4 and 5.5) led Hughes (1974) to propose that myrcene was a precursor of these pheromone components in *Ips*. Exposure of males of *I. paraconfusus* to myrcene vapor resulted in their production of compounds with GC retention times identical to ipsenol and ipsdienol (Hughes, 1974). Byers et al. (1979) confirmed these identifications using gas chromatography and mass spectrometry (GC-MS) and behavioral assays, and reported the male-specific quantitative relationships between precursor vapor concentration and pheromone products. Hendry et al. (1980) radiolabeled myrcene and proved it could be directly converted to the pheromone components (Fig. 5.8). Earlier, Hughes (1974) had proposed that ipsdienol was directly converted to ipsenol, since topical application of ipsdienol on males resulted in ipsenol production. This was confirmed by deuterium-labeled ipsdienol (64% *d*) being converted to labeled ipsenol (25% *d*) (Fish et al., 1979). Intermediates such as ipsdienone and ipsenone (Fig. 5.8) also have been suggested in the biosynthetic pathways (Fish et al., 1979; Byers and Birgersson, 1990).

Another host monoterpene, (−)-α-pinene, in the vapor phase is converted to (*S*)-*cis*-verbenol in both sexes of *I. paraconfusus* (Fig. 5.9) (Renwick et al. 1976), and the relationship between vapor concentration and pheromone component production was quantified (Byers 1981b). The production of (−)-*cis*-verbenol and (+)-*trans*-verbenol in *I. typographus* and *I. amitinus* from the (−)- and (+)-enantiomers of α-pinene was shown to have a similar relationship by Klimetzek and Francke (1980). The ratio of *cis-/trans*-verbenols produced by *I. typographus* was consistent with the ratio of enantiomers of α-pinene in the host tree, which differed between trees of different regions and genotypes (Klimetzek and Francke, 1980; Lindström et al., 1989).

Based on the above studies, a paradigm was established that *I. paraconfusus*, and probably most other *Ips* species, use myrcene and (−)-α-pinene in their host tree as precursors to ipsenol and ipsdienol or *cis*-verbenol, respectively. The evolution of host selection behavior by *Ips* bark beetles could be influenced by the amounts of α-pinene and myrcene in the tree (Elkinton et al. 1980). The

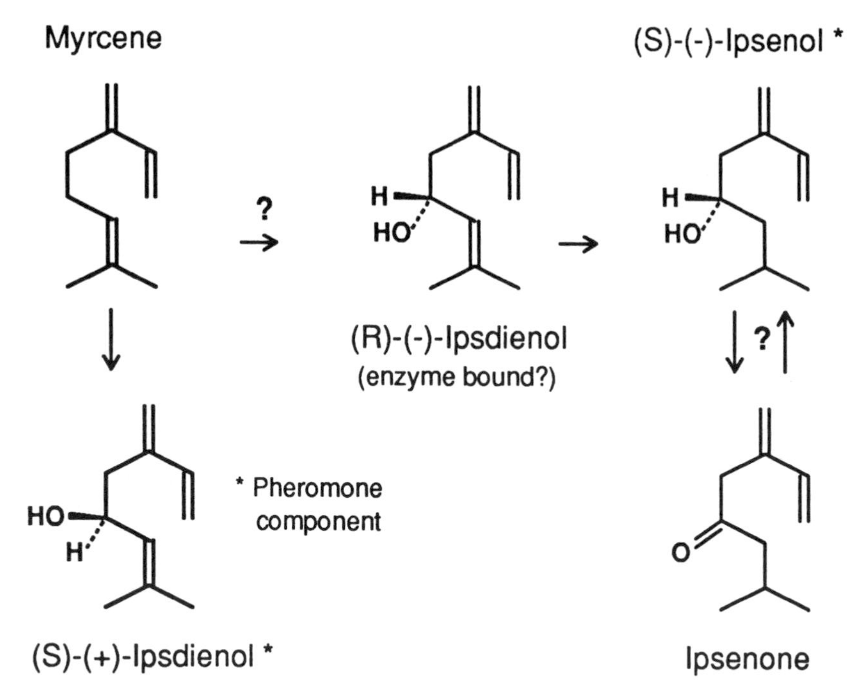

Figure 5.8. Proposed scheme for the conversion of the host-tree compound, myrcene, to the pheromone components (*S*)-(−)-ipsenol and (*S*)-(+)-ipsdienol in *Ips paraconfusus* based on radiolabeling experiments and enantiomers found in the male (Silverstein et al., 1966; Wood et al., 1968; Hughes, 1974; Renwick et al., 1976; Byers et al., 1979; Fish et al., 1979; Hendry et al., 1980; Byers, 1981; Byers and Wood, 1981). Conversion arrows with question marks have not been proven. (*R*)-(−)-ipsdienol does not accumulate in the hindgut but may occur as an enzyme-bound intermediate. However, the myrcene used in the biosynthetic pathways does not appear to come directly from the tree since males feeding in four species of pine with different titers of myrcene produced the same amounts of ipsenone, (*S*)-(−)-ipsenol, and (*S*)-(+)-ipsdienol as did males in *Pinus sabiniana*, which did not contain any detectable myrcene (Byers and Birgersson, 1990).

variation of monoterpenes among trees within a species (Mirov 1961; Smith 1964, 1967) might allow beetles to have evolved preferences for trees that had large amounts of precursor for use in biosynthesis of aggregation pheromone components. A second hypothesis is that tree genotypes may have evolved through natural selection which are lower in pheromone precursor monoterpenes as a means of resistance to bark beetles (Byers, 1989a). It is still not known whether selection pressure by bark beetle predation is severe enough to cause the evolution of trees with lower amounts of α-pinene as a means of resistance.

In contrast to the paradigm above, recent studies question the importance of myrcene precursor quantities within the host tree in the coevolution of aggregation

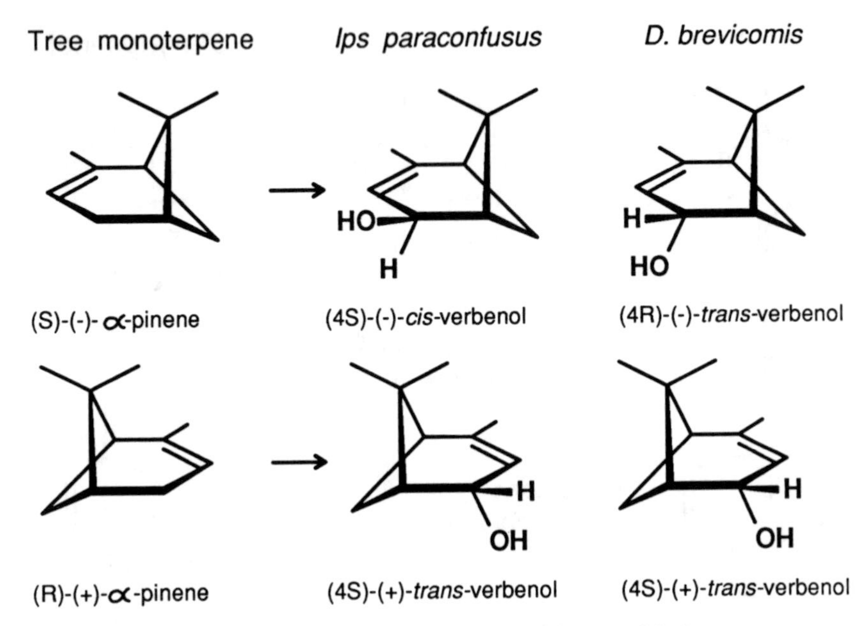

Figure 5.9. Stereoselective conversion of the two enantiomers of the host-tree monoterpene, α-pinene, to pheromone components (4*S*)-*cis*-verbenol in *Ips paraconfusus* and (4R)-*trans*-verbenol in *D. brevicomis*.

pheromone components in *Ips paraconfusus*. This is because males of *I. paraconfusus* produced almost identical amounts of the pheromone components ipsenol and ipsdienol when feeding in five different host pine species, regardless of the large differences in concentrations of myrcene in these trees (Byers and Birgersson, 1990). In fact, digger pine, *P. sabiniana*, had so little myrcene that it could not be detected by GC-MS, and it was calculated that a beetle would need to eat at least eight times its weight in toxic and repellent oleoresin in order to have any possibility of obtaining the required amounts of myrcene. Apparently in this case the expected energetic advantages of using a precursor from the host that is structurally similar to the pheromone components are outweighed by the need for the beetle to retain control over the ecologically critical pheromone system. The beetle could maintain control by means of either *de novo* biosynthesis or use of a simpler host precursor found in all potential hosts.

Streptomycin, a 70S ribosome-active antibiotic, when fed to *I. paraconfusus* inhibited their ability to produce ipsenol and ipsdienol, but the antibiotic had no affect on α-pinene conversion to *cis*-verbenol (Byers and Wood, 1981a; Hunt and Borden, 1989). However, *I. paraconfusus* reared axenically (without microorganisms) could still produce reduced amounts of ipsenol and ipsdienol (Conn et al., 1984; Hunt and Borden, 1989). Any symbiotic microorganisms involved in pheromone biosynthesis would have to be transovarially transmitted in order to survive the axenic rearing methods (Hunt and Borden, 1989). Juvenile hormone is also implicated in pheromone biosynthesis: topical application induces ipsenol and ipsdienol production in *I. paraconfusus* in the absence of external myrcene (Hughes and Renwick, 1977). Other pheromones such as verbenone, *exo*-brevicomin, and frontalin are found in emerged *D. brevicomis* before they land on the tree (Byers et al., 1984).

Exposure of *Ips duplicatus* to myrcene vapor causes the beetle to produce small amounts of its pheromone component ipsdienol but not *E*-myrcenol, the other component (Schlyter et al., 1992). Both compounds are expected to be made from myrcene (Pierce et al., 1987). Ivarsson et al. (1993) found that the biosynthesis of both *E*-myrcenol and ipsdienol from mevalonate in *I. duplicatus* could be blocked by the enzyme inhibitor compactin, but the production of *cis*-verbenol from α-pinene in *I. typographus* was not affected by the inhibitor. These studies, and those on *I. paraconfusus* above, indicate that the major pathway in nature for biosynthesis of the pheromone components ipsdienol, ipsenol, and *E*-myrcenol in *Ips* is *de novo* from mevalonate.

5.2. *Precursors of Pheromones and Allomones for Avoiding Intra- and Interspecific Competition*

As mentioned in section 2.4, many prevalent bark beetle species in *Dendroctonus* and *Tomicus* produce verbenone, which affects behavior in their own species as well as competing species in these genera and others (*Ips*, *Pityogenes*) by reducing their response to aggregation pheromone. Exposure of male and female *D. brevicomis* to (+)- and (−)- enantiomers of α-pinene for several hours caused them to produce large amounts of (+)- and (−)-*trans*-verbenol in their hindguts (Fig. 5.9) (Byers, 1983a). However, the biosynthesis of verbenone in these beetles was not affected by exposure to α-pinene enantiomers, even though verbenone is structurally similar to α-pinene (Figs. 5.4 and 5.5) and is found in males landing on trees (Renwick and Vité, 1968, 1970; Byers et al., 1984). The (−)-enantiomer of *trans*-verbenol inhibits female *D. brevicomis* from entering holes and may serve as a signal to arriving females that they should avoid areas colonized by conspecifics (Byers, 1983a).

Both verbenone and *trans*-verbenol are produced by *D. brevicomis* beetles in the greatest amounts early in colonization so it was suggested that they play a role in reducing intraspecific competition (Byers et al., 1984) as well as interspecific competition with *I. paraconfusus* (Byers and Wood, 1980). However, verbenone (and possibly *trans*-verbenol) are also produced increasingly in aging logs infested by bark beetles (Birgersson and Bergström, 1989; Byers et al., 1989c). A common bacterium, *Bacillus cereus*, also isolated from *I. paraconfusus* can make *cis*- and *trans*-verbenol from α-pinene (Brand et al., 1975). Several yeasts from *I. typographus* can interconvert the verbenols, and when grown in a phloem medium they produced the oxygenated monoterpenes α-terpineol, borneol, myrtenol, terpenene-4-ol, and *trans*-pinocarveol, compounds also shown to be released increasingly from bark beetle holes with age of attack (Leufvén et al., 1984, 1988; Birgersson and Bergström, 1989). A mycangial fungus grown in culture media converted alcohol products of α-pinene to verbenone, the end product (Brand et al., 1976). These microorganisms are introduced by bark beetles during colonization and after buildup may release verbenone, thus signalling to flying beetles that immigrating to these bark substrates would entail competition with established bark beetle colonies.

Myrcene and α-pinene may be used as precursors of pheromones and allomones in some species of conifer-feeding bark beetles. *D. brevicomis*, *I. paraconfusus*, and *I. pini* occur sympatrically in California and Oregon and compete for ponderosa pine bark. Myrcene is converted sex-specifically by male *D. brevicomis* to (+)-ipsdienol, also a pheromone component of its competitor *I. paraconfusus* (Hughes 1973; Byers, 1982). (+)-Ipsdienol inhibits response of both *D. brevicomis* and *I. pini* to their synthetic aggregation pheromones (Birch et al., 1980a; Lanier et al., 1980; Byers, 1982). In Europe, ipsdienol, a pheromone component of *I. duplicatus* could act as an allomone to inhibit response of *I. typographus* (Byers et al., 1990; Schlyter et al., 1992). Mated males of *I. typographus* produce small amounts of ipsdienol and ipsenol during colonization that might function in avoiding competition (Birgersson et al., 1984; Birgersson and Leufvén, 1988). Although ipsdienol was earlier thought to be an aggregation pheromone component of *I. typographus* (Bakke et al., 1977), other studies have shown the compound to inhibit attraction of these beetles to *cis*-verbenol and methyl butenol, the most potent components (Schlyter et al., 1987b, 1087c).

6. Secondary Plant Compounds and Bark Beetle-Tree Coevolution

Secondary plant compounds are those chemicals not believed to be necessary for the basic metabolism of plant cells (Whittaker, 1970). Since the compounds occur in some plant families, often of very different lineage, and not in other families, they appear to have evolved independently for the defense against a common enemy such as fungi or insects (Feeny, 1975). Secondary plant compounds include many exotic structures: phenolics, cinnamic acids, coumarins, lactones, quinones, and glycosides of the foregoing, as well as furan and pyranes, anthraquinones, flavonoids, anthocyanin, polymeric phenolics (tannins), monoterpenoids, sesquiterpenoids, di- and triterpenoids, resin acids, steroids, cardiac glycosides, ecdysone analogues, and alkaloids.

Secondary plant compounds important for host finding from long range would necessarily be volatile. Some of the most common volatiles in conifers are the monoterpenes and thus it is not surprising that bark beetles are often attracted to these chemicals even though in high concentrations they are toxic. It could be expected that volatile monoterpenes shown to be important for a species in one geographic region might not be attractive to populations of this species in another region, if the trees in the second region have different proportions of the monoterpenes. The monoterpenes do vary in ponderosa pine throughout the western United States (Smith, 1964, 1966, 1967, 1968, 1969). This could explain why different populations of *D. brevicomis* may have responded differently to myrcene or 3-carene as synergists of pheromone components (Bedard et al., 1969, 1970; Pitman, 1969).

Quantities of volatile compounds such as ethanol and verbenone that are released from killed trees later during colonization by beetles and microorganisms are probably not regulated by the tree's genetics. Thus, trees are not under selection pressure from bark beetles because the tree has no control over how the insects use these compounds (as pheromones, allomones or kairomones) to indicate potential competition or decaying, unsuitable host areas. Kairomones, in contrast to pheromones and allomones, are not used in chemical communication, in which both parties, sender and receiver, must gain a benefit (Burghardt, 1970).

Host-plant specificity by insects may be the result of plant-insect coevolution of chemical defenses and interspecific competition. Those insect species that do not adapt to a genetic change in host-plant chemistry during evolution are not able to eat the plant as effectively and thus are at a competitive disadvantage to those species that adapt first. The consequence is that one or a few species now become prevalent (major pests) because of reduced interspecific competition.

Hughes (1973, 1974) hypothesized that aggregating pheromones in *Dendroctonus* and *Ips* are "waste products from the metabolism of terpenes that have secondarily been utilized as chemical messengers." *D. ponderosae* have an enzyme system that indiscriminately hydroxylates monoterpenes on allylic methyl groups that are *E* to a methylene or vinyl group as a way of detoxifying monoterpenes (Pierce et al., 1987). The products may also be more easily excreted due to the increased solubility in aqueous fluids. Some of the detoxification products of myrcene have been used as pheromones, but there appears to have been a selection for beetles that can synthesize these pheromone components from mevalonate (Byers and Birgersson, 1990; Ivarsson et al., 1993). In *I. paraconfusus* the production of ipsenol and ipsdienol in males is not primarily for detoxification of myrcene since the conversion is sex-specific, inhibited by streptomycin or mating with females, and induced by juvenile hormone. On the other hand, the *cis*-verbenol synthesis is primarily a detoxification process since the compound is not sex-specific (although males produce more *cis*-verbenol) and juvenile hormone, streptomycin or mating have no effect on its production (Hughes and Renwick, 1977; Byers et al., 1979; Byers 1981a, 1981b, 1983b; Byers and Wood, 1981a; Hunt and Borden, 1989). Other than *cis*-verbenol and possibly *trans*-verbenol, no other bark beetle pheromone components (Fig. 5.5) are thought to be primarily a detoxification of monoterpenes, but rather are synthesized *de novo* from acetate or mevalonate (Vanderwel and Oehlschlager, 1987; Lanne et al., 1989; Ivarsson et al., 1993).

A case where coevolution may still occur is in *I. paraconfusus* and *I. typographus* that rely on host tree (−)-α-pinene as the precursor of *cis*-verbenol. However, *I. typographus* is relatively insensitive to large variations in the proportion of *cis*-verbenol to methyl butenol in the pheromone blend (Schlyter et al. 1987c), indicating that most Norway spruce are adequate hosts regardless of their α-pinene content. Thus, a tree genotype that produced less (−)-α-pinene might not gain an advantage since beetles could still produce adequate pheromone to cooperate in killing the tree. Certainly, the large number of beetle generations relative to the tree would allow the genes of beetles to track any changes in the host and compensate in production and/or response. Presumably the benefits of α-pinene for the tree, such as its toxicity to beetles and microorganisms discussed earlier, also would counter any tendency to select trees with a lower titer of α-pinene.

Several authors have suggested that some plants have evolved an indirect tritrophic mechanism of resistance in which they release compounds (synomones) after being fed upon that are attractive to predators or parasites of their herbivores (Turlings et al., 1990; Whitman and Eller, 1990). The plant could gain benefits if their herbivores are not attracted by these same host volatiles. However, the same host compounds that are attractive to the predators and parasites of bark beetles are often attractive to bark beetles as well. For example, *Enoclerus lecontei* (Cleridae) and *Temnochila virescens* (Ostomidae) prey on several bark beetles such as *D. brevicomis* and *I. paraconfusus* in California. The predators are attracted to several monoterpenes or n-heptane (Jeffrey pine) (Mirov, 1961) from host trees of bark beetles (Rice, 1969; Pitman and Vité, 1971). A clerid predator in Europe, *Thanasimus formicarius* preys on *I. typographus* of Norway spruce and *T. piniperda* of Scots pine and is also attracted to α-pinene and other monoterpenes from these conifers (Schroeder, 1988; Schroeder and Lindelöw, 1989). The dipteran parasites *Medetera aldrichii* of *D. pseudotsugae* and *M. bistriata* of *D. frontalis* are attracted to the host-tree monoterpene, α-pinene, presumably aiding in their location of attacked trees (Williamson, 1971; Fitzgerald and Nagel, 1972). These plant monoterpenes are attractive to many bark beetles, or are synergists of attractive pheromone components (see section 2.3). Thus, it seems doubtful that the tree would benefit by producing chemicals that attract bark beetle enemies as well as predators and parasites of these beetles. In this case the host compounds are better thought of as kairomones, i.e., chemicals that are used by receiving individuals to gain advantages. Kairomones are not dispensed with by the plant over evolutionary time because they confer advantages, such as insect resistance, that outweigh any disadvantages of them attracting additional herbivores.

Other parasites and predators of bark beetles are most strongly attracted to volatiles produced by their host bark beetle, although some of the attractive compounds may be derived from monoterpene precursors in the tree. *Temnochila virescens* feeds on *D. brevicomis* and the predator is almost as sensitive to *exo*-brevicomin as is the bark beetle to this pheromone component (Pitman and Vité, 1971; Byers, 1988). However, in Texas *T. virescens* is more attracted to a mixture of *Ips* components (ipsdienol, ipsenol, and *cis*-verbenol) than to *exo*-brevicomin (Billings and Cameron, 1984). Clerid beetles such as *Thanasimus formicarius* are generalist predators of bark beetles, feeding on several species and are attracted to (or perceive) compounds of *Ips* (ipsenol, ipsdienol, *cis*-verbenol), *Trypodendron* (lineatin), and *Dendroctonus/Dryocoetes* (*exo*-brevicomin) (Bakke and Kvamme, 1981; Hansen, 1983; Kohnle and Vité, 1984; Lanne et al., 1987; Tommerås, 1988).

7. Conclusions

Of the 6,000 bark beetle species worldwide, in a particular geographic area there are usually from only a few to some tens of species that colonize any given species of tree, and then only one or a few species that can attack and kill the tree. Each host-tree species has a large variety of chemicals, some of which affect the success of the bark beetle in finding and colonizing its host tree. Bark beetles probably orient to attractive semiochemical sources during flight using odor-modulated anemotaxis, as in moths, but little is known about this process. It is better understood how beetles orient upwind during walking to attractive sources. Most studies have observed chemotaxis in arenas for the purpose of isolating pheromones rather than from the standpoint of basic behavior.

Bark beetles find suitable host trees by orienting to host odors, especially ethanol and monoterpenes, as well as to aggregation pheromones. However, very few studies could be characterized as complete or rigorous because the attractive compounds were discovered usually by screening of semiochemicals previously known for other species. Other studies have selected compounds for testing based on their presence in the insect or related species, but it is likely that many bioactive compounds have been overlooked. Also, blends of compounds have rarely been tested for synergism. Tree volatiles that attract predators and parasites of bark beetles are often the same as those that attract their host, i.e., most of these chemicals have been discovered by chance when testing compounds on bark beetles. Feeding stimulants and deterrents of conifer bark beetles have been isolated in various solvent fractions but not identified. Several compounds that elicit or deter feeding in deciduous bark beetles have been identified, but undoubtedly many behaviorally active compounds remain to be discovered. It is likely that behavioral responses of bark beetles within a species to semiochemicals may vary geographically as well as the semiochemicals produced by the bark beetles and the host trees.

Fractionation of a biological extract by chromatographic methods (usually GC) and then recombination of certain fractions with an additive method has been used to test for synergism among semiochemicals in a behavioral bioassay. This method was used to isolate some of the first multicomponent pheromones of bark beetles (Silverstein et al., 1966, 1967; 1968; Pearce et al., 1975). However, due to the substantial work involved with these classical isolation methods, most studies have discovered semiochemicals by screening or comparative methods which are inherently less rigorous and are dependent on chromatographic resolution. The subtractive method, where each of the fractions is subtracted from the blend and tested such that blends with lowered activity indicate subtracted synergists, should aid in isolation of synergistic semiochemicals that otherwise have been hard to detect (Byers et al., 1990a; Byers, 1992c).

Future studies should be careful to report the release rates of test volatiles, and in many cases these should be adjusted to coincide with natural rates (Byers, 1988). This means that measurements of volatile release of semiochemicals must be done in many bark beetle systems for both host- and beetle-released semiochemicals (Birgersson and Bergström, 1989). When testing semiochemicals in the field, the spatial and temporal variation of responding insect populations with respect to trap placement may lead to erroneous conclusions. To counter this potential problem, relatively numerous trap replications have been previously employed; however, the mechanical slow rotation of a pair of traps (1–2 rph at 6-m separation) can even this catch variation (Byers et al., 1990b).

Resistance of trees has been studied for many years and monoterpenes, such as limonene, are implicated in their resistance to bark beetles and their symbiotic microorganisms (mostly fungi). However, there has been little recent toxicological work and the relative importance of the purported toxins remains to be established. Also, the monoterpenes have been tested at much higher vapor concentrations than those in nature, and they have not been evaluated in diets. Synergism or interactions between various toxins have not been investigated. Also, geographic variation in toxicity of host compounds has been little studied. A correlation between high-limonene trees and historical "predation" by bark beetles has been suggested as an example of host-chemical evolution. However, more studies are needed in stands with ongoing outbreaks of bark beetle to determine if natural selection can alter the genetic frequencies in the population of trees, and at the same time if populations of bark beetle change their tolerance to particular monoterpenes that were initially most toxic.

Host-tree monoterpenes α-pinene and myrcene can be converted by a simple hydroxylation to ipsenol, ipsdienol, *cis*-verbenol and *trans*-verbenol, pheromone components of bark beetles. However, *I. paraconfusus* is able to make the same amounts of ipsenol and ipsdienol regardless of the myrcene titer in the host tree, suggesting the major pathway is *de novo*. Recent studies in *Ips* species also suggest that pheromonal analogues of myrcene may not be derived primarily from myrcene but by synthesis from mevalonate. Although ipsenol/ipsdienol and

E-myrcenol biosynthesis in some species of bark beetle are probably not coevolving with myrcene in the tree, it is possible that *cis*- and *trans*-verbenol biosynthesis may coevolve with α-pinene levels in hosts. Both *cis*-verbenol and *trans*-verbenol appear to be directly produced in bark beetles by conversion of α-pinene enantiomers from the host tree. However, verbenone, an inhibitor of aggregation in many bark beetle species, may not be directly converted from α-pinene. Other bark beetle pheromone components are probably biosynthesized from small molecules into the more complex structures in several or more different biosynthetic pathways.

Evolution of tree chemistry in response to predation by bark beetles is best supported in studies of host compounds that are toxic to bark beetles or that deter feeding. Bark beetles have also coevolved detoxification mechanisms for the toxic monoterpenes, some of which have been secondarily utilized as pheromone components. Volatile host attractants can be termed kairomones, and there is little evidence that trees evolve these compounds to repel herbivores not adapted to this potential host, since the same compounds attract their herbivores. The compounds probably are beneficial in some way to the tree and cannot be dispensed with even though bark beetles (and some of their predators and parasites) have evolved to utilize the compounds as kairomones. Host-tree chemistry affects most aspects of bark beetle biology, moreover, bark beetles probably differentially affect survival of host trees and alter genotypic frequencies and host chemistry both at the microevolutionary scale (cycling of endemic and epidemic insect populations) and at the macroevolutionary level (host-tree selection in response to new species of bark beetle).

Acknowledgments

The chapter was improved by reviews from O. Anderbrant and F. Schlyter, Department of Animal Ecology, Lund University; G. Birgersson and P. Ivarsson, Department of Chemical Ecology, Göteborg University; L.M. Schroeder, A. Lindelöw, and J. Weslien, Department of Forest Entomology, Sweden's Agricultural University, Sweden; K. Hobson, USDA Forest Service, Logan, Utah; and the editors. The review was supported in part by a grant from SJFR (Sweden).

References

Akers, R. P. (1989) Counterturns initiated by decrease in rate of increase of concentration: possible mechanism of chemotaxis by walking female *Ips paraconfusus* bark beetles. *J. Chem. Ecol.* 15: 183–208.

Akers, R. P. and Wood, D. L. (1989a) Olfactory orientation responses by walking female *Ips paraconfusus* bark beetles I. Chemotaxis assay. *J. Chem. Ecol.* 15: 3–24.

Akers, R. P. and Wood, D. L. (1989b) Olfactory orientation responses by walking female

Ips paraconfusus bark beetles II. In an anemotaxis assay. *J. Chem. Ecol.* 15: 1147–1160.

Anderbrant, O. (1985) Dispersal of reemerged spruce bark beetles, *Ips typographus* (Coleoptera, Scolytidae): a mark-recapture experiment. *Z. Angew. Entomol.* 99: 21–25.

Anderson, A. B. (1962) The influence of extractives on tree properties II. Ponderosa pine (*Pinus ponderosa* Dougl.). *J. Inst. Wood Sci.* 10: 29–47.

Anderson, A. B., Riffer, R., and Wong, A. (1969) Monoterpenes, fatty and resin acids of *Pinus ponderosa* and *Pinus jeffreyi*. *Phytochemistry* 8: 873–875.

Anderson, W. W., Berisford, C. W., and Kimmich, R. H. (1979) Genetic differences among five populations of the southern pine beetle. *Ann. Entomol. Soc. Am.* 72: 323–327.

Anderson, W. W., Berisford, C. W., Turnbow, R. H. and Brown, C. J. (1983) Genetic differences among populations of the black turpentine beetle, *Dendroctonus terebrans*, and an engraver beetle, *Ips calligraphus* (Coleoptera: Scolytidae). *Ann. Entomol. Soc. Am.* 76: 896–902.

Atkins, M. D. (1969) Lipid loss with flight in the Douglas-fir beetle. *Can. Entomol.* 101: 164–165.

Austarå, O., Bakke, A., and Midtgaard, F. (1986) Response in *Ips typographus* to logging waste odors and synthetic pheromones. *J. Appl. Entomol.* 101: 194–198.

Baker, B. H., Hostetler, B. B., and Furniss, M. M. (1977) Response of eastern larch beetle (Coleoptera: Scolytidae) in Alaska to its natural attractant and to Douglas-fir beetle pheromones. *Can. Entomol.* 109: 289–294.

Baker, J. E. and Norris, D. M. (1967) A feeding stimulant for *Scolytus multistriatus* (Coleoptera: Scolytidae) isolated from the bark of *Ulmus americana*. *Ann. Entomol. Soc. Am.* 60: 1213–1215.

Baker, J. E. and Norris, D. M. (1968) Further biological and chemical aspects of host selection by *Scolytus multistriatus*. *Ann. Entomol. Soc. Am.* 61: 1248–1255.

Baker, T. C. (1989) Sex pheromone communication in the Lepidoptera: new research progress. *Experientia* 45: 248–262.

Baker, T. C., Hansson, B. S., Löfstedt, C., and Löfqvist, J. (1988) Adaptation of antennal neurones in moths is associated with cessation of pheromone-mediated upwind flight. *Proc. Natl. Acad. Sci. USA* 85: 9826–9830.

Bakke, A. (1975) Aggregation pheromone in the bark beetle *Ips duplicatus* (Sahlberg). *Norw. J. Entomol.* 22: 67–69.

Bakke, A. (1981) Inhibition of the response in *Ips typographus* to the aggregation pheromone; field evaluation of verbenone and ipsenol. *Z. Angew. Entomol.* 92: 172–177.

Bakke, A. (1983) Dosage response of the ambrosia beetle *Trypodendron lineatum* (Oliver) (Coleoptera, Scolytidae) to semiochemicals. *Z. Angew. Entomol.* 95: 158–161.

Bakke, A. and Kvamme, T. (1981) Kairomone response in *Thanasimus* predators to pheromone components of *Ips typographus*. *J. Chem. Ecol.* 7: 305–312.

Bakke, A., Fryen, P., and Skattebl, L. (1977) Field response to a new pheromonal compound isolated from *Ips typographus*. *Naturwissenschaften* 64: 98.

Bedard, W. D., Tilden, P. E., Wood, D. L., Silverstein, R. M., Brownlee, R. G., and Rodin, J. O. (1969) Western pine beetle: field response to its sex pheromone and a synergistic host terpene, myrcene. *Science* 164: 1284–1285.

Bedard, W. D., Silverstein, R. M. and Wood, D. L. (1970) Bark beetle pheromones. *Science* 167: 1638–1639.

Begley, M.J. and Grove, J.F. 1985. Metabolic products of *Phomopsis oblonga* I. 3a,5a,6,7,8,9,9a,9b-octahydro-7,9b-dimethylnaphtho-1,2-c-furan-1-3h-one oblongolide. *J. Chem. Soc. Perkin Trans. I.* 0: 861–864.

Bennett, R. B. and Borden, J. H. (1971) Flight arrestment of tethered *Dendroctonus pseudotsugae* and *Trypodendron lineatum* (Coleoptera: Scolytidae) in response to olfactory stimuli. *Ann. Entomol. Soc. Am.* 64: 1273–1286.

Berenbaum, M. R. and Isman, M. B. (1989) Herbivory in holometabolous and hemimetabolous insects: Contrasts between Orthoptera and Lepidoptera. *Experientia* 45: 229–236.

Berryman, A. A. (1969) Responses of *Abies grandis* to attack by *Scolytus ventralis* (Coleoptera: Scolytidae). *Can. Entomol.* 101: 1033–1041.

Berryman, A. A. (1972) Resistance of conifers to invasion by bark beetle-fungi associations. *Bioscience* 22: 598–602.

Berryman, A. A. and Ashraf, M. (1970) Effects of *Abies grandis* resin on the attack behavior and brood survival of *Scolytus ventralis* (Coleoptera: Scolytidae). *Can. Entomol.* 102: 1229–1236.

Berryman, A. A., Dennis, B., Raffa, K. F., and Stenseth, N. C. (1985) Evolution of optimal group attack with particular reference to bark beetles (Coleoptera: Scolytidae). *Ecology* 66: 898–903.

Billings, R. F. and Cameron, R. S. (1984) Kairomonal responses of Coleoptera: *Monochamus titillator* (Cerambycidae), *Thanasimus dubius* (Cleridae) and *Temnochila virescens* (Trogositidae) to behavioral chemicals of southern pine bark beetles *Dendroctonus frontalis* (Coleoptera: Scolytidae). *Environ. Entomol.* 13: 1542–1548.

Birch, M. C. (1984) Aggregation in bark beetles. In: *Chemical Ecology of Insects* (Bell, W. J. and Cardé, R. T., eds.) pp. 331–353. Chapman and Hall, London.

Birch, M. C., Light, D. M., Wood, D. L., Browne, L. E., Silverstein, R. M., Bergot, B. J., Ohloff, G., West, J. F. and Young, J. C. (1980a) Pheromonal attraction and allomonal interruption of *Ips pini* in California by the two enantiomers of ipsdienol. *J. Chem. Ecol.* 6: 703–717.

Birch, M. C., Svihra, P., Paine, T. D. and Miller, J. C. (1980b) Influence of chemically mediated behavior on host tree colonization by four cohabiting species of bark beetles. *J. Chem. Ecol.* 6: 395–414.

Birgersson, G. and Bergström, G. (1989) Volatiles released from individual spruce bark beetle entrance holes: quantitative variations during the first week of attack. *J. Chem. Ecol.* 15: 2465–2484.

Birgersson, G. and Leufvén, A. (1988) The influence of host tree response to *Ips typographus* and fungal attack on production of semiochemicals. *Insect Biochem.* 18: 761–770.

Birgersson, G., Schlyter, F., Löfqvist, J., and Bergström, G. (1984) Quantitative variation of pheromone components in the spruce bark beetle *Ips typographus* from different attack phases. *J. Chem. Ecol.* 10: 1029–1055.

Birgersson, G., Byers, J. A., Bergström, G., and Löfqvist, J. (1990) Production of pheromone components, chalcogran and methyl (*E,Z*)-2,4-decadienoate, in the spruce engraver *Pityogenes chalcographus*. *J. Insect Physiol.* 36: 391–395.

Bombosch, S., Engler, I., Gossenauer, H., and Herrmann, B. (1985) On the role of pheroprax influencing the settlement of *Ips typographus* on spruce. *Z. Angew. Entomol.* 100: 458–463.

Bordasch, R. P. and Berryman, A. A. (1977) Host resistance to the fir engraver beetle, *Scolytus ventralis* (Coleoptera: Scolytidae) 2. repellency of *Abies grandis* resins and some monoterpenes. *Can. Entomol.* 109: 95–100.

Borden, J. H. (1982) Aggregation pheromones. In: *Bark beetles in North American Conifers: A System for the Study of Evolutionary Biology* (Mitton, J. B. and Sturgeon, K. M., eds.) pp. 74–139. Univ. Texas Press, Austin, Tex.

Borden, J. H. and Wood, D. L. (1966) The antennal receptors and olfactory response of *Ips confusus* (Coleoptera: Scolytidae) to male sex attractant in the laboratory. *Ann. Entomol. Soc. Am.* 59: 253–261.

Borden, J. H., Handley, J. R., McLean, J. A., Silverstein, R. M., Chong, L., Slessor, K. N., Johnston, B. D., and Schuler, H. R. (1980) Enantiomer-based specificity in pheromone communication by two sympatric *Gnathotrichus* species (Coleoptera: Scolytidae). *J. Chem. Ecol.* 6: 445–456.

Borden, J. H., Chong, L., Slessor, K. N., Oehlschlager, A. C., Pierce, Jr., H. D., and Lindgren, B. S. (1981) Allelochemic activity of aggregation pheromones between three sympatric species of ambrosia beetles (Coleoptera: Scolytidae). *Can. Entomol.* 113: 557–563.

Borden, J. H., King, C. J., Lindgren, S., Chong, L., Gray, D. R., Oehlschlager, A. C., Slessor, K. N., and Pierce, Jr., H. D. (1982) Variation in response of *Trypodendron lineatum* from two continents to semiochemicals and trap form. *Environ. Entomol.* 11: 403–408.

Borden, J. H., Hunt, D. W. A., Miller, D. R., and Slessor, K. N. (1986) Orientation in forest Coleoptera: An uncertain outcome of responses by individual beetles to variable stimuli. In: *Mechanisms in Insect Olfaction* (Payne, T. L., Birch, M. C., and Kennedy, C. E. J., eds.) pp. 97–109. Clarendon Press, Oxford, U.K.

Borden, J. H., Pierce, A. M., Pierce, Jr., H. D., Chong, L. J., Stock, A. J., and Oehlschlager, A. C. (1987) Semiochemicals produced by western balsam bark beetle *Dryocoetes confusus* Swaine (Coleoptera: Scolytidae). *J. Chem. Ecol.* 13: 823–836.

Borg, T. K. and Norris, D. M. (1971) Penetration of tritiated catechol: A feeding stimulant into chemo receptor sensilla of *Scolytus multistriatus* (Coleoptera: Scolytidae). *Ann. Entomol. Soc. Am.* 64: 544–547.

Botterweg, P. F. (1982) Dispersal and flight behavior of the spruce bark beetle *Ips typographus* in relation to sex, size and fat content. *Z. Angew. Entomol.* 94: 466–489.

Brand, J. M., Bracke, J. W., Markovetz, A. J., Wood, D. L., and Browne, L. E. (1975) Production of verbenol pheromone by a bacterium isolated from bark beetles. *Nature* 254: 136–137.

Brand, J. M., Bracke, J. W., Britton, L. N., Markovetz, A. J., and Barras, S. J. (1976) Bark beetle pheromones: production of verbenone by a mycangial fungus of *Dendroctonus frontalis*. *J. Chem. Ecol.* 2: 195–199.

Bridges, J. R. (1987) Effects of terpenoid compounds on growth of symbiotic fungi associated with the southern pine beetle. *Phytopathology* 77: 83–85.

Bridges, J. R. and Perry, T. J. (1985) Effects of mycangial fungi on gallery construction and distribution of bluestain in southern pine beetle *Dendroctonus frontalis* infested pine bolts. *J. Entomol. Sci.* 20: 271–275.

Bridges, J. R., Nettleton, W. A., and Connor, M. D. (1985) Southern pine beetle *Dendroctonus frontalis* (Coleoptera: Scolytidae) infestations without the bluestain fungus *Ceratocystis minor*. *J. Econ. Entomol.* 78: 325–327.

Brown, M. W., Nebeker, T. E., and Honea, C. R. (1987) Thinning increases loblolly pine vigor and resistance to bark beetles. *South. J. Appl. For.* 11: 28–31.

Burghardt, G. M. (1970) Defining "communication." In: *Communication by Chemical Signals* (Johnstron, Jr., J. W., Moulton, D. G. and Turk, A., eds.) pp. 5–18. Appleton, New York.

Byers, J. A. (1981a) Effect of mating on terminating aggregation during host colonization in the bark beetle, *Ips paraconfusus*. *J. Chem. Ecol.* 7: 1135–1147.

Byers, J. A. (1981b) Pheromone biosynthesis in the bark beetle, *Ips paraconfusus*, during feeding or exposure to vapours of host plant precursors. *Insect Biochem.* 11: 563–569.

Byers, J. A. (1982) Male-specific conversion of the host plant compound, myrcene, to the pheromone, (+)-ipsdienol, in the bark beetle, *Dendroctonus brevicomis*. *J. Chem. Ecol.* 8: 363–372.

Byers, J. A. (1983a) Bark beetle conversion of a plant compound to a sex-specific inhibitor of pheromone attraction. *Science* 220: 624–626.

Byers, J. A. (1983b) Influence of sex, maturity and host substances on pheromones in the guts of the bark beetles, *Dendroctonus brevicomis* and *Ips paraconfusus*. *J. Insect Physiol.* 29: 5–13.

Byers, J. A. (1983c) Sex-specific responses to aggregation pheromone: Regulation of colonization density by the bark beetle *Ips paraconfusus*. *J. Chem. Ecol.* 9: 129–142.

Byers, J. A. (1984) Nearest neighbor analysis and simulation of distribution patterns indicates an attack spacing mechanism in the bark beetle, *Ips typographus* (Coleoptera: Scolytidae). *Environ. Entomol.* 13: 1191–1200.

Byers, J. A. (1988) Novel diffusion-dilution method for release of semiochemicals: testing pheromone component ratios on western pine beetle. *J. Chem. Ecol.* 14: 199–212.

Byers, J. A. (1989a) Behavioral mechanisms involved in reducing competition in bark beetles. *Holarc. Ecol.* 12, 466–476.

Byers, J. A. (1989b) Chemical ecology of bark beetles. *Experientia* 45: 271–283.

Byers, J. A. (1992a) Attraction of bark beetles, *Tomicus piniperda*, *Hylurgops palliatus*, and *Trypodendron domesticum* and other insects to short-chain alcohols and monoterpenes. *J. Chem. Ecol.* 18: 2385–2402.

Byers, J. A. (1992b) Dirichlet tessellation of bark beetle spatial attack points. *J. Anim. Ecol.* 61: 759–768.

Byers, J. A. (1992c) Optimal fractionation and bioassay plans for isolation of synergistic chemicals: the subtractive-combination method. *J. Chem. Ecol.* 18: 1603–1621.

Byers, J. A. (1993a) Avoidance of competition by spruce bark beetles, *Ips typographus* and *Pityogenes chalcographus*. *Experientia* 49: 272–275.

Byers, J. A. (1993b) Simulation and equation models of insect population control by pheromone-baited traps. *J. Chem. Ecol.* 19: 1939–1956.

Byers, J. A. and Birgersson, G. (1990) Pheromone production in a bark beetle independent of myrcene precursor in host pine species. *Naturwissenschaften* 77: 385–387.

Byers, J. A. and Löfqvist, J. (1989) Flight initiation and survival in the bark beetle *Ips typographus* (Coleoptera: Scolytidae) during the spring dispersal. *Holarc. Ecol.* 12: 432–440.

Byers, J. A. and Wood, D. L. (1980) Interspecific inhibition of the response of the bark beetles, *Dendroctonus brevicomis* and *Ips paraconfusus*, to their pheromones in the field. *J. Chem. Ecol.* 6: 149–164.

Byers, J. A. and Wood, D. L. (1981a) Antibiotic-induced inhibition of pheromone synthesis in a bark beetle. *Science*, 213, 763–4.

Byers, J. A. and Wood, D. L. (1981b) Interspecific effects of pheromones on the attraction of the bark beetles, *Dendroctonus brevicomis* and *Ips paraconfusus* in the laboratory. *J. Chem. Ecol.*, 7, 9–18.

Byers, J. A., Wood, D. L., Browne, L. E., Fish, R. H., Piatek, B., and Hendry, L. B. (1979) Relationship between a host plant compound, myrcene and pheromone production in the bark beetle, *Ips paraconfusus*. *J. Insect Physiol.* 25: 477–482.

Byers, J. A., Wood, D. L., Craig, J., and Hendry, L. B. (1984) Attractive and inhibitory pheromones produced in the bark beetle, *Dendroctonus brevicomis*, during host colonization: Regulation of inter- and intraspecific competition. *J. Chem. Ecol.* 10: 861–877.

Byers, J. A., Lanne, B. S., Schlyter, F., Löfqvist, J., and Bergström, G. (1985) Olfactory recognition of host-tree susceptibility by pine shoot beetles. *Naturwissenschaften* 72: 324–326.

Byers, J. A., Birgersson, G., Löfqvist, J., and Bergström, G. (1988) Synergistic pheromones and monoterpenes enable aggregation and host recognition by a bark beetle. *Naturwissenschaften* 75: 153–155.

Byers, J. A., Anderbrant, O., and Löfqvist, J. (1989a) Effective attraction radius: A method for comparing species attractants and determining densities of flying insects. *J. Chem. Ecol.* 15: 749–765.

Byers, J. A., Högberg, H. E., Unelius, R., Birgersson, G., and Löfqvist, J. (1989b)

Structure-activity studies on aggregation pheromone components of *Pityogenes chalcographus* (Coleoptera: Scolytidae): All stereoisomers of chalcogran and methyl 2,4-decadienoate. *J. Chem. Ecol.* 15: 685–695.

Byers, J. A., Lanne, B. S., and Löfqvist, J. (1989c) Host-tree unsuitability recognized by pine shoot beetles in flight. *Experientia* 45: 489–492.

Byers, J. A., Birgersson, G., Löfqvist, J., Appelgren, M., and Bergström, G. (1990a) Isolation of pheromone synergists of bark beetle, *Pityogenes chalcographus*, from complex insect-plant odors by fractionation and subtractive-combination bioassay. *J. Chem. Ecol.* 16: 861–876.

Byers, J. A., Schlyter, F., Birgersson, G., and Francke, W. (1990b) *E*-myrcenol in *Ips duplicatus*: an aggregation pheromone component new for bark beetles. *Experientia* 46: 1209–1211.

Cade, S. C., Hrutfiord, B. F., and Gara, R. I. (1970) Identification of a primary attractant for *Gnathotrichus sulcatus* isolated from western hemlock logs. *J. Econ. Entomol.* 63: 1014–1015.

Camacho, A. D., Pierce, Jr., H. D. and Borden, J. A. (1993) Geometrical and optical isomerism of pheromones in two sympatric *Dryocoetes* species (Coleoptera: Scolytidae), mediates species specificity and response level. *J. Chem. Ecol.* 19: 2169–2182.

Cates, R. G. (1981) Host plant predictability and the feeding patterns of monophagous, oligophagous, and polyphagous insect herbivores. *Oecologia* 48: 319–326.

Chapman, J. A. (1972) Ommatidia numbers and eyes in scolytid beetles. *Ann. Entomol. Soc. Am.* 65: 550–553.

Chararas, C., Katoulas, M., and Koutroumpas, A. (1982) Feeding preference of *Rugulos-colytus rugulosus* bark beetle parasite of fruit trees. *C. R. Seances Acad. Sci. Ser. III Sci. Vie.* 294: 763–766.

Chénier, J. V. R. and Philogène, B. J. R. (1989) Field responses of certain forest Coleoptera to conifer monoterpenes and ethanol. *J. Chem. Ecol.* 15: 1729–1746.

Choudhury, J. H. and Kennedy, J. S. (1980) Light versus pheromone-bearing wind in the control of flight direction by bark beetles, *Scolytus multistriatus*. *Physiol. Entomol.* 5: 207–214.

Christiansen, E., Waring, R. H., and Berryman, A. A. (1987) Resistance of conifers to bark beetle attack: searching for general relationships. *For. Ecol. Manage.* 22: 89–106.

Claydon, N., Grove, J. F., and Pople, M. (1985) Elm bark beetle boring and feeding deterrents from *Phomopsis oblonga*. *Phytochemistry* 24: 937–944.

Cobb, Jr., F. W., Krstic, M., Zavarin, E. and Barber, Jr., H. W. (1968) Inhibitory effects of volatile oleoresin components on *Fomes annosus* and four *Ceratocystis* species. *Phytopathology* 58: 1327–1335.

Conn, J. E., Borden, J. H., Hunt, D. W. A., Holman, J., Whitney, H. S., Spanier, O. J., Pierce, Jr., H. D., and Oehlschlager, A. C. (1984) Pheromone production by axenically reared *Dendroctonus ponderosae* and *Ips paraconfusus* (Coleoptera: Scolytidae). *J. Chem. Ecol.* 10: 281–290.

Cook, S. P. and Hain, F. P. (1987) Four parameters of the wound response of loblolly and shortleaf pines to inoculation with the blue-staining fungus associated with the southern pine beetle. *Can. J. Bot.* 65: 2403–2409.

Croteau, R., Gurkewitz, S., Johnson, M. A., and Fisk, H. J. (1987) Biochemistry of oleoresinosis. Monoterpene and diterpene biosynthesis in lodgepole pine saplings infected with *Ceratocystis clavigera* or treated with carbohydrate elicitors. *Plant Physiol.* 85: 1123–1128.

David, C. T., Kennedy, J. S., Ludlow, A. R., Perry, J. N., and Wall, C. (1982) A reappraisal of insect flight towards a distant point source of wind-borne odor. *J. Chem. Ecol.* 8: 1207–1215.

Dethier, V. G. (1982) Mechanisms of host-plant recognition. *Entomol. exp. appl.* 31: 49–56.

Dickens, J. C. (1981) Behavioural and electrophysiological responses of the bark beetle *Ips typographus* to potential pheromone components. *Physiol. Entomol.* 6: 251–262.

Dickens, J.C. (1986) Specificity in perception of pheromones and host odours in Coleoptera. In: *Mechanisms in insect olfaction* (Payne, T. L., Birch, M. C., and Kennedy, C. E. J., eds.). pp. 253–261. Clarendon Press, Oxford, U.K.

Dickens, J. C., Gutmann, A., Payne, T. L., Ryker, L. C., and Rudinsky, J. A. (1983) Antennal olfactory responsiveness of Douglas-fir beetle, *Dendroctonus pseudotsugae* Hopkins (Coleoptera: Scolytidae) to pheromones and host odors. *J. Chem. Ecol.* 9: 1383–1395.

Dickens, J. C., Payne, T. L., Ryker, L. C., and Rudinsky, J. A. (1985) Multiple acceptors for pheromonal enantiomers on single olfactory cells in the Douglas-fir beetle, *Dendroctonus pseudotsugae* Hopk. (Coleoptera: Scolytidae. *J. Chem. Ecol.* 11: 1359–1370.

Dickens, J. C., Billings, R. F., and Payne, T. L. (1992) Green leaf volatiles interrupt aggregation pheromone response in bark beetles infesting southern pines. *Experientia* 48: 523–524.

Doke, N. and Tomiyama, K. (1980) Suppression of the hypersensitive response of potato tuber protoplasts to hyphal wall components by water soluble glucans isolated from *Phytophthora infestans*. *Physiol. Plant Pathol.* 16: 177–186.

Doskotch, R. W., Mikhail, A. A., and Chatterji, S. K. (1973) Structure of the water soluble feeding stimulant for *Scolytus multistriatus*: A revision. *Phytochemistry* 12: 1153–1155.

Elkinton, J. S. and Wood, D. L. (1980) Feeding and boring behavior of the bark beetle *Ips paraconfusus* (Coleoptera: Scolytidae) on the bark of a host and non-host tree species. *Can. Entomol.* 112: 797–809.

Elkinton, J. S., Wood, D. L., and Hendry, L. B. (1980) Pheromone production by the bark beetle, *Ips paraconfusus*, in the non-host, white fir. *J. Chem. Ecol.* 6: 979–987.

Elkinton, J. S., Wood, D. L., and Browne, L. E. (1981) Feeding and boring behavior of the bark beetle, *Ips paraconfusus*, in extracts of ponderosa pine phloem. *J. Chem. Ecol.* 7: 209–220.

Ehrlich, P. R. and Raven, P. H. (1965) Butterflies and plants: a study in coevolution. *Evolution* 8: 586–608.

Faucheux, M. J. (1989) Morphology of the antennal club in the male and female bark beetles *Ips sexdentatus* (Boern.) and *Ips typographus* L. (Coleoptera: Scolytidae). *Ann. Sci. Nat. Zool. Biol. Anim.* 10: 231–243.

Feeny, P. (1975) Biochemical coevolution between plants and their insect herbivores. In: *Coevolution of Animals and Plants* (Gilbert, L. E. and Raven, P. H., eds.) pp. 3–19. Univ. Texas Press, Austin, Tex.

Fish, R. H., Browne, L. E., Wood, D. L., and Hendry, L. B. (1979) Pheromone biosynthetic pathways: conversion of deuterium labelled ipsdienol with sexual and enantioselectivity in *Ips paraconfusus* Lanier. *Tetrah. Lett.* 17: 1465–1468.

Fitzgerald, T. D. and Nagel, W. P. (1972) Oviposition and larval bark-surface orientation of *Medetera aldrichii* (Diptera: Dolichopodidae): Response to a prey-liberated plant terpene. *Ann. Entomol. Soc. Am.* 65: 328–330.

Forsse, E. and Solbreck, C. (1985) Migration in the bark beetle *Ips typographus* L.: duration, timing and height of flight. *Z. Angew. Entomol.* 100: 47–57.

Funk, A. (1970) Fungal symbionts of the ambrosia beetle *Gnathotrichus sulcatus*. *Can. J. Bot.* 48: 1445–1448.

Furniss, M. M., Woo, J. Y., Deyrup, M. A., and Atkinson, T. H. (1987) Prothoracic mycangium on pine-infesting *Pityoborus* spp. (Coleoptera: Scolytidae). *Ann. Entomol. Soc. Am.* 80: 692–696.

Gilbert, B. L., Baker, J. E., and Norris, D. M. (1967) Juglone (5-hydroxy-1,4-napthoquinone) from *Carya ovata*, a deterrent to feeding by *Scolytus multistriatus*. *J. Insect Physiol.* 13: 1453–1459.

Goeden, R. D. and Norris, D. M. (1964) Attraction of *Scolytus quadrispinosus* (Coleoptera: Scolytidae) to *Carya* spp. for oviposition. *Ann. Entomol. Soc. Am.* 57: 141–146.

Goeden, R. D. and Norris, D. M. (1965) The behavior of *Scolytus quadrispinosus* (Coleoptera: Scolytidae) during the dispersal flight as related to its host specificities. *Ann. Entomol. Soc. Am.* 58: 249–252.

Goldhammer, D. S., Stephen, F. M., and Paine, T. D. (1991) The effect of the fungi *Ceratocystis minor* Hedgecock-Hunt, *Ceratocystis minor* var. *barrasii* Taylor, and SJB 122 on reproduction of the southern pine beetle *Dendroctonus frontalis* Zimmermann (Coleoptera: Scolytidae). *Can. Entomol.* 122: 407–418.

Gollob, L. (1980) Monoterpene composition in bark beetle-resistant loblolly pine. *Naturwissenschaften* 67: 409–410.

Graham, K. (1959) Release by flight exercise of a chemotropic response from photopositive domination in a scolytid beetle. *Nature* 184: 283–284.

Graham, K. (1968) Anaerobic induction of primary chemical attractancy for ambrosia beetles. *Can. J. Zool.* 46: 905–908.

Gray, B., Billings, R. F., Gara, R. I., and Johnsey, R. L. (1972) On the emergence and initial flight behaviour of the mountain pine beetle, *Dendroctonus ponderosae*, in eastern Washington. *Z. Angew. Entomol.* 71: 250–259.

Gries, G., Nolte, R., and Sanders, W. (1989) Computer simulated host selection in *Ips typographus*. *Entomol. exp. appl.* 53: 211–217.

Groberman, L. J. and Borden, H. J. (1982) Electrophysiological response of *Dendroctonus pseudotsugae* and *Ips paraconfusus* (Coleoptera: Scolytidae) to selected wave length regions of the visible spectrum. *Can. J. Zool.* 60: 2180–2189.

Hain, F. P., Mawby, W. D., Cook, S. P., and Arthur, F. H. (1983) Host conifer reaction to stem invasion. *Z. Angew. Entomol.* 96: 247–256.

Hallberg, E. (1982) Sensory organs in *Ips typographus* (Insecta: Coleoptera)—fine structure of the sensilla of the maxillary and labial palps. *Acta Zool.* 63: 191–198.

Hanover, J. M. and Furniss, M. M. (1966) Monoterpene concentration in Douglas-fir in relation to geographic location and resistance to attack by the Douglas-fir beetle. *U.S. Dep. Agric. For. Serv. Res. Pap. NC 6*, pp. 23–28, Washington, D.C.

Hansen, K. (1983) Reception of bark beetle pheromone in the predaceous clerid beetle, *Thanasimus formicarius* (Coleoptera: Cleridae). *J. Comp. Physiol. A.* 150: 371–378.

Happ, G. M., Happ, C. M., and French, J. R. J. (1976) Ultrastructure of the mesonotal mycangium of an ambrosia beetle *Xyleborus dispar* (Coleoptera: Scolytidae). *Int. J. Insect Morphol. Embryol.* 5: 381–392.

Hendry, L. B., Piatek, B., Browne, L. E., Wood, D. L., Byers, J. A., Fish, R. H., and Hicks, R. A. (1980) *In vivo* conversion of a labelled host plant chemical to pheromones of the bark beetle *Ips paraconfusus*. *Nature* 284: 485.

Himejima, M., Hobson, K. R., Otsuka, T., Wood, D. L., and Kubo, I. (1992) Antimicrobial terpenes from oleoresin of ponderosa pine tree *Pinus ponderosa*: A defense mechanism against microbial invasion. *J. Chem. Ecol.* 18: 1809–1818.

Hobson, K. R., Wood, D. L., Cool, L. G., White, P. R., Ohtsuka, T., Kubo, I., and Zavarin, E. (1993) Chiral specificity in responses by the bark beetle *Dendroctonus valens* to host kairomones. *J. Chem. Ecol.* 19: 1837–1846.

Hodges, J. D. and Lorio, Jr., P. L. (1975) Moisture stress and composition of xylem oleoresin in loblolly pine. *For. Sci.* 21: 283–290.

Hodges, J. D., Elam, W. W., Watson, W. R., and Nebeker, T. E. (1979) Oleoresin characteristics and susceptibility of four southern pines to southern pine beetle (Coleoptera: Scolytidae) attacks. *Can. Entomol.* 111: 889–896.

Hodges, J. D., Nebeker, T. E., DeAngelis, J. D., Karr, B. L., and Blanche, C. A. (1985) Host resistance and mortality: A hypothesis based on the southern pine beetle-microorganism-host interactions. *Bull. Entomol. Soc. Am.* 31: 31–35.

Horntvedt, R. E., Christiansen, H., Solheim, H., and Wang, S. (1983) Artificial inoculation with *Ips typographus*-associated bluestain fungi can kill healthy Norway spruce trees. *Medd. Nor. Inst. Skogforsk.*, 38: 1–20.

Hughes, P. R. (1973) *Dendroctonus*: Production of pheromones and related compounds in response to host monoterpenes. *Z. Angew. Entomol.* 73: 294–312.

Hughes, P. R. (1974) Myrcene: A precursor of pheromones in *Ips* beetles. *J. Insect Physiol.* 20: 1271–1275.

Hughes, P. R. and Renwick, J. A. A. (1977) Neural and hormonal control of pheromone biosynthesis in the bark beetle, *Ips paraconfusus*. *Physiol. Entomol.* 2: 117–123.

Hunt, D. W. A. and Borden, J. H. (1989) Terpene alcohol pheromone production by *Dendroctonus ponderosae* and *Ips paraconfusus* (Coleoptera: Scolytidae) in the absence of readily culturable microorganisms. *J. Chem. Ecol.* 15: 1433–1464.

Hynum, B. G. and Berryman, A. A. (1980) *Dendroctonus ponderosae* (Coleoptera: Scolytidae) pre-aggregation landing and gallery initiation on lodgepole pine. *Can. Entomol.* 112: 185–192.

Ivarsson, P., Schlyter, F., and Birgersson, G. (1993) Demonstration of *de novo* pheromone biosynthesis in *Ips duplicatus* (Coleoptera: Scolytidae): inhibition of ipsdienol and *E*-myrcenol production by compactin. *Insect Biochem. Mol. Biol.* 23: 655–662.

Jactel, H. (1991) Dispersal and flight behavior of *Ips sexdentatus* (Coleoptera: Scolytidae) in pine forest. *Ann. Sci. For.* 48: 417–428.

Jactel, H. and Gaillard, J. (1991) A preliminary study of the dispersal potential of *Ips sexdentatus* Boern (Coleoptera: Scolytidae) with an automatically recording flight mill. *J. Appl. Entomol.* 112: 138–145.

Jones, R. G. and Brindley, W. A. (1970) Tests of eight rearing media for the mountain pine beetle, *Dendroctonus ponderosae* (Coleoptera: Scolytidae), from lodgepole pine. *Ann. Entomol. Soc. Am.* 63: 313–316.

Kajimura, H. and Hijii, N. (1992) Dynamics of the fungal symbionts in the gallery system and the mycangia of the ambrosia beetle *Xylosandrus mutilatus* Blandford (Coleoptera: Scolytidae) in relation to its life history. *Ecol. Res.* 7: 107–117.

Kimmerer, T. W. and Kozlowski, T. T. (1982) Ethylene, ethane, acetaldehyde and ethanol production by plants under stress. *Plant Physiol.* 69: 840–847.

Kinzer, G. W., Fentiman, Jr., A. F., Page, T. F., Foltz, R. L., Vité, J. P., and Pitman, G. B. (1969) Bark beetle attractants: identification, synthesis and field bioassay of a new compound isolated from *Dendroctonus*. *Nature* 211: 475–476.

Klimetzek, D. and Francke, W. (1980) Relationship between the enantiomeric composition of α-pinene in host trees and the production of verbenols in *Ips* species. *Experientia* 36: 1343–1345.

Klimetzek, D., Köhler, J., Vité, J. P., and Kohnle, U. (1986) Dosage response to ethanol mediates host selection by 'secondary' bark beetles. *Naturwissenschaften* 73: 270–272.

Kohnle, U. (1985) Investigations of chemical communication systems in secondary bark beetles. *Z. Angew. Entomol.* 100: 197–218.

Kohnle, U. and Vité, J. P. (1984) Bark beetle predators: Strategies in the olfactory perception of prey species by clerid and trogositid beetles. *Z. Angew. Entomol.* 98: 504–508.

Langor, D. W., Spence, J. R., and Pohl, G. R. (1990) Host effects on fertility and reproductive success of *Dendroctonus ponderosae* Hopkins (Coleoptera: Scolytidae). *Evolution* 44: 609–618.

Långström, B. (1983) Within tree development of *Tomicus minor* (Coleoptera: Scolytidae) in wind thrown scotch pine. *Acta Entomol. Fenn.* 42: 42–46.

Långström, B. and Hellqvist, C. (1991) Shoot damage and growth losses following three years of *Tomicus* attacks in scots pine stands close to a timber storage site. *Silva. Fenn.* 25: 133–145.

Lanier, G. N. (1983) Integration of visual stimuli, host odorants, and pheromones by bark beetles and weevils in locating and colonizing host trees. In: *Herbivorous Insects: Host-Seeking Behavior and Mechanisms* (Ahmad, S., ed.) pp. 161–171. Academic Press, New York.

Lanier, G. N. and Burns, B. W. (1978) Barometric flux. Effects on the responsiveness of bark beetles to aggregation attractants. *J. Chem. Ecol.* 4: 139–147.

Lanier, G. N., Birch, M. C., Schmitz, R. F., and Furniss, M. M. (1972) Pheromones of *Ips pini* (Coleoptera: Scolytidae): Variation in response among three populations. *Can. Entomol.* 104: 1917–1923.

Lanier, G. N., Classon, A., Stewart, T., Piston, J. J., and Silverstein, R. M. (1980) *Ips pini*: The basis for interpopulational differences in pheromone biology. *J. Chem. Ecol.* 6: 677–687.

Lanne, B. S., Schlyter, F., Byers, J. A., Löfqvist, J., Leufvén, A., Bergström, G., Van Der Pers, J. N. C., Unelius, R., Baeckström, P., and Norin, T. (1987) Differences in attraction to semiochemicals present in sympatric pine shoot beetles, *Tomicus minor* and *T. piniperda*. *J. Chem. Ecol.* 13: 1045–1067.

Lanne, B. S., Ivarsson, P., Johnsson, P., Bergström, G., and Wassgren, A. B. (1989) Biosynthesis of 2-methyl-3-buten-2-ol a pheromone component of *Ips typographus* (Coleoptera: Scolytidae). *Insect. Biochem.* 19: 163–168.

Lekander, B., Bejer-Petersen, B., Kangas, E., and Bakke, A. (1977) The distribution of bark beetles in the Nordic countries. *Acta. Entomol. Fenn.* 32: 1–37.

Leufvén, A., Bergström, G., and Falsen, E. (1984) Interconversion of verbenols and verbenone by identified yeasts isolated from the spruce bark beetle *Ips typographus*. *J. Chem. Ecol.* 10: 1349–1361.

Leufvén, A., Bergström, G., and Falsen, E. (1988) Oxygenated monoterpenes produced by yeasts isolated from *Ips typographus* (Coleoptera: Scolytidae) and grown in phloem medium. *J. Chem. Ecol.* 14: 353–362.

Levieux, J., Cassier, P., Guillaumin, D., and Roques, A. (1991) Structures implicated in the transportation of pathogenic fungi by the european bark beetle *Ips sexdentatus* Boerner: Ultrastructure of a mycangium. *Can. Entomol.* 123: 245–254.

Light, D. M. and Birch, M. C. (1982) Bark beetle enantiomeric chemoreception: greater sensitivity to allomone than pheromone. *Naturwissenschaften* 69: 243–245.

Lindelöw, A. and Weslien, J. (1986) Sex-specific emergence of *Ips typographus* L. (Coleoptera: Scolytidae) and flight behavior in response to pheromone sources following hibernation. *Can. Entomol.* 118: 59–67.

Lindelöw, A., Risberg, B., and Sjodin, K. (1992) Attraction during flight of scolytids and other bark and wood-dwelling beetles to volatiles from fresh and stored spruce wood. *Can. J. For. Res.* 22: 224–228.

Lindgren, B. S., Borden, J. H., Chong, L., Friskie, L. M., and Orr, D. B. (1983) Factors influencing the efficiency of pheromone baited traps for three species of ambrosia beetles (Coleoptera: Scolytidae). *Can. Entomol.* 115: 303–314.

Lindström, M., Norin, T., Birgersson, G., and Schlyter, F. (1989) Variation of enantiomeric composition of α-pinene in Norway spruce, *Picea abies*, and its influence on production of verbenol isomers by *Ips typographus* in the field. *J. Chem. Ecol.* 15: 541–548.

Lorio, Jr., P. L. (1986) Growth and differentiation balance: A basis for understanding southern pine beetle *Dendroctonus frontalis* and tree interactions. *For. Ecol. Manage.* 14: 259–274.

Löyttyniemi, K., Heliovaara, K., and Repo, S. (1988) No evidence of a population pheromone in *Tomicus piniperda* (Coleoptera: Scolytidae): A field experiment. *Ann. Entomol. Fenn.* 54: 93–95.

Magema, N., Gaspar, C., and Séverin, M. (1982) Efficacité de l'éthanol dans le piégeage du scolyte *Trypodendron lineatum* (Olivier, 1795)(Coleoptera, Scolytidae) et role des constituants terpeniques de l'epicea. *Ann. Soc. R. Zool. Belg.* 112: 49–60.

Mathre, D. E. (1964) Pathenogenicity of *Ceratocystis ips* and *Ceratocystis minor* to *Pinus ponderosa*. *Contrib. Boyce Thompson Inst.* 22: 363–388.

McLean, J. A. and Borden, J. H. (1977) Attack by *Gnathotrichus sulcatus* (Coleoptera: Scolytidae) on stumps and felled trees baited with sulcatol and ethanol. *Can. Entomol.* 109: 675–686.

McMullen, L. H. and Atkins, M. D. (1962) On the flight and host selection of the Douglas-fir beetle, *Dendroctonus pseudotsugae* Hopk. (Coleoptera: Scolytidae). *Can. Entomol.* 94: 1309–1325.

Meyer, H. J. and Norris, D. M. (1967a) Behavioral responses by *Scolytus multistriatus* (Coleoptera: Scolytidae) to host- (*Ulmus*) and beetle-associated chemotactic stimuli. *Ann. Entomol. Soc. Am.* 60: 642–646.

Meyer, H. J. and Norris, D. M. (1967b) Vanillin and syringaldehyde as attractants for *Scolytus multistriatus* (Coleoptera: Scolytidae). *Ann. Entomol. Soc. Am.* 60: 858–859.

Meyer, H. J. and Norris, D. M. (1974) Lignin intermediates and simple phenolics as feeding stimulants for *Scolytus multistriatus*. *J. Insect Physiol.* 20: 2015–2021.

Miller, D. R. and Borden, J. H. (1990) β-Phellandrene: Kairomone for pine engraver *Ips pini* (Say)(Coleoptera: Scolytidae). *J. Chem. Ecol.* 16: 2519–2531.

Miller, D. R., Borden, J. H., and Slessor, K. N. (1989) Interpopulation and intrapopulation variation of the pheromone ipsdienol produced by male pine engravers *Ips pini* Say (Coleoptera: Scolytidae). *J. Chem. Ecol.* 15: 233–248.

Miller, J. R. and Strickler, K. L. (1984) Finding and accepting host plants. In: *Chemical Ecology of Insects* (Bell, W. J. and Cardé, R. T., eds.) pp. 127–157. Chapman and Hall, London.

Mirov, N.T. (1961) Composition of gum turpentines of pines. *USDA For. Ser. Tech. Bull.* No. 1239, Washington, D.C.

Moeck, H. A. (1970) Ethanol as the primary attractant for the ambrosia beetle *Trypodendron lineatum* (Coleoptera: Scolytidae). *Can. Entomol.* 102: 985–994.

Moeck, H. A. (1981) Ethanol induces attack on trees by spruce beetles *Dendroctonus rufipennis* (Coleoptera: Scolytidae). *Can. Entomol.* 113: 939–942.

Moeck, H. A., Wood, D. L., and Lindahl, Jr., K. Q. (1981) Host selection behavior of bark beetles (Coleoptera: Scolytidae) attacking *Pinus ponderosa*, with special emphasis on the western pine beetle, *Dendroctonus brevicomis*. *J. Chem. Ecol.* 7: 49–83.

Montgomery, M. E. and Wargo, P. M. (1983) Ethanol and other host derived volatiles as attractants to beetles that bore into hardwoods. *J. Chem. Ecol.* 9: 181–190.

Moser, J. C. and Browne, L. E. (1978) A nondestructive trap for *Dendroctonus frontalis* Zimmerman (Coleoptera: Scolytidae). *J. Chem. Ecol.* 4: 1–7.

Mustaparta, H. (1984) Olfaction. In: *Chemical Ecology of Insects* (Bell, W. J. and Cardé, R. T., eds.) pp. 37–70. Chapman and Hall, London.

Mustaparta, H., Angst, M. E., and Lanier, G. N. (1980) Receptor discrimination of enantiomers of the aggregation pheromone ipsdienol, in two species of *Ips*. *J. Chem. Ecol.* 6: 689–701.

Mustaparta, H., Tommerås, B. A., Baeckström, P., Bakke, J. M., and Ohloff, G. (1984) Ipsdienol-specific receptor cells in bark beetles: Structure activity relationships of various analogs and the deuterium-labeled ipsdienol. *J. Comp. Physiol. A.* 154: 591–596.

Namkoong, G., Roberds, J. H., Nunnally, L. B., and Thomas, H. A. (1979) Isozyme variations in populations of southern pine beetles. *For. Sci.* 25: 197–203.

Nijholt, W. W. and Shönherr, J. (1976) Chemical response behavior of scolytids in West Germany and western Canada. *Can. For. Serv. Bi-mon. Res. Notes* 32: 31–32.

Paine, T. D. (1984) Influence of the mycangial fungi of the western pine beetle *Dendroctonus brevicomis* on water conduction through ponderosa pine seedlings. *Can. J. Bot.* 62: 556–558.

Paine, T. D. and Stephen, F. M. (1987) Fungi associated with the southern pine beetle: Avoidance of induced defense response in loblolly pine. *Oecologia* 74: 377–379.

Paine, T. D. and Stephen, F. M. (1988) Induced defenses of loblolly pine, *Pinus taeda*: Potential impact on *Dendroctonus frontalis* within-tree mortality. *Entomol. Exp. Appl.* 46: 39–46.

Paine, T. D., Stephen, F. M. and Cates, R. G. (1988) Moisture stress, tree suitability, and southern pine beetle population dynamics. In: *Integrated Control of Scolytid Bark Beetles* (Payne, T. L. and Saarenmaa, H., eds.) pp. 85–103. Virginia Polytechnic Inst. and State Univ., Blacksburg, V.

Paiva, M. R. and Kiesel, K. (1985) Field responses of *Trypodendron* spp. (Col., Scolytidae) to different concentrations of lineatin and α-pinene. *Z. Angew. Entomol.* 99: 442–448.

Payne, T. L. (1979) Pheromone and host odor perception in bark beetles. In: *Neurotoxicology of Insecticides and Pheromones* (Narahashi, T., ed.) pp. 27–57. Plenum Pub. Co., New York.

Payne, T. L. and Dickens, J. C. (1976) Adaptation to determine receptor system specificity in insect olfactory communication. *J. Insect Physiol.* 22: 1569–1572.

Payne, T. L., Moeck, H. A., Willson, C. D., Coulson, R. N., and Humphreys, W. J. (1973) Bark beetle olfaction -II. Antennal morphology of sixteen species of Scolytidae (Coleoptera). *Int. J. Insect Mor. Emb.* 2: 177–192.

Payne, T. L., Richerson, J. V., Dickens, J. C., West, J. R., Mori, K., Berisford, C. W., Hedden, R. L., Vité, J. P., and Blum, M. S. (1982) Southern pine beetle: olfactory receptor and behavior discrimination of enantiomers of the attractant pheromone frontalin. *J. Chem. Ecol.* 8: 873–881.

Payne, T. L., Klimetzek, D., Kohnle, U. and Mori, K. (1983) Electrophysiological and field responses of *Trypodendron*-spp to enantiomers of lineatin. *Z. Angew. Entomol.* 95: 272–276.

Pearce, G. T., Gore, W. E., Silverstein, R. M., Peacock, J. W., Cuthbert, R. A., Lanier, G. N., and Simeone, J. B. (1975) Chemical attractants for the smaller European elm bark beetle, *Scolytus multistriatus* (Coleoptera: Scolytidae). *J. Chem. Ecol.* 1: 115–124.

Phillips, T. W. (1990) Responses of *Hylastes salebrosus* to turpentine, ethanol and pheromones of *Dendroctonus* (Coleoptera: Scolytidae). *Fla. Entomol.* 73: 286–292.

Phillips, T. W., Wilkening, A. J., Atkinson, T. H., Nation, J. L., Wilkinson, R. C. and Foltz, J. L. (1988) Synergism of turpentine and ethanol as attractants for certain pine-infesting beetles (Coleoptera). *Environ. Entomol.* 17: 456–462.

Pierce, Jr., H. D., Conn, J. E., Oehlschlager, A. C., and Borden, J. H. (1987) Monoterpene metabolism in female mountain pine beetles, *Dendroctonus ponderosae* Hopkins attacking ponderosa pine. *J. Chem. Ecol.* 13: 1455–1480.

Pitman, G. B. (1969) Pheromone response in pine bark beetles: Influence of host volatiles. *Science* 166: 905–906.

Pitman, G. B. and Vité, J. P. (1969) Aggregation behavior of *Dendroctonus ponderosae* (Coleoptera: Scolytidae) in response to chemical messengers. *Can. Entomol.* 101: 143–149.

Pitman, G. B. and Vité, J. P. (1971) Predator-prey response to western pine beetle attractants. *J. Econ. Entomol.* 64: 402–404.

Pitman, G. B. and Vité, J. P. (1974) Biosynthesis of methylcyclohexenone by male Douglas-fir beetle. *Environ. Entomol.* 3: 886–887.

Pitman, G. B., Renwick, J. A. A., and Vité, J. P. (1966) Studies on the pheromone of *Ips confusus* (LeConte). IV. Isolation of the attractive substance by gas-liquid chromatography. *Contrib. Boyce Thompson Inst.* 23: 243–250.

Pitman, G. B., Hedden, R. L., and Gara, R. I. (1975) Synergistic effects of ethyl alcohol on the aggregation of *Dendroctonus pseudotsugae* (Col., Scolytidae) in response to pheromones. *Z. Angew. Entomol.* 78: 203–208.

Raffa, K. F. and Berryman, A. A. (1979) Flight responses and host selection by bark beetles. In: *Dispersal of Forest Insects: Evaluation, Theory and Management Implications* (Berryman, A. A. and Safranyik, L., eds.) pp. 213–233. Proc. second IUFRO conf., Canad. and USDA Forest Service, Washington State Univ., Pullman, W.

Raffa, K. F. and Berryman, A. A. (1982a) Accumulation of monoterpenes and associated volatiles following fungal inoculation of grand fir with a fungus vectored by the fir engraver *Scolytus ventralis* (Coleoptera: Scolytidae). *Can. Entomol.* 114: 797–810.

Raffa, K. F. and Berryman, A. A. (1982b) Gustatory cues in the orientation of *Dendroctonus ponderosae* (Coleoptera: Scolytidae) to host trees. *Can. Entomol.* 114: 97–104.

Raffa, K. F. and Berryman, A. A. (1982c) Physiological differences between lodgepole pines resistant and susceptible to the mountain pine beetle and associated microorganisms. *Environ. Entomol.* 11: 486–492.

Raffa, K. F. and Berryman, A. A. (1983) Physiological aspects of lodgepole pine wound responses to a fungal symbiont of the mountain pine beetle, *Dendroctonus ponderosae* (Coleoptera: Scolytidae). *Can. Entomol.* 115: 723–734.

Raffa, K. F. and Berryman, A. A. (1987) Interacting selective pressures in conifer-bark beetle systems: A basis for reciprocal adaptations? *Am. Nat.* 129: 234–262.

Raffa, K. F., Berryman, A. A., Simasko, J., Teal, W., and Wong, B. L. (1985) Effects of grand fir, *Abies grandis* monoterpenes on the fir engraver, *Scolytus ventralis* (Coleoptera: Scolytidae) and its symbiotic fungus. *Environ. Entomol.* 14: 552–556.

Raffa, K. F., Phillips, T. W., and Salom, S. M. (1993) Strategies and mechanisms of host colonization by bark beetles. In: *Beetle-Pathogen Interactions in Conifer Forests* (Schowalter, T. D. and Filip, G. M., eds.) pp. 103–128. Academic Press, London.

Ramisch, H. (1986) Host location by *Trypodendron domesticum* and *Trypodendron lineatum* (Coleoptera: Scolytidae). *Z. Angew. Zool.* 73: 159–198.

Reed, A. N., Hanover, J. W., and Furniss, M. M. (1986) Douglas-fir and western larch: chemical and physical properties in relation to Douglas-fir bark beetle attack. *Tree Physiol.* 1: 277–288.

Reid, R. W., Whitney, H. S. and Watson, J. A. (1967) Reactions of lodgepole pine to attack by *Dendroctonus ponderosae* Hopkins and blue stain fungi. *Can. J. Bot.* 45: 1115–1126.

Renwick, J. A. A. and Vité, J. P. (1968) Isolation of the population aggregating pheromone of the southern pine beetle. *Contrib. Boyce Thompson Inst.* 24: 65–68.

Renwick, J. A. A. and Vité, J. P. (1969) Bark beetle attractants: Mechanisms of colonization by *Dendroctonus frontalis*. *Nature* 224: 1222--1223.

Renwick, J. A. A. and Vité, J. P. (1970) Systems of chemical communication in *Dendroctonus*. *Contrib. Boyce Thompson Inst.* 24: 283–292.

Renwick, J. A. A., Hughes, P. R., and Krull, I. S. (1976) Selective production of *cis*- and *trans*-verbenol from (−)-and (+)-α-pinene by a bark beetle. *Science* 191: 199–201.

Rice, R. E. (1969) Response of some predators and parasites of *Ips confusus* (LeC.)(Coleoptera: Scolytidae) to olfactory attractants. *Contrib. Boyce Thompson Inst.* 24: 189–194.

Richeson, J. S., Wilkinson, R. C., and Nation, J. L. (1970) Development of *Ips calligraphus* on foliage based diets. *J. Econ. Entomol.* 63: 1797–1799.

Rose, W. F., Billings, R. F. and Vité, J. P. (1981) Southern pine bark beetles *Ips calligraphus*: Evaluation of nonsticky pheromone trap designs for survey and research. *Southwest. Entomol.* 6: 1–9.

Rudinsky, J. A. (1966) Host selection and invasion by the Douglas-fir beetle, *Dendroctonus pseudotsugae* Hopkins, in coastal Douglas-fir forests. *Can. Entomol.* 98: 98–111.

Rudinsky, J. A., Novak, V., and Svihra, P. (1971) Attraction of the bark beetle *Ips typographus* L. to terpenes and a male-produced pheromone. *Z. Angew. Entomol.* 67: 179–188.

Rudinsky, J. A., Furniss, M. M., Kline, L. N., and Schmitz, R. F. (1972) Attraction and repression of *Dendroctonus pseudotsugae* (Coleoptera: Scolytidae) by three synthetic pheromones in traps in Oregon and Idaho. *Can. Entomol.* 104: 815–822.

Rudinsky, J. A., Morgan, M. E., Libbey, L. M., and Putnam, T. B. (1974) Antiaggregative-rivalry pheromone of the mountain pine beetle, and a new arrestant of the southern pine beetle. *Environ. Entomol.* 3: 90–98.

Rudinsky, J. A., Ryker, L. C., Michael, R. R., Libbey, L. M., and Morgan, M. E. (1976) Sound production in Scolytidae: Female sonic stimulus of male pheromone release in two *Dendroctonus* beetles. *J. Insect Physiol.* 22: 1675–1681.

Salom, S. M. and Mclean, J. A. (1989) Influence of wind on the spring flight of *Trypodendron lineatum* Olivier (Coleoptera: Scolytidae) in a second-growth coniferous forest. *Can. Entomol.* 121: 109–120.

Salom, S. M. and Mclean, J. A. (1991) Environmental influences on dispersal of *Trypodendron lineatum* (Coleoptera: Scolytidae). *Environ. Entomol.* 20: 565–576.

Salonen, K. (1973) On the life cycle, especially on the reproduction biology of *Blastophagus piniperda* L. (Col., Scolytidae). *Acta For. Fenn.* 127: 1–72.

Schlyter, F. and Löfqvist, J. (1986) Response of walking spruce bark beetles *Ips typographus* to pheromone produced in different attack phases. *Entomol. exp. appl.* 41: 219–230.

Schlyter, F., Birgersson, G., Byers, J. A., Löfqvist, J., and Bergström, G. (1987a) Field response of spruce bark beetle, *Ips typographus*, to aggregation pheromone candidates. *J. Chem. Ecol.* 13: 701–716.

Schlyter, F., Byers, J. A., and Löfqvist, J. A. (1987b) Attraction to pheromone sources of different quantity, quality, and spacing: Density-regulation mechanisms in bark beetle *Ips typographus*. *J. Chem. Ecol.* 13: 1503–1523.

Schlyter, F., Löfqvist, J., and Byers, J. A. (1987c) Behavioural sequence in the attraction of the bark beetle *Ips typographus* to pheromone sources. *Physiol. Entomol.* 12: 185–196.

Schlyter, F., Birgersson, G., Byers, J. A. and Bakke, A. (1992) The aggregation pheromone of *Ips duplicatus* and its role in competitive interactions with *I. typographus* (Coleoptera: Scolytidae). *Chemoecology* 3: 103–112.

Schroeder, L. M. (1987) Attraction of the bark beetle *Tomicus piniperda* to Scots pine trees in relation to tree vigor and attack density. *Entomol. exp. appl.* 44: 53–58.

Schroeder, L. M. (1988) Attraction of the bark beetle *Tomicus piniperda* and some other bark- and wood-living beetles to the host volatiles α-pinene and ethanol. *Entomol. exp. appl.* 46: 203–210.

Schroeder, L. M. (1990) Duct resin flow in scots pine in relation to the attack of the bark beetle *Tomicus piniperda* L. (Coleoptera: Scolytidae). *J. Appl. Entomol.* 109: 105–112.

Schroeder, L. M. (1992) Olfactory recognition of nonhosts aspen and birch by conifer bark beetles *Tomicus piniperda* and *Hylurgops palliatus*. *J. Chem. Ecol.* 18: 1583–1593.

Schroeder, L. M. and Eidmann, H. H. (1987) Gallery initiation by *Tomicus piniperda* (Coleoptera: Scolytidae) on Scots pine trees baited with host volatiles. *J. Chem. Ecol.* 13: 1591–1599.

Schroeder, L. M. and Lindelöw, A. (1989) Attraction of scolytids and associated beetles by different absolute amounts and proportions of α-pinene and ethanol. *J. Chem. Ecol.* 15: 807–818.

Schuh, B. A. and Benjamin, D. M. (1984) The chemical feeding ecology of *Neodipron dubiosus* Schedl, *N. rugifrons* Midd., and *N. lecontei* (Fitch) on Jack pine (*Pinus banksiana* Lamb.). *J. Chem. Ecol.* 10: 1071–1079.

Scriber, J. M. (1984) Host-plant suitability. In: *Chemical Ecology of Insects* (Bell, W. J. and Cardé, R. T., eds.) pp. 159–202. Chapman and Hall, London.

Shain, L. (1967) Resistance of sapwood in stems of loblolly pine to infection by *Fomes annosus*. *Phytopathology* 57: 1034–1045.

Shrimpton, D. M. (1973) Extractives associated with the wound response of lodgepole pine attacked by the mountain pine beetle and associated microorganisms. *Can. J. Bot.* 51: 527–534.

Silverstein, R. M., Rodin, J. O., and Wood, D. L. (1966) Sex attractants in frass produced by male *Ips confusus* in ponderosae pine. *Science* 154: 509–510.

Silverstein, R. M., Rodin, J. O., and Wood, D. L. (1967) Methodology for isolation and identification of insect pheromones with reference to studies on California five-spined *Ips*. *J. Econ. Entomol.* 60: 944–949.

Silverstein, R. M., Brownlee, R. G., Bellas, T. E., Wood, D. L., and Browne, L. E. (1968) Brevicomin: Principal sex attractant in the frass of the female western pine beetle. *Science* 159: 889–891.

Smith, L. V. and Zavarin, E. (1960) Free mono- and oligosaccharides of some California conifers. *Tech. Assoc. Pulp Pap. Ind.* 43: 218–221.

Smith, M. T., Busch, G. R., Payne, T. L. and Dickens, J. C. (1988) Antennal olfactory responsiveness of three sympatric *Ips* species [*Ips avulsus* (Eichhoff), *Ips calligraphus* (Germar), *Ips grandicollis* (Eichhoff)], to intra- and interspecific behavioral chemicals. *J. Chem. Ecol.* 14: 1289–1304.

Smith, R. H. (1961) The fumigant toxicity of three pine resins to *Dendroctonus brevicomis* and *D. jeffrei*. *J. Econ. Entomol.* 54: 365–369.

Smith, R. H. (1964) Variation in the monoterpenes of *Pinus ponderosa* Laws. *Science* 143: 1337–1338.

Smith, R. H. (1965a) A physiological difference among beetles of *Dendroctonus ponderosae* (=*D. monticolae*) and *D. ponderosae* (=*D. jeffreyi*). *Ann. Entomol. Soc. Am.* 58: 440–442.

Smith, R. H. (1965b) Effect of monoterpene vapors on the western pine beetle. *J. Econ. Entomol.* 58: 509–510.

Smith, R. H. (1966) The monoterpene composition of *Pinus ponderosa* xylem resin and of *Dendroctonus brevicomis* pitch tubes. *For. Sci.* 12: 63–68.

Smith, R. H. (1967) Variations in the monoterpene composition of the wood resin of Jeffrey, Washoe, Coulter and lodgepole pines. *For. Sci.* 13: 246–252.

Smith, R. H. (1968) Intratree measurements of the monoterpene composition of ponderosa pine xylem resin. *For. Sci.* 14: 418–419.

Smith. R. H. (1969) Local and regional variation in the monoterpenes of ponderosa pine xylem resin. *USDA For. Ser. Res. Pap. PSW–56 p. 1–10*, Berkeley, CA.

Städler, E. (1984) Contact chemoreception. In: *Chemical Ecology of Insects* (Bell, W. J. and Cardé, R. T., eds.) pp. 3–35. Chapman and Hall, London.

Stephen, F. M. and Paine, T. D. (1985) Seasonal patterns of host tree resistance to fungal associates of the southern pine beetle. *Z. Angew. Entomol.* 99: 113–122.

Stock, M. W. and Amman, G. D. (1980) Genetic differentiation among mountain pine beetle populations from lodgepole pine and ponderosa pine in northeast Utah. *Ann. Entomol. Soc. Am.* 73: 472–478.

Stock, M. W., Pitman, G. B., and Guenther, J. D. (1979) Genetic differences between Douglas-fir beetles (*Dendroctonus pseudotsugae*) from Idaho and coastal Oregon. *Ann. Entomol. Soc. Am.* 72: 394–397.

Struble, G. R. 1957. The fir engraver, a serious enemy of western true firs. *U.S. Dep. Agric. Prod. Res. Rep. II.*, Washington, D. C.

Sturgeon, K. B. (1979) Monoterpene variation in ponderosa pine xylem resin related to western pine beetle predation. *Evolution* 33: 803–814.

Sturgeon, K. B. and Mitton, J. B. (1986) Allozyme and morphological differentiation of mountain pine beetles *Dendroctonus ponderosae* (Coleoptera: Scolytidae) associated with host tree. *Evolution* 40: 290–302.

Teale, S. A., Webster, F. X., Zhang, A., and Lanier, G. N. (1991) Lanierone: a new pheromone component from *Ips pini* (Coleoptera: Scolytidae) in New York. *J. Chem. Ecol.* 17: 1159–1176.

Thompson, S. N. and Bennett, R. B. (1971) Oxidation of fat during flight of male Douglas-fir beetles, *Dendroctonus pseudotsugae. Insect Physiol.* 17: 1555–1563.

Tilden, P. E., Bedard, W. D., Lindahl, Jr., K. Q., and Wood, D. L. (1983) Trapping *Dendroctonus brevicomis*: Changes in attractant release rate, dispersion of attractant, and silhouette. *J. Chem. Ecol.* 9: 311–321.

Tommerås, B. A. (1988) The clerid beetle *Thanasimus formicarius* is attracted to the pheromone of the ambrosia beetle *Trypodendron lineatum. Experientia* 44: 536–537.

Tommerås, B. A., Mustaparta, H., and Gregoire, J. C. (1984) Receptor cells in *Ips typographus* and *Dendroctonus micans* specific to pheromones of the reciprocal genus. *J. Chem. Ecol.* 10: 759–769.

Tuomi, J. and Augner, M. (1993) Synergistic selection of unpalatability in plants. *Evolution* 47: 668–672.

Turlings, T. C. J., Tumlinson, J. H., and Lewis, W. J. (1990) Exploitation of herbivore-induced plant odors by host-seeking parasitic wasps. *Science* 250: 1251–1253.

Vanderwel, D. and Oehlschlager, A. C. (1987) Biosynthesis of pheromones and endocrine regulation of pheromone production in Coleoptera. In: *Pheromone Biochemistry* (Prestwich, G. D. and Blomquist, G. J., eds.) pp. 175–215. Academic Press, New York.

Vité, J. P. (1961) The influence of water supply on oleoresin exudation pressure and resistance to bark beetle attack in *Pinus ponderosa*. *Contrib. Boyce Thompson Inst.* 21: 37–66.

Vité, J. P. and Bakke, A. (1979) Synergism between chemical and physical stimuli in host selection by an ambrosia beetle. *Naturwissenschaften* 66, 528–529.

Vité, J. P. and Pitman, G. B. (1969) Insect and host odors in the aggregation of the western pine beetle. *Can. Entomol.* 101: 113–117.

Vité, J. P., Bakke, A., and Renwick, J. A. A. (1972) Pheromones in *Ips* (Coleoptera: Scolytidae): Occurrence and production. *Can. Entomol.* 104: 1967–1975.

Vité, J. P., Volz, H. A., Paiva, M. R., and Bakke, A. (1986) Semiochemicals in host selection and colonization of pine trees by the pine shoot beetle *Tomicus piniperda*. *Naturwissenschaften* 73: 39–40.

Volz, H. A. (1988) Monoterpenes governing host selection in the bark beetles *Hylurgops palliatus* and *Tomicus piniperda*. *Entomol. exp. appl.* 47: 31–36.

Wagner, M. R., Benjamin, D. M., Clancy, K. L., and Schuh, B. A. (1983) Influence of diterpene resin acids on feeding and growth of larch sawfly, *Pristphora erichsonii* (Hartig). *J. Chem. Ecol.* 9: 119–127.

Webb, J. L. (1906) The western pine destroying bark beetle. *U.S. Dep. Agric. Bur. Entomol. Bul.* 58, Pt. II, 30 pp., Washington, D.C.

Webb, J. W. and Franklin, R. T. (1978) Influence of phloem moisture on brood development of the southern pine beetle (Coleoptera: Scolytidae). *Environ. Entomol.* 7: 405–410.

White, J. D. (1981) A bioassay for tunneling responses of southern pine beetles to host extractives. *J. Georgia Entomol. Soc.* 16: 484–492.

Whitehead, A. T. (1981) Ultrastructure of sensilla of the female mountain pine beetle *Dendroctonus ponderosae* (Coleoptera: Scolytidae). *Int. J. Insect Morphol. Embryol.* 10: 19–28.

Whitman, D. W. and Eller, F. (1990) Parasitic wasps orient to green leaf volatiles. *Chemoecology* 1: 69–75.

Whitney, H. S. (1982) Relationships between bark beetles and symbiotic organisms. In: *Bark Beetles in North American Conifers: A System for the Study of Evolutionary Biology* (Mitton, J. B. and Sturgeon, K. B., eds.) pp. 183–211. Univ. Texas Press, Austin, Tex.

Whitney, H. S. and Spanier, O. J. (1982) An improved method for rearing axenic mountain pine beetles *Dendroctonus ponderosae* (Coleoptera: Scolytidae). *Can. Entomol.* 114: 1095–1100.

Whittaker, R. H. (1970) The biochemical ecology of higher plants. In: *Chemical Ecology* (Sondheimer, E. and Simeone, J. B., eds.) pp 43–70. Academic Press, New York.

Williamson, D. L. (1971) Olfactory discernment of prey by *Medetera bistriata* (Diptera: Dolichopodidae). *Ann. Entomol. Soc. Am.* 64: 586–589.

Witanachchi, J. P. and Morgan, F. D. (1981) Behavior of the bark beetle, *Ips grandicollis*, during host selection. *Physiol. Entomol.* 6: 219–223.

Wollerman, E. H. (1979) Dispersion and invasion by *Scolytus multistriatus* in response to pheromone. *Environ. Entomol.* 8: 1–5.

Wong, B. L. and Berryman, A. A. (1977) Host resistance to the fir engraver beetle. 3. Lesion development and containment of infection by resistant *Abies grandis* inoculated with *Trichosporium symbioticum*. *Can. J. Bot.* 55: 2358–2365.

Wood, D. L. (1962) Experiments on the interrelationship between oleoresin exudation pressure in *Pinus ponderosa* and attack by *Ips confusus* (LeC.)(Coleoptera: Scolytidae). *Can. Entomol.* 94: 473–477.

Wood, D. L. (1982) The role of pheromones, kairomones, and allomones in the host selection and colonization behavior of bark beetles. *Annu. Rev. Entomol.* 27: 411–446.

Wood, D. L. and Vité, J. P. (1961) Studies on the host selection behavior of *Ips confusus* (LeConte)(Coleoptera: Scolytidae) attacking *Pinus ponderosa*. *Contrib. Boyce Thompson Inst.* 21: 79–96.

Wood, D. L., Browne, L. E., Bedard, W. D., Tilden, P. E., Silverstein R. M., and Rodin, J. O. (1968) Response of *Ips confusus* to synthetic sex pheromones in nature. *Science* 159: 1373–1374.

Wood, S. L. (1982) The bark and ambrosia beetles of North and Central America (Coleoptera: Scolytidae), a taxonomic monograph. *Great Basin Naturalist Memoirs*, Brigham Young Univ., Provo, Utah.

Wright, L. E., Berryman, A. A., and Gurusiddaiah, S. (1979) Host resistance to the fir engraver beetle, *Scolytus ventralis* (Coleoptera: Scolytidae). 4. Effect of defoliation on wound monoterpenes and inner bark carbohydrate concentrations. *Can. Entomol.* 111: 1255–1261.

6

Host Plant Choice in *Pieris* Butterflies

F.S. Chew
Department of Biology, Tufts University, Medford, MA 02155

J.A.A. Renwick
Boyce Thompson Institute, Tower Road, Ithaca, NY 14853

1. Introduction

Pieris specificity for crucifers was recorded as early as 1660 by John Ray (Mickel, 1973) and the chemical affinity of glucosinolates produced by plants in the four major *Pieris* host plant families—Cruciferae, Tropaeolaceae, Capparaceae, and Resedaceae—was established by Guignard in the 1890s [reviewed by Feltwell, 1982]. Verschaffelt (1910) demonstrated that *P. brassicae* larvae would feed on paper saturated with plant sap expressed from *Bunias orientalis* (Cruciferae). Thorsteinson (1953) tested behavioral responses of these larvae to glucosinolates, and David and Gardiner (1962) showed that *Pieris brassicae* females would lay eggs on nonhost substrates dipped in glucosinolate solutions. *Pieris* butterflies thus provided some of the earliest and most compelling evidence for the biochemical mediation of insect-plant associations.

In addition to the association with Cruciferae, major lineages of pierid butterflies are associated with the Leguminosae, as well as with phytochemically diverse families including Loranthaceae, Ericaceae and Pinaceae (Ehrlich and Raven, 1964). Thus host associations within the Pieridae exhibit both (1) radical gaps that do not appear to involve shared plant chemistry of potential hosts with current hosts; and (2) variation in host specificity among butterflies that feed within groups sharing strong chemical affinities, such as the Cruciferae.

Mechanisms to account for major differences among pierid lineages are unresolved. Based on a suggestion by Singer et al. (1971) that close ecological associations between butterflies and nonfood plants might provide an opportunity for evolution of host association with a novel resource unrelated to the ancestral food, Chew and Robbins (1984) speculated that plants in the Loranthaceae, a family of plant parasites and hemiparasites, might serve as ecological bridges by providing opportunities for butterfly exposure to phytochemically disparate mistletoe hosts. The observation that host-produced quinolizidine alkaloids are transferred to a *Pedicularis* hemiparasite (Schrophulariaceae) from its lupine host (Leguminosae) (Stermitz et al., 1989) suggests a mechanism by which plant parasites could provide a biochemical bridge for butterflies to evolve new host specificity. However, recent cladistic analysis of morphological characters shows that the Loranthaceae-feeding pierid butterflies are unlikely to be ancestral to either legume-feeding or crucifer-feeding lineages (Adrienne Venables, personal communication), so it is now doubtful that the parasitic Loranthaceae provided a bridge between these phytochemically disparate hosts. Further, although legume-feeding pierids generally refuse crucifers and vice versa, Shapiro (1986) reports that Argentinian *Tatochila distincta* feeds on *Astragalus* (Leguminosae) in nature but can be reared on crucifers in the laboratory. *Tatochila* is a large, South American genus, probably polyphyletic, containing both legume-feeding and crucifer-feeding lineages (Shapiro, 1991).

Within each major association between pierid lineage and plant family, diet breadth varies among butterfly species. For example in the Pieridae, crucifer-feeding butterflies in a North African community range from monophagous to broadly oligophagous species whose diets include Resedaceae as well as Cruciferae (Chew and Courtney, 1991). The chemical determinants of these differences, which are based at least partly on presumed affinities and differences among the host plants, are the focus of this paper. Other aspects of the extensive work on ecology of pierines, particularly population dynamics and community ecology, are addressed by Courtney (1986).

The genus *Pieris* historically has included two dozen species. More recently the subgenus *Artogeia,* containing *Pieris rapae, P. napi,* and related species was elevated to generic status along with other subgenera of *Pieris*. Under this revision, in a strict sense *Pieris* would now include only *P. brassicae* and its subspecies (Bernardi, 1947; Kudrna, 1974). However, there is no evidence that the group called *Artogeia* is monophyletic and therefore *Pieris* should continue to be used in a broad sense until we have better evidence to determine phylogenetic relationships within the genus (Feltwell and Vane-Wright, 1982; Robbins and Henson, 1986). We treat *Pieris* in its broad sense in this paper. *P. brassicae, P. rapae* and *P. napi* represent three lineages whose phylogenetic relations are unresolved (R.K. Robbins, personal communication).

2. Sensory Basis of Specificity

Most work on behavioral responses to plant chemicals and sensory physiology in *Pieris* butterflies has involved European strains of *P. brassicae*. More recent work includes various subspecies and strains of *P. rapae* and *P. napi*. These *Pieris* all oviposit on leaf tissue of the hosts where their larvae feed, in contrast to the use of inflorescences by many other pierid butterflies whose behavioral and physiological responses to plant chemicals are as yet unexamined.

Like other butterflies, *Pieris* use visual and chemical cues to locate host plants. During the sequence of behaviors leading to oviposition [reviewed by Renwick and Chew, 1994], chemical cues become critical determinants of acceptance or rejection.

Because glucosinolates in leaf tissue do not yield volatile products until cells are disrupted, contact chemoreception (gustation) has been considered to have primary importance. But the variety of potential volatile glucosinolate-derived products is large [reviewed by Fenwick et al. (1983) and Chew (1988a)]. Glucosinolate-derived volatiles, green-plant volatiles, and glucosinolates characteristic of rapidly growing parts have received increased attention recently, partly as a result of improved analytical methods in the past decade.

2.1. Adults

Although the role of volatiles in host-selection behavior of pierid butterflies is not entirely clear, chemoreceptors on the antennae of several species have been described. Adult antennae of female *Pieris brassicae* bear four types of sensilla on each of 31 distal flagellar segments (excluding six proximal segments), arranged along the flagellum in a groove that faces forward during flight (Behan and Schoonhoven, 1978; Den Otter et al., 1980). These sensilla include mechanosensitive setae and densely packed chemosensory structures of three types: thin trichoid, blunt basiconic, and coeloconic sensilla. Scanning electron micrographs of antennae of both sexes show that both bear these structures [reviewed by Feltwell (1982)]. These observations for *P. brassicae* are similar to findings for the legume-feeding pierids *Colias eurytheme* and *C. philodice* (Grula and Taylor, 1980). Morphological appearance of antennal sensilla of both sexes and both species was nearly identical [Grula and Taylor, 1980; reviewed by Douglas (1986)].

Electrophysiological characterization of antennal response revealed that no cells or sensilla were specific for odors of eggs or cabbage leaf components. Cluster analysis of responses failed to reveal clear patterns, but structurally related compounds showed relatively small differences in response profiles or odor spectra compared to structurally dissimilar compounds (Den Otter et al., 1980). These similarities suggest that receptor cell membranes contain several acceptor sites for specific molecular structures—for which structurally similar compounds may compete; different types of acceptor sites may occur in varying proportions in the membranes of different cells (Den Otter et al., 1980).

Olfactory sensitivity spectra are similar though not identical for female *P. brassicae* and *P. rapae* and may reflect their similar host-plant preferences (van Loon et al., 1992b). Comparisons with sensitivity rank-order data of Behan and Schoonhoven (1978) and Topazzini et al. (1990) led van Loon et al. (1992b) to suggest that the olfactory sensitivity spectra are characteristic at the species level rather than at the level of specific laboratory colonies.

Antennal receptor sensitivities to many compounds changed only slightly after larval and adult experience with *Sinapis arvensis* and *Brassica oleracea*. However, for two of the four crucifer-specific compounds tested—phenylacetonitrile and benzylisothiocyanate (one of three isothiocyanates tested), *P. brassicae* females became more sensitive after larval and adult exposure to *S. arvensis* while *P. rapae* did not (van Loon et al., 1992b). Possibly such effects of experience on female antennal receptors are correlated with effects of experience on oviposition behavior (Traynier and Truscott, 1991), but very little is known about what plant volatiles or concentrations are encountered by insects in nature. Oviposition may proceed without antennae (Ma and Schoonhoven, 1973). Furthermore, antennae function to detect species-specific pheromones that regulate mating behavior in some pierid species [*Colias:* Grula and Taylor (1980)] and that are secreted in the androconial scales of male *P. brassicae, P. rapae,* and *P. napi* (Bergström and Lundgren, 1973).

Sensory physiology of contact chemoreception by adults involves receptors on the proboscis and foretarsi [reviewed by Feltwell (1982) and Chew and Robbins (1984)] and ovipositor valves (Klijnstra and Roessingh, 1986). In female *Pieris brassicae,* foretarsal chemosensory hairs [B-type chemosensory hairs (Ma and Schoonhoven, 1973)] are associated with spines on proximally adjacent tarsomeres. A similar association of spines with sensilla is also found in female Nymphalidae (Calvert and Hanson, 1983) and *Papilio polyxenes* (Roessingh et al., 1991). This anatomical arrangement, found only in females, was the basis of Fox's (1966) suggestion that the spines might function to abrade leaves to make plant compounds accessible to foretarsal chemoreceptors.

One general finding, detailed in Section 3, is that the active forms of contact stimulants as well as deterrents appear to be glucosides. This observation suggests that there may be glycosidic receptor sites for which these compounds may compete. Both adults and larvae are more sensitive to glycosides than to their corresponding aglycones. Given the slow rate of glucosinolate hydrolysis, which would occur for example, only after feeding larvae initiated glucosinolate-thioglucosidase contact by damaging plant cells [reviewed by Chew (1988a)], this observation is not surprising.

However, this finding raises concerns about how ovipositing adults perceive glucosides in intact plants. Spines associated with sensilla on the foretarsi of female butterflies might be capable of piercing plant cell membranes but microscopic examination of leaves following drumming behavior has failed so far to reveal callose deposition or any other signs of cell wounding (Traynier and Hines, 1987). Glucosinolates that stimulate oviposition can be isolated by dipping the leaves in various nonpolar solvents followed by various polar solvents (Renwick et al., 1992; van Loon et al., 1992a). Because the solvent dips were very brief (2 s), and no appreciable color was extracted with the glucosinolates, it is unlikely that compounds were extracted from disrupted cells. These observations combined with the lack of evidence for cell wounding (Traynier and Hines,

1987) suggest that the glucosinolates and other glycosides are in fact present at the leaf surface, but must be tightly bound. Free sugars and amino acids can be isolated from the leaf surface by cold water washes [reviewed by Städler and Roessingh (1990) and Städler (1992)], but these compounds are not active in stimulating oviposition by *Pieris*. One possibility is that glucosinolates are present in the hydrated cell wall just below the leaf cuticle. Glucosinolates accumulate in vacuoles [reviewed by Chew (1988a)], but when crucifers are grown in liquid media (Elliot and Stowe, 1971) or agar (Glenn et al., 1988), the media also accumulate glucosinolates from crucifer root cells. In leaves, hydrated cell wall microstructures might similarly accumulate glucosinolates from leaf cells via the cell membrane. Because cell wall hydration is very resistant to removal (Salisbury and Ross, 1992), accumulated glucosinolates just below the leaf cuticle might require not only removal of waxes and other cuticular components with nonpolar solvents, but also extreme treatments with polar solvents such as boiling water (Renwick et al., 1992) or methanol washes (van Loon et al., 1992a), to be removed. Such occurrence would be consistent with the finding that glucosinolates in water solutions are most stimulatory (Traynier and Truscott, 1991). Possibly the foretarsal spines facilitate contact between sensilla and the hydrated cell wall microstructures by piercing the cuticle during drumming behavior, but not all crucivorous insects that respond to crucifer leaf surfaces exhibit drumming behavior (Städler, 1992).

2.2. Larvae

Gustation in larval *P. brassicae* was the focus of much pioneering work by Schoonhoven and colleagues [reviewed by Feltwell (1982)]. Studies of olfactory and gustatory coding by lepidopterous larvae suggest that the number of receptors is small, and that sensitivity spectra for specific compounds depends on central nervous system integration of input from receptor cells that are differentially sensitive to compounds with specific molecular structures (Schoonhoven and Dethier, 1966). Electrophysiological studies on both *P. rapae* and *P. brassicae* have recently demonstrated a good correlation between sensory and behavioral responses of late instar larvae (van Loon, 1990). Responses to five phenolic acids and seven flavonoids were tested, and certain structural requirements for activity were suggested. Chemoreception of amino acids by these *Pieris* species has also been investigated, and differences between responses of the species as well as activities of the different amino acids were noted (van Loon and van Eeuwijk, 1989). However, no attempt was made to relate these electrophysiological responses to behavior.

3. *Pieris* Behavioral Responses to Plant Compounds

3.1. Adult Oviposition

Recognition of host plants for oviposition by *Pieris* species is known to depend on both visual and chemical cues. Vision appears to be the predominant sensory modality in the orientation (precontact) phase of host finding, and alightment on a potential host depends to a large extent on the reflectance spectrum of the leaf surface (Ilse, 1937; Kolb and Scherer, 1982). In the case of *P. rapae,* it has been shown that other visual cues such as shape and size of foliage do not play a role in host discrimination (Renwick and Radke, 1987). Electroantennogram responses have been obtained to plant volatiles for both *P. rapae* and *P. brassicae* (van Loon et al., 1992b), but these were not related to behavior. No clear evidence for the involvement of volatiles in attracting butterflies to hosts has been reported. Nevertheless, volatile compounds released from damaged plants may result in avoidance of these plants (Renwick and Radke, 1983) and electrophysiological responses may reflect this behavior. The most critical chemical cues responsible for triggering oviposition behavior appear to be contact stimuli which are perceived at the leaf surface after landing (Renwick and Radke, 1983, 1987).

Glucosinolates have long been thought to play a role in stimulating oviposition by *Pieris* species, and butterflies have been induced to lay eggs on treated artificial substrates or on leaves of nonhosts that were immersed in a glucosinolate solution (David and Gardiner, 1962; Terofal, 1965; Ma and Schoonhoven, 1973). However, most early studies relied on the use of sinigrin as a representative glucosinolate, and confusing or contradictory results led to questions about the validity of previous assumptions about the key role of glucosinolates (Renwick and Radke, 1983; Traynier, 1984; Chew, 1988b). Subsequent studies have shown that sinigrin is not typical of the glucosinolates in the behavior it elicits from *P. rapae* adults, and that different glucosinolates vary dramatically in their stimulatory activity (Fig. 6.1). An additional reason for the confusing results obtained with sinigrin is that individual *P. rapae* females differ in their behavior. This variation may well reflect individual variation and inconsistency observed at the electrophysiological level in *P. brassicae* (Schoonhoven, 1977). In behavioral experiments, a majority of individuals failed to oviposit on a substrate treated with sinigrin, and those individuals that did respond, tended to lack the ability to discriminate between treated and control substrates (Renwick, 1988).

Recent advances in the methodology for glucosinolate analyses have resulted in the discovery that certain specific glucosinolates are powerful stimulants that induce oviposition by females of *Pieris* species. Independent studies in The Netherlands and the United States have identified 3-indolylmethyl glucosinolate (glucobrassicin) as the major stimulant responsible for oviposition by *P. brassicae* and *P. rapae* on cabbage plants (van Loon et al., 1992a; Renwick et al. 1992). Other glucosinolates present in cabbages—sinigrin and glucoiberin—had greatly reduced or no activity as stimulants for *P. rapae* (Renwick et al., 1992). Traynier and Truscott (1991) have also shown that glucobrassicin more effectively strengthens associative learning behavior in *P. rapae*. Further experiments have been conducted to determine how *P. rapae* recognize host plants that do not produce glucobrassicin. Benzyl glucosinolate (glucotropaeolin) in *Tropaeolum majus,* allyl glucosinolate (sinigrin) in *Brassica juncea*, and 1-sulfonyl-3-indolyl-glucosi-

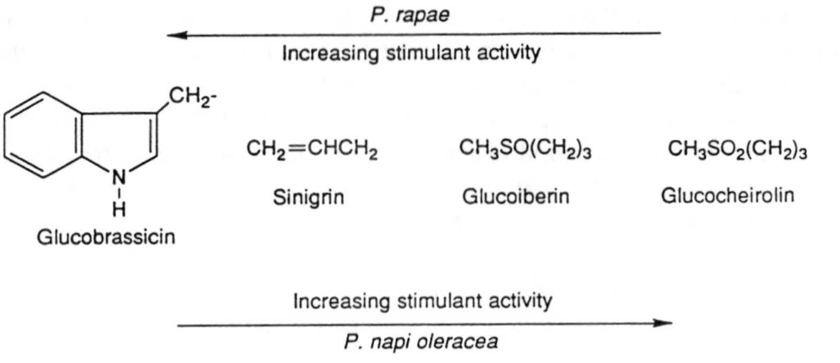

P. rapae

← Increasing stimulant activity

Glucobrassicin (indol-3-yl CH_2-)	$CH_2{=}CHCH_2$	$CH_3SO(CH_2)_3$	$CH_3SO_2(CH_2)_3$
Glucobrassicin	Sinigrin	Glucoiberin	Glucocheirolin

Increasing stimulant activity →

P. napi oleracea

The structures shown are for the R-group in the general formula for glucosinolates:

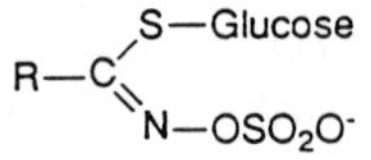

$$R{-}C(S{-}Glucose){=}N{-}OSO_2O^-$$

Figure 6.1. Stimulant activity of glucosinolates for *Pieris rapae* and *P. napi oleracea* oviposition.

nolate in *Isatis tinctoria* stimulate oviposition on these plants (Sachdev-Gupta et al., 1992). Eight glucosinolates have now been shown to stimulate oviposition by *Pieris* butterflies (Table 1), but a high degree of specificity exists. *P. napi oleracea* and *P. rapae* respond most strongly to distinctly different groups of glucosinolates (Fig. 6.1).

Avoidance or rejection of nonhost plants for oviposition by *Pieris* butterflies can be attributed to the presence of oviposition deterrents in these plants. Early experiments showed that ground tissue or extracts of nonhosts can deter oviposition by *Pieris* species (Lundgren, 1975). Later work by Renwick and Radke (1985) demonstrated the presence of oviposition deterrents to *P. rapae* in a range of nonhosts that included two crucifers, *Erysimum cheiranthoides* and *Capsella bursa-pastoris*. The major deterrent compounds from *E. cheiranthoides* were recently identified as the cardenolides erysimoside and erychroside (Renwick et al., 1989; Sachdev-Gupta et al., 1990). When extracts containing these cardenolides were applied to cabbage plants in the field, a significant reduction in oviposition by a natural population of *P. rapae* occurred, at least over a short period of time (Dimock and Renwick, 1991). Similar studies in Europe with *P. brassicae* have shown that a strophanthidin-based cardenolide is responsible for deterrent activity in the Siberian wallflower, *Cheiranthus allionii* (Rothschild et al., 1988). *Iberis amara* is another crucifer that is generally rejected by *P. rapae* adults, and oviposition deterrents in this plant have been identified as 2-O-β-D-glucosyl cucurbitacin I and 2-O-β-D-glucosyl cucurbitacin E (Huang et al., 1993a). The

Table 6.1. Oviposition stimulants affecting host selection by Pieris *species.*

Stimulants	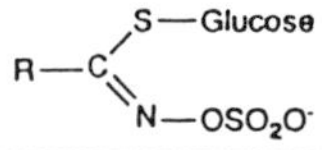 $R{-}C(S{-}Glucose){=}N{-}OSO_2O^-$ R-Group	*Pieris* species	Plant Source	Reference
Glucobrassicin	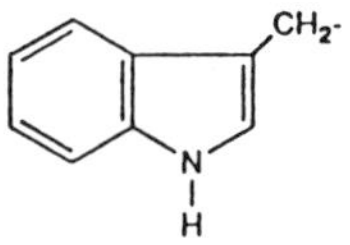 CH_2- N H	*rapae* *brassicae*	Cabbage	1, 2
Sinigrin	$CH_2{=}CHCH_2$	*rapae* *napi oleracea* *brassicae*	many crucifers	1, 3, 4, 6
Glucoiberin	$CH_3SO(CH_2)_3$	*rapae* *napi oleracea*	*Iberis amara*	3
Glucocheirolin	$CH_3SO_2(CH_2)_3$	*napi oleracea*	*Erysimum cheiranthoides*	5
Glucotropaeolin	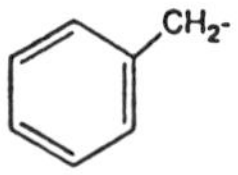 CH_2-	*rapae*	*Tropaeolum majus* & *Carica papaya*	6
Glucobrassicin-1-sulfonate	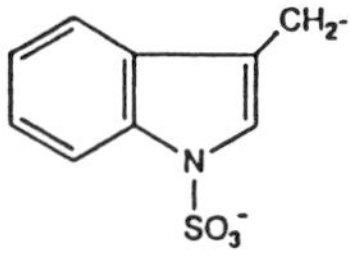 CH_2- N SO_3^-	*rapae*	*Isatis tinctoria*	6
(2R)-Glucobarbarin (Glucosibarin)	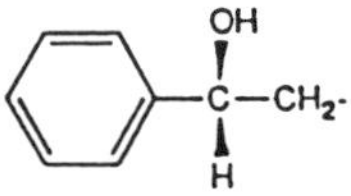 OH C—CH_2- H	*rapae* *napi oleracea*	*Barbarea vulgaris*	7
(2S)-Glucobarbarin	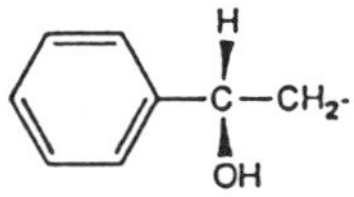 H C—CH_2- OH	*rapae* *napi oleracea*	*Barbarea vulgaris*	7

References: 1. Renwick *et al.*, 1992; 2. van Loon *et al.*, 1992a; 3. Schoonhoven, 1972; 4. Huang *et al.*, 1993b; 5. Huang *et al.*, 1993a; 6. Sachdev-Gupta *et al.*, 1992; 7. Huang *et al.*, 1993c.

cucurbitacin E glycoside was already identified as a feeding deterrent to *P. rapae* larvae (Sachdev-Gupta et al., 1993b). A list of known oviposition deterrents isolated from crucifers is presented in Table 6.2.

The presence of both oviposition stimulants and deterrents in a plant has been demonstrated in the case of *P. rapae* and *Erysimum cheiranthoides* (Renwick and Radke, 1987). The positive and negative contact stimuli were separated by means of solvent partitioning between water and n-butanol. This protocol has provided a means for detecting the presence of stimulants and deterrents in a wide range of plants. Comparative studies on *Pieris rapae* and *P. napi oleracea* have recently shown that differential preferences of the two species for potential host plants can be explained by their differential sensitivities to stimulants and deterrents in the plants (Huang and Renwick, 1993) (Table 6.3). Subsequent examination of the responses of the two butterflies to *Erysimum cheiranthoides* and its active constituents has revealed that *P. napi oleracea* oviposition is stimulated by glucocheirolin and glucoiberin. The acceptance of *E. cheiranthoides* by *P. napi* is further explained by its relative lack of sensitivity to the cardenolides that deter *P. rapae* (Huang et al., 1993a). In a similar study with *Iberis amara, P. napi oleracea* was insensitive to the deterrent cucurbitacin glycoside and was stimulated by glucoiberin and sinigrin (Huang et al., 1993a). Structure-activity studies on the responses of *P. rapae* and *P. napi oleracea* to a range of glucosinolates suggest that *P. rapae* is most sensitive to indoles, whereas *P. napi* respond more strongly to the sulfinylalkyl and sulfonylalkyl glucosinolates (Fig. 6.1) (Huang and Renwick, unpublished data).

Oviposition by *Pieris* butterflies may also be influenced by nonplant factors. Rothschild and Schoonhoven (1977) demonstrated the existence of a mechanism whereby *P. brassicae* avoids oviposition on plants that have a high egg load or where larvae are feeding. This behavior was found to be mediated by both visual and chemical cues. Visual cues with oviposition-deterring functions in the Pieridae are reviewed by Shapiro (1981). For *P. brassicae,* perception of chemical cues involves both olfactory and contact chemoreceptors (Behan and Schoonhoven, 1978). An oviposition-deterring pheromone [ODP, also termed host marking pheromone (Städler, personal communication)] can be extracted into water from the eggs, and appears to be stable and relatively nonvolatile (Schoonhoven et al., 1981). Subsequent experiments with egg washes from *P. brassicae* and *P. rapae* indicated that the ODP from one species was also active in deterring oviposition by the other. Also, the tarsal receptors and antennae of both species responded to ODPs from their own as well as the other's eggs (Klijnstra and Roessingh, 1986; Schoonhoven et al., 1990a). Contact chemoreceptors on the ovipositor valves of *P. brassicae* also responded electrophysiologically to ODP (Klijnstra, 1982). Field studies on *P. rapae* have caused some doubt about the importance of the ODP in oviposition behavior and egg distribution (Ives, 1978; Root and Kareiva, 1984). However, Schoonhoven et al. (1990b) used *P. rapae* from Europe and North America to show that both populations

Table 6.2. Oviposition deterrents from crucifers affecting host selection by Pieris *species.*

Deterrents	*Pieris* species	Plant Source (Reference)
Erysimoside 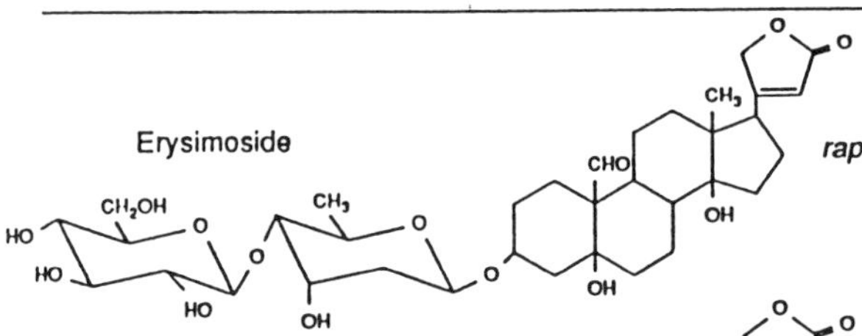	*rapae*	*Erysimum cheiranthoides* (1)
Erychroside 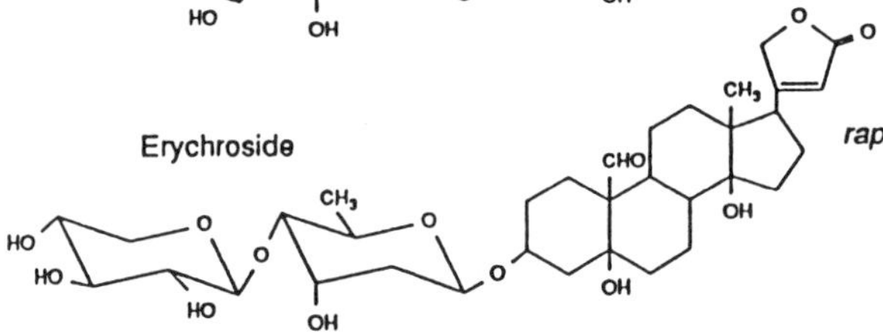	*rapae*	*Erysimum cheiranthoides* (1)
2-O-β-D-glucosyl cucurbitacin E 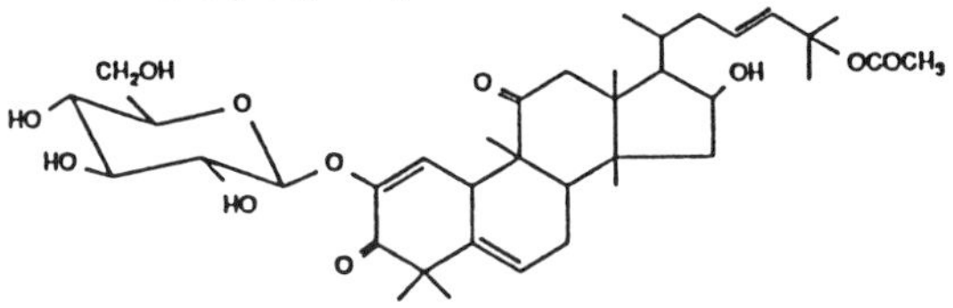	*rapae*	*Iberis amara* (2)
2-O-β-D-glucosyl cucurbitacin I 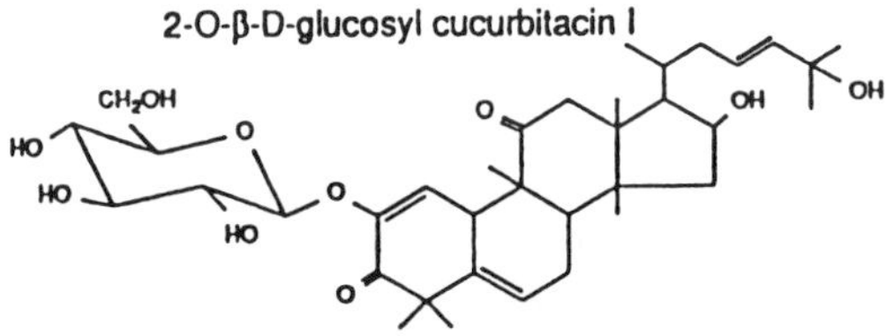	*rapae*	*Iberis amara* (2)
A strophanthidin glycoside	*brassicae*	*Cheiranthus allionii* (3)

References: 1. Sachdev-Gupta *et al.*, 1990; 2. Huang *et al.*, 1983b; 3. Rothschild *et al.*, 1988.

Table 6.3. Presence of oviposition stimulants and deterrents in hosts or potential hosts of P. rapae *and* P. napi oleracea[1]

Plant species	*P. rapae*		*P. napi oleracea*	
	Stimulant	Deterrent	Stimulant	Deterrent
Alyssum saxatile	X		X	X
Cleome spinosa	X		X	X
Conringia orientalis	X		X	X
Erysimum cheiranthoides		X	X	
Iberis amara	X	X	X	X
Lunaria annua		X	X	X
Brassica juncea	X		X	
Tropaeolum majus	X	X		X
Isatis tinctoria	X	X		X

[1]Based on data from Huang and Renwick (1992).

produced and responded, behaviorally and electrophysiologically, to egg washes and gland extracts in laboratory assays. Chemical studies on the ODP have subsequently resulted in the identification of three active compounds from *P. brassicae* and *P. rapae* as novel cinnamic acid derivatives (J.J.A. van Loon, personal communication).

Gravid female butterflies are obviously confronted with an array of opposing positive and negative chemical cues as they approach or make contact with a plant. The insect's behavioral response is likely to depend on the balance of these sensory inputs (Miller and Strickler, 1984). *P. rapae* butterflies refuse to lay eggs on *Erysimum cheiranthoides,* despite the presence of a stimulant, because of the overwhelming effect of deterrents in the plant (Renwick and Radke, 1987). Other plants are quite acceptable hosts despite the presence of deterrents (Renwick and Radke, 1985). Thus subtle changes in the relative concentrations of stimulants and deterrents in a plant may affect its acceptability as a host. Such changes may occur as a result of nutritional or other environmental factors. Preliminary studies on *P. rapae* suggest that nitrogen fertilization influences the ratios of oviposition stimulants and deterrents at the surface of *E. cheiranthoides* leaves (U. Hugentobler and J.A.A. Renwick, unpublished results). Experiments to alter the balance of stimulants and deterrents in a plant or in individual leaves suggest that the degree of acceptance or rejection of a plant is in fact dependent on the balance of positive and negative signals at the leaf surface (Hugentobler and Renwick, unpublished results).

3.2. Larval Feeding

Many chemical compounds have been tested on *Pieris brassicae* as feeding stimulants or deterrents, based on both behavioral and electrophysiological responses (Schoonhoven, 1972). Sinigrin was active as a stimulant, and several alkaloids, ecdysterone, and quinine proved to be deterrents. David and Gardiner (1966) observed that nine glucosinolates stimulated *P. brassicae* larval feeding on semisynthetic diet. Of these, methyl and benzyl glucosinolates were the most effective feeding stimulants. However, most of these compounds were chosen because of their activity for other insects or because of their association with host plants. Only recently was a systematic approach to isolate and identify deterrents from nonhost plants adopted to explain rejection of these plants by larvae. Dimock et al. (1991) showed that *Erysimum cheiranthoides* contains butanol-soluble compounds that deter feeding by *P. rapae*. These were characterized as cardenolides, but differed from the compounds that were previously identified as oviposition deterrents (Sachdev-Gupta et al., 1990). Subsequent chemical studies on the feeding deterrents resulted in the identification of digitoxigenin-based cardenolides as active constituents (Sachdev-Gupta et al., 1993b). These results suggest that cardenolides in *E. cheiranthoides* have differential effects on the larvae and adults of *P. rapae*. Cardenolides have been isolated (although not identified) from *Erysimum asperum* (Rodman et al. 1982) and may account for unconditional refusal of this plant by both ovipositing adults and feeding larvae of *P. napi macdunnoughii* (Rodman and Chew, 1980) and *P. occidentalis* (Chew, 1975). The refusal of *P. rapae* to feed on *Iberis amara* has also been investigated, and two cucurbitacin glycosides were isolated and identified from deterrent extracts of foliage. However, only one of these, 2-O-β-D-glucopyranosyl cucurbitacin E, proved to be an effective feeding deterrent (Sachdev-Gupta et al., 1993a). A list of *Pieris* larval feeding deterrents identified from crucifers is presented in Table 6.4.

4. Ecological Context of *Pieris* Behaviors

Adult female and larval perception of stimuli at the plant surface depends primarily on chemoreception and sensory coding integrated by the central nervous system. We know little about how sensory input is transduced into behavioral responses. However, the two life stages are under quite different ecological constraints to accept or reject host plants. Female oviposition behavior represents a compromise between accurately assessing host suitability for larvae and laying all of the egg complement before dying [reviewed by Courtney (1986)]. By contrast, larvae with limited mobility are under selection to accept whatever plants their mothers choose for oviposition. Also at issue are the different scales at which each life stage operates. Ovipositing females are able to sample individual plants and different species growing in a community over short periods (minutes to hours), while *Pieris* larvae gain sufficient mobility to move successfully between plants only after they are more than half-grown—a period of days to weeks (Chew, 1975, 1980; Davies and Gilbert, 1985).

Our current knowledge about structure-function relationships and differential

Table 6.4. Feeding deterrents to Pieris rapae *from potential host plants.*

Deterrents	Plant Source	Reference
Erysimoside 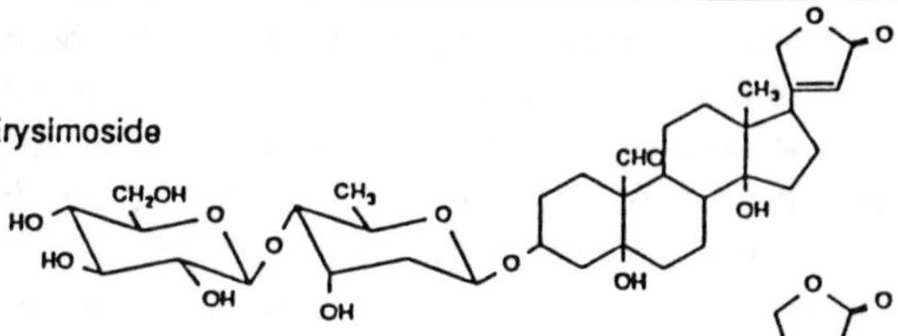	*Erysimum cheiranthoides*	1
Erychroside 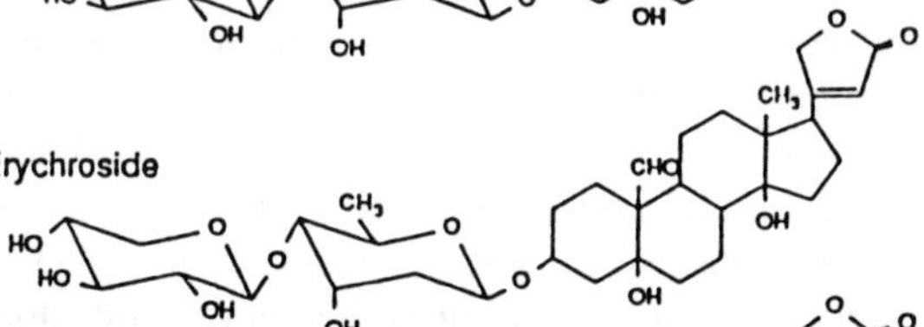		
Digitoxigenin 3-O-β-D-glucoside 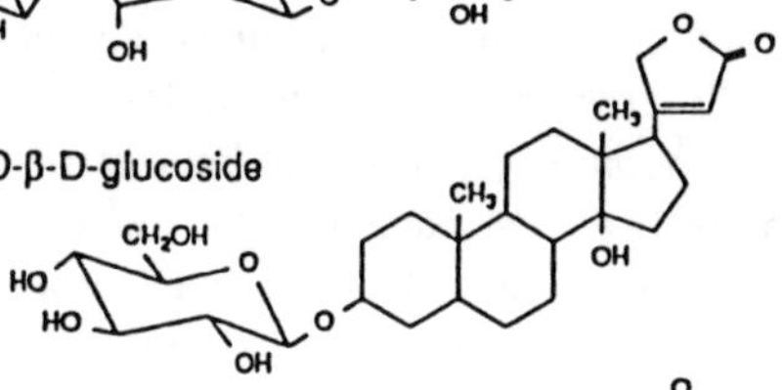		
Glucodigigulomethyloside 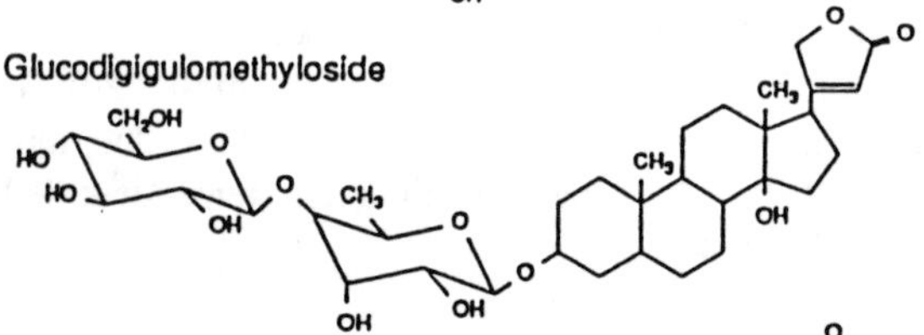		
Glucodigifucoside 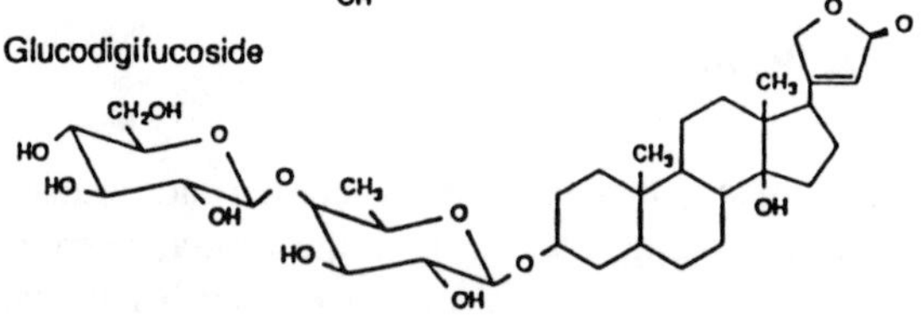		
2-O-β-D-glucopyranosyl cucurbitacin E 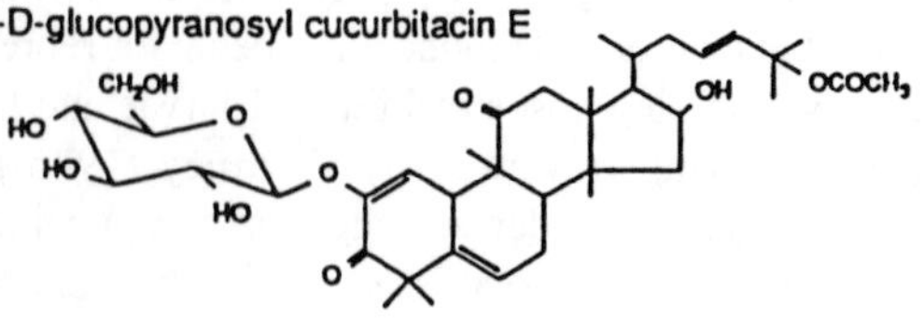	*Iberis amara*	2

References: 1. Sachdev-Gupta *et al.*, 1993b; 2. Sachdev-Gupta *et al.*, 1993a.

sensitivity of *Pieris* species to specific glucosinolates and deterrent glycosides now accounts for some of the behavioral specificity observed among pierids in natural populations. Based on studies showing differential sensitivity of *P. rapae* and *P. napi* to both cardenolide deterrents and specific glucosinolates, we can now reject the speculation (Rodman and Chew, 1980) that *Erysimum asperum* was refused by ovipositing *P. napi macdunnoughii* because its leaves contain 3-methylsulfinylpropyl glucosinolate. Although we do not yet know how *P. napi macdunnoughii* compares to *P. n. oleracea* in its sensitivities, at worst the 3-methylsulfinylpropyl glucosinolate is likely to be a very weak stimulant rather than a deterrent. It seems more likely that the unidentified cardenolides in this plant (Rodman et al., 1982) will have deterrent activity, and occur in sufficiently high concentrations to outweigh the stimulatory effect that the glucosinolate may have on this *P. napi* subspecies.

Recent findings suggest that for the *Pieris*-crucifer association, lack of a stimulant alone would not confer protection against specialist herbivores such as *Pieris* butterflies. Many crucifers that are refused contain very low foliar concentrations of glucosinolates or they contain glucosinolates for which the insect has low sensitivity (Rodman and Chew, 1980; Renwick and Radke, 1985; Huang et al. 1993a, 1993b). Hence these plants contain reduced levels of attractive or stimulant compounds and may be candidates for ecological escape from *Pieris* and other crucivorous insects (Feeny, 1975). But bioassay-guided fractionation of foliar extracts has revealed the additional presence of deterrent compounds in the few such species examined so far. Moreover, unapparent crucifers with reduced pierid encounter rates (sensu Feeny, 1975) are unconditionally refused by both ovipositing adults and pierine larvae including *P. rapae* and *P. brassicae* (Courtney, 1986; Courtney and Chew, 1987). Because newly hatched *Pieris napi* and *P. rapae* larvae will often eat semisynthetic diets lacking crucifer material or glucosinolates (Webb and Shelton, 1988; Chew et al., unpublished observations), this unconditional refusal suggests these crucifers contain deterrents. It now appears probable that crucifers do not produce very low levels of stimulatory glucosinolates without also producing other compounds with deterrent activity.

In natural situations, *Pieris* are extremely adept in locating crucifers. However, *P. napi* in North America (at least *P.n. oleracea* and *P.n. macdunnoughii* as two exemplars) sometimes make apparently maladaptive choices, laying eggs on crucifers that do not support complete larval development (Chew, 1975, 1977a, 1977b, 1981). The greater sensitivity of *P. napi oleracea* to sinigrin and other alkyl glucosinolates compared to *P. rapae* suggests a possible mechanism for these mistakes in *P. napi macdunnoughii,* since the maladaptive choices in that plant community involve sinigrin in contrast to branched alkyl glucosinolates (Rodman and Chew, 1980). In the case of *Pieris virginiensis* and *P. napi oleracea* ovipositing on *Alliaria petiolata* (Bowden, 1971), sinigrin present in leaf tissue (Chew, unpublished results; Renwick, unpublished results) may also be implicated. By contrast, *P. rapae* has never been reported to accept non-suitable plants under natural conditions, even though this now-cosmopolitan species has undoubtedly encountered many crucifers not found in its native Palearctic region. These observations suggest some alternative hypotheses about evolution of *Pieris* behavior toward potential host plants.

1. Adults and larvae may have similar sensory constraints, at least for glucosidic compounds. For example, glucobrassicin is the most stimulatory glucosinolate tested for both *Pieris brassicae* and *P. rapae* adults and it is the most effective stimulus for the larval receptor that responds to glucosinolates (Schoonhoven, 1972). Sensory coding in adults and larvae may be constrained so that adults and larvae exhibit similar, though not necessarily identical, sensitivities [as demonstrated by differential activity of different classes of cardenolides on *P. rapae* adults and larvae (Sachdev-Gupta et al. (1993b)]. Such similar sensory constraints on larval and adult stages of individual insects might explain why *P. napi* often accepts novel crucifers as both adult and larva, while *P. rapae* larvae do not feed on crucifers refused by their ovipositing mothers.

Work on other pierid butterflies demonstrates genetic variation among females for oviposition preference [*Colias:* Tabashnik et al. (1981)]. *Pieris napi* oviposition behavior may eventually evolve to more accurately reflect host suitability for larvae. Van Loon et al. (1992b) showed that *P. brassicae* and *P. rapae* females had similar—though apparently species-specific—electroantennogram responses to volatiles, but whether behavioral specificity is a function of differential sensitivity at the sensory level or differential processing at the central nervous system level is not known.

2. *P. rapae* may be under stronger selection than *P. napi* to oviposit accurately. *P. rapae* is often multivoltine and may be nearly continuously brooded in warmer regions of its now-cosmopolitan distribution. By comparison *P. napi* subspecies are strongly seasonal insects throughout their extensive northern hemisphere range. Where adult flight seasons are sharply defined, females may gain no increase in fitness by choosing larval hosts that support rapid growth because adult offspring emerging asynchronously with the rest of the population will not easily find mates (Wiklund et al., 1991). In strongly bivoltine *P. napi,* host plant choice may be less critical as long as larvae develop successfully [cf. Scriber and Lederhouse (1992)]. By contrast, in *P. rapae* in the northeast United States, where there are overlapping generations and multiple broods, selection may favor rapidly developing larvae (Gilbert, 1986; van der Reijden and Chew, 1992). However, host selection is not the only mechanism for growth acceleration. This acceleration might equally be achieved through thermal microsite selection by ovipositing females, as exhibited by *P. rapae* in Japan (Ohsaki, 1980, 1986).

3. Alkyl glucosinolates may correlate poorly with *P. rapae* host suitability. *Pieris rapae* may have evolved reduced sensitivity to alkyl glucosinolates in favor of greater sensitivity for indolyl glucosinolates, because the latter were better indicators of host suitability than the former. By contrast, for *P. napi* the

occurrence of alkyl glucosinolates among North American crucifers may have been better correlated with highly suitable hosts (Rodman and Chew, 1980). Some alkyl glucosinolates such as sinigrin and biosynthetically related compounds such as the thioalkyl, sulfinylalkyl and sulfonylalkyl glucosinolates are very widely distributed among the Cruciferae and figure prominently in the glucosinolate profiles of many seeds [reviewed by Rodman (1981) and Daxenbichler et al. (1991)] and leaves (Cole, 1976) from species of Paleartic origin. [Whether sinigrin and related glucosinolates are more widespread than the indolyl glucosinolates in ecologically relevant plant parts of Palearctic crucifers or any other crucifers awaits systematic surveys using green plant material. Despite "fair predictive value" in correspondence between glucosinolate patterns in seeds and leaves of cabbage cultivars (Tookey et al., 1980), older surveys used methods that would not have detected indolyl glucosinolates and new methods have been available only in the past decade (Truscott et al., 1983)]. If sinigrin and related compounds are particularly prevalent among crucifers from Palearctic centers of diversity for the Cruciferae (Hedge, 1976), they may be associated with both suitable and unsuitable host species, and are therefore poor correlates of host suitability. Many crucifer species refused by Palearctic *P. brassicae* and *P. rapae* (Courtney and Chew, 1987) contain sinigrin and other alkylglucosinolates in leaves (Cole, 1976). In North America, weedy crucifers of Palearctic origin became naturalized not more than a few hundred years ago (Munz and Keck, 1968; Fernald, 1950; Harrington, 1954; Chew, 1977a, 1981).

North American pierids may be particularly sensitive to alkyl glucosinolates as a function of prior selection for sensitivity in plant communities where the most highly suitable hosts contain alkyl glucosinolates (Rodman and Chew, 1980). Alternatively, sensitivity to alkyl glucosinolates may be an ancient Holarctic pierid trait, conserved in the North American fauna in the absence of selection favoring other sensitivities. These pierids may then make mistakes in encounter with recently naturalized unsuitable Palearctic crucifers containing alkyl glucosinolates. Similar oviposition "mistakes" by several pierids including euchloeine pierids, *P. napi oleracea, P. napi marginalis,* and *P. virginiensis*, on sinigrin-containing *Alliaria petiolata* (Bowden, 1971; Chew, 1977b, and unpublished observations) suggest that *P. napi's* sensitivity to alkyl glucosinolates may be shared by other pierid lineages in the North American fauna. However, oviposition mistakes by various pierids on *Barbarea vulgaris,* whose leaves do not contain sinigrin (Chew, 1977, 1981; Renwick, unpublished; J. Nielsen, personal communication), suggest that the plant stimuli may be more complex than presence or absence of sinigrin.

4. Indolyl glucosinolates may correlate especially well with *P. brassicae* host suitability and availability. *P. brassicae* and *P. rapae* may also have evolved increased sensitivity to indolyl glucosinolates because these occur in particularly suitable host plants (cf. van Loon et al., 1992a). In Europe, the geographic source for *P. brassicae* and the European *P. rapae* colonies used in the studies

reviewed in this paper, *Brassica* species are the most widely cultivated crucifers [reviewed by Hedge (1976)], and indolyl glucosinolates are characteristic of green parts of many *Brassica* cultivars (Truscott et al., 1983; Bodnaryk, 1992), *Isatis tinctoria* (Truscott et al., 1983), *Capparis spinosa* (Schraudolf, 1989) and *Moricandia arvensis* (Belkhiri and Lockwood, 1990). Did increased sensitivity to indolyl glucosinolates permit these butterflies to attain pest status or did sensitivity evolve after pest status was attained? Available evidence points to both possibilities. Adults in indigenous, nonagricultural North African populations of these two butterflies are associated with crucifer populations—not necessarily *Brassica*—that are dense and relatively stable from year to year (Chew and Courtney, 1991). So it is hardly surprising that these butterflies become pests in crucifer agroecosystems where host plants are dense and plant population density is stable from year to year (Chew, in press). But some North African plants, such as *Isatis tinctoria* and *Capparis spinosa*, both exhibit dense, stable populations of large plants (Chew and Courtney, 1991) and contain indolyl glucosinolates. Whether sensitivity to indolyl glucosinolates was a prerequisite to or consequence of pest status on *Brassica* crops will be resolved only by additional data on the distribution of indolyl glucosinolates and sensitivities of North African—versus European—*P. brassicae* and *P. rapae* to them.

5. Concluding Remarks

Specificity of adult oviposition behavior in *Pieris* can now be explained by differential sensitivity to stimulant and deterrent compounds perceived on contact with a plant. Compared, for example to chemical cues determining oviposition behavior in *Papilio* butterflies [reviewed by Nishida (1993)] where synergistic interactions may involve as many as ten compounds, the surprising simplicity of oviposition stimuli in the *Pieris*-crucifer association may provide an excellent model system for exploring structure-function relationships among the compounds. In addition, it may provide another character for examining relationships among lineages within the genus *Pieris*.

Our knowledge of larval specificity is not so clear. In addition to the glucosinolates, other compounds may be involved in stimulating *Pieris* feeding behavior (van Loon and van Eeuwijk, 1989; van Loon, 1990). The involvement of other compounds in host-related behaviors in other crucivorous insects [reviewed by Städler (1992)] cautions us against premature generalization.

We know very little about the chemical stimuli that butterflies and larvae confront. Despite our ability to explain several features of host specificity in these few species of butterflies, there are additional complexities. For example, parasitoids and predators have been suggested as important agents in evolution of host specificity (Bernays and Graham, 1988). Plant odors, insect effects on plant odors and combinations of insect and plant products are implicated in parasitoid searching for hosts and parasitoid learning (Keller, 1990; Turlings et al., 1990; reviewed by Vet and Dicke, 1992; Vet et al., 1994). Headspace volatiles are different when cabbage plants are damaged by *P. brassicae* or by *P. rapae* (van Loon, personal communication). Flea beetle damage can induce increases in indolyl glucosinolates of *Brassica* cultivars (Koritsas et al., 1991; Bodnaryk, 1992). It would be extremely interesting to examine whether this increase renders such a plant more attractive to ovipositing *P. brassicae* or *P. rapae*. Such increase, if accompanied by increased susceptibility to *Pieris* oviposition, may account for the ecological observation that individuals of *Isatis tinctoria* were simultaneously attacked by crucifer specialists belonging to several guilds (Chew and Courtney, 1991).

Sensory perception of plant compounds sets limits on host utilization by *Pieris* species and most other herbivorous insects (Jermy, 1993). But habitat and host plant utilization patterns of *Pieris* adults also constrain the range of plants—and presumably the diversity of compounds—to which larvae are exposed (Courtney, 1986; Chew and Courtney, 1991). Further, visual stimuli and plant community diversity affect *Pieris* ability to locate host plants [reviewed by Courtney (1986)]. We are optimistic that detailed laboratory observations, combined with observations to identify ecologically relevant plants and compounds will enable us to determine how sensory and ecological factors interact to generate differential host utilization patterns among different *Pieris* species.

Acknowledgments

This work was supported in part by NSF grants IBN-9107322 and IBN-9108987, USDA grant 91-37302-6198, and the Arabis Fund (Tufts University). We thank Adrienne Venables and R.K. Robbins for discussion regarding pierid phylogeny, G.S. Ellmore for discussion of plant cell wall hydration, and Erich Städler and Jens Kvist Nielsen for discussion and comments on an early version. FSC dedicates this paper to the memory of Adrienne Venables, who died while this paper was in the final stages of preparation.

References

Behan, M. and Schoonhoven, L.M. (1978) Chemoreception of an oviposition deterrent associated with eggs in *Pieris brassicae*. *Entomol. exp. appl.* 24: 163–179.

Belkhiri, A. and Lockwood, G.B. (1990) An indole derivative and glucosinolates from *Moricandia arvensis*. *Phytochemistry* 29: 1315–1316.

Bergström, G. and Lundgren, L. (1973) Androconial secretion of three species of butterflies of the genus *Pieris*. *Zoon* (Suppl.) 1: 67–75.

Bernardi, G. (1947) Révision de la classification des espèces holoarctique des genres *Pieris* Schr. et *Pontia* Fabr. *Miscnea entomol.* 44: 65–79.

Bernays, E.A. and Graham, M. (1988) On the evolution of host specificity in phytophagous arthropods. *Ecology* 69: 886–892.

Bodnaryk, R.P. (1992) Effects of wounding on glucosinolates in the cotyledons of oilseed rape and mustard. *Phytochemistry* 31: 2671–2677.

Bowden, S.R. (1971) American white butterflies and English foodplants. *J. Lepid. Soc.* 25: 6–12.

Calvert, W.H. and Hanson, F.E. (1983) The role of sensory structures and preoviposition behavior in oviposition by the patch butterfly, *Chlosyne lacinia*. *Entomol. exp. appl.* 33: 179–187.

Chew, F.S. (1975) Coevolution of pierid butterflies and their cruciferous foodplants. I. The relative quality of available resources. *Oecologia* 20: 117–127.

Chew, F.S. (1977a) Coevolution. . .II. the distribution of eggs on potential foodplants. *Evolution* 31: 568–579.

Chew, F.S. (1977b) The effects of introduced mustards (Cruciferae) on some native North American cabbage butlerflies (Lepidoptera: Pieridae). *Atala* 5: 13–19.

Chew, F.S. (1980) Foodplant preferences of *Pieris* caterpillars. *Oecologia* 46: 347–353.

Chew, F.S. (1981) Coexistence and local extinction in two pierid butterflies. *Am. Nat.* 118: 655–672.

Chew, F.S. (1988a) Biological effects of glucosinolates. *Am. Chem. Soc. Symp.* 380: 155–181.

Chew, F.S. (1988b) Searching for defensive chemistry in the Cruciferae. In: *Chemical Mediation of Coevolution*, (Spencer, K.C., ed.), pp. 81–112. Academic Press, San Diego, Calif.

Chew, F.S. (1995) From weeds to crops: Changing habitats of *Pieris* butterflies. *J. Lepid. Soc.* (in press).

Chew, F.S. and Courtney, S.P. (1991) Plant apparency and evolutionary escape from insect herbivory. *Am. Nat.* 138: 729–750.

Chew, F.S. and Robbins, R.K. (1984) Egg-laying in butterflies. *Symp. R. Entomol. Soc. Lond.* 11: 65–79.

Cole, R.A. (1976) Isothiocyanates, nitriles, and thiocyanates as products of autolysis of glucosinolates in Cruciferae. *Phytochemistry* 15: 759–762.

Courtney, S.P. (1986) The ecology of pierid butterflies: Dynamics and interactions. *Adv. Ecol. Res.* 15: 51–131.

Courtney, S.P. and Chew, F.S. (1987) Coexistence and host use by a large community of pierid butterflies: Habitat is the templet. *Oecologia* 71: 210–220.

David, W.A.L. and Gardiner, B.O.C. (1962) Oviposition and the hatching of eggs of *Pieris brassicae* (L.) in a laboratory culture. *Bull. Entomol. Res.* 53: 91–109.

David, W.A.L. and Gardiner, B.O.C. (1966) Mustard oil glucosides as feeding stimulants for *Pieris brassicae* larvae in a semi-synthetic diet. *Entomol. exp. appl.* 9: 247–255.

Davies, C.R. and Gilbert, N. (1985) A comparative study of the egg-laying behavior and larval development of *Pieris rapae* L. and *P. brassicae* L. on the same host plants. *Oecologia* 67: 278–281.

Daxenbichler, M.E., Spencer, G.F, Carlson, D.G., Rose, G.B., Briker, A.M., and Powell, R.G. (1991) Glucosinolate composition of seeds from 297 species of wild plants. *Phytochemistry* 30: 2623–2638.

Den Otter, C.J., Behan, M. and Maes, F.W. (1980) Single cell responses in female *Pieris brassicae* to plant volatiles and conspecific odours. *J. Insect Physiol.* 26: 465–472.

Dimock, M.B. and Renwick, J.A.A. (1991) Oviposition by field populations of *Pieris rapae* (Lepidoptera: Pieridac) deterred by an extract of a wild crucifer. *Environ. Entomol.* 20: 802–806.

Dimock, M.B., Renwick, J.A.A., Radke, C.D., and Sachdev-Gupta, K. (1991) Chemical constituents of an unacceptable crucifer, *Erysimum cheiranthoides,* deter feeding by *Pieris rapae. J. Chem. Ecol.* 17: 525–533.

Douglas, M.M. (1986) *The Lives of Butterflies.* Univ. of Michigan, Ann Arbor.

Ehrlich, P.R. and Raven, P.H. (1964) Butterflies and plants: A study in coevolution. *Evolution* 18: 586–608.

Elliot, M.C. and Stowe, B.B. (1971) Distribution and variation of indole glucosinolates in woad *(Isatis tinctoria* L.). *Plant Physiology* 48: 498–503.

Feeny, P. (1975) Biochemical coevolution between plants and their insect herbivores, in *Coevolution of Animals and Plants* (Gilbert, L.E. and Raven, P.H., eds.), pp. 3–19. Univ. of Texas, Austin.

Feltwell, J. (1982) *The Large White Butterfly: the Biology, Biochemistry, and Physiology of Pieris brassicae* (Linnaeus). Dr. W. Junk Publishers, The Hague.

Feltwell, J. and Vane-Wright, R.I. (1982) Classification of *Pieris brassicae.* In: *The Large White Butterfly: The Biology, Biochemistry, and Physiology of Pieris brassicae* (Linnaeus), (J. Feltwell, ed). Dr. W. Junk Publishers, The Hague.

Fenwick, G.R., Heaney, R.K., and Mullin, W.J. (1983) Glucosinolates and their breakdown products in food and food plants. *CRC Crit. Rev. Fd. Sci. Nut.* 18, 123–201.

Fernald, M.L. (1950) *Gray's Manual of Botany*, American Book Co., New York.

Fox, R.M. (1966) The forelegs of butterflies. I. Introduction: Chemoreception. *J. Res. Lepid.* 5: 1–12.

Gilbert, N. (1986) Control of fecundity in *Pieris rapae.* IV. Patterns of variation and their ecological consequences. *J. Anim. Ecol.* 55: 317–329.

Glenn, M.G., Chew, F.S., and Williams, P.H. (1988) Influence of glucosinolate content of *Brassica* (Cruciferae) roots on growth of vesicular-arbuscular mycorrhizal fungi. *New Phytol.* 110: 217–225.

Grula, J.W. and Taylor, O.R., Jr. (1980) A micromorphological and experimental study of the antennae of the sulfur butterflies, *Colias eurytheme* and *C. philodice* (Lepidoptera, Pieridae). *J. Kans. Entomol. Soc.* 52: 476–484.

Harrington, H.D. (1954) *Manual of the Plants of Colorado*, Swallow Press, Chicago.

Huang, X.P. and Renwick, J.A.A. (1993) Differential selection of host plants by two

Pieris species: The role of oviposition stimulants and deterrents. *Entomol. exp. appl.* 36: 59–69.

Huang, X.P., Renwick, J.A.A. and Sachdev-Gupta, K. (1993a) A chemical basis for differential acceptance of *Erysimum cheiranthoides* by two *Pieris* species. *J. Chem. Ecol.* 19: 195–210.

Huang, X.P., Renwick, J.A.A. and Sachdev-Gupta, K. (1993b) Oviposition stimulants and deterrents regulating differential acceptance of *Iberis amara* by *Pieris rapae* and *P. napi oleracea*. *J. Chem. Ecol.* 19: 1645–1663.

Hedge, I.C. (1976) A systematic and geographical survey of the Old World Cruciferae, in *The Biology and Chemistry of the Cruciferae*. (Vaughan, J. G., MacLeod, A.J., and Jones, B.M.G., eds.), pp. 1–45, Academic Press, London.

Ilse, D. (1937) New observations on responses to colours in egg-laying butterflies. *Nature* 140: 544–545.

Ives, P.M. (1978) How discriminating are cabbage butterflies? *Aust. J. Ecol.* 3: 261–276.

Jermy, T. (1993) Evolution of insect-plant relationships—a devil's advocate approach, *Entomol. exp. appl.* 55: 3–12.

Keller, M. (1990) Responses of the parasitoid *Cotesia rubecula* to its host *Pieris rapae* in a flight tunnel. *Entomol. exp. appl.* 57: 243–249.

Klijnstra, J.W. (1982) Perception of the oviposition deterrent pheromone in *Pieris brassicae, Proc. 5th* Intl. Symp. Insect-Plant Relationships (Visser, J.H. and Minks, A.K., eds.), pp. 145–51. Pudoc, Wageningen.

Klijnstra, J.W. and Roessingh, P. (1986) Perception of the oviposition deterring pheromone by tarsal and abdominal contact chemoreceptors in *Pieris brassicae*. *Entomol. exp. appl.* 40: 71–79.

Kolb, G. and Scherer, C. (1982) Experiments on wavelength specific behavior of *Pieris brassicae* L. during drumming and egg-laying. *J. Comp. Physiol.* 149: 325–332.

Koritsas, V.M., Lewis, J.A. and Fenwick, G.R. (1991) Glucosinolate responses of oilseed rope, mustard and kale to mechanical wounding and infestation by cabbage stem flea beetle *(Psylliodes chrysocephala)*. *Ann. Appl. Biol.* 118: 209–221.

Kudrna, O. 1974. *Artogeia* Verity, 1947. Gen. rev. for *Papilio napi* Linnaeus (Lep., Pieridae). *Ent. Gaz.* 25: 9–12.

Lundgren, L. (1975) Natural plant chemicals acting as oviposition deterrents on cabbage butterflies [*Pieris brassicae* (L.), *P. rapae* (L.), and *P. napi* (L.)]. *Zool. Scripta* 4: 253–258.

Ma, W.C. and Schoonhoven, L.M. (1973) Tarsal chemosensory hairs of the large white butterfly *Pieris brassicae* and their possible role in oviposition behavior. *Entomol. exp. appl.* 16: 343–357.

Mickel, C.R. (1973) John Ray: Indefatiguable student of nature. *Annu. Rev. Entomol.* 18: 1–16.

Miller, J.R. and Strickler, K.L. (1984) Finding and accepting host plants. In: *Chemical*

Ecology of Insects (Bell, W.J. and Cardé, R.T., eds.), pp. 127–155. Chapman and Hall, New York.

Munz, P.A. and Keck, D.D. (1968) *A California Flora*. University of California Press, Berkeley, California.

Nishida, R. (1994) Oviposition stimulants of swallowtail butterflies. In: *Swallowtail Butterflies: Their Ecology and Evolutionary Biology*. (Scriber, J.M., Tsubaki, Y., and Lederhouse, R.C., eds.) Scientific Publishers, Inc., Gainesville, Florida, in press.

Ohsaki, N. (1980) Comparative population studies of 3 *Pieris* butterflies, *P. rapae*, *P. melete*, and *P. napi*, living in the same area. II. Utilization of patchy habitats by adults through migratory and non-migratory movements. *Res. Popul. Ecol.* 22: 163–183.

Ohsaki, N. (1986) Body temperature and behavioral thermoregulation strategies of three *Pieris* butterflies in relation to solar radiation. *J. Ethology* 4 (1): 1–9.

Renwick, J.A.A. and Radke, C.D. (1983) Chemical recognition of host plants for oviposition by the cabbage butterfly, *Pieris rapae* (Lepidoptera: Pieridae). *Environ. Entomol.* 12: 446–450.

Renwick, J.A.A. and Radke, C.D. (1985) Constituents of host- and non-host plant deterring oviposition by the cabbage butterfly, *Pieris rapae*. *Entomol. exp. appl.* 39: 21–26.

Renwick, J.A.A. and Radke, C.D. (1987) Chemical stimulants and deterrents regulating acceptance or rejection of crucifers by cabbage butterflies. *J. Chem. Ecol.* 13: 1771–1776.

Renwick, J.A.A. (1988) Plant constituents as oviposition deterrents to lepidopterous insects. Am. Chem. Soc. Symp. 380: 378–385.

Renwick, J.A.A., Radke, C.D. and Sachdev-Gupta, K. (1989) Chemical constituents of *Erysimum cheiranthoides* deterring oviposition by the cabbage butterfly, *Pieris rapae*. *J. Chem. Ecol.* 15: 2161–2169.

Renwick, J.A.A., Radke, C.D., Sachdev-Gupta, K. and Städler, E. (1992) Leaf surface chemicals stimulating oviposition by *Pieris rapae* (Lepidoptera: Pieridae) on cabbage. *Chemoecology* 3: 33–38.

Renwick, J.A.A. and Chew, F.S. (1994) Oviposition behavior in Lepidoptera. *Annu. Rev. Entomol.* 39: 377–400.

Robbins, R.K. and Henson, P.M. (1986) *Pieris rapae* is a better name than *Artogeia rapae* (Pieridae). *J. Lepid. Soc.* 40: 79–92.

Rodman, J.E. (1981) Divergence, convergence, and parallelism in phytochemical characters: The glucosinolate-myrosinase system. In: *Phytochemistry and Angiosperm Phylogeny*, (Young, D.A. and Seigler, D.A., eds.), pp. 43–79. Praeger, New York.

Rodman, J.E. and Chew, F.S. (1980) Phytochemical correlates of herbivory in a community of native and naturalized Cruciferae. *Biochem. Syst. Ecol.* 8: 43–50.

Rodman, J.E., Brower, L. and Frey, J. (1982) Cardenolides in North American *Erysimum* (Cruciferae, a preliminary chemotaxonomic report. *Taxon* 3: 507–516.

Roessingh, P., Städler, E., Schoni, R., and Feeny, P. (1991) Tarsal contact chemoreceptors of the black swallowtail butterfly, *Papilio polyxenes:* Responses to phytochemicals from host and non-host plants. *Physiol. Entomol.* 16: 485–495.

Root, R.B. and Kareiva, P.M. (1984) The search for resources by cabbage butterflies *(Pieris rapae):* Ecological consequences and adaptive significance of Markovian movements in a patchy environment. *Ecology* 65: 147–165.

Rothschild, M. and Schoonhoven, L.M. (1977) Assessment of egg-load by *Pieris brassicae* (Lepidoptera: Pieridae). *Nature* 266: 352–355.

Rothschild, M., Alborn, H., Stenhagen, G and Schoonhoven, L.M. (1988) A strophanthidin glycoside in Siberian wallflower: A contact deterrent for the large white butterfly. *Phytochemistry* 27: 101–108.

Sachdev-Gupta, K., Renwick, J.A.A. and Radke, C.D. (1990) Isolation and identification of oviposition deterrents to cabbage butterfly, *Pieris rapae,* from *Erysimum cheiranthoides. J. Chem. Ecol.* 16: 1059–1067.

Sachdev-Gupta, K., Radke, C.D., and Renwick, J.A.A. (1992) Chemical recognition of diverse hosts by *Pieris rapae* butterflies. In: *Proc. 8th Intl. Symp. Insect-Plant Relationships,* (Mencken, S.B.J., Visser, J.H., and Harrewijn, P., eds.), pp. 136–138. Kluwer, Dordrecht, The Netherlands.

Sachdev-Gupta, K., Radke, C.D., and Renwick, J.A.A. (1993a) Antifeedant activity of curcubitacins from *Iberis amara* against larvae of *Pieris rapae. Phytochemistry* 33: 1385–1388.

Sachdev-Gupta, K, Radke, C.D., Renwick, J.A.A., and Dimock, M.B. (1993b) Cardenolides from *Erysirnum cheiranthoides;* feeding deterrents to *Pieris rapae* larvae. *J. Chem. Ecol.* 33: 1385–1369.

Salisbury, F.B. and Ross, C.W. (1992) *Plant Physiology,* 4th Ed. Wadsworth, Belmont, Calif.

Schoonhoven. L.M. (1972) Secondary plant substances and insects. *Rec. Adv. Phytochem.* 5: 197–224.

Schoonhoven, L.M. (1977) Individuality of insect feeding behavior. *Proc. Sci. K. Akad. Wet. Amsterdam. Ser. C* 80: 341–350.

Schoonhoven, L.M. and Dethier, V.G. (1966) Sensory aspects of host plant discrimination by ledpiopterous larvae. *Arch. Neerland. Zool.* 16: 497–530.

Schoonhoven, L.M., Beerling, E.A.M., Braaksma, R., and van Vugt, Y. (1990a) Does the imported cabbageworm, *Pieris rapae,* use an oviposition deterring pheromone? *J. Chem. Ecol.* 16: 1649–1655.

Schoonhoven, L.M., Beerling, E.A.M., Klijnstra, J.W., and van Vugt, Y. (1990b) Two related butterfly species avoid oviposition near each other's eggs. *Experientia* 46: 526–528.

Schoonhoven, L.M., Sparnaay, T., van Wissen, W., and Meerman, J. (1981) Seven-week persistence of an oviposition-deterrent pheromone. *J. Chem. Ecol.* 7: 583–588.

Schraudolf, H. (1989) Indole glucosinolates of *Capparis spinosa. Phytochemistry* 28: 259–260.

Scriber, J.M. and Lederhouse. R.C. (1992) The thermal environment as a resource dictating geographic patterns of feeding specialization of insect herbivores. In: *Effects of resource distribution on animal-plant interactions* (Hunter, M.R., Ohguishi, T., and Price, P.W., eds.), pp. 429–466. Academic Press, New York.

Shapiro, A.M. (1981) The pierid red-egg syndrome. *Am. Nat.* 117: 276–294.

Shapiro, A.M. (1986) The natural history of *Tatochila distincta distincta,* a rare butterfly from the *puna* of northwestern Argentina. *J. New York Entomol. Soc.* 94: 526–530.

Shapiro, A.M. 1989 (91). The zoogeography and systematics of Argentine Andean and Patagonian pierid fauna. *J. Res. Lepid.* 28 (3): 137–238.

Singer, M.C., Ehrlich, P.R., and Gilbert, L.E. (1971) Butterfly feeding on lycopsid. *Science* 172: 1341.

Städler, E. (1992) Behvaioral responses of insect to plant secondary compounds. In: *Herbivores: Their Interactions with Secondary Plant Metabolites,* 2nd Ed., Vol. II, Ecological and Evolutionary processes (Rosenthal, G.A. and Berenbaum, M.R., eds.), pp. 45–88. Academic Press, New York.

Städler, E. and Roessingh, P. (1990) Perception of surface chemicals by feeding and ovipositing insects. In: *Proc. 7th Intl. Symp. Insect-Plant Relationships,* Symp. Biol. Hung. 39 (Szentesi, A. and Jermy, T., eds.), pp. 71–86. Akadémia Kiadó, Budapest.

Stermitz, F.R., Belofsky, G.N., Ng, D., and Singer, M.C. (1989) Quinolizidine alkaloids obtained by *Pedicularis semibarbarta (Schrophulariaceae)* from *Lupinus falcratus* (Leguminosae) fail to influence the specialists herbivore, *Euphydryas editha* (Lepidoptera). *J. Chem. Ecol.* 15: 2521–2530.

Tabashnik, B., Wheelock, H., Rainboldt, J.D., and Watt, W.B. (1981) Individual varation in oviposition preferences in the butterfly *Colias eurytheme. Oecologia* 50: 225–230.

Terofal, F. (1965) Zum Problem der Wirtsspezifitat bei Pienden (Lep.) *Mitt. Munch. Ent. Ges.* 55: 68–74.

Thorsteinson, A.J. (1953) The chemotactic responses that determine host specificity in an oligophagous insect (*Plutella maculipennis.* (Curt.) Lepidoptera). Can. J. Zool. 31: 53–72.

Tookey, H.L., Daxenbichler, M.E., VanEtten, C.H., Kwolek, W.F., and Williams, P.H. (1980) Cabbage glucosinolates: Correspondence patterns in seeds and leafy heads. *J. Amer. Soc. Hort. Sci.* 105: 714–717.

Topazzini, A., Mazza, M., and Pelosi, P. (1990) Electroantennogram responses of five Lepidoptera species to 26 general odourants. *J. Insect Physiol.* 36: 619–624.

Traynier, R.M.M. (1984) Associative learning in the oviposition behavior of the cabbage butterfly, *Pieris rapae. Physiol. Entomol.* 9: 465–472.

Traynier, R.M.M. and Hines, E.R. (1987) Probes by aphids indicated by stain induced fluorescence in leaves. *Entomol. exp. appl.* 45: 198–201.

Traynier, R.M.M. and Truscott, R.J.W. (1991) Potent natural egg-laying stimulant for cabbage butterfly *Pieris rapae. J. Chem. Ecol.* 17: 1371–1380.

Truscott, R.J.W., Johnstone, P.K., Minchinton, I.R., and Sang, J.P. (1983) Indole glucosinolates in swede *(Brassica napobrassica* L. Mill.). *J. Agric. Food Chem.* 31: 863–867.

Turlings, T.C.J., Tumlinson, J.H., and Lewis, W.J. (1990) Exploitation of herbivore-induced plant odors by host-seeking parasitic wasps. *Science* 250: 1251–1253.

van der Reijden, E.D. and Chew, F.S. (1992) Assessing host-plant suitability in caterpillars: Is the weight worth the wait? In: *Proc. 8th Intl. Symp. Insects-plant Relationships.* (Mencken, S.B.J., Visser, J.H., and Harrewijn, P., eds.), pp. 69–70, Kluwer, Dordrecht, The Netherlands.

van Loon, J.J.A. (1990) Chemoreception of phenolic acids and flavonoids in larvae of two species of *Pieris*. *J. Comp. Physiol. A* 166: 889–899.

van Loon, J.J.A. and van Eeuwijk, F.A. (1989) Chemoreception of amino acids in larvae of two species of *Pieris*. *Physiol. Entomol.* 44: 459–469.

van Loon, J.J.A., Frentz, W.H. and can Eeuwijk, F.A. (1992b) Electroantennogram responses to plant volatiles in two species of *Pieris* butterflies. *Entomol. exp. appl.* 62: 253–260.

van Loon, J.J.A., Blaakemeer, A., Griepink, F.C., van Beek, T.A., Schoonhoven, L.M. and de Groot, A. (1992a) Leaf surface compound from *Brassica oleracea* (Cruciferae) induces oviposition by *Pieris brassicae* (Lepidoptera: Pieridae). *Chemoecology* 3: 39–44.

Verschaffelt, E. (1910) The cause determining the selection of food in some herbivorous insects. *Proc. K. Ned. Akad. Wet.* 13: 536–542.

Vet, L.E.M. and Dicke, M. (1992) Ecology of infochemical use by natural enemies in a tritrophic context. *Annu. Rev. Entomol.* 37: 141–172.

Vet, L.E.M., Lewis, W.J., and Cardé, R.T. (1995) Parasitoid foraging and learning. In: *Chemical Ecology of Insects II* (Cardé, R.T. and Bell, W.J., eds) Chapman and Hall, New York.

Webb, A. and Shelton, A. (1988) Laboratory rearing of the imported cabbageworm. *New York Food Life Sci. Bull.* 122: 1–5.

Wiklund, C., Nylin, S., and Forsberg, J. (1991) Sex-related variation in growth rate as a result of selection for large size and protandry in a bivoltine butterfly, *Pieris napi*. *Oikos* 60: 241–250.

Insect-Insect Interactions

7

Trail and Territorial Communication in Social Insects

James F. A. Traniello and Simon K. Robson
Department of Biology, Boston University

1. Introduction

The social properties of insect colonies are sometimes described in seemingly contradictory terms. As pinnacles of biological complexity they are superorganisms and their emergent, colony-level characteristics are often referred to in terms of their elaborate and sophisticated nature. Yet the mechanisms that mediate social interactions and group phenomena, after empirical or theoretical analysis, are simple and parsimonious. This complexity-mediated-by-simplicity paradigm provides a heuristic approach to the analysis of the basic behavioral characteristics of the individual members of an insect society and the regulatory mechanisms of cooperative response, which are the fundamental elements from which colony-level behavior is derived. Inevitably, the dissection and reconstruction of insect social organization involves semiochemicals, because the principal sensory modality of integration, social coordination, and assembly of colony-level patterns is olfaction.

The ecological contexts and mechanisms of chemical communication in the social insects are well known if not all equally well understood, and the adaptiveness of communication seems obvious. Communication is necessary to direct workers to places outside the nest where labor is required. The essentially fixed placement of a social insect colony determines its trophic and competitive environment, because each colony is one point in an overdispersed, clumped, or random matrix of conspecific and interspecific nests (Hölldobler and Lumsden, 1980; Levings and Adams, 1984; Levings and Traniello, 1981). Nestmates must be mobilized cooperatively to harvest food resources or combat encroachments on territorial borders or nest areas. Foraging behavior, colony dispersion and space use patterns, competition, resource partitioning and other community-level interactions are either partially or totally dependent upon semiochemicals. Correlated with these ecologically significant processes are the behavioral mechanisms that constitute recruitment, orientation, and intercolony communication, mechanisms that are based on the structural characteristics of trail and territorial pheromones and the responses they induce (Fig. 7.1).

Many milestones have been passed since the early observations of Carthy (1950) on trail communication in ants, but progress in all groups of social insects has not been equal. The first behavioral bioassays of a social insect trail pheromone (Wilson, 1959, 1962) identified the glandular source of recruitment chemicals and demonstrated the extreme reliance on pheromones for the organization of mass communication during foraging. The 1960s and 1970s subsequently witnessed a mushrooming of studies in which the glandular sources, chemical structures, and levels of specificity of social insect semiochemicals were identified, including the first isolation and chemical characterization of ant and termite trail pheromones (Matsumura et al., 1968; Moore, 1966; Tumlinson et al., 1971). Descriptive studies of exocrine gland histology revealed previously undiscovered and often functionally unknown semiochemical sources (Downing, 1991; Hölldobler and Engel, 1978; Jeanne et al., 1983), providing new phylogenetic insights. A great diversity of forms of recruitment and trail communication was described, which led to the creation of ethological models of the evolution of mass communication [reviewed in Hölldobler, 1978 and Wilson, 1971], as well as field research on the ecological significance of chemical signaling and the behavioral mechanisms of territoriality [e.g., Adams, 1990a; Adams and Traniello, 1981; Hölldobler, 1976; Hölldobler and Lumsden, 1980; Hölldobler et al., 1978 and Traniello, 1983, 1987]. And most recently, principles of self-organization have been applied to foraging strategies in an attempt to understand how the ecologically important

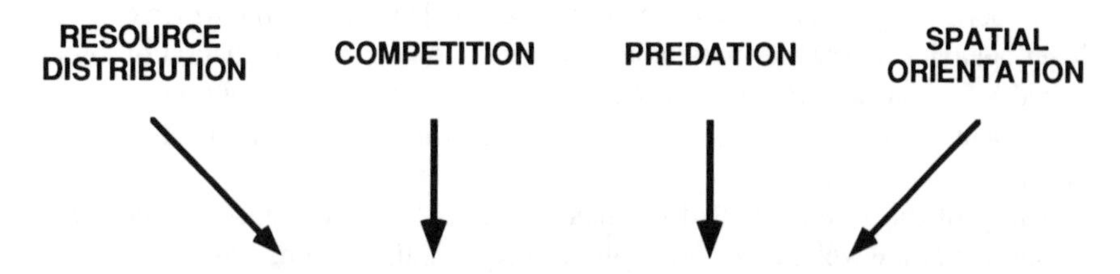

Figure 7.1. Ecological influences on the evolution of trail and territorial communication systems in social insects.

collective properties of social insect colonies are derived from trail pheromone-mediated behavior (Goss and Deneubourg, 1989; Goss et al., 1990).

In this chapter we present the basic elements of trail and territorial communication, focusing primarily on ants and termites, the groups that have been studied most intensively, while reviewing the literature on bees and wasps. We concentrate on those areas of ethochemistry and behavioral ecology that have not been the subject of current, comprehensive reviews [e.g., Attygalle and Morgan, 1985; Hölldobler, 1987; Hölldobler and Lumsden, 1980; Hölldobler and Wilson, 1990 and Howse 1984].

2. Glandular sources of trail and territorial pheromones in the social insects

The anatomy and histology of exocrine glands in social insects have been described in detail for a number of species. Several reviews (Billen, 1986; Downing, 1991; Hölldobler and Wilson, 1990; Noirot, 1969) provide excellent summaries. In the following section we treat exocrine gland nomenclature, structure, and physiology in a limited way to introduce the reader to terminology that will appear in the present review.

In ants, the accessory glands to the sting (the Dufour's gland and the poison gland), the pygidial glands and sternal glands, the hindgut, the rectal gland, and the tibial and tarsal glands are the sources of trail substances and territorial markings (Hölldobler and Wilson, 1990; Hölldobler et al., 1992). The metapleural gland is known to produce antimicrobial secretions but the gland has secondarily involved a defensive function in some species (Maschwitz, 1974) and is reported to produce a territorial pheromone in *Solenopsis geminata* (Jaffe and Puche, 1984) and *Pseudomyrmex* (Jaffe et al., 1986).

Trail and territorial communication in the social wasps has received relatively little attention; recruitment to food sources is uncommon and does not appear to involve trail-laying behavior. But during recruitment to a new nest site, gaster dragging behavior is used to apply secretions of the sternal glands to the substrate and this gland has been shown to produce a trail pheromone in *Polybia sericea* (Downing, 1991; Jeanne, 1981). Male *Polistes dominulus* drag their gasters and legs while defending territories at leks, apparently marking the area with sternal and leg tegumental gland pheromones (Beani and Calloni, 1991). And recruiting bees in the meliponine genus *Trigona* employ trail communication, discharging mandibular gland scent marks at intervals on the ground (Lindauer, 1961).

Only one source of trail pheromones, the sternal gland, has been described in termites. The number, location and morphology of sternal glands varies in different genera (Ampion and Quennedey, 1981; Noirot, 1969). In *Mastotermes*, sternal glands are located on abdominal sternites II, III, and IV; in *Hodotermes*, a single gland is found on segment III, and in the higher termites (Termitidae) as well as the Rhinotermitidae and Kalotermitidae there is a single sternal gland on segment V.

3. Trail-laying behavior

The structure of the active space of a pheromone trail is due in part to the method of deposition of glandular secretions. Observations of trail-laying postures and patterns indicate that a wide variety of mechanisms can be used to create a chemical space for attraction and orientation. In termites in the genera *Reticulitermes*, *Zootermopsis*, and *Nasutitermes*, the sternal gland is pressed against the substrate. Campaniform sensilla most likely monitor contact between sternites and the substrate to regulate pheromone deposition (Stuart and Satir, 1968). The harvester termite *Hodotermes mossambicus* marks areas around the nest entrance and a food source by dotting the tip of the abdomen and lays directional trails with the sternal gland (Heidecker and Leuthold, 1984). Dotting the gaster tip has also been observed in ants, including *Lasius neoniger* (Traniello, 1983), *Formica schaufussi* (Traniello and Beshers, 1991), and *Myrmica sabuleti* (Cammaerts and Cammaerts, 1980). In *L. neoniger* and *F. schaufussi* the hindgut is the source of trail pheromone, which is emitted through the acidopore; in *M. sabuleti*, Dufour's gland secretion is deposited by gaster dotting. Chemical trails in these species are a series of point sources or 'sign posts' deposited between the nest and the target area, sometimes in conjunction with other secretions. In some cases cuticular modifications associated with exocrine glands function as applicators for the secretions of the glands beneath. In *Pachycondyla laevigata*, the trail-laying posture is unusual in that the secretions of the pygidial gland, which is located dorsally beneath the 7th tergite, are applied to the substrate by curling the gaster forward ventrally and dragging the cuticular applicator (Hölldobler and Traniello, 1980a).

The significance of these different trail-laying behaviors is only partly understood. For example, in the fire ant *Solenopsis* information about food quality is conveyed to colony members through pheromone concentration, which is related to the degree of continuity of the sting trail (Hangartner, 1969). But why is gaster or abdominal dotting employed as opposed to dragging the gaster, the sting or an exocrine gland? Deposition by dotting may be a mechanism of modulating colony response by controlling trail pheromone release and the structure of the pheromone's active space or it may be a method of providing orientation cues. Few studies have correlated the frequency and pattern of point sources of secretions with foraging organization. Ambient light conditions appear to be important in some cases (Cammaerts and Cammaerts, 1980; Heidecker and Leuthold, 1984).

4. Trail pheromone bioassays

The identification of glandular sources of trail substances and their chemical characterization represent two analytical phases of the study of recruitment communication that require careful, detailed, and biologically relevant methodology. The accuracy and interpretation of results hinges completely upon the bioassay's nature and conditions. Although investigators repeatedly have cautioned that bioassays must be designed critically (Hall and Traniello, 1985; Howard et al., 1976; Stuart, 1969; Vander Meer et al., 1988), the glandular sources of trail pheromones have been misidentified and the roles that chemically characterized secretions play in recruitment communication misunderstood.

Stuart (1969) noted that "the only trustworthy assay (of a termite trail pheromone) must use termites which are in as natural a condition as possible." Howard et al. (1976) described five criteria necessary to the authentic identification of trail pheromones; among these criteria are localization in a glandular source, active deposition, and correspondence between the stored chemical and the deposited pheromone in both structure and biological activity. Hall and Traniello (1985) emphasized that the bioassays used in trail pheromone isolation and identification must distinguish between recruitment and orientation, which are experimentally discernible behavioral responses involved in trail communication and foraging organization in both ants and termites.

During recruitment, nestmates are induced to leave the nest to perform work at a target area. In the systems of mass communication found in ant and termite species, a chemical signal, without any other modulatory stimulus, is capable of eliciting the full behavioral response of recruitment. Individuals previously inactive or engaged in some other task change their behavior and leave the nest as part of a newly-mobilized group. To accurately measure the recruitment property of a trail pheromone it is critically important that the individuals are *undisturbed* during the bioassay. Testing undisturbed individuals dates back to the initial trail pheromone bioassays on fire ants and termites (Stuart, 1969; Wilson, 1959) and has generally been adopted in other studies of trail communication in ants [reviewed in Hölldobler and Wilson, 1990] and termites (Hall and Traniello, 1985; Stuart, 1981; Traniello, 1982; Traniello and Busher, 1985). Other bioassays use a Y-configuration of trails; individuals moving along a naturally established trail [e.g., Heidecker and Leuthold, 1984; and Oloo, 1981] or an artificial extract trail [e.g., Hall and Traniello, 1985; and Traniello and Busher, 1985] or in an olfactometer (Vander Meer et al., 1988) are tested for responsiveness to specific trail properties such as recruitment, orientation, and species- or colony-specific components. Many bioassays of the chemical and behavioral characteristics of trail pheromones in termites offer experimental conditions in which only orientation is tested (Birch et al., 1972; Grace et al., 1989; Tokoro et al., 1991, 1992), because the bioassay involves placing a termite on an artificial test trail. But it is well-established that termite and ant trail pheromones have more than one behavioral effect (Kaib et al., 1982; Runcie, 1987; Traniello, 1982; Vander Meer et al., 1988, 1990). Behavioral assays must be able to discern among the behavioral subcategories regulated by trail pheromones, and investigators must design a bioassay that is relevant and appropriate to the range of behavioral effects induced by multicomponent pheromone systems.

5. Recruitment and orientation properties of trail pheromones

Recruitment is defined as "communication that brings nestmates to some point in space where work is required" (Hölldobler and Wilson, 1990). Trail communication thus involves stimulating nestmates to leave the nest or change task performance; excited (alerted) individuals are then led to a target area where their cooperation is required. In a number of ant species such as *Camponotus socius* and *Formica fusca*, alerting nestmates requires behavioral stimulation in the form of a motor display from a recruiter (Hölldobler, 1971, 1978). This motor display is a modulatory signal that either lowers the response threshold or provides a switching mechanism for trail-following behavior. Vibrational and tactile stimuli may affect trail-following behavior in the dampwood termite *Zootermopsis* (Stuart, 1969). In the ants *Camponotus socius* and *Formica fusca* the trail substance originates in the hindgut; artificial trails do not have recruitment (alerting) effects on nestmates, but such trails are used as orientation guides if a recruited ant first contacts the behavioral displays of a recruiter. Yet in many ponerine, myrmicine, dolichoderine, and formicine ants, trail pheromone alone can elicit excitation and orientation: an artificial trail drawn out from the nest entrance can completely reproduce the natural response [reviewed in Hölldobler and Wilson, 1990].

A trail pheromone, in its most basic form, is thus a chemical substance that after deposition on the substrate diffuses to form an active space in which the concentration of pheromone molecules is at or above the threshold required for orientation (Bossert and Wilson, 1963). Hangartner (1967), using the ant *Lasius fuliginosus* as a model, described the spatial processing of the information encoded in an odor trail as a type of tropotactic orientation. Olfactory receptors on the antennae monitor pheromone concentration and lateral movements of the antennae and the body maintain an ant oriented within the pheromone's active space as it travels along a trail. During trail orientation the paired antennae sample the pheromone's active space and detect its boundaries. Theoretically, the active space of the trail substance is semiellipsoidal (Bossert and Wilson, 1963). The molecular gradient created by chemical diffusion yields an active space in which pheromone concentration is highest at the point of application and symmetrically decreases laterally. By continuously and simultaneously sampling this gradient and sensing the lower concentration at its periphery (or no pheromone), compensatory (opposing) movements can be made during locomotion to maintain position within the active space. This mechanism also occurs in the harvester termite *Schedorhinotermes lamanianus*, but in *Hodotermes mossambicus* orientation on a chemical trail is klinotactic (Leuthold, 1975; Leuthold et al., 1976). Trail orientation may be accompanied by guideline and phototactic orientation (Hangartner, 1967; Jander and Daumer, 1974; Klotz and Reid, 1992). Trail orientation has also been modeled by Calenbuhr and Deneubourg (1992).

5.1. Chemical composition and behavioral effects of trail substances

A myriad of studies on a wide variety of ants and termites show that for many species it is not realistic to refer to the secretion of a single exocrine gland or one pheromone component as being *the* chemical responsible for trail communication. Even if a trail pheromone has been chemically identified, artificial trails prepared from gland extracts may be more effective than trails of an authentic chemical (Attygalle and Morgan, 1985; Van Vorhis Key and Baker, 1982). More than one gland may contribute to the chemical structure of a trail, and a given gland may synthesize and secrete more than one pheromone product. Trail communication may thus involve a system of multiple sources in which the secretions of different exocrine glands are added to a trail, or secretions may comprise several components, often chemical homologues (Attygalle and Morgan, 1985; Vander Meer at al. 1988, 1990). In either case, trail signaling may involve a pheromone blend as has been documented in alarm communication. And finally, the organization of foraging and defensive behaviors by social insects may depend upon tactile or acoustical sensory input to modulate responsiveness to trail substances, forming a communicative system based on several modalities (Hölldobler and Wilson, 1990).

The isolation and biochemical characterization of trail substances preceded the recognition of the subtleness of the responses induced by semiochemicals and the development of behavioral bioassays sensitive enough to permit these responses to be separated and assigned to a chemical homologue or glandular constituent of a trail substance. As noted by Vander Meer et al. (1988), the isolation of the chemical components of the Dufour's gland trail pheromones of fire ants relied upon a bioassay that was restricted to measuring worker orientation, which only measures the tendency to *follow* a trail as opposed to being drawn out of the nest, i.e., *recruited* by a trail. Yet the process of mass recruitment in fire ants requires greater behavioral complexity than simple orientation; in fact, different Dufour's gland chemicals regulate functionally different behavioral responses ('subcategories' of trail-following behavior; Vander Meer et al., 1988, 1990).

During foraging, nest emigration, territorial encounters, and alarm/recruitment in ants and termites, chemical stimuli mediate specific behaviors. These behaviors include attraction and aggregation, measured by a 'point source' bioassay in ants

(Wilson, 1962) and termites (Stuart, 1981; Traniello, 1981). Trail substances also elicit recruitment and have the ability to induce nestmates to depart from the nest (Hölldobler and Wilson, 1990; Stuart, 1969; Traniello, 1982). The chemistry regulating behavior may involve recruitment primers, synergists and orientation inducers (Vander Meer et al., 1988, 1990). The primer and inducer substances are necessary to elicit the full recruitment and orientation responses. Movement of individuals between the nest and the target area is regulated by orientation pheromones (Hölldobler and Wilson, 1990; Oloo and Leuthold, 1979; Traniello, 1982; Vander Meer et al., 1990). These substances do not elicit true recruitment, but seem to solely mediate tropotactic or klinotactic orientation.

5.2. Pheromone blends

Fire ants, *Solenopsis* spp., provide an instructive model system in which the behavioral correlates of exocrine gland chemistry have been elucidated successfully. Mass recruitment is organized by the secretion of the Dufour's gland, which produces a collection of pheromones that play different roles in trail communication. A Dufour's gland extract bioassayed in an olfactometer for recruitment at a concentration of one worker equivalent elicited a 79.5% response; the response to (*Z*, *E*)-α-farnesene, the principal trail orientation component isolated from the Dufour's gland was lower, 63% (Vander Meer et al., 1988). However, an additional Dufour's gland fraction present at about 75 pg/worker (the C-1 fraction) induced a 60% response when offered alone but gave a 76.5% response when presented in combination with (*Z*, *E*)-α-farnesene. Two homofarnesenes present in the Dufour's gland secretion did not increase the attractiveness of the C-1 component; their role is unknown. An orientation inducer present in the secretion provides the chemical equivalent of behavioral or acoustical displays which also increase trail following (Vander Meer et al., 1990).

Homofarnesenes present in the Dufour's gland of *Myrmica* sp. may be added to 3-ethyl-2,5-dimethyl pyrazine (EDMP), the poison gland trail pheromone, as part of a multicomponent trail system (Jackson et al., 1991). In the dolichoderine ant *Linepithema humile* [= *Iridomyrmex humilis*], the presence of a pheromone fraction similar in function to one fire ant trail pheromone constituent may augment the attractiveness of (Z)-9-hexadecenal, the major purification product obtained from Pavan's gland (Cavill et al., 1980) since synthetic trails are not as active as gaster-extract trails (Attygalle and Morgan, 1985; Van Vorhis Key and Baker, 1982). Similarly, EDMP has been isolated from the poison gland of the myrmicine *Pheidole pallidula* and induces trail following, but bioassays of this chemical alone show low trail activity in comparison to artificial trails made from poison gland extracts (Ali et al., 1988). These results again suggest the importance of synergists, primers, or inducers as part of a pheromone blend.

In the myrmicine ant *Tetramorium caespitum* the trail pheromone is a 70:30

synergistic mixture of 2,5-dimethylpyrazine and EDMP. Pharaoh's ant, *Monomorium pharaonis*, uses the main constituent of the Dufour's gland (6*E*,10*Z*)-3,4,7,11-tetramethyl-6,10-tetradecadienal (faranal) as a recruitment pheromone, but nitrogen heterocycles present in the gland may be responsible for worker attraction (Ritter et al., 1977). Morgan et al. (1990) have suggested that in the ant *Tetramorium impurum* methyl 2-methoxy-6-methyl benzoate has a slight synergistic effect when combined with methyl 6-methyl salicylate, the major component of the trail pheromone.

Pheromone blends exist in other species of ants but their role is not well understood. Decanoic, dodecanoic, heptanoic, hexanoic, nonanoic, and octanoic acids have been identified in the hindgut trail pheromone of the ant *Lasius fuliginosus* (Huwyler et al., 1975) and other fatty acid collections have been found in *Pristomyrmex pungens* (Hayashi and Komae, 1977), but their function is unknown. Traniello (1980) speculated that different blends of fatty acids may be responsible for the colony specificity of the trail substance of *Lasius neoniger*.

In termites, sternal gland pheromones mediate recruitment and orientation behavior (Kaib et al., 1982; Oloo and Leuthold, 1979; Runcie, 1987; Traniello, 1982), but only single chemicals such as neocembrene-A (*E*-6-cembrene A, nasutene) (Birch et al., 1972) and dodecatrienol (Matsumura et al., 1968; Tokoro et al., 1991, 1992) have been isolated from whole body extracts and bioassayed. It seems evident that termite sternal gland secretions represent a pheromonal blend system, given the distinct behaviors different substances control, but the secretions have not been tested with this perspective in mind. The sternal gland secretions of *Nasutitermes costalis* depending on context induce recruitment and orientation and serve as arrestants (Traniello, 1981, 1982; Traniello and Busher, 1985). Gas chromatography profiles of sternal gland secretions of *N. costalis* indicate the presence of numerous constituents (Prestwich and Traniello, unpublished), but their functions have not been studied.

Pheromone blends also may result from the simultaneous deposition of the contents of more than one exocrine gland. In the formicine ant *Camponotus pennsylvanicus* artificial trails of hindgut fluid and poison gland secretions are most effective as recruitment signals (Traniello, 1977), and pygidial gland and venom gland secretions mixed as a trail pheromone are responsible for organizing group-predatory behavior in the ponerine ant *Leptogenys chinensis* (Maschwitz and Schönegge, 1977, 1983). Furthermore, compounds of low and high volatility present in the venom gland of this species may regulate orientation and trail age discrimination, respectively. In *L. diminuta* the poison gland and pygidial gland produce (3R,4S)-4-methyl-3-heptanol and isogeraniol, respectively, to regulate orientation and recruitment (Attygalle et al., 1988). Species of *Pogonomyrmex* and *Daceton armigerum* also use the pheromones of different glands in combination to provide short-lived recruitment signals and persistent orientation cues (Hölldobler and Wilson, 1970; Hölldobler et al., 1990).

5.3. Orientation trails and pheromone persistence

The orientation components of some termite and ant trail pheromones are remarkably persistent. In the Neotropical termite *Nasutitermes costalis*, chemicals extracted from four-year-old trails are capable of serving as pheromone guides to soldier and worker termites (Traniello, 1982). Additional tests with cembrene-A, which has been identified as a trail pheromone in *N. exitiosus* (Birch et al., 1972; Moore, 1966) and *Trinervitermes* (McDowell and Oloo, 1984) suggest that this molecule may be responsible for what appears to be a general orientation response in higher termites. Runcie (1987) offered behavioral data indicating the presence of a functionally similar orientation chemical in the rhinotermitid *Reticulitermes flavipes*, and Oloo and Leuthold (1979) noted the existence of short-lived 'recruitment' and more durable 'basic' trails in *Trinervitermes bettonianus*.

In termites, the chemicals regulating recruitment and orientation are produced in the sternal gland, but in ants they may originate in the same or different glands. Hölldobler and Wilson (1970) showed that artificial trails prepared from extracts of the poison gland of the harvester ant *Pogonomyrmex badius* had a recruitment effect that lasted approximately 20 minutes, whereas artificial trails prepared form Dufour's gland secretions and aged for longer periods of time could elicit orientation responses. The hindgut secretions of the ant *Lasius neoniger* are composed of an ephemeral recruitment regulator and a persistent, colony-specific orientation cue (Traniello, 1980, 1989a). Persistent pheromones may serve ecological functions. For example, *L. neoniger* colonies often occur in densely packed arrays in which neighboring entrances of their polydomous nests may be as close as 5 cm (Traniello, 1989a; Traniello and Levings, 1986). Trail systems channel foragers away from neighboring nests; their persistent orientation cues may permit territorial recognition cues or serve as initial guides for naive foragers. The trunk trail systems of *P. badius*, *P. rugosus*, and *P. barbatus* are marked with Dufour's gland secretions that are composed of blends of hydrocarbons that impart species, population- and colony-level specificity, perhaps playing a similar role (Hölldobler, 1987; Regnier et al., 1973).

In other ant species different glands may produce trail chemicals with different behavioral effects. In *Daceton armigerum*, for example, the sternal gland secretes a short-lived recruitment pheromone; poison gland trails are far more durable (Hölldobler et al., 1990). And in the ant species *Myrmica rubra* and *M. scabrinodis* poison gland secretions have recruitment properties whereas Dufour's gland contents appear to provide more persistent territorial markings (Cammaerts et al., 1978).

6. Signal Specificity in Trail and Territorial Communication

Hölldobler and Carlin (1987) note that there has been a 'double standard' applied to insect and mammalian chemical communication that unrealistically accentuates the fine-tuned nature and specificity of signal/response patterns in the former group and the varied complexity of pheromone-induced behaviors in the latter. Although many social insect semiochemicals are in fact chemical blends that often rival the complexity of vertebrate pheromonal signals, species specificity has nevertheless been emphasized in many studies [reviewed in Attygalle and Morgan, 1985; Hölldobler and Wilson, 1990; and Prestwich, 1984].

The identification of the species-specific sex pheromone, bombykol, of the silk moth *Bombyx mori* led to the premature and inaccurate assumption that reproductive isolation in insects was solely achieved through the molecular design of attractants. Similarly, the first research on trail pheromone specificity in ants (Wilson and Pavan, 1959) found that these substances were highly species-specific, assuring 'privacy' during trail communication and serving as useful biochemical traits to distinguish species within a genus. But additional research did not support such a general rule. Species in at least six different genera of ants in the subfamily Myrmicinae (*Myrmica*, *Manica*, *Atta*, *Tetramorium*, *Pheidole*, and *Messor*) use EDMP as a trail pheromone (Ali et al., 1988; Attygalle and Morgan, 1985; Attygalle et al., 1986; Jackson et al., 1989). Moreover, fourteen species of *Myrmica* distributed in Europe and North America all have similar quantities of EDMP as the major constituent of their poison gland secretion and follow trails made from gland extracts or the authentic chemical. Dodecatrienol, identified as a trail pheromone in the termites *Reticulitermes virginicus*, *R. speratus* and *Coptotermes formosanus* and a product of wood decay by the brown-rot fungus *Gleophyllum trabeum* (Matsumura et al., 1968; Tokoro et al., 1991, 1992) serves as a nonspecific orientation guide in these species as well as in other species of *Reticulitermes* and representatives of a diverse group of geographically dispersed termites in different families such as *Pseudacanthotermes*, *Odontotermes*, *Trinervitermes*, *Schedorhinotermes*, and *Amitermes* (Bordereau et al., 1990; Kaib et al., 1982; McDowell and Oloo, 1984; Prestwich, 1984). The diterpene hydrocarbon (*E*)-6-cembrene A, an orientation pheromone isolated from whole-body extracts of *Nasutitermes exitiosus*, also elicits orientation responses in *N. walkeri* and *N. gravelosus* (Birch et al., 1972), *Trinervitermes bettonianus* and *T. gratiosus* (McDowell and Oloo, 1984), as well as *Schedorhinotermes*, *Amitermes*, *Reticulitermes*, and weaker following in *Macrotermes* (Kaib et al., 1982). In this latter study only *Hodotermes mossambicus* failed to respond to test trails generated by other species. Specificity was demonstrated, however, in the same study in a binary choice test. Using this bioassay, *Schedorhinotermes lamanianus*, *Trinervitermes bettonianus*, *T. gratiosus*, and *Amitermes unidentatus* were able to discriminate their trails from those of other species. Conspecific trails were consistently preferred to intraspecific test trails, and congeneric trails were preferred if conspecific trails were not offered, suggesting the existence of a multi-component trail pheromone [see also Howard et al., 1976].

These results are as interesting as they are difficult to reconcile. For example, Kaib et al. (1982) showed that the African nasutitermitine *T. bettonianus* would follow trails of *T. gratiosus*, *T. togoensis*, *T. geminatus*, *T. occidentalis*, *Nasutitermes kempai*, *Amitermes messinae* and *A. unidentatus*. *T. bettonianus* would also follow trails of the African rhinotermitid *Schedorhinotermes lamanianus* and the European rhinotermitid *Reticulitermes lucifugus*. The orientation component of the trail pheromone of the nasutitermitine species is (*E*)-6-cembrene-A, but it is dodecatrienol in the rhinotermitids. Can a diterpene hydrocarbon and a C12 alcohol both be the basic component of the trail pheromones of species representative of both lower and higher termite subfamilies? Molecular shape and the presence of a primary alcohol and (*Z*)-3-alkene are required for trail following in *Reticulitermes* (Prestwich, 1984), but these structural properties are absent in (*E*)-6-cembrene A. In termites a broad range of molecules seem capable of eliciting general orientation, and many results may be of little value in determining meaningful patterns of signal specificity. Combined with a lack of appropriate behavioral bioassays for termite trail pheromones (Hall and Traniello, 1985), interpretation of the significance of specificity studies on termites is difficult and results can be misleading.

Among myrmicine ants, numerous studies have indicated great variety in the degree of specificity of trail pheromones. In the fungus-growing ants of the genera *Atta*, *Acromyrmex*, and *Trachymyrmex* interspecific trail following is common, but there are differences in the degree of response to the poison gland secretions of *Atta* and *Acromyrmex* owing to the specificity of pyrrole and pyrazine constituents present in the exocrine organs of species in these genera [reviewed in Attygalle and Morgan, 1985]. In *Daceton armigerum*, ephemeral trails are produced from substances in the sternal gland, whereas trails made from the contents of the poison gland are more persistent (Hölldobler et al., 1990). These authors demonstrated that workers of *Atta cephalotes* will follow artificial trails prepared from the poison gland (but not the Dufour's gland) of *D. armigerum* and that *D. armigerum* Dufour's gland trails will elicit trail-following behavior by workers of *Solenopsis invicta*. Although these results suggested that the Dufour's gland of *D. armigerum* may contain sesquiterpenes, the major orientation component of the Dufour's gland of this dacetine species may contain the pyrrole common in the poison gland of attine species. In fact, the poison gland of *D. armigerum* contains a mixture of pyrazines and the Dufour's gland contents include 9-tricosene, tricosane, and 9-pentacosene together with small quantities of sesquiterpene aldehydes (Morgan et al., 1992). Additional trail-following bioassays are needed to fully understand the specificity of responses in these species.

In *Tetramorium*, species specificity is found in the trails of *T. guineense*, *T. caespitum*, and *T. impurum* (Morgan et al., 1990). For *T. caespitum* and *T. impurum*, the specificity is due to a 70:30 mixture of EDMP in the former, and methyl 6-methyl salicylate in the latter. Trails of *T. guineense* are followed by *Atta texana* and *Trachymyrmex septentrionalis*, suggesting that the poison gland of *T. guineense* may contain the pyrrole found in the attine species. Other examples of the presence, absence, and chemistry of trail pheromone specificity are summarized in Attygalle and Morgan (1985).

7. Ecology and the evolution of trail pheromone specificity

The origin of the biochemical variation serving as the proximate mechanism underlying species specificity may be the result of metabolic end products (Blum, 1974), dietary differences (Traniello, 1980), or genetics. In some taxonomic groups, such as the myrmicine ants, pyrazines and farnesenes are shared by different species and genera. Species may synthesize chemicals common to the subfamily in different exocrine glands; some constituents present in small quantities that play no role in one species may have a prominent role in a congener.

The properties that impart specificity have been analyzed, but studies of the degree of specificity in trail communication generally beg the question of its *significance*. Conceptual arguments have been made about the 'promotion of privacy' during communication (Hölldobler and Wilson, 1990) and selection for functional specificity from the 'chemical noise' in an ancestral anonymous signal through the process of chemical ritualization. In this process substances previously used in a noncommunicative context acquire a signal function through evolution (Hölldobler and Carlin, 1987). There are several studies on the behavioral ecology of chemical trail systems that suggest trail discrimination may enhance colony fitness, but there has been little research on the ecological significance of trail pheromone specificity. The existing literature provides some insight into the nature of the behavioral and ecological processes that might select for adaptive signal specificity. Four aspects of ecological interaction could be associated with the origin of specificity in trail signaling; current studies provide tests of the following hypotheses.

- *Inter- and intraspecific competition has produced divergences in trail communication systems.*

Divergences among sympatric ant species in recruitment communication systems have been correlated with patterns of resource use and competitive interactions (Davidson, 1977; Hölldobler, 1976; Traniello, 1987). Collective and individual foraging behavior in desert harvester ants (*Pogonomyrmex*, *Aphaenogaster* [= *Novomessor*] and *Pheidole*; Hölldobler, 1976; Hölldobler and Möglich, 1980; Hölldobler et al., 1978) and North Temperate generalist ants (Traniello, 1987) is organized (respectively) directly and indirectly by pheromone-based recruitment communication systems that differ among species. Different foraging systems are considered to be adaptations to exploiting resources with different density distributions, facilitating resource partitioning (Davidson, 1977) or enhancing resource holding ability and speed of retrieval to minimize interference competition (Adams and Traniello, 1981; Hölldobler et al., 1978; Traniello, 1983, 1987;

Traniello and Beshers, 1991). But are changes in the behavioral mechanisms of food recruitment accompanied by changes in trail pheromone chemistry and response specificity?

The sympatric harvester ants *Pogonomyrmex rugosus* and *P. barbatus* have virtually identical ecological requirements and have trunk trail systems that divert groups of foragers away from each other (Hölldobler, 1976, 1987). The use of new seed patches by workers of *P. barbatus* and *P. rugosus* is regulated by short-lived recruitment pheromones and persistent orientation cues discharged from the poison and Dufour's glands, respectively (Hölldobler, 1976). The poison gland pheromones are anonymous, as are the Dufour's gland secretions. However, trunk routes may be colony-specific. Hölldobler (1987) argues that the lack of species specificity may be the behavioral mechanism that is responsible for the interchangeability of *P. rugosus* and *P. barbatus* in territorial interactions. Colony-specific Dufour's gland signatures may represent biochemical divergence due to competitive interactions; their high degree of specificity may give an advantage in territorial defense.

It seems unusual, however, that the poison gland secretions of *Pogonomyrmex* lack specificity. We envision that if selection is acting to ensure privacy of communication, then 'eavesdropping' would play a critical role during the early phase of foraging organization, soon after a new food source is discovered [see for example Adams, 1990b]. Specificity in the recruitment component of the poison gland trail pheromone would thus be predicted, because it would ensure that detection of the recruitment signal by a congener did not lead to interspecific contest competition at the food source, which can result in intense aggression. This specificity does not occur. It may be that trunk route systems provide such an effective spatial partitioning of foraging area that foragers from neighboring colonies rarely search for seed patches in the same sector. Foragers from different colonies may search concurrently the terminal areas of trunk route branches, and thus be sufficiently separated spatially, or it may not be possible to encode a high level of specificity in a relatively low molecular weight pheromone that is a volatile attractant.

The ant *Lasius neoniger* provides a potential example of the evolution of biochemical divergence in trail pheromone structure (Traniello, 1980; 1989a). Colonies of *L. neoniger* occur in dense populations and show strong intraspecific territoriality (Traniello and Levings, 1986). Trail communication includes a colony-specific orientation pheromone used to mark trunk routes; this specificity appears to be due to intraspecific competition.

Colony specificity in the trail pheromones of termites has rarely been examined. Oloo (1981) analyzed the preference of *Trinervitermes bettonianus* using pheromone trails generated by termites and a Y-maze bioassay. Termites did not discriminate their own trails from those of neighboring colonies. However, *T. bettonianus*, *Odontotermes* sp. and *Macrotermes michaelseni* workers preferentially followed trails of their own species. These three species occur sympatrically, and results thus support the hypothesis that trail communication systems should diverge.

- *Pheromone specificity increases the accuracy of orientation of homing foragers and facilitates nest recognition.*

Foragers returning to the nest use celestial, landmark, and odor cues as orienting stimuli during homing (Wehner, 1983). Chemical marks along trunk trails and at the nest entrance may facilitate accurate homing by offering odors that can be used as complementary cues to visual stimuli, or used under conditions in which visual cues are not available. In the latter situation, a colony-specific pheromone may provide a cue critical to nest discrimination during the final stage of homing (Traniello, 1989a). Colony-specific markings permit nest recognition in *Pogonomyrmex badius* (Hangartner et al., 1970); other species of ants also mark their nest entrances (Hölldobler and Wilson, 1990).

But in some species of social insects visual cues appear to play the dominant role in orientation. For example, in the ants *Camponotus pennsylvanicus* and *Leptothorax unifasciatus* visual cues quickly replace the use of chemical trails in orientation during foraging (Aron et al., 1988; Traniello, 1977). In fact, in *L. unifasciatus* visual orientation predominates after only one trip to a food source (Aron et al., 1988). In *Lasius neoniger*, foragers home using visual cues preferentially when visual and chemical cues are experimentally placed in competition with each other (Traniello, 1989a). The harvester termite *Hodotermes mossambicus* uses both optical and chemical orientation during above ground foraging activity; optical orientation is more precise than orientating by chemical cues when distinct light cues are available (Leuthold et al., 1976).

Given the precision of homing by visual cues, it may seem questionable that foragers returning to the nest would require additional signals to increase the accuracy of orientation. Nevertheless, specificity in nest markings could serve as a chemical check on colony identity or a homing cue if visual cues are unavailable; this seems to be the function of the nest exit pheromones of *Atta cephalotes* (Hölldobler and Wilson, 1986).

Surprisingly, the chemical trails of some ants may also encode information about the identity of the individual that laid the trail. Individually specific trail markings are used during nest emigration by workers of *Pachycondyla tesserinoda* (Jessen and Maschwitz, 1986). Because of its individual specificity, the origin of the trail substance could not be determined. A similar system of individually specific trails has been described in *Leptothorax affinis*; in this case the trails are used for food recruitment (Maschwitz et al., 1986). *Paraponera clavata* and *Leptothorax unifasciatus* may offer additional examples of individuality superimposed upon colony-specific pheromones (Aron et al., 1988; Breed and Harrison, 1988). Maschwitz et al. (1986) hypothesized that individuality in a trail pheromone could provide a finely tuned mechanism of directional discrimination that would not be possible with only visual acuity. We hypothesize that individual specificity could play a role in maintaining the path fidelity of individual foragers. The role of trail individuality in the orientation hierarchy requires further study, as does its ecological role.

Mechanisms used by other hymenopteran species may give insight into the relative importance of olfactory and visual cues in homing and nest recognition. Altering the landmark panorama of a nest of an African carpenter bee (*Xylocopa*) does not affect nest locating ability (Anzenberger, 1986). Recent research by Hefetz (1992) demonstrated that *X. pubescens* uses individualistic olfactory cues in nest recognition. Pheromones originating in the Dufour's gland, mandibular gland, and rectal fluid may be involved. Bumblebees (*Bombus occidentalis*) also mark nest entrances with colony-specific odors (Foster and Gamboa, 1989), as does the halictid bee *Lasioglossum malachurum* (Ayasse, 1990). In the anthophorid bee *Eucera palestinae*, Dufour's gland secretions may play a role in nest odor recognition (Shimron et al., 1985), and in the eusocial wasp *Polistes metricus* lipids on the nest surface may serve as recognition factors (Espelie et al., 1990).

- *Pheromone specificity permits territory advertisement and recognition and may yield an advantage in territorial conflicts.*

A territory is an area used exclusively by an individual or a group; current occupation of a territory is advertised in social insects by scent marking. Both the African weaver ant *Oecophylla longinoda* and the green tree ant *O. smaragdina* mark their arboreal territories with colony-specific pheromones present in rectal fluid (Hölldobler, 1979, 1983; Hölldobler and Wilson, 1977). Recognition of these pheromones appears to regulate agonistic interactions between neighboring colonies in the field. The colony-specific persistent fractions of trail pheromones of the ants *Lasius neoniger* and *Pogonomyrmex* spp. may also be involved in recognition and defense of foraging area (Hölldobler, 1976; Traniello, 1980), but the exact role of the specificity of these substances has yet to be determined. It is possible that recognition of these pheromones results in enhanced aggressiveness when individuals are "fighting on their own territory" (Hölldobler and Wilson, 1978). Workers of *Atta laevigata* are considered to deposit colony-specific territorial pheromones that increase aggressive behavior in nestmates and submissive behavior in alien conspecifics (Salzemann and Jaffe, 1990), but the existence of territorial pheromones in *Atta* has been disputed (Hölldobler and Wilson, 1986).

A high degree of territoriality in ants need not be associated with territory marking. *Azteca trigona* workers, for example, strongly defend arboreal territories by actively patrolling boundaries rather than passively advertising occupancy through the use of pheromones (Adams, 1987, 1990a).

Some termite species are known to defend space and recognize intra- and interspecific competitors (Adams and Levings, 1987; Thorne and Haverty, 1991;

Traniello and Beshers, 1985). Individuals deposit rectal fluid and sternal gland secretions on trails; this *pavé* (Grassé and Noirot, 1951) may be responsible for colony specificity, but the phenomena of territory marking in termites remains to be demonstrated.

- *Interspecific trail-following behavior evolves as a mechanism to exploit information about the location of food sources.*

Experimental laboratory studies are numerous, but interspecific trail following is uncommon in nature. Ants in parabiotic associations provide the only example of trail sharing, which may represent weak parasitism, mutualism, or commensalism (Wilson, 1965). *Camponotus beebei* foragers lay and follow their own chemical trails and are able to interpret the trail pheromones of *Azteca chartifex* (Wilson, 1965). In another parabiotic association, Neotropical *Crematogaster* nest with the dolichoderine *Monacis debilis* and the formicine *Camponotus femoratus*, and these latter two species exploit the scouting and quick food-finding ability of *Crematogaster* by following their scent trails (Swain, 1980). Additionally, Adams (1990b) has shown that workers of *Zacryptocerus maculatus* follow the trails laid by scouts of *Azteca trigona* to find new food sources. If trail composition in these formicine, myrmicine and dolichoderine species is similar to what has been described in other representatives of these subfamilies, which do not appear to share common chemistry, it is likely that selection has favored generalist chemoreceptors to permit such information exploitation.

Trail following between species of different genera also occurs in several socially parasitic ant species (Lenoir et al., 1992).

8. Phylogeny, foraging ecology, and the evolution of trail communication

8.1. Ethological models

In ants, tandem running is a form of recruitment communication in which a single nestmate is led ‘in tandem’ from the nest to a new nest or food source (Wilson, 1971). The signals that maintain a communicative tie between the leader and follower involve motor displays, surface pheromones, and glandular secretions (Hölldobler, 1978; Hölldobler and Traniello, 1980b). Because of its presence in the Ponerinae, tandem running has been viewed as a possible evolutionary antecedent of chemical trail communication, in which the secretions of various glands are deposited singly or in combination to regulate nestmate recruitment and orientation (Hölldobler and Wilson, 1990; Wilson, 1971).

According to ethological models of the evolution of chemical communication, the most derived form of trail signaling is chemical mass communication. At this organizational level, information regulating recruitment is wholly encoded in the structure of the trail pheromone, and the concentration of trail pheromone mediates communication between groups of individuals (Wilson, 1962). This type of trail communication is characteristic of many species in the majority of ant subfamilies and in the termites it appears to be the only method of trail communication (Stuart, 1969; Heidecker and Leuthold, 1984; Prestwich, 1984; Traniello and Busher, 1985). Trail-laying behavior has been described in some species of wasps (Downing, 1991; Jeanne, 1991), but the behavioral effects of the pheromones alone have not been assessed.

During the past two decades, an ecological perspective on the evolution of recruitment communication has augmented traditional ethological analyses of signals used to direct the work of nestmates. In ants, the ability to correlate mechanisms of information transfer with specific environmental circumstances such as competition, nest spacing and food distribution has permitted recruitment systems and their underlying regulatory processes to be analyzed as social components of community-level interaction. This represents an important addition to the more narrowly circumscribed ethological approaches to the evolution of recruitment communication that emphasize the role of phylogeny and views the evolution of behavior as a series of systems of graded complexity [reviewed in Hölldobler, 1978; Wilson, 1971]. But characterizing recruitment communication techniques as 'primitive' or 'advanced,' which seems to have developed its own inertia independent of its biological relevance, masks the complex relationship of phylogeny and ecology as determinants of trail communication. Also, the claim that morphologically 'primitive' species such as those in the Ponerinae have less 'sophisticated' recruitment communication than species in the advanced subfamilies such as the Formicinae and Myrmicinae is questionable since some 'primitive' species may use more signals involving tactile and chemical modalities that have high levels of specificity (Jessen and Maschwitz, 1986; Maschwitz et al., 1986).

Mechanistic ethological studies, however, are not "suspect as a basis for understanding the evolution and ecological efficiency of observed foraging patterns" (Carroll and Janzen, 1973). It is highly unlikely that modes of foraging organization observed in the laboratory are only artifacts of that environment. Rather, laboratory studies of communication mechanisms reveal the details of signal structure and function. By their nature such studies are limited in providing ecological insight, but that is clearly not their focus. As components of a combined ecological and ethological approach they are powerful tools to examine the details of communication during foraging.

8.2. Resource distribution and recruitment communication

The ecological influences on the evolution of communication and the relative role of phylogeny are apparent when a variety of species having no recruitment communication are surveyed. A lack of any ability to direct nestmates to food sources greater in size than one worker load is shown by the ants *Cataglyphis* and *Ocymyrmex* (Formicinae) (Schmid-Hempel, 1987; Wehner, 1987) and *Neoponera* (= *Pachycondyla*) *apicalis* (Ponerinae) (Fresneau, 1985). Phylogeny, therefore, does not seem to be an important correlate of the absence of recruitment communication. The distribution patterns of the resources (arthropod prey) collected by these species, however, are similar, suggesting that among these species the major selective force for individual retrieval has been random, widely dispersed food items. Foragers of *Cataglyphis* scavenging in the arid, food-impoverished Saharan shrub deserts, live dangerously, foraging at soil surface temperatures of up to 70° C. As is apparently true for some other thermophilic ants, the reason that *Cataglyphis* foragers 'live on the edge' is that their primary food is arthropods that have died due to desiccation from heat stress (Wehner et al., 1992). Such temporally and spatially unpredictable food items, each equal to a single forager load, provides no selection for recruitment or cooperative retrieval because individual retrieval maximizes the rate of food delivery to the colony.

The absence of cooperative foraging does not appear to be due to the lack of preadaptive traits. For example, although *Pachycondyla apicalis* and *P. obscuricornis* show no food recruitment, each species uses communication by tandem running during nest emigration (Traniello and Hölldobler, 1984), suggesting that cooperative foraging may evolve given the appropriate selection pressure. And within the Dacetini, species of *Smithistruma*, *Neostruma* and *Strumigenys* are specialist, solitary predators (Brown and Wilson, 1959) but *Orectognathus versicolor* exhibits chemical trail communication during nest emigration (Hölldobler, 1981). Similarly, the Ponerinae are often cited as exhibiting primitive communication systems, but recruitment behavior, including chemical mass communication, is common in species that utilize patchily distributed or clumped food resources. Examples include *Megaponera foetens* (Longhurst et al., 1979), *Leptogenys* (Maschwitz and Schönegge, 1977, 1983; Maschwitz and Steghaus-Kovac, 1991), *Paraponera clavata* (Breed and Bennett, 1985), *Pachycondyla* (= *Termitopone*) *laevigata* (Hölldobler and Traniello, 1980a), *Paltothyreus tarsatus* (Hölldobler, 1984), *Ectatomma ruidum* (Lachaud, 1985; Pratt, 1989), and the ambyloponines *Prionopelta amabilis* (Hölldobler et al., 1992), *Onychomyrmex* spp. (Hölldobler et al., 1982), and *Ambylopone australis* (Hölldobler and Palmer, 1989). Also, the aneuretine *Aneuretus simoni* feeds on patchily distributed fallen fruits and has well-developed trail communication (Traniello and Jayasuriya, 1981a,b).

Species in the same myrmicine genus vary according to resource distribution in their use of chemical communication during foraging. *Pogonomyrmex rugosus* and *P. barbatus* occur sympatrically and feed on seed clumps and have well-developed trail communication. In contrast, *P. maricopa* collects scattered seeds, primarily through individual retrieval, has a comparatively narrow diet breadth, and has a relatively weak response to seed patches (Hölldobler, 1976; see also Kugler, 1984). Strong intrageneric variation in recruitment behavior is seen in *Camponotus* (Hölldobler, 1971; Hölldobler et al., 1974; Traniello, 1977), but its ecological significance is unclear.

The role of food distribution and the involvement of pheromones and their preadaptive condition in the evolution of foraging behavior are neatly illustrated by a comparison of the biology of *Pachycondyla obscuricornis* and *P*. (= *Termitopone*) *laevigata* (Hölldobler and Traniello, 1980a, 1980b; Traniello and Hölldobler, 1984). *P. obscuricornis* foragers individually hunt arthropod prey and show no apparent form of recruitment communication. During nest emigration, however, tandem running and worker transport are used to change the colony's location. Communication between the tandem pair leader and her follower is mediated by secretions from the pygidial gland. The congeneric *P. laevigata* is an obligate predator of termites. A raiding column is organized after a scout discovers a single termite, and the pygidial gland secretion alone, acting as a mass recruitment trail pheromone, stimulates nestmates to leave the nest and orients them to the termite find. It is therefore evident that the dietary specialization of *P. laevigata* selected for changes in recruitment behavior and pygidial gland secretion chemistry and/or the sensory perception and response to the secretion. Because of the probable economy of these changes, a gland characteristic of the genus readily became incorporated into a new repertoire of foraging behavior associated with the exploitation of clumped resources.

Recent attention in studies of territoriality have focused on trail communication as a behavioral mechanism underscoring the social organization of territoriality and foraging and its significance in community-level interaction (Hölldobler 1987). Trail communication not only regulates cooperative foraging, prey retrieval, and prey selection, but also affects nest dispersion, the use of foraging space and orients homing foragers [reviewed in Traniello, 1989b]. Ants that use trunk routes or trail systems that are marked with chemicals are widely scattered in several subfamilies and include species of *Formica*, *Atta*, *Lasius*, *Pheidole*, *Pogonomyrmex*, *Aneuretus*, *Leptogenys* and *Camponotus*, among others. The stability and temporal persistence of the trail systems of different species are variable. Functionally, trail systems represent a finer level of spatial segregation of foragers that may have a number of functions including (1) Adjusting search to areas of high food availability; (2) adjusting search away from neighboring nests and reduction of injury or loss of workers; (3) directing the search patterns of foragers to converge at one area to facilitate short-or long-range recruitment of prey and thus successful prey retrieval; (4) serving as initial orientation guides for naive foragers [reviewed in Traniello, 1989a, 1989b].

8.3. Communication, competition, and cooperative retrieval

Perhaps the principal reason for the infrequent evolution of polymorphism in the worker caste is that tasks can be performed by groups. A monomorphic species' diet may be expanded to include food items that exceed a single forager's load size limit and thus simulate the retrieval ability of a larger forager caste (Oster and Wilson, 1978). Foragers of *Lasius neoniger* are small and monomorphic;

individual retrieval of prey is restricted to small prey, but cooperative foraging enables workers of *L. neoniger* to collect larger prey, increasing diet breadth 30-fold (Traniello 1983). Foraging costs associated with cooperative retrieval appear to be lower than those of individual retrieval, and there appears to be an energetic advantage to cooperative foraging, reflected in reduced search and retrieval costs. The plasticity of recruitment communication also allows this species to efficiently acquire patchily distributed resources that occur in units of different sizes.

Success in competition should vary with the number of workers recruited to a contested resource. As a result, factors that affect the recruitment response of a colony will alter the way in which resources are partitioned between species. First, the strength of an ant colony's recruitment response varies with temperature, and each species has a temperature optimum and a range of temperature tolerances that overlap with those of its competitors (Hölldobler and Wilson, 1990). Competitive ability is enhanced within the range of optimal foraging temperatures. Second, recruitment response may be adjusted to food source size leading to a greater number of workers foraging cooperatively at a large food item. This socially regulated foraging process may result in an inadequate recruitment rate to small- or intermediate-size prey and consequently, the loss of that prey to competitors. For example, in *Monomorium minimum*, higher temperature tolerance and the abilities to recruit rapidly, to produce a chemical repellent and to dissect prey constitute a strategy for the retrieval of large food sources (Adams and Traniello, 1981; see also Chapter 12, this volume). The recruitment response of *M. minimum* is adjusted to the size of a food source, so that large items induce the recruitment of more workers than small ones. In effect, the pheromone-mediated recruitment system fails to bring workers fast enough and in sufficient numbers to defend small prey. Thus, in communities of ants, the properties of species that determine how resources are partitioned are properties of groups of workers and their recruitment systems.

Recruitment makes possible the exploitation of items too large for individual retrieval, but it is time consuming and leaves opportunity for discovery by competing ant species (Oster and Wilson, 1978). The probability of interference can increase from 0% to 100% when items become too large for a single forager to carry (Adams and Traniello, 1981). Species which utilize prey that are large relative to the body size of workers can be expected to evolve methods to decrease losses due to interference during recruitment. In *L. neoniger*, short-range recruitment appears to decrease the time during which a food item is unoccupied and therefore accessible to intra- and interspecific competitors. Prey movement is also critical, and as is the case with *Aphaenogaster* (= *Novomessor*); prey must be moved before competitor mass-recruiting species arrive (Hölldobler, 1982; Hölldobler et al., 1978; Traniello and Beshers, 1991). In contrast, these small-bodied species are unable to move prey, but have rapid recruitment and use chemical repellents to defend against competitors. Workers of these small-bodied repellent strategists dissect prey effectively. We have observed a single colony of *Monomorium minimum* dissect and retrieve over a four-day period the body of an approximately 8-gm shrew that had died near the nest entrance. Because of the effective chemical defenses of *M. minimum* this extremely rich food source could not have been utilized by sympatric species. By contrast, larger workers of other species dissect unevenly and rely more upon individual foraging and cooperative transport to retrieve foods (Traniello, 1987; Traniello and Beshers, 1991).

9. Defensive recruitment responses to competitors and predators

The competitive and predatory interactions among many social insects may lead to decreased fitness in colonies of sympatric species. In some communities, social insects are among the most important predators of other social insects. Army ants, for example, exert considerable predation pressure on social wasps and ground-dwelling ant species; both groups have adapted to their predatory raids by detecting and rapidly responding to the presence of *Eciton* and *Neivamyrmex* (Chadab, 1979; Droual, 1983, 1984).

Competitor recognition occurs in some ant species. *Pheidole dentata* is an ant with a completely dimorphic worker caste. Occurring sympatrically are species of the fire ant, *Solenopsis*, which are fierce competitors and potential predators of *P. dentata*. When minor workers of *P. dentata* contact fire ants, they quickly return to the nest and recruit major workers which are adapted for defense (Wilson, 1976). Because this caste-restricted recruitment to a competitor was observed only in response to *Solenopsis*, this phenomenon was termed *enemy specification*. Neotropical arboreal ants (*Azteca*, *Camponotus*) also show specific responses to territory intruders (Adams, 1987).

Defensive recruitment systems have been reported as well in termites (Traniello and Beshers, 1985). *Nasutitermes costalis* has both large and small worker castes, which chiefly forage, construct the nest, and provide brood care; predator defense is accomplished mainly by nasute soldiers through sternal and frontal gland secretions (Traniello, 1981; Traniello and Busher, 1985). Soldiers search for food and communicate its location to workers by laying sternal gland trails, which initially have a recruitment effect only on other soldiers. As pheromone concentration builds, workers are recruited. Soldier-organized foraging appears to be an adaptation to deter predation during the early phases of foraging when termites are exposed to ants. A similar system occurs in *Schedorhinotermes lamanianus* (Kaib, 1990).

N. costalis also responds defensively to other termites. The response to other species of *Nasutitermes* primarily involves the recruitment of soldiers, but when one soldier or worker of an alien colony of *N. costalis* is detected, large third-instar workers are mobilized. These workers typically are not aggressive, but when recruited to an area where other *N. costalis* have been found, they are truculent defenders, attacking the intruders with their mandibles. Sternal gland pheromones most likely are employed in this enemy-specific recruitment process, but the details of the mechanism of caste-specific recruitment are not known. Similar recognition and recruitment systems exist in other termite species (Lesniak and Traniello, unpublished; Thorne and Haverty, 1991).

10. Chemical recruitment communication, foraging decisions and self-organization

Social insect colonies allocate foraging effort according to food profitability (Breed et al., 1987; Hangartner, 1969; Jaffe and Howse, 1979; Taylor, 1977; Traniello, 1983; Traniello and Beshers, 1991; reviewed in Traniello, 1989b), foraging risk (Nonacs and Dill, 1990, 1991) and competition (Hölldobler, 1976; Traniello, 1989a,b). Chemical communication provides the mechanistic basis for such decisions, but how are decisions implemented at the colony level?

In recent years the concept of self-organization has been used to explain the emergence and maintenance of complex stable states from simple component systems. Developed primarily in the physical sciences to explain the origin of complexity in random systems, it argues that complexity at one level (colony-level foraging patterns) can emerge from the interactions of the simple units that make up the system (individual workers), reacting only to local situations (trail pheromone concentrations) (Deneubourg and Goss, 1989; Nicolis and Prigogine, 1977).

Studies applying self-organization concepts to trail communication have emphasized mass-recruiting species (Beckers et al., 1992a; de Biseau et al., 1991; Calenbuhr and Deneubourg, 1992; Calenbuhr et al., 1992). The relatively large colony sizes, simple individual behaviors and chemical communication systems mediated by the physical processes of evaporation and diffusion of trail substances allow researchers to adopt organizational principles that are perhaps more readily applied to large physicochemical systems (Deneubourg and Goss, 1989). Such an approach is considered fundamentally different from the pervasive view that insect colonies are comprised of "genetically preprogrammed specialists" (Deneubourg et al., 1990). Through a combination of computer simulations, mathematical formulations, and experimental manipulations, the organization of colony foraging decisions from simple individual responses to local trail pheromone concentrations can be described. These demonstrations are particularly important given that the decentralized hierarchical systems of social insects make it difficult to intuitively predict the influence of individual changes on group-level behavior (Pasteels et al., 1987; Resnick, 1990, 1991; Wilson and Hölldobler, 1988).

During the past decade, interest has grown in applying these concepts to social insect biology. The major conclusions of self-organization studies of trail communication and foraging dynamics follow.

- *Colony-level foraging patterns can arise through the collective actions of numerous relatively simple individuals reacting only to the local pheromone environment.*

It has been demonstrated repeatedly that simple individual behaviors alone can produce complex group-level exploratory and foraging patterns (Fig. 7.2a) (Aron et al., 1989; Franks et al., 1991; Goss et al., 1989), minimize colony

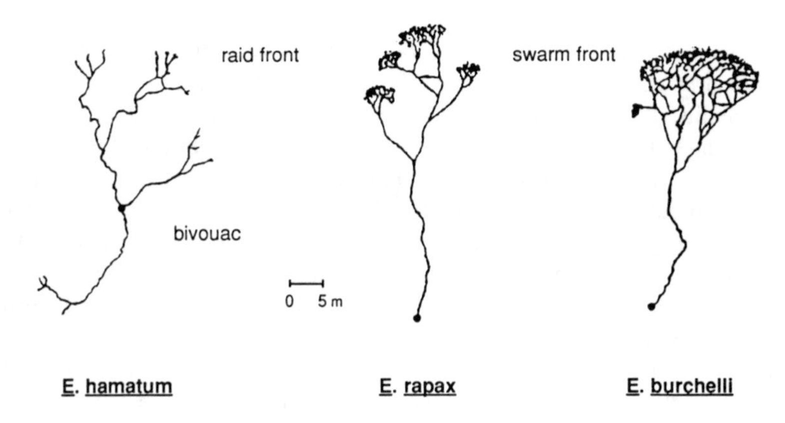

Figure 7.2. (a) Foraging patterns of three army ant species, *Eciton hamatum*, *E. rapax* and *E. burchelli* [modified after Rettenmeyer, 1963, and Burton and Franks, 1985], (b) Monte Carlo simulations of *Eciton* raid patterns under varying food densities. From left to right, the probability and number of food items at each site in the simulation are: 0.01/400 food items; 0.5/one food item; and 0/0 [modified after Deneubourg et al., 1989, and Franks et al., 1991].

foraging path lengths both between colonies (Aron et al., 1990a) and between a colony and food sources (Aron et al., 1990b; Goss et al., 1989; Strickland et al., 1992), as well as influencing the pattern of trail selection and resource utilization when a colony is presented with multiple food sources (Beckers et al., 1990; de Biseau et al., 1991; Pasteels et al., 1987). For example, when presented simultaneously with two nondepleting food sources of equal quality and distance from the colony, mass-recruiting ant species such as *Tetramorium caespitum*, *Monomorium viridum*, *Pheidole hyatti* and *Liometopum apiculatum* initially recruit in equal numbers to both sources, but then switch and continue to recruit to a single source only, with recruitment to the second decreasing significantly (Pasteels et al., 1987; Robson, unpublished data). The second food source can be discovered up to fifteen minutes after the first by *M. viridum* workers yet still attracts the greater numbers of foragers, presumably as a result of individual variance in both the amount of time foragers spend at the food source and the number of ants they recruit (Fig. 7.3a) (Robson, unpublished data).

The original explanation of this phenomenon was based on the analysis of the behavior of groups of individuals, describing how the typical pattern of exponential recruitment to a single food source can produce an unequal allocation of workers when a colony is faced with two identical food sources (Pasteels et al., 1987). To understand the individual behavior that may underlie this phenomenon, we used the StarLogo computer language (Resnick, 1991) operating on the Connection Machine CM-2 (Thinking Machines Corp.) to program the perceptual capabilities and decision rules of individuals placed in a computer environment, similar to that of a laboratory arena, to see if similar foraging patterns emerged. The hierarchical architecture of massively parallel computers such as the Connection Machine makes them well suited to the study of self-organizing phenomena arising in similarly organized systems such as insect societies (Resnick, 1990; Villa, 1992). In this simulation, individuals move randomly, modifying their behavior on the basis of whether the site they occupy is empty or contains food, pheromone, or nest material. Individuals that encounter food will feed for a variable period of time, return to the nest laying a recruitment and orientation pheromone, and recruit individuals from the nest. Nestmates (and any other searching individuals that come across a trail) examine the pheromone concentration at 45° to their immediate left and right and move into the area of highest concentration, with a small error in accuracy. The pheromone itself evaporates and diffuses through the environment with each time step, and each parameter is calibrated (using only a single food source) to allow an individual to discover the food, return to the nest and initiate successful recruitment. By modifying the simulation environment to include a second food source of similar quality at the same distance from the colony, and allowing individuals to continue to operate under the same rules, we were able to produce the same differential recruitment

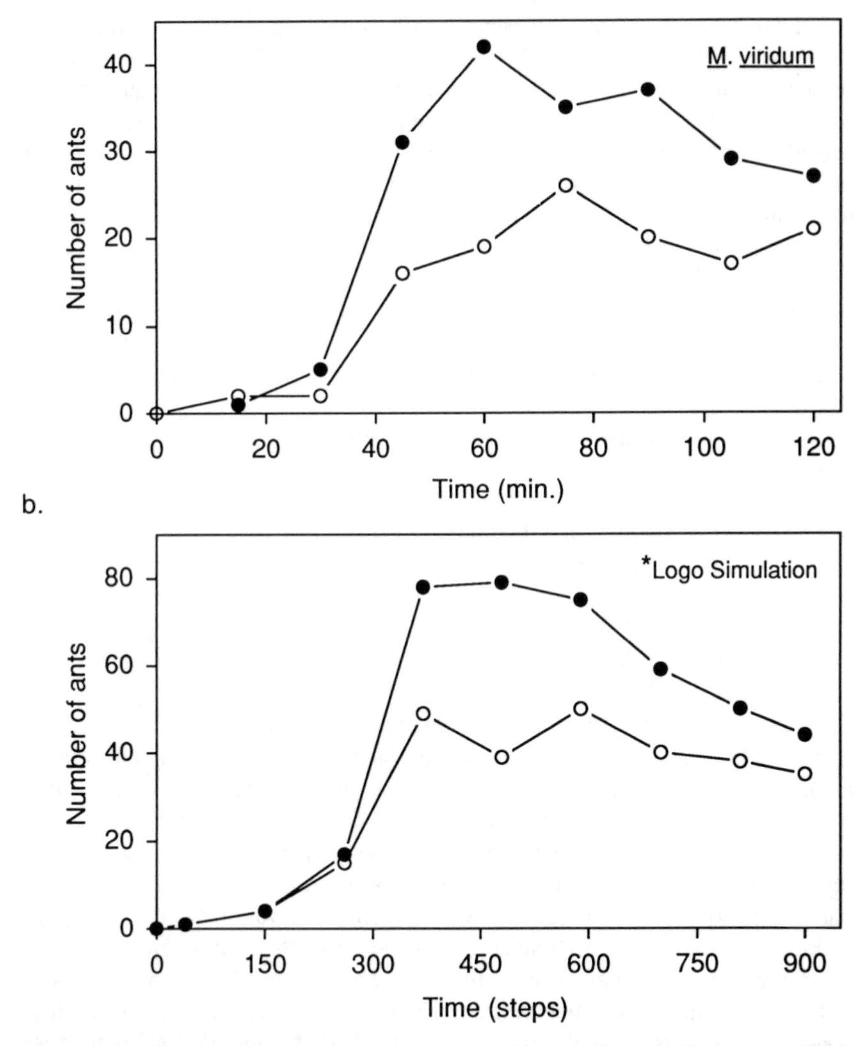

Figure 7.3. (a) The number of *Monomorium viridum* workers feeding at two food sources (○ and ●) of identical quality and distance from the colony entrance. The source that received the greatest recruitment (●) was discovered 15 minutes after the first, (b) StarLogo simulation of the number of ants either feeding at or returning to the colony from two food sources (○ and ●) of identical quality and distance from the colony entrance. Both sources were discovered within two time steps of each other.

found in studies of ants (Figs. 7.3a, 7.3b and 7.4). Workers initially recruited equally to both food sources, then switched and concentrated their efforts at a single source.

The organization of this 'bifurcation' phenomena is thus explained entirely by the simple responses of individuals to pheromone concentrations as regulated by the physical properties of pheromone evaporation and diffusion. Once a difference in the concentration of pheromone on either trail arises (through stochastic processes), it becomes amplified due to the greater number of individuals that follow and in turn lay trails on this path. Pheromone evaporation also contributes to the reduced attractiveness of the remaining path. The differential recruitment therefore, in this simulation at least, is a consequence of the characteristic of the mass-communication system itself, rather than a decision by individuals aware of the resource environment to concentrate on a single source. Hence, such foraging complexity at the colony level need not imply that the behavior of any individual worker is itself complex. This is in contrast to the more typical

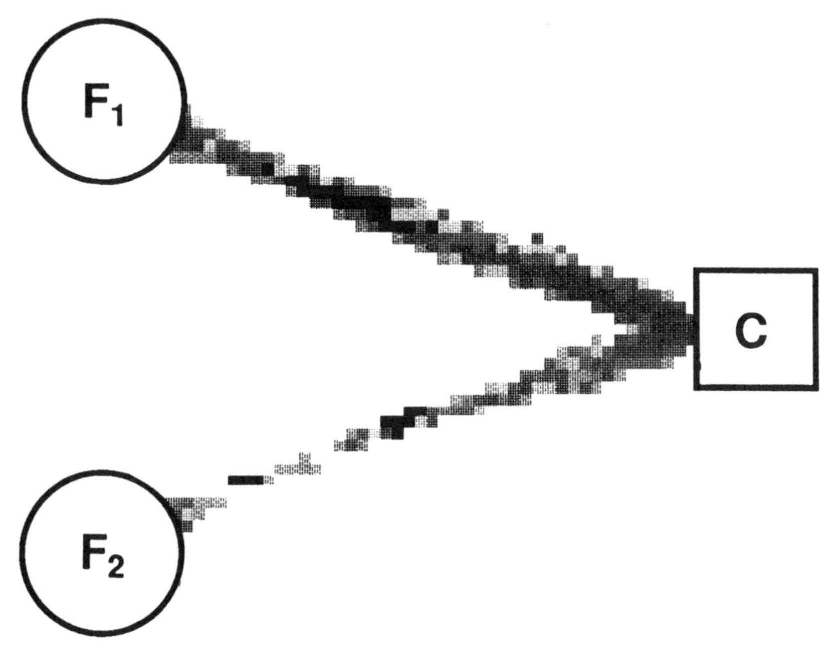

Figure 7.4. StarLogo simulation of an ant colony foraging in an arena containing two identical food sources (after 350 time steps). 'Ants' (■) are moving between the 'colony' on the right (C) and the two 'food sources' on the left (F_1, F_2). The intensity of the pheromone trail is indicated by the shading intensity. Having initially recruited equally to both food sources, the recruitment effort is now concentrating on F_1 and the strength of the pheromone trail and the number of ants recruited to F_2 is diminishing.

approaches to the organization of complexity, in which group-level complexity is based on individual complexity (Resnick, 1990).

- *Different chemically organized colony-level foraging patterns may depend upon the resource environment more than different individual rules per se.*

The influence of food source distribution on the form of colony behavior has been demonstrated for a variety of mass-recruiting ant species. A single column of foragers typically rotates slowly around the nest entrance of the ant *Messor pergandei*; the period of rotation can vary from one to three weeks, with rotation rate decreasing with increasing food abundance (Bernstein, 1975; Rissing and Wheeler, 1976). Monte Carlo simulations whose parameters include the rate of food arrival, the rate at which foragers leave the colony, the rate of pheromone evaporation and the rules by which individuals deposit and follow a pheromone trail produce qualitative results in accordance with field observations (Goss and Deneubourg, 1989). At low food densities, individuals forage randomly around the colony entrance. However, with increasing food density a single foraging trail forms, and through stochastic factors food is depleted more on one side of the trail than the other. As a consequence of pheromone evaporation and a greater tendency for individuals to lay trails in areas where they discover food, the trail slowly rotates away from the area of food depletion, decreasing and eventually stopping if food densities and depletion rates are sufficiently high. The significance of this simulation is that even though the rules governing individual behavior in this model do not change, single trail formation and rotation still arises as a consequence of resource distribution patterns and the communication system.

The species-specific patterns of foraging by army ants, *Eciton*, have also been analyzed within this framework. Through computer simulations, Deneubourg et al. (1989) demonstrated how similar individual behaviors operating within different resource environments can produce fundamentally distinct raid patterns. When the behaviors of individuals within the model were held constant (i.e., the rules governing pheromone deposition and running speed), varying food item size and distributions in the simulation environment produced qualitatively similar foraging patterns to those of different *Eciton* species (Fig. 7.2a, 7.2b). Subsequent field manipulations of the food resources available to *E. burchelli* colonies support these results (Franks et al., 1991).

The column-raid foraging pattern of *E. hamatum*, an army ant that specializes on the clumped but relatively rare and rich resource of social wasp nests, is predominantly formed by a trunk trail with a minimally diffuse foraging front, while the foraging pattern of *E. burchelli* feeding on small, relatively abundant arthropods shows an extensive arborization at the front of the raid. The diet of *E. rapax* is intermediate between that of *E. hamatum* and *E. burchelli*, as is the form of its raid (Fig. 7.2a). By manipulating the density and size of prey items around *E. burchelli* colonies in the field, Franks et al. (1991) induced foraging patterns similar to those species of *Eciton* that feed on prey of a similar nature. Such interspecific variation therefore may represent subtle differences among species in the food recognition capabilities of individuals and the distribution of food in the environment. Individual ants need not perceive the overall resource distribution pattern, or, at the colony level, decide on the form of their foraging pattern.

The environmental parameters influencing colony behavior may also include internal parameters of the colony and individuals themselves, such as colony size and individual satiation, motivation and experience as well as external environmental modifications arising from the previous behavior of individuals (colony patterns of pheromone deposition). For example, the differential recruitment that can occur when mass-recruiting species exploit two identical food sources appears to require a critical density of individuals (Detrain et al., 1991). When the pool of foragers is experimentally reduced, there is an insufficient number of ants present to allow the pheromone concentration of one trail to eventually dominate the other. The foraging patterns of army and Argentine ants and the organization of trails between Argentine ant colonies are mediated by individual responses to pheromone concentrations (Aron et al., 1990a; Deneubourg et al., 1989), and hence should also demonstrate density-dependent (i.e., intrinsic) effects.

- *Simple and variable individual behavioral responses to local pheromone concentrations are sufficient to produce adaptive (sensu Deneubourg and Goss, 1989) colony-level foraging patterns.*

Theoretical analyses of self-organizing systems and behavioral observations of social insects show that the mechanism of colony response need not be based upon individuals possessing sophisticated decision-making capabilities. The existence of simple individuals with a level of stochasticity in their behavioral decisions is sufficient to produce what are considered "adaptive" colony responses (Deneubourg and Goss, 1989). For example, the foraging trails of mass-recruiting ants exploiting a large food source tend to change from the initial trail, reflecting the original path of the discovering ant, to a more direct route between food and colony as the recruitment process becomes established. The ability of the colony to minimize path length appears to depend on the existence of variability in the behavior of individuals, in particular the rate at which individuals lose the original trail, incorrectly discriminating between areas of high and low pheromone concentration [e.g., Aron et al., 1990b]. Previous observations have noted the influence of individual trail-following variability on the overall foraging pattern of the colony [e.g., Pasteels et al., 1987; and Wilson, 1962], and its significance and requirement to simulation models has been investigated in detail [e.g., Beckers et al. 1992b; Calenbuhr and Deneubourg, 1992; Pasteels et al., 1982; 1987; and

Millonas, 1992]. The rate at which individuals lose a pheromone trail has even been considered an 'adaptive error,' increasing colony foraging efficiency by allowing recruited ants to discover and exploit new food sources (Deneubourg et al., 1983; Pasteels et al., 1987).

As well as simplifying mathematical descriptions of self-organizing processes, individual simplicity is considered to allow the colony to achieve a greater variety and flexibility in its behavior than could be achieved through a society based on many complex individuals. This represents an extension of the idea that large decentralized systems operating in parallel are more reliable and collectively show a greater flexibility to novel situations [e.g., Herbers, 1981]. Any limitation in system efficiency related to individual simplicity is considered to be more than overcome by the greater reliability and flexibility such a system can provide (Aron et al., 1990a). Indeed, the positive correlation between mass recruitment systems and colony size has been used to argue that the greater complexity of larger social insect societies is most efficiently achieved through a system of self-organization based on individual simplicity (Beckers et al., 1989; Jaffe and Deneubourg, 1992).

11. Self-organization and trail communication in retrospect and prospect

Despite these significant findings, the theory of self-organization has been slowly accepted within the study of social insect foraging behavior as a whole. One possible reason is that the basic elements comprising a self-organizing system have been previously described, albeit under a different terminology. Individual social insects have long been known to possess a relatively simple behavioral repertoire and base their decisions predominantly on local information, with the consequence that social insect groups collectively display a variety of emergent behaviors in the face of variable environmental conditions (Hangartner, 1969; Sudd, 1963; Wilson, 1962, 1971). 'Collective decision making' seems conceptually equivalent to the 'electorate response' described previously by Wilson (1962); such similarity may obscure the novel findings of self-organization theory.

A further potential problem with the application of self-organization theory is that these studies seem to consistently consider the behaviors described as being adaptive or efficient in the sense of contributing to colony fitness. Although it is possible to envisage selective advantages for the various behaviors described above, the adaptive significance of these behaviors can only be determined in reference to their ecological context, and should not be assessed on the basis of their having arisen through a process of self-organization alone. Until appropriate studies have been made, such observations seem to demonstrate the unavoidable consequence of the perceptions and behaviors of individuals to emergent group-level foraging patterns. Our own StarLogo simulations reveal that the ultimate form of the group response to two identical food sources is sensitive to parameters such as the relative level of pheromone deposition and individual pheromone sensitivity. Variation in these parameters may provide a mechanistic explanation for interspecific differences (Robson, unpublished data).

While recognizing the limits of adaptive assessments, we suggest that the potential for the processes arising through self-organization to limit behavior at the group level should also be considered. The ability of mass-recruiting ants to 'choose' between food sources and to minimize foraging trail lengths illustrates this point.

At least six species of mass-recruiting ants, *Tetramorium caespitum* (Pasteels et al., 1987), *Linepithema humile* and *Pheidole pallidula* (Beckers et al., 1990), and *Liometopum apiculatum*, *Pheidole hyatti* and *Monomorium viridum* (Robson, unpublished observations) are known to concentrate foraging effort at a single food source when simultaneously offered two identical food sources. Most ant species will also switch to a second higher quality food source (1M vs. 0.1M sucrose), even if already using a food source of lesser value. Such behavior has been interpreted as demonstrating the adaptive consequence of a self-organizing system by allowing the appropriate collective decision to emerge [e.g., Beckers et al., 1990; Pasteels et al., 1982]. However, recent research has shown that not all mass-recruiting species consistently allocate workers predominantly to a single source when offered two sources of identical quality; *Pheidole hyatti* is such a species.

Significant variation in the response to food resources of unequal quality and foraging paths of unequal length also exists. In the ant *Lasius niger* recruitment does not switch to the higher quality food source if recruitment to the first is well established (Beckers et al., 1990). Similarly, colonies of the Argentine ant *Linepithema humile* are capable of allocating workers to the shortest of two foraging routes only if both are available simultaneously; they cannot switch to a shorter route once recruitment along the initial path has progressed. The pattern of worker allocation when two foraging paths of identical length to the same food source are present is also variable between species. *Lasius flavus* and *Linepithema humile* eventually concentrate their effort on a single path, whereas *Myrmica sabuleti* workers continue to fluctuate their effort between paths (de Biseau et al., 1992). *Lasius niger* was originally considered to always concentrate on a single path (de Biseau et al., 1992) but now is only considered to "nearly always do so" (Beckers et al., 1992b). It is clear that the colony-level behavior arising through self-organization need not always confer the ability to select the highest-quality food source or shortest foraging path. The same processes that result in colonies concentrating on the richest of two food sources appear to constrain the colony from switching to a third, more profitable source if recruitment to the first is already well established. When compared between species, the response of mass-recruiting ants to similar situations clearly differs, but it is not obvious why a particular pattern of resource utilization or path selection may be adaptive for one species but not for another. We consider this interspecific variability in colony-level foraging patterns (Beckers et al., 1990; Beckers et al., 1992b; de Biseau et al., 1992) to further argue against *a priori* adaptive interpretations of the consequences of self-organization.

Another difficulty with self-organization theory is that the majority of studies appear to reject the need to consider natural selection as a significant factor influencing the organization of colony behavior. Rather, the process of self-organization itself is seen as the source of these behaviors, which arise as an unavoidable consequence of the interaction between individuals (Goss and Deneubourg, 1989). However, independent of whether selectionist arguments can be rejected, the problem is compounded by an attempt to nevertheless interpret results within an evolutionary context by using a loosely defined concept of efficiency to describe a variety of colony behaviors, or by citing the proposed advantages of a self-organization system as being the reason for the widespread evolution of sociality itself [e.g., Beckers et al., 1990; Goss and Deneubourg, 1989]. If the goal is to explore the optimal way for individual units to be constructed and interact to maximize the efficiency of the system itself [i.e., to maximize reliability, flexibility and fault tolerance (Aron et al., 1990a)], then an approach that ignores external factors such as natural selection may be valid. Indeed, this may explain why similar approaches are being actively used in a variety of nonbiological fields, such as systems design and 'complexity engineering' in which selection has no relevance [e.g., Brooks, 1991; Wolfram, 1986]. However, it is erroneous to evaluate the efficiency of a particular recruitment system, which is rooted in a species' natural history, on the basis of such design principles alone, and then use this information to compare the efficiency of various strategies across species. For example, it is possible to measure both the rate and accuracy of recruitment associated with various foraging strategies, but such measures cannot be combined into a meaningful index of efficiency that provides insight into the ecological significance or phylogenetic distribution of such systems.

The application of evolutionary biology without critical evolutionary analysis thus blurs the significance of the results of self-organization studies. For example, collective exploration may be "more efficient than individual exploration" (Detrain et al., 1991) and minimizing total path lengths between subcolonies at the cost of increased travel time may be of benefit to Argentine ants (Aron et al., 1990a). But these hypotheses require testing in light of current models of the energetics of ant foraging. Similarly, the proposed adaptive value of a colony concentrating on a single source in order to keep the other "in reserve" (Pasteels et al., 1987) needs to be measured in terms of the impact of such a strategy on competitive ability. While these studies may provide insight into the organization of behavior, their relevance to colony fitness, a parameter that must be estimated before adaptive assessments can be made, is less clear. Can observations of the mechanisms by which only some species recruit unequally to two identical food sources provide insight into the fitness correlates of this behavior? Does the rate at which an ant loses a trail (the "adaptive error") (Deneubourg et al., 1982) really result in a greater food intake by the colony than would occur if all recruited ants successfully reached the particular food item? Can ants that have lost the trail really be considered likely to discover new food items? Social insect colonies operate within ecological communities of varying competitor intensity and food distributions. Self-organization theory should benefit from an infusion of behavioral ecology and field study.

The significance of the concept of complexity in clarifying our understanding of the organization of social insect colonies is also not well developed. Indeed, its definition and application within a variety of fields remains unclear, despite recent attempts to address this concept explicitly [e.g., Bennett, 1990]. Biological groups appear to have a greater behavioral repertoire than that of any single individual. Yet in studies of social organization, the use of the term complexity seems to imply that the behavioral capabilities of the colony are greater than would be expected from knowledge of the behavior of the simple individuals alone, and that this greater capability arises through the process of self-organization. But this argument confuses the concepts of complexity, emergence, and synergy, and by describing all colony behaviors as complex it fails to define the exact nature of the relationship between individual and colony behavior. Terms such as simplicity and complexity appear to reflect the field of origin of this approach and the growing theories of 'complexity' and 'order' [e.g., Prigogine, 1980; Prigogine and Stengers, 1984; Nicolis and Prigogine, 1989] rather than illuminate our understanding of biological phenomena within an evolutionary framework. Until such concepts are clearly delineated, and a more rigorous system of their acceptance or rejection is developed, the use of such qualitative and value-laden terminology may be confusing.

Rather than representing a fundamentally different approach to the study of social insect organization [e.g., Deneubourg and Goss, 1989] self-organization provides a useful methodological framework (nonlinear differential equations and computer simulations) demonstrating that a diversity of group-level behaviors may arise more as a consequence of the organizational characteristics of the group itself, rather than as a result of a change in the behavior of individuals. In particular, by examining a variety of foraging behaviors these studies have elucidated many nonintuitive relationships between individual and colony behavior and have suggested that individual behavioral variance can have a far greater 'creative' role than previously thought [e.g., Deneubourg and Goss, 1989]. The perception of the goal of a particular colony behavior can be quite different from the actual behavior of the individuals, and under certain circumstances groups of individuals cannot help but display a greater behavioral repertoire at the colony level. These concepts may have great significance for our understanding of the chemical organization and evolution of insect societies.

12. Conclusions

Traditionally, research on trail communication has emphasized exocrine gland systems and pheromone chemistry. But these are only two aspects of the chemical ecology of social insects. The organizational levels of communication have been correlated with foraging ecology, suggesting the functional significance of different modes of chemical recruitment behavior. Other research has concentrated on the dynamics of the organization of such group behaviors, arguing that there is a significant association between intrinsic factors and foraging complexity.

We are still far from fully integrating behavioral and chemical ecology in the analysis of social insect trail and territorial communication. Considerable research on the ecology of foraging and territorial behavior is required, and the mechanistic details of communication systems need to be evaluated with respect to the community-level processes that have directed their evolutionary course. Detailed comparative research is often lacking. The fitness consequences of communication systems must be determined, most likely through measures of energetic efficiency, colony growth dynamics, and alate production. Currently, a qualitative scale of efficiency is too often used to characterize foraging communication, and is too loosely applied to provide in-depth insight into the adaptiveness of the behavior it describes. Theories of foraging and territorial strategy should be aligned with communication system analyses to generate testable hypotheses.

Acknowledgments

We thank Rob Page and Jacques Pasteels for helpful discussion and E. Adams, B. Hölldobler, R. Vander Meer and T. Strickland for reprint updates. M. Resnick, R. Putnam (Thinking Machines Corp.) and the staff of the Boston University Center for Computational Science assisted greatly with the computer simulations. Mary Lesniak provided indispensable assistance with literature searches. Thanks also to the National Science Foundation and the Whitehall Foundation for supporting J.F.A.T. and the Fulbright Foundation for supporting S.K.R.

References

Adams, E.S. (1987) The behavioral ecology of territory defense in a community of neotropical arboreal ants. Ph.D. Thesis, University of California, Berkeley, Calif.

Adams, E.S. (1990a) Boundary disputes in the territorial ant *Azteca trizona*: Effects of asymmetries in colony size. *Anim. Behav.* 39: 321–328.

Adams, E.S. (1990b) Interaction between the ants *Zacryptocerus maculatus* and *Azteca trigona*: Interspecific parasitization of information. *Biotropica* 22: 200–206.

Adams, E.S., Levings, S.C. (1987) Territory size and population limits in mangrove termites. *J. Anim. Ecol.* 56: 1069–1081.

Adams, E.S., Traniello, J.F.A. (1981) Chemical interference competition by *Monomorium minimum*. *Oecologia* 51: 270–275.

Ali, M.F., Morgan, E.D., Detrain, C., Attygalle, A.B. (1988) Identification of a component of the trail pheromone of the ant *Pheidole pallidula* (Hymenoptera: Formicidae). *Physiol. Entomol.* 13: 257–265.

Ampion, M., Quennedey, A. (1981) The abdominal epidermal glands of termites and their phylogenetic significance, in *Biosystematics of Social Insects* (P.E. Howse and J.L. Clément, eds). Academic Press, N.Y.

Anzenberger, G. (1986) How do carpenter bees recognize the entrance of their nests? An experimental investigation in a natural habitat. *Ethology* 71: 54–62.

Aron, S., Deneubourg, J.L., Pasteels, J.M. (1988) Visual cues and trail-following idiosyncrasy in *Leptothorax unifasciatus*: An orientation process during foraging. *Insectes Soc.* 35: 355–366.

Aron, S., Pasteels, J.M., Deneubourg, J.L. (1989) Trail-laying behavior during exploratory recruitment in the Argentine ant, *Iridomyrmex humilis* (Mayr). *Biol. Behav.* 14: 207–217.

Aron, S., Deneubourg, J.L., Goss, S., Pasteels, J.M. (1990a) Functional self-organization illustrated by inter-nest traffic in ants: The case of the Argentine ant, in *Biological Motion. Lecture Notes in Biomathematics* (eds, W. Alt and W. Hoffman). Springer Verlag, Heidelberg. pp. 533–547.

Aron, S., Pasteels, J.M., Goss, S., Deneubourg, J.L. (1990b) Self-organizing spatial patterns in the Argentine ant *Iridomyrmex humilis* (Mayr), in *Applied Myrmecology: a world perspective* (eds, R.K. Vander Meer, K. Jaffe and R. Cedeno). Westview Press, Boulder, CO. pp. 438–451.

Attygalle, A.B., Morgan, E.D. (1985) Ant trail pheromones. *Adv. Insect Physiol.* 18: 1–30.

Attygalle, A.B., Cammaerts, M.C., Cammaerts, R., Morgan, E.D., Ollet, D.G. (1986) Chemical and ethological studies of the trail pheromone of the ant *Manica rubida* (Hymenoptera: Formicidae). *Physiol. Entomol.* 11: 125–132.

Attygale, A.B., Vostrowsky, O., Bestmann, H.J., Steghaus-Kovac, S., Maschwitz, U. (1988) (3R,4S)-4-Methyl-3-heptanol, the trail pheromone of the ant *Leptogenys diminuta*. *Naturwissenschaften* 75: 315–317.

Ayasse, M. (1990) Odor based interindividual and nest recognition in the sweat bee *Lasioglossum malachurum* (Hymenoptera: Halictidae), in *Social Insects and The Environment* Proc. 11th Int. Congr. IUSSI (eds. G.K. Vereesh, B. Malik, and C.A. Viraktamath). Oxford and IBH Publishing, New Delhi, pp. 511–512.

Beani, L., Calloni, C. (1991) Leg tegumental glands and male rubbing behavior at leks in *Polistes dominulus* (Hymenoptera: Vespidae). *J. Insect Behav.* 4: 449–462.

Beckers, R., Deneubourg, J.L., Goss, S. (1992a) Trail laying behavior during food recruitment in the ant *Lasius niger* (L.). *Insectes Soc.* 39: 59–72.

Beckers, R., Deneubourg, J.L., Goss, S. (1992b) Trails and U-turns in the selection of a path by the ant *Lasius niger*. *J. theor. Biol.* 159: 397–415.

Beckers, R., Deneubourg, J.L., Goss, S., Pasteels, J.M. (1990) Collective decision making through food recruitment. *Insectes Soc.* 37: 258–267.

Beckers, R., Goss, S., Deneubourg, J.L., Pasteels, J.M. (1989) Colony size, communication and ant foraging strategy. *Psyche* 96: 239–256.

Bennett, C.H. (1990) How to define complexity in physics, and why, in *Complexity, Entropy, and the Physics of Information* SFI Studies in the Sciences of Complexity, vol. VIII (ed. W.H. Zurek), Addison-Wesley, N.Y., pp. 137–148.

Bernstein, R.A. (1975) Foraging strategies of ants in response to variable food density. *Ecology* 56: 213–219.

Billen, J.P.J. (1986) Comparative morphology and ultrastructure of the Dufour gland in ants (Hymenoptera: Formicidae). *Entomol. General.* 11: 165–181.

Birch, A.J., Brown, W.V., Corrie, J.E.T., Moore, B.P. (1972) Neocembrene-A, a termite trail pheromone. *J. Chem. Soc. Perkin Trans. I* 1972, 2653–2658.

Blum, M.S. (1974) Pheromonal sociality in the Hymenoptera, In *Pheromones*, (ed. M.C. Birch), North-Holland, Amsterdam, pp. 222–249.

Bordereau, C., Robert, A., Bonnard, O., Le Quere, J.L. (1990) Cis-(3Z,6Z,8E)-3,6,8-dodecatrien-1-ol: Sex pheromone in a higher fungus-growing termite, *Pseudacanthotermes spiniger* (Isoptera, Macrotermitinae). *J. Chem. Ecol.* 17: 2177–2191.

Bossert, W.H., Wilson, E.O. (1963) The analysis of olfactory communication among animals. *J. theor. Biol.* 5: 443–469.

Breed, M.D., Bennett, B. (1985) Mass recruitment to nectar sources in *Paraponera clavata*: a field study. *Insectes Soc.* 32: 198–208.

Breed, M.D., Harrison, J. (1988) Individually discriminable trails in a ponerine ant. *Insectes Soc.* 34: 22–26.

Breed, M.D., Fewell, J.H., Moore, A.J., Williams, K. (1987) Graded recruitment in a ponerine ant. *Behav. Ecol. Sociobiol.* 20: 407–411.

Brooks, R.A. (1991) Intelligence without representation. *Artificial Intelligence* 47: 139–160.

Brown, W.L., Wilson, E.O. (1959) The evolution of the dacetine ants. *Quart. Rev. Biol.* 34: 278–294.

Burton, J.L., Franks, N. R. (1985) The foraging ecology of the army ant *Eciton rapax*: An ergonomic enigma? *Ecol. Entomol.* 10: 131–141.

Calenbuhr, V., Deneubourg, J.L. (1992) A model for osmotropic orientation (I). *J. theor. Biol.* 158: 359–393.

Calenbuhr, V., Chrétein, L., Deneubourg, J.L., Detrain, C. (1992) A model for osmotropic orientation (II). *J. theor. Biol.* 158: 395–407.

Cammaerts, M.C., Cammaerts, R. (1980) Food recruitment strategies of the ants *Myrmica sabuleti* and *Myrmica ruginodis*. *Behav. Proc.* 5: 251–270.

Cammaerts, M.C., Inwood, M.R., Morgan, E.D., Parry, K., Tyler, R.C. (1978) Comparative study of the pheromones emitted by workers of the ants *Myrmica rubia* and *Myrmica scabrinodis*. *J. Insect Physiol.* 24: 207–214.

Carroll, C.R., Janzen, D.H. (1973) Ecology of foraging by ants. *Ann. Rev. Ecol. Syst.* 4: 231–257.

Carthy, J.D. (1950) Odour trails of *Acanthomyops fuliginosus*. *Nature* 166: 154.

Cavill, G.W.K., Davies, N.W., McDonald, F.J. (1980) Characterization of aggregation factors and associated compounds from the Argentine ant, *Iridomyrmex humilis*. *J. Chem. Ecol.* 6: 371–384.

Chadab, R. (1979) Early warning cues for social wasps attacked by army ants. *Psyche* 86: 115–124.

Davidson, D.W. (1977) Foraging ecology and community organization in seed-eating ants. *Ecology* 58: 725–737.

de Biseau, J.C., Deneubourg, J.L., Pasteels, J.M. (1991) Collective flexibility during mass recruitment in the ant *Myrmica sabuleti* (Hymenoptera: Formicidae). *Psyche* 98: 323–336.

de Biseau, J.C., Deneubourg, J.L., Pasteels, J.M. (1992) Mechanisms of food recruitment in the ant *Myrmica sabuleti*: An experimental and theoretical approach, in *Biology and Evolution of Social Insects* (ed. J.Billen). Leuven University Press, Leuven, Belgium. pp. 438–451.

Deneubourg, J.L., Goss, S. (1989) Collective patterns and decision-making. *Ethol. Ecol. Evol.* 1: 295–311.

Deneubourg, J.L., Aron, S., Goss, S., Pasteels, J.M. (1990) The self-organizing exploratory pattern of the Argentine ant. *J. Insect Behav.* 3: 159–168.

Deneubourg, J.L., Pasteels, J.M., Verhaeghe, J.C. (1983) Probabalistic behavior in ants: a strategy of errors? *J. theor. Biol.* 105: 259–271.

Deneubourg, J.L., Goss, S., Franks, N., Pasteels, J.M. (1989) The blind leading the blind: Modelling chemically mediated army ant raid patterns. *J. Insect Behav.* 2: 719–725.

Detrain, C., Deneubourg, J.L., Goss, S., Quinet, Y. (1991) Dynamics of collective exploration in the ant *Pheidole pallidula*. *Psyche* 98: 21–31.

Downing, H. (1991) The Function and Evolution of Exocrine Glands, in *The Social Biology of Wasps* (eds. K.G. Ross and R.W. Matthews), Cornell University Press, NY, pp. 540–569.

Droual, R. (1983) The organization of nest evacuation in *Pheidole desertorum* Wheeler and *P. hyatti* Emery (Hymenoptera: Formicidae). *Behav. Ecol. Sociobiol.* 2: 203–238.

Droual, R. (1984) Anti-predator behaviour in the ant *Pheidole desertorum*: the importance of multiple nests. *Anim Behav.* 32: 1054–1058.

Espelie, K.E., Wenzel, J.W., Chang, G. (1990) Surface lipids of the social wasp *Polistes metricus* Say and its nest and nest pedicel and their relation to nestmate recognition. *J. Chem. Ecol.* 16: 2229–2241.

Foster, R.L, Gamboa, G.J. (1989) Nest entrance marking with colony-specific odors by the bumblebee *Bombus occidentalis* (Hymenoptera: Apidae). *Ethology* 81: 273–278.

Franks, N.R., Gomez, N., Goss, S., Deneubourg, J.L. (1991) The blind leading the blind in army ant raid patterns: Testing a model of self-organization (Hymenoptera: Formicidae). *J. Insect Behav.* 4: 583–607.

Fresneau, D. (1985) Individual foraging and path fidelity in a ponerine ant. *Insectes Soc.* 32: 109–116.

Goss, S., Deneubourg, J.L. (1989) The self-organizing clock-pattern of *Messor pergandei* (Formicidae, Myrmicinae). *Insectes Soc.* 36: 339–376.

Goss, S., Aron, S., Deneubourg, J.L., Pasteels, J.M. (1989) Self-organized shortcuts in the Argentine ant. *Naturwissenschaften* 76: 579–581.

Goss, S., Beckers, R., Deneubourg, J.L., Aron, S., Pasteels, J.M. (1990) How trail laying and trail following can solve foraging problems for ant colonies, In *Behavioural Mechanisms of Food Selection* NATO ASI Series, Vol. G 20. (ed. R.N. Hughes). Springer-Verlag, Heidelberg, pp. 661–677.

Grace, J.K., Wood, D.L., Frankie, G.W. (1989) Trail-following behavior of *Reticulitermes hesperus* Banks (Isoptera: Rhinotermitidae). *J. Chem. Ecol.* 14: 653–667.

Grassé, P.-P., Noirot, C. (1951) Orientation et routes chez les termites: le "balisage" des pistes. *L'Année Psychol.* 50: 273–280.

Hall, P., Traniello, J.F.A. (1985) Behavioral bioassays of termite trail pheromones. Recruitment and orientation effects of cembrene-A in *Nasutitermes costalis* (Isoptera: Termitidae) and discussion of factors affecting termite response in experimental contexts. *J. Chem. Ecol.* 11: 1503–1513.

Hangartner, W. (1967) Spezifität und Inaktivierung des Spurpheromons von *Lasius fuliginosus* Latr. und Orientierung der Arbeiterinnen in Duftfeld. *Zeit. vergl. Physiol.* 57: 103–136.

Hangartner, W. (1969) Structure and variability of the individual odor trail in *Solenopsis geminata* Fabr. (Hymenoptera: Formicidae). *Zeit. vergl. Physiol.* 62: 111–120.

Hangartner, W., Reichson, J.M., Wilson, E.O. (1970) Orientation to nest material by the ant *Pogonomyrmex badius* (Latrielle). *Anim. Behav.* 18: 331–334.

Hayashi, N., Komae, H. 1977. The trail and alarm pheromones of the ant *Pristomyrmex pungens* Mayr. *Experientia* 33: 424–425.

Hefetz, A. (1992) Individual scent marking of the nest entrance as a mechanism for nest recognition in *Xylocopa pubescens* (Hymenoptera: Anthophoridae). *J. Insect Behav.* 5: 763–772.

Heidecker, J.L., Leuthold, R.H. (1984) The organization of collective foraging in the harvester termite *Hodotermes mossambicus* (Isoptera). *Behav. Ecol. Sociobiol.* 14: 195–202.

Herbers, J. M. (1981) Reliability theory and foraging by ants. *J. theor. Biol.* 89: 175–189.

Hölldobler, B. (1971) Recruitment behavior in *Camponotus socius* (Hym. Formicidae). *Zeit. vergl. Physiol.* 75: 123–142.

Hölldobler, B. (1976) Recruitment behavior, home range orientation and territoriality in harvester ants, *Pogonomyrmex*. *Behav. Ecol. Sociobiol.* 1: 3–44.

Hölldobler, B. (1978) Ethological aspects of chemical communication in ants. *Adv. Study Behav.* 8: 75–115.

Hölldobler, B. (1979) Territories of the African weaver ant (*Oecophylla longinoda* [Latreille]): A field study. *Zeit. Tierpsychol.* 51: 201–213.

Hölldobler, B. (1981) Trail communication in the dacetine ant *Orectognathus versicolor* (Hymenoptera: Formicidae). *Psyche* 88: 245–257.

Hölldobler, B. (1982) Interference strategy of *Iridomyrmex pruinosum* (Hymenoptera: Formicidae) during foraging. *Oecologia* 52: 208–213.

Hölldobler, B. (1983) Territorial behavior in the green tree ant (*Oecophylla smaragdina*). *Biotropica* 15: 241–250.

Hölldobler, B. (1984) Communication during foraging and nest relocation in the African stink ant, *Paltothyreus tarsatus* Fabr. (Hymenoptera, Formicidae, Ponerinae). *Zeit. Tierpsychol.* 65: 40–52.

Hölldobler, B. (1987) Communication and competition in ant communities, in *Evolution and Coadaptation in Biotic Communities* Proceedings of the Second International Symposium held in conjunction with the International Prize for Biology, Tokyo, 1986 (eds. S. Kadwano, J.H. Connell, and T. Hidaka). Tokyo University Press, Tokyo, pp. 95–124.

Hölldobler, B., Carlin, N.F. (1987) Anonymity and specificity in the chemical communication signals of social insects. *J. Comp. Physiol. A.* 161: 567–581.

Hölldobler, B., Engel, H. (1978) Tergal and sternal glands in ants. *Psyche.* 85: 285–330.

Hölldobler, B., Lumsden, C.J. (1980) Territorial strategies in ants. *Science* 210: 732–739.

Hölldobler, B., Möglich, M. (1980) The foraging system of *Pheidole militicida* (Hymenoptera: Formicidae). *Insectes Soc.* 27: 237–264.

Hölldobler, B., Palmer, J. (1989) Footprint glands in *Amblyopone australis* (Formicidae, Ponerinae). *Psyche* 96: 111–121.

Hölldobler, B., Traniello, J.F.A. (1980a) The pygidial gland and chemical recruitment communication in *Pachycondyla* (= *Termitopone*) *laevigata*. *J. Chem. Ecol.* 6: 883–893.

Hölldobler, B., Traniello, J.F.A. (1980b) Tandem running pheromone in ponerine ants. *Naturwissenschaften* 67: 360.

Hölldobler, B., Wilson, E.O. (1970) Recruitment trails in the harvester ant *Pogonomyrmex badius*. *Psyche* 77: 385–399.

Hölldobler, B., Wilson, E.O. (1977) Colony-specific territorial pheromone in the African weaver ant *Oecophylla* (Latreille). *Proc. Natl. Acad. Sci. USA* 74: 2072–2075.

Hölldobler, B., Wilson, E.O. (1978) The multiple recruitment systems of the African weaver ant *Oecophylla longinoda* (Latreille) (Hymnenoptera: Formicidae). *Behav. Ecol. Sociobiol.* 3: 19–60.

Hölldobler, B., Wilson, E.O. (1986) Nest area exploration and recognition in leafcutter ants (*Atta cephalotes*). *J. Insect Physiol.* 32: 143–150.

Hölldobler, B., Wilson, E.O. (1990) *The Ants*. Belknap Press, Cambridge, MA.

Hölldobler, B., Engel, H., Taylor, R.W., (1982) A new sternal gland in ants and its function in chemical communication. *Naturwissenschaften* 69: 90–91.

Hölldobler, B., Möglich, M., Maschwitz, U. (1974) Communication by tandem running in the ant *Camponotus sericeus*. *J. Comp. Physiol.* 90: 105–127.

Hölldobler, B., Obermayer, M., Wilson, E.O. (1992) Communication in the primitive cryptobiotic ant *Prionopelta amabilis*. *J. Comp. Physiol. A.* 170: 9–16.

Hölldobler, B., Palmer, J.M., Moffett, M.W. (1990) Chemical communication in the dacetine ant *Daceton armigerum* (Hymenoptera: Formicidae). *J. Chem. Ecol.* 16: 1207–1219.

Hölldobler, B., Stanton, R.C., Markl, H. (1978) Recruitment and food-retrieving behavior in *Novomessor* (Formicidae, Hymenoptera), I: Chemical signals. *Behav. Ecol. Sociobiol.* 4: 163–181.

Howard, R.W., Matsumura, F., Coppel, H.C. (1976) Trail-following pheromones of the Rhinotermitidae: Approaches to their authentication and specificity. *J. Chem. Ecol.* 2: 147–166.

Howse, P.E. (1984) Sociochemicals of termites, in *Chemical Ecology of Insects* (eds. W.J. Bell and R.T. Cardé). Sinauer, Sunderland, MA, pp. 475–519.

Huwyler, S., Grob, K., Viscontini, M. (1975) The trail pheromone of the ant *Lasius fuliginosus*: Identification of six components. *J. Insect Physiol.* 21: 299–304.

Jackson, B.D., Wright, P.J., Morgan, E.D. (1989) 3-Ethyl-2,5-dimethylpyrazine, a component of the trail pheromone of the ant *Messor bouvieri*. *Experientia* 45: 487–489.

Jackson, B.D., Keegans, S.J., Morgan, E.D., Clark, W.H., Blom, P.E. (1991) Chemotaxonomic study of an undescribed species of *Myrmica* ant from Idaho. *J. Chem. Ecol.* 17: 335–342.

Jaffe, K., Deneubourg, J.L. (1992) On foraging, recruitment systems and optimum number of scouts in eusocial colonies. *Insectes Soc.* 39: 201–213.

Jaffe, K., Howse, P.E. (1979) The mass recruitment system of the leaf-cutting ant *Atta cephalotes* (L.) *Anim. Behav.* 27: 930–939.

Jaffe, K., Puche, H. (1984) Colony-specific territorial marking with the metapleural gland secretion in the ant *Solenopsis geminata* (Fabr.) *J. Insect Physiol.* 30: 265–270.

Jaffe, K., Lopez, M.E., Aragort, W. (1986) On the communication systems of the ant *Pseudomyrmex termitarius* and *P. triplarinus*. *Insectes Soc.* 33: 105–117.

Jander, R., Daumer, K. (1974) Guide-line and gravity orientation of blind termites foraging in the open (Termitidae: *Macrotermes*, *Hospitalitermes*). *Insectes Soc.* 21: 45–64.

Jeanne, R.L. (1981) Chemical communication during swarm emigration in the social wasp *Polybia sericea* (Olivier). *Anim. Behav.* 29: 102–113.

Jeanne, R.L. (1991) The swarm founding Polistinae, in *The Social Biology of Wasps* (eds. K.G. Ross and R.W. Matthews). Cornell University Press, NY, pp. 191–231.

Jeanne, R.L., Downing, H.A., Post, D.C. (1983) Morphology and function of sternal glands in polistine wasps (Hymenoptera: Vespidae). *Zoomorphology* 103: 149–164.

Jessen, K., Maschwitz, U. (1986) Orientation and recruitment behavior in the ponerine ant *Pachycondyla tesserinoda* (Emery): laying of individual-specific trails during tandem running. *Behav. Ecol. Sociobiol.* 19: 151–155.

Kaib, M. (1990) Multiple functions of exocrine secretions in termite communication exemplified by *Schedorhinotermes lamanianus*, In *Social Insects and the Environment* Proc. 11th Int. Congr. IUSSI (eds. G.K. Vereesh, B. Malik, and C.A. Viraktamath), Oxford and IBH Publishing, New Delhi, pp. 37–38.

Kaib, M., Bruinsma, O., Leuthold, R.H. (1982) Trail-following in termites: Evidence for a multicomponent system. *J. Chem. Ecol.* 8: 1193–1205.

Klotz, J., Reid, B. (1992) The use of spatial cues for structural guideline orientation in *Tapinoma sessile* and *Camponotus pennsylvanicus* (Hymenoptera: Formicidae). *J. Insect Behav.* 5: 71–82.

Kugler, C. (1984) Ecology of the ant *Pogonomyrmex mayri*: foraging and competition. *Biotropica* 16: 227–234.

Lachaud, J.P. (1985) Recruitment by selective activation: an archaic type of mass recruitment in a ponerine ant (*Ectatomma ruidum*). *Sociobiology* 11: 133–142.

Lenoir, A., Detrain, C., Barbazanges, N. (1992) Host trail following by the guest ant *Formicoxenus provancheri*. *Experentia* 48: 94–97.

Leuthold, R.H. (1975) Orientation mediated by pheromones in social insects, in *Pheromones and Defensive Secretions in Social Insects* (eds. C. Noirot, P.E. Howse and J. Le Masne). Univeristy of Dijon Press, Dijon, pp. 197–211.

Leuthold, R.H., Bruinsma, O., Huis, A. van (1976) Optical and pheromonal orientation and memory for homing distance in the harvester termite *Hodotermes mossambicus* (Hagen). *Behav. Ecol. Sociobiol.* 1: 127–139.

Levings, S.C., Adams, E.S. (1984) Intra- and interspecific territoriality in *Nasutitermes* (Isoptera: Termitidae) in a Panamanian Mangrove forest. *J. Anim. Ecol.* 53: 705–714.

Levings, S.C., Traniello, J.F.A. (1981) Territoriality, nest dispersion, and community structure in ants. *Psyche* 88: 265–319.

Lindauer, M. (1961) *Communication Among Social Bees*. Harvard University Press, Cambridge, MA.

Longhurst, C., Baker, R., Howse, P.E. (1979) Termite predation by *Megaponera foetens* (Fab.) (Hymenoptera: Formicidae): Coordination of raids by glandular secretions. *J. Chem. Ecol.* 5: 703–725.

Maschwitz, U. (1974) Vergleichende Untersuchungen zur Funktion der Ameisenmetathorakaldruse. *Oecologia* 16: 303–310.

Maschwitz, U., Schönegge, P. (1977) Recruitment gland of *Leptogenys chinensis*. *Naturwissenschaften* 64: 589.

Maschwitz, U., Schönegge, P. (1983) Forage communication, nest moving recruitment, and prey specialization in the oriental ponerine *Leptogenys chinensis*. *Oecologia* 57: 175–182.

Maschwitz, U., Steghaus-Kovac, S. (1991) Individualismus versus kooperation. *Naturwissenschaften* 78: 103–113.

Maschwitz, U., Lenz, S., Buschinger, A. (1986) Individual specific trails in the ant *Leptothorax affinus* (Formicidae: Myrmicinae). *Experentia* 42: 1173–1174.

Matsumura, F., Coppel, H.C., Tai, A. (1968) Isolation and identification of termite trail-following pheromone. *Nature* 219: 963–964.

McDowell, P.G., Oloo, G.W. (1984) Isolation, identification and biological activity of the trail-following pheromone of the termite *Trinervitermes bettonianus*. *J. Chem. Ecol.* 10: 835–851.

Millonas, M.M. (1992) A connectionist type model of self-organized foraging and emergent behavior in ant swarms. *J. theor. Biol.* 159: 529–552.

Moore, B.P. (1966) Isolation of the scent-trail pheromone of an Australian termite. *Nature* 211: 746–747.

Morgan, E.D., Hölldobler, B., Vaisar, T., Jackson, B.D. (1992) Contents of poison apparatus and their relation to trail-following in the ant *Daceton armigerum*. *J. Chem. Ecol.* 18: 2161–2168.

Morgan, E.D., Jackson, B.D., Ollett, D.G., Sales, G.W. (1990) Trail pheromone of the ant *Tetramorium impurum* and model compounds: Structure-activity comparisons. *J. Chem. Ecol.* 16: 3493–3510.

Nicolis, G., Prigogine, I. (1977) *Self-Organization in Non-Equilibrium Systems*. Wiley, N.Y.

Nicolis, G., Prigogine, I. (1989) *Exploring Complexity*. Freeman & Co., N.Y.

Noirot, C. (1969) Glands and Secretions, In *The Biology of Termites* (eds K. Krishna and F.M. Weesner). Academic Press, N.Y., pp. 89–104.

Nonacs, P., Dill, L.M. (1990) Mortality risk vs. food quality trade-offs in a common currency: Ant patch preferences. *Ecology* 71: 1886–1892.

Nonacs, P., Dill, L.M. (1991) Mortality risk versus food quality trade-offs in ants: Patch use over time. *Ecol. Entomol.* 16: 73–80.

Oloo, G.W. (1981) Specifity of termite trails: analysis of natural trails of *Trinervitermes*, *Macrotermes*, and *Odontotermes* from sympatric populations. *Entomol. exp. appl.* 29: 162–168.

Oloo, G.W., Leuthold, R.H. (1979) The influence of food on trail-laying behavior in *Trinervitermes bettonianus* (Termitidae: Nasutitermitinae). *Entomol. exp. appl.* 26: 267–278.

Oster, G., Wilson, E.O. (1978) *Caste and Ecology in the Social Insects*. Princeton Univ. Press, N.J.

Pasteels, J.M., Deneubourg, J.L., Goss, S. (1987) Self-organization mechanisms in ant societies (I): Trail recruitment to newly discovered food sources, in *From Individual to Collective Behavior in the Social Insects* (eds. J.M. Pasteels and J.L. Deneubourg). Birkhauser, Basel, pp. 155–174

Pasteels, J.M., Verhaeghe, J.C., Deneubourg, J.L. (1982) The adaptive value of probabalistic behaviour during food recruitment in ants: Experimental and theoretical approaches, In *The Biology of Social Insects* (eds, M.D. Breed, C.D. Michener and H.E. Evans). Westview Press, Colorado. pp. 297–301.

Pratt, S.C. (1989) Recruitment and other communication behavior in the ponerine ant *Ectatomma ruidum*. *Ethology* 81: 313–331.

Prestwich, G.D. (1984) Chemical systematics of termite exocrine secretions. *Annu. Rev. Ecol. Syst.* 14: 287–311.

Prigogine, I. (1980) *From Being to Becoming. Time and Complexity in the Physical Sciences*. W.H. Freeman and Company, San Francisco.

Prigogine, I., Stengers, I. (1984) *Order out of Chaos. Man's New Dialogue with Nature*. Bantam Books, Toronto.

Regnier, F.E., Nieh, M., Hölldobler, B. (1973) The volatile Dufour's gland components of the harvester ants *Pogonomyrmex rugosus* and *P. barbatus*. *J. Insect Physiol.* 19: 981–992.

Resnick, M. (1990) Overcoming the centralized mindset: towards an understanding of emergent phenomena. *M.I.T. Media Lab. Epistemology & Learning Group Memo #* 11. Massachusetts Institute of Technology, Cambridge, Mass.

Resnick, M. (1991) Animal simulations with *Logo: massive parallelism for the masses, in *From Animals to Animats* Proc. 1st Int. Conf. on Simulation of Adaptive Behavior (eds J-A. Meyer and S.W. Wilson). MIT Press, Cambridge, MA. pp 534–539.

Rettenmyer, C.W. (1963) Behavioural studies of army ants. *Univ. Kansas Sci. Bull.* 44: 281–465.

Rissing, S.W., Wheeler, J. (1976) Foraging responses of *Veromessor pergandei* to changes in seed production. *Pan-Pac. Entomol.* 52: 63–72.

Ritter, F.J., Brüggemann-Rotgans, I.E.M., Verwiel, P.E.J., Persoons, C.J., Talman, E. (1977) Trail pheromone of the Pharaoh's ant, *Monomorium pharaonis*: Isolation and identification of faranol, a terpenoid related to juvenile hormone II. *Tetrahedron Lett.* 30: 2617–2618.

Runcie, C.D. (1987) Behavioral evidence for multicomponent trail pheromone in the termite *Reticulitermes flavipes* (Kollar) (Isoptera: Rhinotermitidae). *J. Chem. Ecol.* 9: 1967–1978.

Salzemann, A., Jaffe, K. (1990) On the territorial behavior of field colonies of the leaf-cutting ant *Atta laevigata* (Hymenoptera: Myrmicinae). *J. Insect Physiol.* 36: 133–138.

Schmid-Hempel, P. (1987) Foraging characteristics of the desert ant *Cataglyphis*, in *From Individual to Collective Behavior in the Social Insects,* (eds. J.M. Pasteels and J.L. Deneubourg) Birkhauser, Basel, pp. 43–61.

Shimron, O., Hefetz, A., Tengo, J. (1985) Structural and communicative functions of Dufour's gland secretion in *Eucera palestinae* (Hymenoptera: Anthophoridae). *Insect Biochem.* 15: 635–638.

Strickland, T.R., Tofts, C.M.N., Franks, N.R. (1992) A path choice algorithm for ants. *Naturwissenschaften* 79: 567–572.

Stuart, A.M. (1969) Social behavior and communication, In *The Biology of Termites* (eds. K. Krishna and F.M. Weesner). Academic Press, N.Y., pp. 193–232.

Stuart, A.M. (1981) The role of pheromones in the initiation of foraging, recruitment, and defense by the soldiers of a tropical termite *Nasutitermes corniger* Motschulsky. *Chem. Senses* 6: 409–420.

Stuart, A.M., Satir, P. (1968) Morphological and functional aspects of an insect epidermal gland. *J. Cell Biol.* 36: 527–549.

Sudd, J. (1963) How insects work in groups. *Discovery* 24: 15–19.

Swain, R. (1980) Trophic competition among parabiotic ants. *Insectes Soc.* 27: 377–390.

Taylor, F. (1977) Foraging behavior of ants: experiments with two species of myrmicine ants. *Behav. Ecol. Sociobiol.* 2: 147–167.

Thorne, B.L., Haverty, M.I. (1991) A review of intracolony, intraspecific, and interspecific agonism in termites. *Sociobiology* 19: 115–145.

Tokoro, M., Takahashi, M., Yamaoka, R. (1992) Identification of the trail pheromone of the subterranean termite *Coptotermes formosanus* Shiraki. *Mokuzai Gakaishi* 38: 294–300.

Tokoro, M., Takahashi, M., Tsunoda, K., Yamaoka, R., *Hayashiya*, K. (1991) Isolation and identification of the trail pheromone of the subterranean termite *Reticulitermes speratus* (Kolbe). *Wood Res.* 78: 1–14.

Traniello, J.F.A. (1977) Recruitment behavior, orientation, and the organization of foraging in the carpenter ant *Camponotus pennsylvanicus* DeGeer. *Behav. Ecol. Sociobiol.* 2: 61–79.

Traniello, J.F.A. (1980) Colony specificity in the trail pheromone of an ant. *Naturwissenschaften* 67: 360.

Traniello, J.F.A. (1981) Enemy deterrence in the recruitment strategy of a termite: Soldier-organized foraging in *Nasutitermes costalis*. *Proc. Natl. Acad. Sci. USA* 78: 1976–1979.

Traniello, J.F.A (1982) Recruitment and orientation components in a termite trail pheromone. *Naturwissenschaften* 69: 343–344.

Traniello, J.F.A. (1983) Social organization and foraging success in *Lasius neoniger* (Hymenoptera: Formicidae): Behavioral and ecological aspects of recruitment communication. *Oecologia* 59: 94–100.

Traniello, J.F.A. (1987) Comparative foraging ecology of North Temperate ants: the role of worker size and cooperative foraging in prey selection. *Insectes Soc.* 34: 118–130.

Traniello, J.F.A. (1989a) Chemical trail systems, orientation, and territoriality in the ant *Lasius neoniger*. *J. Insect Behav.* 2: 339–354.

Traniello, J.F.A. (1989b) Foraging strategies of ants. *Annu. Rev. Entomol.* 34: 191–210.

Traniello, J.F.A., Beshers, S.N. (1985) Species-specific alarm/recruitment responses in a Neotropical termite. *Naturwissenschaften* 72: 491–492.

Traniello, J.F.A., Beshers, S.N. (1991) Maximization of foraging efficiency and resource defense by cooperative retrieval in the ant *Formica schaufussi*. *Behav. Ecol. Sociobiol.* 29: 283–289.

Traniello, J.F.A., Busher, C. (1985) Chemical regulation of polyethism during foraging in the Neotropical termite *Nasutitermes costalis*. *J. Chem. Ecol.* 11: 319–332.

Traniello, J.F.A., Hölldobler, B. (1984) Chemical communication during tandem running in *Pachycondyla obscuricornis* (Hymenoptera: Formicidae). *J. Chem. Ecol.* 10: 783–794.

Traniello, J.F.A., Jayasuriya, A.K. (1981a) The sternal gland and recritment communication in the primitive ant *Aneuretus simoni*. *Experientia* 37: 46.

Traniello, J.F.A. and Jayasuriya, A.K. (1981b) Chemical communication in the primitive ant *Aneuretus simoni*: The role of the pygidial and sternal glands. *J. Chem. Ecol.* 7: 1023–1033.

Traniello, J.F.A., Levings, S.C. (1986) Intra- and intercolony patterns of nest dispersion in the ant *Lasius neoniger*: Correlations with territoriality and foraging ecology. *Oecologia* 69: 413–419.

Tumlinson, J.H., Silverstein, R.M., Moser, J.C., Brownlee, R.G., Ruth, J.M. (1971) Identification of the trail pheromone of a leaf-cutting ant, *Atta texana*. *Nature* 234: 348–349.

Vander Meer, R.K., Alvarez, F., Lofgren, C.S. (1988) Isolation of the trail recruitment pheromone of *Solenopsis invicta*. *J. Chem. Ecol.* 14: 825–838.

Vander Meer, R.K., Lofgren, C.S., Alvarez, F.M. (1990) The orientation inducer pheromone of the fire ant *Solenopsis invicta*. *Physiol. Entomol.* 15: 483–488.

Van Vorhis Key, S.E., Baker, T.C. (1982) Specificity of trail-following by Argentine ant *Iridomyrmex humilis* (Mayr) to (Z)-9-hexadecenal, analogs, and gaster extracts. *J. Chem. Ecol.* 8: 1057–1063.

Villa, F. (1992) New computer architectures as tools for ecological thought. *TREE* 7: 179–182.

Wehner, R. (1983) Celestial and terrestrial navigation: human strategies-insect strategies, in *Neuroethology and Behavioral Physiology* (eds. F. Huber and H. Markl). Springer, Heidelberg, pp. 366–381.

Wehner, R. (1987) Spatial organization of foraging behavior in individually searching desert ants, *Cataglyphis* (Sahara Desert) and *Ocymyrmex* (Namib Desert), in *From Individual to Collective Behavior in the Social Insects* (eds. J.M. Pasteels and J.L. Deneubourg). Birkhauser, Basel, Switzerland, pp. 15–42.

Wehner, R., Marsh, A.C., Wehner, S. (1992) Desert ants on a thermal tightrope. *Nature* 357: 586–587.

Wilson, E.O. (1959) Source and possible nature of the odor trail of fire ants. *Science* 129: 643–644.

Wilson, E.O. (1962) Chemical communication among workers of the fire ant *Solenopsis saevissima* (Fr. Smith), 1: The organization of mass-foraging; 2: An information analysis of the odor trail; 3: The experimental induction of social responses. *Anim. Behav.* 10: 134–147, 148–158; 159–164.

Wilson, E.O. (1965) Trail sharing in ants. *Psyche* 72: 2–7.

Wilson, E.O. (1971) *The Insect Societies*. Harvard University Press, Cambridge, Mass.

Wilson, E.O. (1976) The organization of colony defense in the ant *Pheidole dentata* Mayr (Hymenoptera: Formicidae). *Behav. Ecol. Sociobiol.* 1: 63–81.

Wilson, E.O., Hölldobler, B. (1988) Dense heterarchies and mass communication as the basis of organization in ant colonies. *TREE* 3: 65–68.

Wilson, E.O., Pavan, M. (1959) Glandular sources and specificity of some chemical releasers of social behavior in dolichoderine ants. *Psyche* 66: 70–76.

Wolfram, S. (1986) Approaches to complexity engineering, in *Evolution, Games and Learning. Models for Adaptation in Machines and Nature* (eds. D. Farmer, A. Lapedes, N. Packard, B. Wendroff). North-Holland, New York., pp 385–399.

8

The Chemical Basis for Nestmate Recognition and Mate Discrimination in Social Insects

Brian H. Smith
Department of Entomology, Ohio State University

Michael D. Breed
Department of Environmental, Population, and Organismic Biology, University of Colorado

1. Introduction

In this chapter we discuss the chemical basis for nestmate recognition in social insects. Animals that live in family groups are often able to discriminate family members from non-family members. For example, humans and other mammals can use a variety of cues—visual, auditory, and perhaps olfactory—to learn the identities of family members. Birds, on the other hand, rely primarily on auditory cues (Beecher, 1988). While any phenotypic trait that is associated with group membership could be utilized in nestmate recognition, studies of social insects have shown that they depend solely on olfactory cues to discriminate family members from other conspecifics.

In addition to discrimination of nestmates from non-nestmates, which is the primary topic of this chapter, an interesting, but controversial, additional form of nestmate recognition may take place in social insects when there is discrimination among subgroups within colonies. A possible example is between patrilines in honey bees (Getz et al., 1982; Frumhoff and Schneider, 1987; Noonan, 1986) or among matrilines in polygynous ants (Bennett, 1989).

2. Recognition Chemistry

While there has been interest in the chemistry of nestmate recognition for some time, there is still relatively little knowledge of the chemical identity of recognition cues [recent reviews that consider the chemistry of nestmate recognition signals include Michener and Smith (1987), Smith and Wenzel (1988), Breed and Bennett (1987), Getz and Page (1991) and Getz (1991)]. The three systems in which substantial progress has been made are the primitively eusocial bee, *Lasioglossum zephyrum,* the highly eusocial bee, *Apis mellifera*, and the primitively eusocial wasp, *Polistes*. Interesting insight has also been gained into the chemistry of nestmate recognition in ants through the study of "guests", i.e. myrmecophiles, in ant colonies (see Howard and Akre, this volume). Cuticular hydrocarbons have been implicated as recognition cues in a number of ants, but it has yet to be shown that these compounds can be perceived as signals (Obin, 1986; Bonavita-Cougourdan et al., 1989). In most social insect species we have only rudimentary knowledge of the source of recognition cues (e.g. genetic versus environmental) and no detailed knowledge of their chemical nature.

Two factors have slowed the study of recognition chemistry. First, the quantities of compounds available for analysis are often small. This is a factor held in common with many systems of pheromonal communication, but advances in chemical techniques over the last two decades have made recognition chemistry far more accessible. Second, virtually any compound produced by the insects, or available in the insects' environment, could be used for nestmate recognition. Unlike many pheromonal signaling systems, in which species-level stereotypy of signals is important for effective signaling, nestmate recognition systems rely on intraspecific diversity of signals. Honey bees and at least some species of ant are capable of employing cues acquired from food or other materials that are brought into the colony from the surrounding environment. This multiplicity of candidate compounds makes chemical analyses and associated bioassays complicated and difficult.

3. Production and Reception in Nestmate Recognition

All models of recognition processes must have at least two components (Hölldobler and Michener, 1980; Breed and Bennett, 1987; Getz, 1991). A *production* component encodes and transmits the signal to a conspecific. A *reception* component must decode the signal, determine its meaning, and select an appropriate behavioral response. The production component varies considerably among species and is discussed in detail for *A. mellifera* and *L. zephyrum* below.

The reception component is more uniform among species and can be discussed in broader terms. In general, recognition cues are learned by newly emerged adult workers. These cues can be learned from nesting material (*Polistes* wasps, Gamboa et al., 1986a, 1986b; halictine bees, Buckle and Greenberg, 1981; honey bees, Breed et al., 1988a, 1988b), from other individuals (halictine bees; Buckle and Greenberg, 1981; Greenberg, 1979; Smith, 1983; honey bees, Breed, 1983) or from self (honey bees, Getz and Smith, 1986). There is evidence for learning recognition cues as larvae in one ant (Isigrini et al., 1985), but tests have not been conducted to determine if this mechanism is widespread in social insects. There is no evidence for genetically encoded reception of information; all known mechanisms for nestmate recognition in social insects involve some form of learning.

Phenotype matching and familiarity (Bekoff, 1981; Blaustein, 1983; Reeve,

1989, discussed in more detail below) both postulate recognition templates that are learned. Breed and Bennett (1987) proposed a number of different types of templates that might be formed. The most parsimonious of these is the "common feature" template, in which the members of the colony fix on a chemical cue carried by all members of the colony, which is presumably absent or present in different concentrations in other colonies, and use this common feature to make nestmate/non-nestmate discriminations. Other mechanisms for template formation involve averaging of the cues held by colony members, so the template is reflective of a colony characteristic, but individual colony members could vary substantially from the template. Breed and Julian (1992) argue in favor of the common feature template, but much work remains to be done to gain an understanding of how templates are formed. Common feature cues are probably easily transferred among workers in the colony, so that individual distinctiveness is lost.

The chemical mechanisms for subgroup discrimination, if it takes place, may be somewhat different than nestmate recognition. In particular, individual distinctiveness must be retained so that a worker can classify, by group membership, other workers with which it interacts. This implies compounds with lower volatility and a higher affinity for the surface of the insect. The common feature model clearly does not work in subgroup discrimination and must be replaced by a model that invokes a template for the worker's own subgroup and perhaps templates for the other subgroups in the colony.

No experiments have manipulated the acceptance threshold (Reeve, 1989). This is the level of dissimilarity between a memory template and the phenotypic cues being matched with the template that will induce a given action (e.g., attempt to mate, reject from nest). This component to the discrimination process is clearly important and may relate to different degrees of acceptance under different conditions such as, for example, mating versus nest entry or discrimination by individuals in different physiological conditions. Presumably, there is an optimal balance between the probabilities of error by being too permissive or too restrictive in accepting or rejecting a match between template and phenotype (Breed and Bennett, 1987).

The nature of the recognition context can be critical for establishing allelic diversity at loci responsible for cue production (Ratnieks, 1991). Actions that favor high degrees of similarity between template and phenotype (e.g., kin discrimination during queen rearing in honey bees) may reduce allelic diversity. This hypothesis might explain the increased importance of environmentally derived cues in honey bees versus halictine bees. Other actions that favor high degrees of dissimilarity may generate allelic diversity at recognition loci by favoring rare alleles.

4. Criteria for Identifying Recognition Cues

For a compound to be effective as a recognition cue, it must meet the following criteria: (1) the insects must be able to perceive the compound, (2) they should be able to discriminate the compound from other compounds, (3) the concentration of the compound or the nature of a blend of compounds must vary among colonies, (4) the concentration of the compound in a colony must be relatively stable over time, and (5) the use of the compound in recognition should not interfere or conflict with other pheromonal uses.

The necessity for perception of the compound is obvious. One class of candidate compounds for recognition cues, the higher molecular weight hydrocarbons of the cuticle, may not meet this criterion. Similarly, if all compounds in a functional class or compounds in a broad molecular weight range are perceived as being the same, then the set of compounds presents limited utility as recognition cues. In general, the narrower the range of compounds that are perceived as being similar, the larger the number of independent signals that can be used in making nestmate-nonnestmate discriminations. Intercolonial variability is, of course, a necessary prerequisite for making intercolonial discriminations. Temporal changes in concentration could be accommodated by flexibility in learning cues or by an ability of older individuals to lose or acquire a cue, but stable systems are clearly more parsimonious. Finally, certain kinds of compounds may meet all of the above criteria but have other critical communicatory functions. An example would be the components of the alarm pheromone (Collins and Blum, 1983), which vary genetically among colonies but which are probably unsuitable for use in a recognition because of possible confusion with their alarm function.

These criteria allow for a broad range of compounds to be used as recognition cues; in our experience it is not possible to predict on an *a priori* basis which compounds may have actual activity in recognition systems. The mere fact that a compound meets the criteria does not prove that it is used as a recognition cue. Empirical testing of compounds in an actual recognition context is required to demonstrate such use. Many studies contain the flawed argument that because compounds meet criterion 3 (intercolonial variability) they serve as recognition cues. In our experience meeting criterion 3 is only suggestive of function; not only must the other criteria be met, but many compounds that meet all five criteria do not show activity in recognition bioassays.

Assays for biological activity of specific compounds in recognition systems have been conducted in only honey bees (*Apis mellifera*) (Breed and Stiller, 1992) and a halictine bee, *Lasioglossum zephyrum* (Smith et al., 1985). There are two reasons for this. First, obtaining chemical identifications of candidate compounds has been problematic. Second, difficulties in designing effective bioassays has impeded progress in work on some species. Ultimately, bioassays will need to be performed on compounds that meet the five criteria in order to determine which compounds are actually used by the insects in question.

5. Sources of Recognition Cues

Compounds that can serve as recognition cues can be classified by source (Table 1). This provides a useful framework for analysis of recognition systems and

Table 8.1. A classification, by source of the compounds that may serve as recognition cues in social insects.

A. Compounds synthesized (as metabolic products) by the insect
   1. Compounds retained in or on the individual (e.g., cuticular hydrocarbons)
   2. Compounds used in nest structures (e.g., bee's was)
   3. Compounds spread directly from one individual to another (e.g., pheromones produced by the queen)

B. Compounds present in or collected from the environment
   1. Floral oils (for social insects that forage on flowers)
   2. Nest construction materials (e.g., propolis, a mixture of plant resins used by honey bees)
   3. Compounds associated with other types of food
   3. Compounds from miscellaneous other sources

for, ultimately, forming hypotheses concerning the evolution of recognition systems. The fundamental dichotomy in the Table, between compounds that are endogenous (metabolic products) to the insects and compounds that are exogenous (acquired from the environment), reflects empirical experience that either sort of compound may be active in recognition systems.

6. The Evolution of Recognition Systems

Members of the worker caste in social insect colonies forego their individual reproductive potential and work to maximize the reproductive output of the colony's queen(s). Biologists beginning with Darwin recognized that worker sterility posed an interesting evolutionary problem, as any mutation that conferred sterility might not be easily passed to succeeding generations. One way of overcoming this difficulty is to postulate that the sterile workers are enhancing the reproduction of related individuals; this inclusive fitness argument was developed by Hamilton (1964).

One of the consequences of this theoretical line of reasoning is the hypothesis that colony members should be able to discriminate between related and unrelated conspecifics. Recognition could occur by recognition alleles [the "green beard" model of Dawkins, (1982)], by learning the characteristics of surrounding individuals [the familiarity model of Bekoff (1981)], or by phenotypic matching of self versus surrounding individuals. The recognition allele model has not been conclusively demonstrated in any system (Blaustein, 1983) and seems unlikely to be relevant in considerations of social insect colonies. Familiarity and phenotypic matching models, on the other hand, are not mutually exclusive [i.e., they both involve generation of a memory template; Reeve (1989)] and seem to explain most of the phenomena associated with nestmate recognition in social insects.

The well-established role of learning in social insect nestmate discrimination indicates that in most cases physical proximity with colony members is correlated with genetic relationship. [See Chapter 10, Howard and Akre, for a discussion of exceptions, such as slavery, to this generalization]. Nestmate discrimination seems, in most cases, to function effectively in ensuring that the activities of the sterile caste aid close relatives, usually the mother or sisters.

Evolved mechanisms rarely arise *de novo*; rather they are nearly always derived from an existing structure, behavior, or metabolic process. There are two leading candidate mechanisms for the evolutionary source of nestmate recognition, cues used in nest recognition and cues used in assortative mating. Both of these mechanisms may occur in solitary bees and wasps, as well as the cockroach-like predecessors of termites, and strong logical arguments can be invoked in support of either mechanism.

Gamboa and his associates (1986a, 1986b) have demonstrated that nesting materials in *Polistes* wasps serve as an avenue for transfer of recognition cues to wasps in the colony. The actual cues seem to be metabolic products of the wasps, which are incorporated into the nesting material as the nest is constructed (in these wasps the nest is formed of masticated plant fiber). Such odor cues in the nesting materials could aid solitary wasps in locating their nest, or in discriminating their nest from nearby nests of conspecifics.

In honey bees, *Apis mellifera*, Breed et al. (1988b) showed that bees acquire recognition cues from the colony environment, and that cues are specifically acquired from the wax combs in the colony. While the wax is a metabolic product of the bees, the comb is also highly absorbent of compounds that are brought into the colony with food or plant resin (termed propolis, this is used to seal the nest structure). Again, one can postulate a solitary bee that constructs a nest of plant resin (as some of the Euglossine bees) and then uses odors associated with the nest to locate it or discriminate it from neighboring nests.

Arguments concerning nest discrimination as an evolutionary precedent for nestmate discrimination gain support from the fact that many solitary wasp or bee species nest in dense aggregations. Under conditions in which neighboring nests are close and the substrate is relatively uniform, visual cues may be inadequate for individuals in locating nests and olfactory cues could be called upon. Olfactory cues could either be environmentally derived, perhaps as food odors, or be derived from materials deposited by the insect in the nest, e.g. as waterproofing. Examples of olfactory identification of nests or nesting materials by social insects include *Polistes* (Ferguson et al., 1987), *Apis* (Breed and Stiller, 1992), *Lasioglossum* (Kukuk et al., 1977), *Dolichovespula* (Ferguson et al., 1987), and *Bombus* (Gamboa et al., 1987; Foster and Gamboa, 1989).

While the argument for nestmate recognition as a derivative of assortative mating mechanisms is logically sound, there is less empirical support for this mechanism. The haplo-diploid mechanism of sex determination in the Hymenoptera makes inbred matings highly costly—diploid individuals that are homozygous at the sex-determining locus are sterile. Thus there should be strong selective pressure for disassortative mating.

Unfortunately there have been few investigations of mating preferences in Hymenoptera. In honey bees queen recognition is mediated by different cues than worker recognition (Breed et al., 1992b; see below); the cues used in queen recognition might be useful in disassortative mating, but this hypothesis has not yet been tested conclusively. In sweat bees, *Lasioglossum zephyrum*, males are less attracted to females with whom they have been associated in prior, unsuccessful, mating attempts (Barrows et al., 1975; see below). Investigations of the behavior and chemistry of mate choice in Hymenoptera should prove to be very interesting, and may ultimately give insight into the evolution of nestmate recognition.

7. Recognition Chemistry in Halictine Bees

The hymenopteran family Halictidae contains several thousand species. They occur in practically all geographic regions in which bees are found and display a wide array of life histories (Michener, 1974; Moure and Hurd, 1987; Westrich, 1989). The majority of species have a solitary lifestyle in which the sole female overwinters and rears one or more generations of brood the following year. Some species are obligate cleptoparasites (Michener 1978). Levels of social behavior range among communal, semisocial, and eusocial (Michener, 1974); most of these types of colonies exhibit an annual colony cycle, but at least one species has a perennial colony cycle (Plateaux-Quénu, 1959). Social behavior has probably arisen several different times within evolving clades of Halictidae, and some species have become secondarily solitary (Packer, 1991); that is, some solitary species have most likely evolved from a social ancestor. Because of this wide variability in lifestyle, this family of bees represents an array of possibilities for those interested in the conditions underlying the evolution of sociality in general, including the types of recognition systems that may be involved in social interactions.

A typical annual life cycle begins in the spring or early summer, during which the overwintered female provisions cells from which the next generation of bees emerges. In solitary species, this next generation overwinters and completes the cycle by establishing new nests the following spring. In communal and social species, many individuals of the next generation continue to occupy the same nest. Each bee in a communal colony works independently to establish and maintain its own brood, whereas most first generation bees from social species become workers and collect food to provision cells in which the queen lays eggs. If the founding female dies, which happens frequently in some species, one of her daughters assumes the role of queen. In this case, the colony becomes semisocial because the queen and workers are derived from the same generation (Michener, 1974). In at least one social species, some females in the first brood become workers whereas others begin diapause and emerge as foundresses the following year (Yanega, 1988).

Studies of olfactory-based kin recognition focus on several species in the genus

Lasioglossum. [Most publications refer to the genus as *Lasioglossum*, although some refer to the subgenera *Evylaeus* and *Dialictus* as genera (Moure and Hurd, 1987); we use *Lasioglossum* to maintain consistency with most authors]. Most of the species studied to date nest in aggregations of a few to thousands of colonies. Two recent reviews have extensively summarized the literature on behavioral studies of kin recognition (Michener and Smith, 1987) and the learning mechanisms involved in the mate recognition process (Smith, 1992). We refer briefly to studies cited in those reviews and focus instead on a slightly different subject; identification and bioassay of putative recognition chemicals.

7.1. Candidate Chemicals for Recognition Cues

Studies of semiochemicals in *Lasioglossum* meet only a subset of the criteria for identification of recognition cues. In no case does a series of studies conclusively establish the chemical basis for recognition cues. The first tactic has been to chemically identify putative recognition cues produced by the insects themselves. The second tactic correlates variation in chemical secretions to important biological variables. Finally, in the final two subsections below, evidence will be presented that halictine bees can detect and utilize those cues in specific behavioral contexts.

Because of its large size, studies have focused exclusively on the contents of the Dufour's gland (de Lello, 1971). This gland takes up a large part of the abdominal cavity and, along with the venom gland, empties through a pore at the base of the sting. Several studies have shown that the predominant contents of this gland in adult females are macrocyclic lactones bearing an even number of from 16 to 26 carbons (Hefetz et al., 1978; Bergström and Tengö, 1979; Smith et al., 1985). Major components are typically saturated and unbranched, although unsaturated and methyl branched homologs almost always occur in lower amounts and may not be identified in chemical analyses unless proper extraction and separation procedures are employed. Recently, more extensive analyses have identified several saturated and unsaturated hydrocarbons as well as isopentenyl and ethyl esters as also being characteristic components of this secretion in *Lasioglossum* (Smith et al., 1985; Ayasse et al., 1993).

The Dufour's gland secretion serves primarily as the protective lining of brood cells (Hefetz et al., 1979; Cane, 1981). Females tunnel in soil or in rotting wood, the typical nesting substrates for halictine bees, and periodically excavate teardrop shaped brood cells extending laterally from the tunnel. Once a cell is complete, the female rotates around her longitudinal axis within the cell while pressing the opening of this gland to the cell wall. She then licks the walls with her mouthparts. This process takes approximately an hour, after which the cell is provisioned and a single egg is laid in the cell. Chemical analyses of the waxy lining that is formed during this process on the cell wall has shown that it consists of a polyester that is derived from the macrocyclic lactones of the Dufour's gland

(Hefetz et al., 1979). Presumably the esterification process occurs with the aid of an unidentified secretion from one of the cephalic glands. This lining protects the larva and pupa from dehydration and fungal invasion.

Several studies have established correlations between the composition of the Dufour's gland secretion and genetic variation. Species differences, familial membership, and individual differences all correlate with this secretion (Hefetz et al., 1986; Smith and Wenzel, 1988). For example, Smith and Wenzel (1988) performed principal components analysis on the absolute amount of macrocyclic lactones from *L. (Dialictus) zephyrum)* females collected from different nests in two nesting aggregations in midsummer, just prior to production of gynes destined to overwinter (FIG. 8.1). The most variation (42.5%) was captured on the first

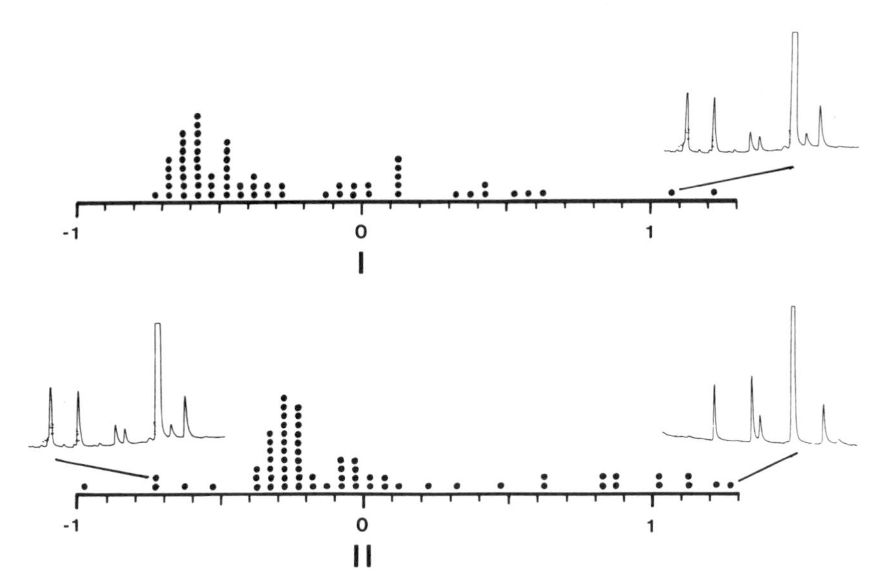

Figure 8.1. Distribution of factor scores for individual female *L. (D.) zephyrum* for the first two principal components from analyses by Smith and Wenzel (1988). The first factor (I) describes interindividual variation in regard to overall amount of lactones present; larger values indicate larger overall amounts of lactones extracted in headspace samples. A representative chromatogram from one such female is shown to the right. The largest peak is an internal standard used to calculate amounts of lactones. The five remaining peaks indicate from left to right lactones of increasing molecular weight (carbon numbers in parentheses); octadecanolide (18), eicosanolide (20), docosanolide (22), docosenolide (22), tetracosanolide (24), tetracosenolide (24), respectively. Females possessing low values on this axis had barely detectable amounts of these lactones. The second factor (II) describes varying ratios of the 18- and 22-carbon lactones. Representative chromatograms are from two females that possessed widely varying mixtures of these substances; the 18-carbon lactone was not detected in the chromatogram to the right. The third factor (not shown) describes varying amounts of the 24-carbon lactones.

principal component, which describes differences in absolute amounts of lactones extracted from different females. The second and third axes account for 31.5% and 18.0% of the variability in the sample, respectively, and describe differences in ratios among component lactones. Factor scores on the second and third axes for females collected from the same colony are significantly more similar than those for females from different colonies. No such nestmate correlation exists for the first axis. Genetic analysis of the same females demonstrated that nestmates were genetically more similar, on average, than females from different colonies (Crozier et al., 1987). These data demonstrate that the lactones could encode information on genealogical relationship, although it has not been determined whether the similarity is due to individual genotype or to transfer of lactones among females within a colony.

Thus, for encoding kinship, the Dufour's gland secretion contains what might be described as *redundant* information. That is, two or more independently varying groups of components (corresponding to the principal components axes) are capable of carrying kinship information. A similar pattern also exists in regard to patterns of cuticular hydrocarbons in the honey bee (Getz and Page, 1991). This redundancy may represent signal amplification. Multiple, independent communication channels may enhance the probability of correct reading of the signal.

If this hypothesis is correct, then two testable predictions emerge. First, bees must have the capacity to process the sensory channels for the different odorants independently. Second, in cases such as mate recognition it is in the interest of both sender and receiver to accurately transmit information and perhaps avoid inbreeding. In contrast, there may be cases in which it may not be to the advantage to both parties to transmit species or kinship identification. For example, in dominance interactions it may be more in the interest of the worker than the queen to determine kinship. If average within-colony genealogical relationship is low, the queen might be expected to hide kinship if possible. Queens in social insects may evolve other chemical signals that mask or swamp out the kinship information. Neither hypothesis has been tested to date.

Recent studies have demonstrated, through similar correlation analyses, other types of biological information that could be transmitted by the Dufour's gland secretion. The total amount of lactone, which accounted for most of the variation in the study by Smith and Wenzel (1988), is significantly correlated with the queen size (Smith, Weller and Tengö, unpublished); the larger queens have more extractable lactone in the Dufour's gland. Furthermore, ontogenetic factors dramatically affect the composition of the secretion (Ayasse et al., 1993). Isopentenyl esters make up a larger relative proportion of the secretion of young, unmated queens than of older, mated queens or workers; macrocyclic lactones dominated the secretions of the latter.

To date studies have used correlation analyses to establish a potential link between chemical cues and biological recognition factors. Biological assays of these chemicals are now required to test the third criterion for recognition cues cited above; whether individuals can perceive and act on this information. These studies have been initiated in several behavioral contexts.

7.2. Recognition During Mating

Several studies have documented the potential use of kinship information in mate recognition in halictine bees (Barrows, 1975a; Smith, 1983; Smith and Ayasse, 1987; Wcislo, 1987; see review of natural history and learning mechanisms in Smith, 1992). In the presence of female odor males are attracted to objects that approximate the size and shape of females (Barrows, 1975b). Frequency of contact with those objects, either models or actual females, is initially high but tapers off to near background (no odor) levels within a minute or two. When the female is removed and immediately returned (as a control), there is no change in rate of contact. However, when a different female is placed into the cage or in the vicinity of the males, the contact frequency and or mating attempts return to a level that is equal to that elicited during the initial exposure to the first female's odor. This latter result indicates that the males' sensory system is not adapting to the odor, but instead, the decline in responsiveness over time is more likely due to one or more learning processes (Smith, 1992). Furthermore, if the initial female is replaced by a close relative, then the frequency of contact by the males increases in inverse proportion to genealogical relationship (Smith, 1983). Therefore, males are capable of using kinship information encoded in the odor signal emitted by the female.

The strong attraction of males to odor signals that are *unlike* those that they have experienced indicates that kin recognition during mating may be a means of maintaining outbreeding in the usually small nesting populations of halictine bees (Bateson, 1983). Field studies support this contention. Smith and Ayasse (1987) collected females from nesting aggregations and froze them. The next day the thawed females were pinned at one of several sites within the aggregation. Mating attempts of patrolling males could then be readily observed. Males displayed the same learning curve in this field study that was obtained in the laboratory (Smith, 1983; Wcislo, 1987, 1993); initially attraction was high and tapered off over a few minutes (FIG. 8.2). Replacement with the same female or with a female from the same colony did not increase the attraction as much as replacement with a female from a different colony (i.e., females from different colonies are less likely to be closely related). Finally, females collected from an aggregation other than the one in which they were tested elicited more male activity than did females from the same aggregation.

Wcislo (1987; 1993) was not able to replicate Smith and Ayasse's (1987) finding of increased attractiveness when bees were switched among aggregations. Other factors, such as dissipation of odors, site-specific learning, decay of motivation and male-produced repellents have little or no detectable effect (Wcislo,

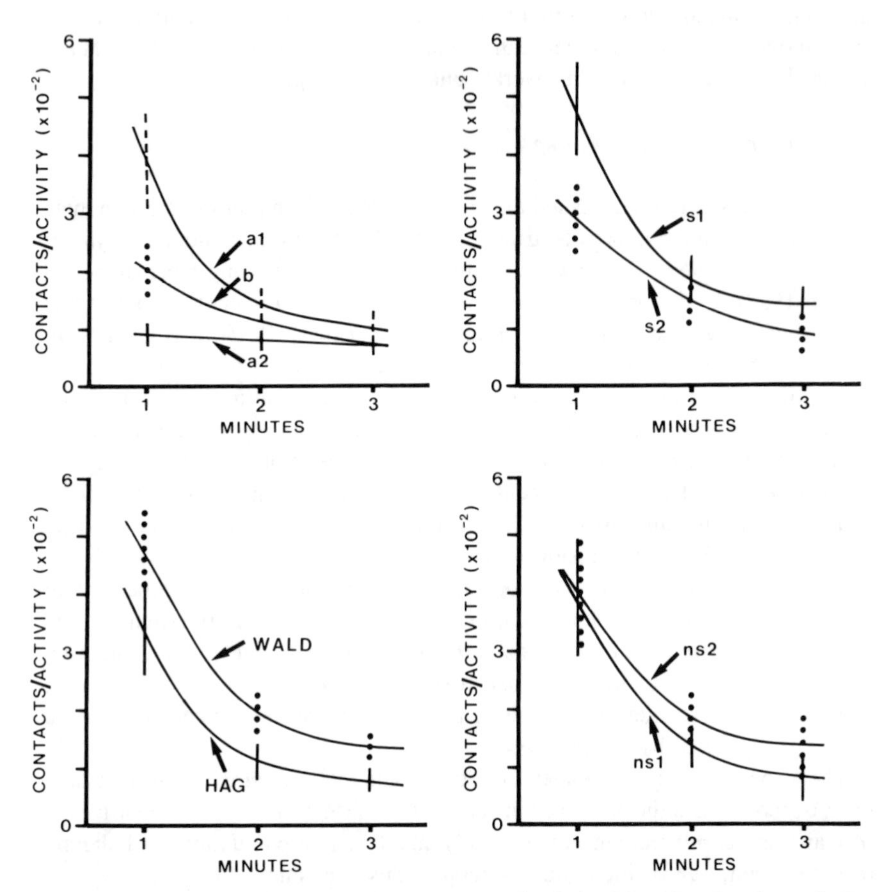

Figure 8.2. Field studies of learning and generalization of female odors by male *L. (Evylaeus) malachurum*. Curves show decrement in copulatory activity over three successive one-minute intervals. Activity is measured by number of contacts males flying in the test area of the aggregation made with a female. Because differing numbers of males were flying on different days, or at different times of day, this activity was adjusted by dividing flight activity into the number of contacts [see Smith and Ayasse (1987)]. Top left: A female (a1) was pinned in the test area for three minutes before removal. She was replaced into the same area (a2) or into a test area where different marked males were patrolling (b). Top right: A female (s1) was tested and then replaced by a female that had been collected from the same colony (s2), which presumably was a close relative. Bottom right A female (ns1) was tested and then replaced with a female from a different colony (ns2), which was presumably less related than a female from the same colony. Bottom left: Mean response on first tests by males from one aggregation (termed "HAG") to females from HAG and to females from a different ("WALD") aggregation several kilometers away [reprinted with permission from Smith (1992)].

1993). He therefore argued that his results support a hypothesis of males attempting to avoid individual, unreceptive females with which they have had experience, rather than outbreeding avoidance *per se*.

Smith (1992) reviews several hypotheses that need to be tested in more detail in order to resolve these differences. For example, the acceptance/rejection thresholds of males may be modified by state-dependent variables such as age and overall availability of females (Reeve, 1989). All of the field studies have utilized population samples composed of males of uncontrolled, variable physiological states. Furthermore, females vary significantly in regard to attractiveness of males (Wcislo, 1993; Ayasse et al., 1993), which may reflect their reproductive and/or mating status. More controlled conditions under which *individual* males and females can be manipulated and tracked are required to determine whether males avoid specific females or avoid inbreeding.

The macrocyclic lactones from the Dufour's gland secretion serve as signals in mate finding. Smith et al. (1985) applied synthetic lactones to female models in field studies. This treatment increased attraction of males to the model. Synthetic mixtures of all four of the saturated lactones found in the natural secretion produced activity equal to that of glandular extracts. Individual lactones and solvent controls elicited significantly less activity. The role of the lactones in kin recognition still needs to be tested.

Recent work has identified the potential importance for mate identification of other components of the Dufour's secretion (Ayasse et al., 1990; 1993). Newly emerged gynes (destined to become queens) are more attractive than older, mated females. That the former have relatively larger quantities of isopentenyl esters in the Dufour's gland secretion may indicate that this set of compounds has an important signaling function in addition to the lactones.

7.3. Recognition Among Females

Nestmate recognition in semisocial and eusocial colonies has been extensively studied in *L. (D.) zephyrum* (see review in Michener and Smith, 1987). In a colony containing a queen and several (2–20) workers, one or a few bees specialize behaviorally as guards (Brothers and Michener, 1974). They typically remain at or near the nest entrance and engage any bee/intruder attempting to enter the nest. Early studies showed that guards would preferentially allow their nestmates to enter while repelling other conspecifics. Furthermore, the recognition cues are learned by frequent contact with nestmates (Kukuk et al., 1977). There is a significant heritable component of the variability in the recognition cues; guards allow close relatives of their nestmates to enter even when they had no prior contact with those relatives (Greenberg, 1979). Furthermore, the recognition cues maintain their genetic integrity, i.e., are not transferred among bees, when unrelated bees live together in the same colony (Buckle and Greenberg, 1981).

Kinship recognition also functions in interactions between queens and workers within colonies (Smith, 1987). Colonies composed either of sisters (i.e., r = 0.75 or higher in inbred lines) or of unrelated females were established and observed over several weeks. Behavioral acts correlated with dominance status decline equally over time in the two nest types. Aggressive interactions between queens and workers are stable over time but are significantly more frequent in colonies composed of unrelated bees. However, workers in colonies of unrelated bees have more ovarian development than workers in colonies of related bees.

To date no behavioral studies have been performed to test specific odors for biological activity in any of the female-female social interactions just reviewed. However, several studies have eliminated cues from other sensory modalities (Bell, 1974; Barrows et al., 1975) as possible sources for kinship cues as well as indicated the potential importance of environmentally derived food odors (Buckle and Greenberg, 1981). However, a study of conflict over nest ownership among *L. malachurum* queens during the solitary phase of the nesting cycle demonstrated that lactones can function as signals (Smith and Weller, 1989). During the spring, when the queens are solitary, females often become nestless due to disease, natural factors such as severe rainfall, or nest usurpation. These females then attempt to usurp nests of other foundresses. When the original nest owner returns from a foraging trip to find a usurper in her nest, the usurper either departs immediately or a fight ensues. During such fights, combatants can be injured (e.g., loss of a leg or antenna) or even killed. Several behaviors are employed during a fight, including use of the mandibles to grasp an opponent and use of the sting, which is bent around underneath the mandibles in the C-posture characteristic of halictine bees (Bell and Hawkins, 1974). Kaitala et al. (1990) have modeled this behavior to put forward hypotheses regarding conditions under which this fighting would be expected to emerge.

Similar fights can be induced under laboratory conditions (Smith and Weller, 1989). Under those conditions, the larger of the two bees displays aggressive postures (including the C-posture) more frequently than the smaller queen (FIG. 8.3). It was hypothesized that the contents of the Dufour's gland might be expelled during such encounters and thus be a source of size information. Two experiments support this contention. First, when a small amount of synthetic lactone (a fraction of what normally can be found in the gland) is applied to the thorax of the smaller queen, both queens show behavior patterns characteristic of small individuals. Solvent-treated (control) smaller queens had no such effect on the larger queens. Second, as noted above, the amount of lactone contained in a queen's Dufour's gland is positively correlated with size (Smith, Weller, and Tengö, unpublished). However, this signaling function can only occur if use of the Dufour's gland contents is in some way costly such that smaller females cannot cheat and use more of it than they can afford. This contention has not been tested to date, but the use of the contents in lining of brood cells, and the gland's large size relative to the abdominal cavity, indicate that it is a reasonable hypothesis.

As with data on mating behavior, no finding on interactions among females

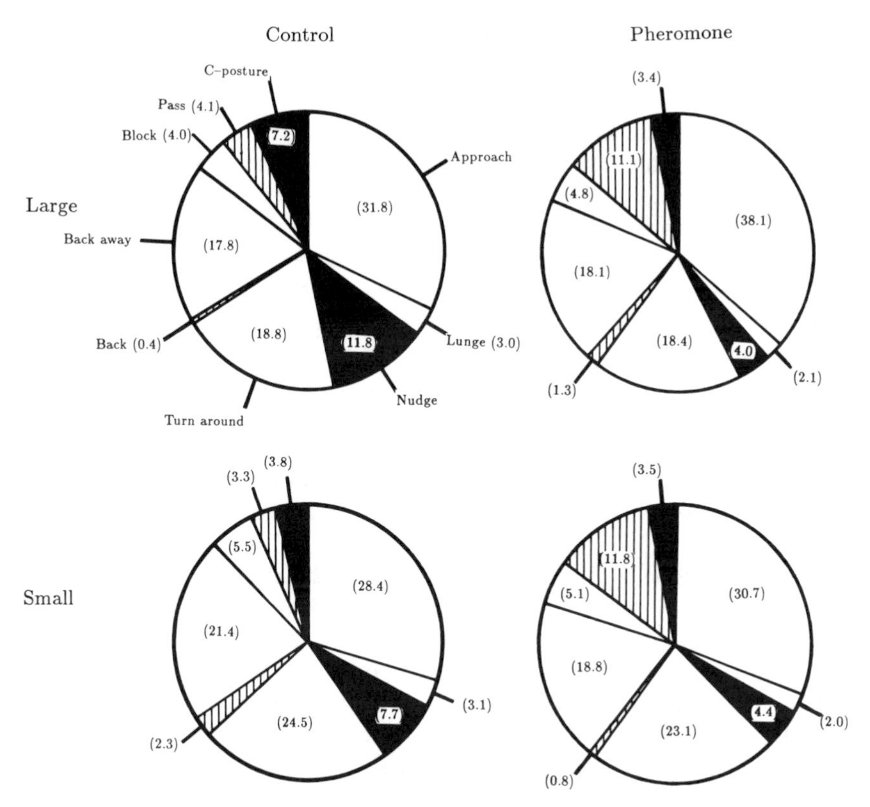

Figure 8.3. Distribution of behaviors by pairs of *L. (E.) malachurum* queens in circle tubes [data from Smith and Weller (1989)]. The left two diagrams indicate behaviors for the larger (top) and smaller (bottom) queens from "control" pairs in which the smaller of the two individuals had been treated with an application of acetone. In these pairs the probability of showing aggressive behaviors such as nudging and Cposturing [see Smith and Weller (1989) and Bell and Hawkins (1974) for further descriptions of behaviors] was significantly higher for the larger queen. The right two diagrams show behavior distributions for similar pairs in which the smaller queen had been treated with acetone containing a small amount of the synthetic lactones characteristic for this species. The larger and smaller queens no longer differ in distribution of behaviors and show characteristic differences between control and pheromone treatments. [See Smith and Weller 1989 for a detailed statistical analysis of these data].

conclusively links any of the Dufour's gland components to the kin recognition system. There are only tantalizing indications that the components can function as signals in such contexts, and much work remains to be done.

8. Recognition Chemistry of *Polistes* Wasps

Gamboa and his students pioneered the study of nestmate recognition in *Polistes* wasps (Pfennig et al., 1983; Gamboa et al., 1986a). Recently Espelie and his students have followed up on Gamboa's work with studies of the cuticular hydrocarbons of social wasps (Espelie and Hermann, 1990; Espelie et al., 1990a, 1990b; Singer et al., 1992a, 1992b; Butts et al., 1993) and their function in nestmate recognition (Singer and Espelie, 1992).

Polistes wasps learn their colony's recognition template from the nest; wasps that have not had contact with the nest structure do not express tolerance of nestmates (Gamboa et al. 1986a). Espelie et al. (1990b) washed nests of *Polistes metricus* with hexane and found that the wasps then were not able to discriminate their own nest. This finding implicated hexane soluble hydrocarbons in nest and nestmate recognition. The major (greater than 10% of the total composition) compounds found in these hexane extracts were heptacosane, nonacosane, methylhentriacontane and methyltritriacontane. In a factor analysis the first factor reflected differences in chain lengths (Espelie et al. 1990b). In a similar analysis of *Polistes exclamans* (Singer et al. 1992a) extracts from workers, the nest, and the nest pedicel were separated in discriminant analysis, and workers from the same nest clustered together. Worker and nest extracts, however, were not similar.

Singer and Espelie (1992) showed that the hydrocarbons in hexane extracts are used for nestmate recognition. They first replicated Gamboa's finding that the development of tolerance for nestmates depends on exposure to the nest. This experiment was then repeated, but exposure was to hexane washed nests. Wasps with exposure to hexane washed nests did not develop nestmate tolerance. Finally, the hydrocarbons were placed back on the nests; wasps exposed to nests treated in this manner developed nestmate tolerance. This conclusively demonstrates that the compounds contained in the hexane extract play a critical role in nestmate recognition in *Polistes metricus*.

None of the compounds identified from *Polistes* nests have been bioassayed individually for sensory responses or for individual roles in nest or nestmate recognition. This will be an interesting next step. The compounds identified are relatively large and are solid at ambient temperatures; it will be particularly interesting if these compounds, which presumably have low volatilities and may be difficult to transfer between a surface and a chemoreceptor, are involved in nestmate recognition.

9. Recognition Chemistry of the Honey Bee

Four sources of recognition cues have been identified in honey bees—surface compounds on bees (cuticular hydrocarbons), comb wax, floral or resin scents, and queen pheromones (Breed, 1987). The range of type and number of different compounds available from these sources is truly enormous.

9.1. Cuticular Hydrocarbons

Cuticular hydrocarbons are dominated by alkanes and alkenes, with other hydrocarbons present in smaller proportions (Carlson and Bolton, 1984: Francis et al.,

1989, Getz et al., 1989; Page et al., 1991). The predominant cuticular hydrocarbons are odd-chained alkanes, ranging from C-23 to C-33 (Page et al., 1991). Cuticular hydrocarbons vary in proportion among families of bees (Page et al., 1991). Getz and Page (1991) argued that the ratio between relatively short chain and relatively long-chain hydrocarbons may play a key role in nestmate recognition, but there is no evidence for this assertion. Given the volatile nature of the cues (Moritz and Southwick, 1987) it is probable that the relatively long-chain hydrocarbons are unimportant, but this remains to be tested. Breed (1983) showed that cues on the surface of bees could be used in nestmate discrimination if other cues, such as those from comb, were absent. Presumably these cues are cuticular hydrocarbons, but there is no direct evidence for this. Breed et al. (1985) found that these cues are not transferred among bees in a social group, so that genetic subgroups in a colony retain their differentiation.

9.2. Comb Wax

Comb wax has a composition similar to the cuticular hydrocarbons, but also contains alcohols, acids, esters, and aromatic compounds (Tulloch, 1980; Hepburn, 1986). Comb wax is produced by the wax glands on the venter of the abdomen of workers, but is modified with glandular secretions from the head during manipulation into comb (Hepburn, 1986). The wax can be a source of recognition cues, but it can also absorb odors brought into the colony from the surrounding environment. These environmental odors are considered separately, below.

There is intercolonial variation in the composition of comb wax (Breed, Hibbard, Bjostad, and Page, in press), so it meets at least one of the criteria for sources of recognition cues. Colonies typically vary in the relative amounts of each compound, rather than in the presence or absence of compounds. Gas chromatograms of the hydrocarbon component of wax typically reveal 20–30 major peaks. The larger peaks correspond to a series of odd-chained alkanes; the smaller peaks are primarily even-chained compounds, alkenes, and alkadienes. More polar compounds, such as acids are also present but may not be resolved using the same chromatographic column.

The acids are of interest because they are the primary products of the mandibular glands, and may be applied to the wax as the bee chews it and forms it into comb. Crewe (1982) found interpopulational differences in the acid contents of the mandibular glands, so this set of compounds also may meet the criterion of intercolonial variation. The acid component of wax, once it has been formed, has not been analyzed in detail.

Breed and Stiller (1992) bioassayed a small number of hydrocarbons for activity as recognition cues and found that hexadecane and octadecane are used as recognition cues, but that tetradecane, pentadecane, heptadecane, and eicosane are not (FIG. 8.4). Hexadecane and octadecane are usually not reported as

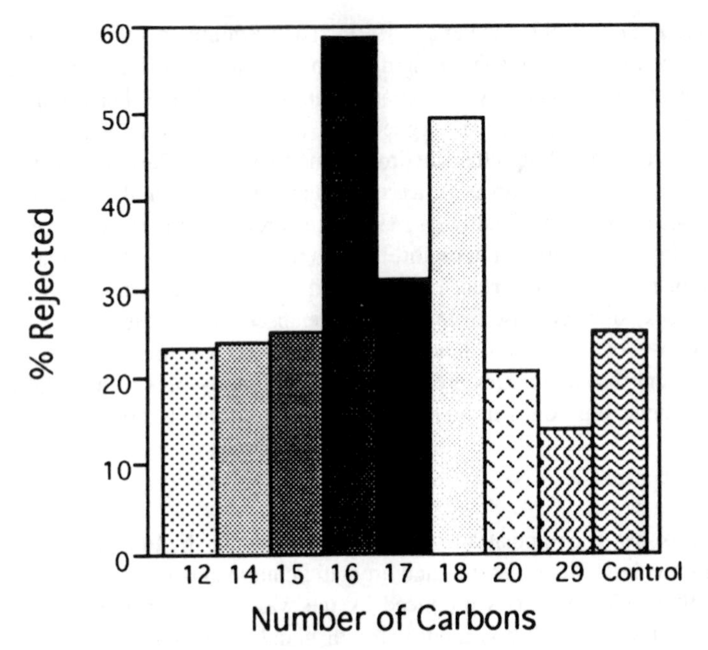

Figure 8.4. Tests for activity of hydrocarbons as recognition cues in the honey bee, *Apis mellifera*. Worker bees were maintained for five days in a container with 100 m g of a compound. After this exposure each bee was placed in a container with its sister bees that had not been exposed. The responses of the sisters were observed; the introduced bee was scored as rejected if the sisters bit or attempted to sting it. The x-axis represents the treatment, showing the number of carbons in the compound. The last bar is the control, in which the introduced bees were not treated. The y-axis is the percentage of the introduced bees that were rejected. Only hexadecane (16 carbons) and octadecane (18 carbons) differ significantly from the control.

components of the cuticular hydrocarbon mixture (e.g., Francis et al., 1989) but may occur in low concentrations (Hibbard and Bjostad, personal communication). These two compounds do occur in comb wax (Hibbard and Bjostad, personal communication). Interestingly, Getz and Smith (1987) found that bees could discriminate between compounds that differed by as little as two carbons in length using a tongue extension bioassay, but in the nestmate recognition bioassay employed by Breed and Stiller (1992) the bees could not discriminate between hexadecane and octadecane.

In addition to the alkanes, Breed and Stiller (1992) also performed bioassays on a number of organic acids and methyl esters. Two compounds, methyl docosanoate and tetracosanoic acid, gave positive results. Tetracosanoic acid is the predominant fatty acid of bee's wax (Hepburn, 1986). Bees could discriminate these compounds from hexadecane, but could not discriminate between the two compounds. In sum, of 18 compounds assayed 4 showed activity but only 2 distinct cues were found. While this is a small sample of the total number of alkanes, acids and esters, between C-10 and C-30, the result indicates a lack of predictability in which compounds might be used as recognition cues.

9.3. Floral Scents

Floral scents also can serve as recognition cues. Early researchers who investigated nestmate recognition in honeybees felt that odors brought into the colony with food were the primary source of recognition cues (Kalmus and Ribbands, 1952). Each colony was hypothesized to gain a distinctive set of odors by foraging on a different blend of flowers than other colonies. Floral scents are used by bees to locate flowers, so the criterion of perception is met for these compounds.

There are two major problems in the use of floral scents as recognition cues. First, a bee discovering a novel food source might be rejected from a colony (Breed et al., 1988b). In fact, bees carrying food loads are generally accepted by colony guards—this is most apparent in the "drift" of foragers in apiaries (Free, 1959)—so new floral scents may not be problematic. Second, and more difficult, is the fact that the suite of floral resources in an area remains relatively constant through time. Bees may occupy a nest for several years or even decades, continuously using comb for brood rearing and food storage. Under these conditions it is reasonable to hypothesize that the floral scents for all of the colonies in a local area would converge; intercolonial variation would be damped by the effects of averaging over time.

Knudsen et al. (1993) review the compounds associated with floral scents. The analyses were based on taking samples of air from the headspace around the flowers and identifying the compounds present using gas chromatography and mass spectrometry. Knudsen et al. (1993) list more than 700 compounds from 441 species in 60 families of plants. Notably, there are no analyses presented of mints (family Labiatae) or Umbelliferae; these two families are favored by bees and produce many of the distinctive odors used by humans in flavoring food. Thus the total number of compounds presented by the floral environment probably greatly exceeds that listed in the review.

The chemical structures of the compounds listed by Knudsen et al. (1993) spans virtually the entire range of organic natural products. The most common compounds are benzenoids and terpenoids, but many other types are possible. It is not clear just how many of these compounds are actually perceived by bees and used in orientation to flowers, but logic indicates that the vast majority of these compounds would have this function.

There are two approaches to obtaining floral compounds for testing for a role in nestmate discrimination. One is to employ extracts from plants; these are available as food flavorings for human consumption and can be easily obtained for a range of plants. This has the disadvantage of yielding a mixture of compounds of uncontrolled relative concentration. Alternatively individual compounds that have been identified as floral odors can be purchased or synthesized. Concentrations can be easily controlled in this case. Ironically, however, many of these compounds are lethal in high concentrations and must be delivered in oil or wax media to facilitate transfer of effective, but non-lethal, quantities to bees. We have used compounds from both sources in our assays.

Breed et al. (1988a) found that anise oil was used by bees as a cue when the oil was dissolved in paraffin wax. Anise oil is derived from *Pimpinella anisum* (Umbelliferae) and the predominant component of the odor is anethole. Paraffin alone did not have this effect, nor did the oil when delivered in the bee's sugar candy food. These results led us to conclude that anise oil can act as a recognition cue, but that a waxy medium is necessary for the transfer of the oil to the bees. As an alternative to wax as a medium, we have in more recent experiments placed the compounds to be bioassayed on filter paper and allowed the bees to walk on the filter paper for a period of time (usually five days). In some cases the compound being tested is dissolved in 100 ml oil and the solution is applied to the filter paper. The waxy (or oily) medium is not necessary in all cases, but can be useful in facilitating transfer of compounds to bees.

Unpublished experiments with anethole (Bowden and Breed, in press) support this conclusion; quantities large enough to transfer directly to bees are lethal, very small amounts placed on filter paper are not adequately transferred, but the same small amounts dissolved in an oil medium are transferred and used as recognition cues.

Bowden and Breed (in preparation) tested several plant extracts (anise—to replicate the previous result, orange, lemon, and peppermint) and found that all of these scents can be used as recognition cues. The primary odorant of lemon is limonene, of peppermint is menthol, and of anise is anethole. These floral scents were transferred to the bees from filter paper without the use of a oil medium. As predicted, a wide variety of floral oils seem to be used by honey bees as recognition cues. Bowden and Breed (in preparation) also tested a number of synthetic compounds. Of these, linalool, geraniol, and anethole all had significant effects in our bioassay for recognition effects.

In addition to floral scents bees may bring odors to the colony in resin collected from plants. This resin is used to plug gaps and cracks in the nest structure and, when used in this way, is termed propolis. The role of scents associated with propolis in nestmate recognition has not been explored.

9.4. *Queen Pheromones*

There is a final type of compound that may come into play in honey bee recognition systems. Honey bee queens can be individually discriminated by workers; hypothetically, queen recognition pheromones could be transfered to the workers (Breed, 1981; Page and Erickson, 1986; Moritz and Crewe, 1988). Virgin queen honey bees produce large quantities of esters in their feces (Blum et al., 1983;

Blum and Fales, 1988). The acid group of these compound ranges from C-8 to C-14 and the alcohol group from C-6 to C-16. Breed et al. (1992b) showed that many of these esters can serve as recognition cues. In discrimination tests among esters bees could discriminate between some ester pairs and not others, but Breed et al. (1992b) could find no predictive factors that would indicate which pairs of esters would be distinguishable. It seems likely that these esters are involved in the discrimination of queens, but not of workers, although more investigation will be required to confirm this point.

9.5. Use of Comb-Derived Cues

In the preceding sections we have shown that at least some bee's wax compounds can serve as recognition cues and that a wide variety of plant-derived compounds are potential recognition cues. In order to understand how these cues might be used we have investigated the role of exposure to the hive environment, and to wax, in the acquisition of recognition cues.

Guard bees sit at the entrance of a honey bee colony and regulate admission to the colony (Breed et al., 1992a). The cuticular hydrocarbon cues are not sufficient for a bee to gain entrance past the guards to its natal nest (Breed et al., 1988b). Exposure to the hive environment, even for very short periods of time (Breed et al., 1988b) modifies the recognition cues carried by bees, so that bees with this exposure can gain entrance. Exposure to comb in the laboratory also modifies the recognition cues of bees; the effect is not as strong as exposure to the hive environment but is quite significant (Breed et al., 1988b).

Cues can be acquired from comb in five minutes (FIG. 8.5) (Breed, Garry, Pearce, Hibbard, Bjostad, and Page, in press) and can be transferred among bees by physical contact among the bees (cuticular hydrocarbons are not transferred, see above). Bees can discriminate among combs from different hives in a way that indicates that the comb conveys information about the genetic background of the bees that produced it, and cues acquired by bees from the comb communicates the same genetic information (Breed, Garry, Pearce, Hibbard, Bjostad, and Page, in press).

The combs used in these experiments were obtained from field hives, so they could possibly carry environmental cues, but the genetic effect is also clearly present. While Breed, Hibbard, Bjostad and Page (in press) have shown that there is a genetic component to the variation among combs from different colonies, Breed and Stiller (1992) found only a few compounds of the types that are commonly found in comb to be active in recognition. Thus there is a major gap in the linkage between our knowledge of the effects of comb and the components of comb that are active.

Comb, of course, is a perfect waxy medium for the absorption and release of floral scents. While no experiment that establishes a direct link between food

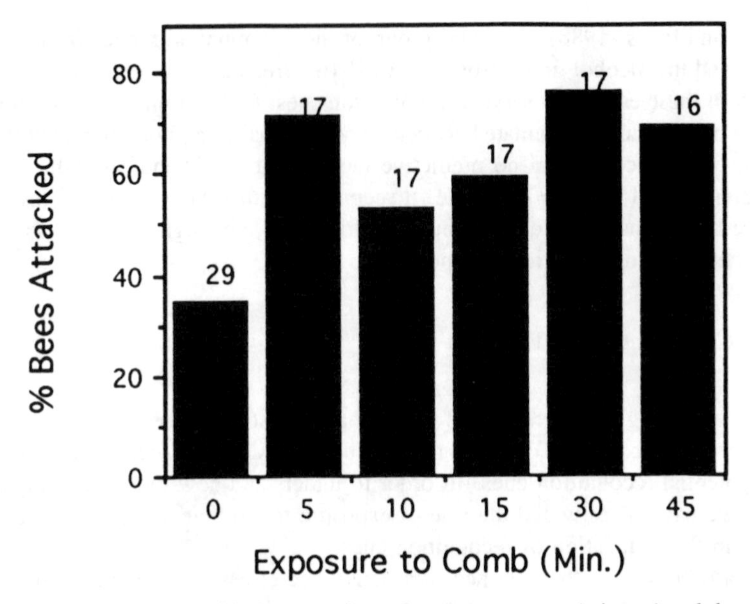

Figure 8.5. The relationship between time of exposure to wax (minutes) and the effect of the wax on acceptability of bees to unexposed sister bees. The method is as described in the legend for Fig. 4 except in this case the introduced bees were exposed to comb for controlled periods of time. Sample sizes appear at the tops of the bars. The 0 minute exposure is the control. The 5 minute, 30 minute, and 45 minute tests are significantly different from the control.

and odors in comb has been performed, it is very reasonable to hypothesize such a link. If this is the case, then the scented candle that Gamboa et al. (1986a) discuss for *Polistes* wasps clearly applies to honey bees.

Comb, then, could serve as a source of two very different types of cues—genetic cues that result from the glandular production of wax and environmental cues that result from the absorption of floral scents into the comb. In either case these cues could be transferred directly from the comb to bees or from bee to bee; empirical evidence supports both mechanisms (Breed, Garry and Pearce, in preparation). How this information might be integrated is discussed below.

9.6. How Do Honeybees Integrate the Available Cues?

Breed and Julian (1992) argued that honey bees have rules for prioritizing cues. They interpreted this as support for a common feature model of nestmate recognition, in which a decision process (not necessarily cognitive) results in members of a colony using a parsimonious cue or set of cues to discriminate nestmates from non-nestmates. This was based on the finding that hexadecane is used in preference over methyl docosanoate when both cues are present. Getz (1993) challenged this conclusion, arguing that tests would need to be conducted over a broad range of ratios of the cues in order to demonstrate prioritization; his challenge represents a strongly held belief in a ratio-based cue template (Getz and Smith, 1987). Breed (1993) replied to Getz, presenting data that tested for ratio and concentration effects.

This dispute aside, our knowledge of how information is processed in honey bee nestmate recognition is very poorly developed. All bees in a colony have the information available from a variety of sources to make nestmate/non-nestmate discriminations, although not all express the behavior associated with the discrimination (Breed et al., 1992a). The exact nature of the template, the degree of tolerance for variation from the template, and the relative importance of environmental and genetic cues all remain unknown. Whether all bees even follow the same decision rules in forming their individual template has not been tested.

Perhaps the only conclusive statement that can be made about cue integration is that cues from the comb and the colony environment have priority in nestmate recognition over the cues (possibly cuticular hydrocarbons) that are produced by bees and which can be used if the other cues are absent. Following the argument of Breed and Julian (1992) it is likely that a set of decision rules guides bees to use a relatively limited number of cues that are likely to vary among colonies in forming their recognition template. This is clearly an intriguing area for further inquiry.

10. Recognition Chemistry of Ants

Given the large number of ant species and the diversity of colony life cycles and colony genetic structure in ants it would be surprising if all species of ant employed the same mechanism of nestmate recognition (Carlin and Hölldobler, 1986; Bennett, 1989; Crosland, 1989). Carlin and Hölldobler (1986) implicate the queen as an important source of recognition cues. Other studies [e.g., Bennett (1989)] have shown strong environmental effects on nestmate recognition. Cues that are produced by the workers are also common (Crosland, 1989) and are generally thought to be cuticular hydrocarbons (Bonavita-Cougourdan et al. 1989).

Chemical explorations of nestmate recognition in ants have focused entirely on cuticular hydrocarbons (Bonavita-Cougourdan et al. 1989). This approach was stimulated by the finding that parasites and enslaved individuals often have cuticular hydrocarbon profiles that match their hosts [e.g. Franks et al. (1990)]. These profiles are acquired after entry into the host nest. Studies within species have shown intercolonial variation in cuticular hydrocarbons that could be used to make discriminations among colonies (Clement et al., 1987; Morel and Vander Meer, 1987; Vander Meer et al., 1989). Bioassays of cuticular extracts and identified compounds from these extracts are now required to determine if these correlations have functional significance.

11. Discussion

The two main evolutionary sources for the derivation of nestmate recognition abilities—nest recognition and assortative mating cues—have been at least partially investigated in *Lasioglossum* and *Apis*. The role of nest recognition has been thoroughly investigated in Polistes. There is an intriguing parallel among these three systems in the possible use of nest materials as recognition cues. In *Lasioglossum* the candidate compounds are Dufour's gland products whose major functional purpose is lining brood cells. In *Apis* one set of candidate compounds are waxes produced by the wax glands and used in comb construction. There is a rather remarkable similarity in Polistes wasps, in which hydrocarbons applied to nesting materials play a key role in nestmate recognition. These three taxa are from quite different evolutionary lines and make it tempting to draw a broad conclusion about the importance of nesting materials in nestmate recognition.

Ants present a very different picture. The nesting biology of ants is very different from that of bees and wasps and no study to date has demonstrated any special role of nest structures. Less is known about the recognition chemistry of any ant than *Lasioglossum* or *Apis*; studies to date have focused on cuticular hydrocarbons. Perhaps further investigation will show that ants fit the generality that seems to be developing from studies of other eusocial Hymenoptera, but the diversity of nesting habits in ants and lack, in many cases, of modification of the nesting cavity, beyond simple excavation, make this seem less likely.

Even in *Lasioglossum* and *Apis* the picture is far from complete. The possible use of Dufour's gland products in assortative mating in *Lasioglossum* makes it possible that we will never discriminate between the two evolutionary hypotheses. We do not yet know if honey bees assortatively mate, but the queen pheromones certainly raise this possibility.

In the honey bee, nests may be recognized using environmentally-derived cues as well as glandular products; environmental cues are also used as nestmate recognition cues in some ants [e.g., Obin (1986)]. Nest recognition in *Lasioglossum zephyrum* can take place in the absence of differences in environmental cues among nests (Kukuk et al. 1977) but there has been no satisfactory test of the possibility that environmental cues, when present, have a role in nest or nestmate recognition.

Unlike most pheromonal signaling systems, in which stereotypy is important for efficient signaling and interindividual variation in pheromones should be minimized, nestmate recognition systems rely on a certain degree of variation among colonies. It is not surprising that glandular products and the environment both may have a role in characterizing a colony population. This complexity makes the chemistry, sensory biology, learning aspects, and evolution of nestmate recognition systems fascinating areas of study.

Acknowledgements. We thank Louis Bjostad, Bruce Hibbard, and Karl Espelie for their comments on a draft of this paper. This work was supported in part by grant IBN–9209016 to M.D.B.

References

Ayasse, M., Engels, W., Hefetz, A., Lübke, G., and Franke, W. (1990) Ontogenetic patterns in amounts and proportions of Dufour's gland volatile secretions in virgin and nesting queens of *Lasioglossum malachurum* (Hymenoptera: Halictidae). *Z. Naturforsch.* 45c: 709–714.

Ayasse, M., Engels, W., Hefetz, A., Tengö, J., Lübke, G., and Franke, W. (1993) Ontogenetic patterns of volatiles identified in Dufour's gland extracts from queens and workers of the primitively eusocial halictine bee, *Lasioglossum malachurum* (Hymenoptera: Halictidae). *Insectes Soc.* 40: 41–58.

Barrows, E. M. (1975a) Individually distinctive odors in an invertebrate. *Behav. Biol.* 15: 57–64.

Barrows, E. M. (1975b) Mating behavior in halictine bees (Hymenoptera: Halictidae): III. Copulatory behavior and olfactory communication. *Insectes Soc.* 22: 307–331.

Barrows, E. M., Bell, W. J., and Michener, C. D. (1975) Individual odor differences and their social functions in insects. *Proc. Nat. Acad. Sci.* USA 72: 2824–2828.

Bateson, P. P. G. (1983) Optimal outbreeding. In: *Mate Choice* (Bateson, P.P.G., ed.). pp 367377. Cambridge U. Press, Cambridge. U.K.

Beecher, M. D. (1988) Kin recognition in birds. *Behav. Genet.* 18: 465–482.

Bekoff, M. (1981) Mammalian sibling interactions. Genes, facilitative environments, and the coefficient of familiarity. In: *Parental Care In Mammals* (Gubernick, D. J., and Klopfer, P. H.eds.). pp. 307–346. Plenum, New York.

Bell, W.J. (1974) Recognition of resident and non-resident individuals in intraspecific nest defense of a primitively eusocial halictine bee. *J. Comp. Physiol.* 93: 195–202.

Bell, W. J., and Hawkins, W. A. (1974) Patterns of intraspecific agonisitc interactions involved in nest defense of a primitively eusocial halictine bee. *J. Comp. Physiol.* 93: 183–193.

Bennett, B. (1989) Nestmate recognition systems in a monogynous-polygynous species pair of ants. Parts I and II. *Sociobiol.* 16: 121–147.

Bergström, G. and Tengö, J. (1979) C24, C22, C20, and C18 macrocyclic lactones in halictid bees. *Acta Chem. Scand.* B33: 390.

Blaustein, A. R. (1983) Kin recognition mechanisms: Phenotype matching or recognition alleles. *Am. Nat.* 121: 749–754.

Blum, M. S. and Fales, H. M. (1988) Eclectic chemisociality of the honeybee: A wealth of behaviors, pheromones, and exocrine glands. *J. Chem. Ecol.* 14: 2099–2107.

Blum, M. S., Fales, H. M., Jones, T. H., Rinderer, T. E. and Tucker, K. W. (1983) Caste-specific esters derived from the queen honey bee sting apparatus. *Comp. Biochem. Physiol.* 75B:237–238.

Bonavita-Cougourdan, A., Clement, J-L., and Lange, C. (1989) The role of cuticular hydrocarbons in recognition of larvae by workers of the ant *Camponotus vagus*: Changes in the chemical signature in response to social environment. *Sociobiol.* 16: 49–74.

Bowden, R.M. and Breed, M.D. (in press) The effects of floral extracts and oils on nestmate recognition in the honey bee. *J. Insect Behav.*

Breed, M. D. (1981) Individual recognition and learning of queen odors by worker honeybees (*Apis mellifera*). *Proc. Natl. Acad. Sci. USA* 78: 2635–2637.

Breed, M. D. (1983) Nestmate recognition in honey bees. *Anim. Behav.* 31: 86–91.

Breed, M. D. (1987) Multiple inputs in the nestmate discrimination system of the honey bee. In: *The Chemistry and Biology of Social Insects* (Eder, J., and Rembold, H., eds). pp. 461–462. Verlag J. Peperny: Munich, Germany.

Breed, M. D. (1993) Odour detection in bees. *Nature* 362: 120.

Breed, M. D. and Bennett, B. (1987) Kin recognition in highly eusocial insects. In: *Kin Recognition In Animals* (Fletcher, D.J.C. and Michener, C.D., eds.). pp. 243–285. John Wiley, Chichester, U.K.

Breed, M.D., Garry, M.F., Pearce, A.N., Hibbard, B.E., Bjostad, L.B., and Page, R.E. (in press) The role of wax comb in honey bee nestmate recognition: Genetic effects on comb discrimination, acquisition of comb cues by bees, and passage of cues among individuals. *Anim. Behav.*

Breed, M.D., Hibbard, B.E., Bjostad, L.B., and Page, R.E. (in press) Genetic component of variation in comb wax hydrocarbons produced by honey bees. *J. Chem. Ecol.*

Breed, M. D. and G. Julian. (1992) Honey bee nestmate recognition: Simple rules do not apply. *Nature*. 357: 685–686.

Breed, M. D., Smith, T. A., and Torres, A. (1992a) Guard honey bees: role in nestmate recognition and replacement. *Annals Entomol. Soc. Am.* 85: 633–637.

Breed, M. D., Butler, L., and Stiller, T. M. (1985) Kin recognition by worker honey bees in genetically mixed groups. *Proc. Natl. Acad. Sci. USA* 82: 3058–3061.

Breed, M. D., Fewell, J. H. and Williams, K. R. (1988a) Comb wax mediates the acquisition of nest-mate recognition cues in honey bees. *Proc. Natl. Acad. Sci. USA* 85: 8766–8769.

Breed, M. D. and Stiller, T. M. (1992) Honey bee, *Apis mellifera*, nestmate discrimination: hydrocarbon effects and the evolutionary implications of comb choice. *Anim. Behav.* 43: 875–883.

Breed, M. D., Stiller, T. M., and Moor, M. J., (1988b) The ontogeny of kin discrimination cues in the honey bee, *Apis mellifera*. *Behav. Genet.* 18: 439–448.

Breed, M. D., Stiller, T. M., Blum, M. S., and Page, R. E., Jr. (1992b) Honey bee nestmate recognition: effects of queen fecal pheromones. *J. Chem. Ecol.* 18: 1633–1640.

Brothers, D. J., and Michener, C. D. (1974) Interactions in colonies of primitively social bees, III. Ethometry of division of labor in *Lasioglossum zephyrum*. *J. Comp. Physiol.* 90: 129–168.

Buckle, G. R., and Greenberg, L. (1981) Nestmate recognition in sweat bees (*Lasioglos-*

sum zephyrum): Does an individual recognize its own odour or only odours of its nestmates? *Anim. Behav.* 29: 802–809.

Butts, D. P., Camann, M. A., and Espelie, K. E. (1993) Discriminant analysis of cuticular hydrocarbons of the baldfaced hornet, *Dolichovespula maculata. Sociobiol.* 21: 193–201.

Cane, J. H. (1981) Dufour's gland secretion in the cell linings of bees (Hymenoptera: Apoidea). *J. Chem. Ecol.* 7: 403–410.

Carlin, N. F. and Hölldobler, B. (1986) The kin recognition system of carpenter ants (*Camponotus spp.*) I. Hierarchical cues in small colonies. Behav. Ecol. *Sociobiol.* 19: 123–134.

Carlson, D. A. and Bolton, A. B.. (1984) Identification of Africanized and European honey bees using extracted hydrocarbons. *Bull. Entomol. Soc. Am.* 30: 32–35.

Clement, J-L., Bonavita-Cougourdan, A. and Lange, C. (1987) Nestmate recognition and cuticular hydrocarbons in *Camponotus vagus*. In: *The Chemistry and Biology of Social Insects* (Eder, J. and Rembold, H., eds.). pp. 473–474. Verlag J. Peperny, Munich, Germany.

Collins, A. M. and Blum, M. S. (1983) Alarm responses caused by newly identified compounds derived from the honey bee sting. *J. Chem. Ecol.* 9: 57–65.

Crewe, R. M. (1982) Compositional variability, the key to the social signals produced by honey bee mandibular glands. In: *The Biology of Social Insects* (Breed, M.D., Michener, C.D. and Evans, H.E., eds.). pp. 318–322. Westview Press, Boulder, Colorado.

Crosland, M. W. J. (1989) Kin recognition in the ant *Rhytidoponera confusa*. II. Gestalt odour. *Anim. Behav.* 37: 920–926.

Crozier R. H., Smith B. H., and Crozier Y. C. (1987) Relatedness and population structure of the primitively eusocial bee *Lasioglossum zephyrum* in Kansas. *Evolution* 41: 902–910.

Dawkins, R. 1982. The extended phenotype. Freeman: Oxford.

de Lello, E. 1971) Adexnal glands of the sting apparatus of bees: Anatomy and histology. II. (Hymenoptera: Halictidae). *J. Kansas Entomol. Soc.* 44: 14–20.

Espelie, K. E., and Hermann, H. R. (1990) Surface lipids of the social wasp *Polistes annularis* (L.) and its nest and nest pedicel. *J. Chem. Ecol.* 16: 1841–1852.

Espelie, K. E., Butz, V. M., and Dietz, A. (1990a) Decyl-decanoate—A major component of the tergite glands of honeybee queens. *J. Apic. Res.* 29: 15–19.

Espelie, K. E., Wenzel, J. W. and Chang, G. (1990b) Surface lipids of social wasp *Polistes metricus* Say and its nest and nest pedicel and their relation to nestmate recognition. *J. Chem. Ecol.* 16: 2229–2241.

Ferguson, I. D., Gamboa, G. J. and Jones, J. K. (1987) Discrimination between natal and non-natal nests by the social wasps *Dolichovespula maculata* and *Polistes fuscatus*. *J. Kansas Entomol. Soc.* 60: 65–69.

Foster, R. and Gamboa, G. (1989) Nest entrance marking with colony specific odors by the bumble bee *Bombus occidentalis*. *Ethology* 81: 273–278.

Francis, B. R., Blanton, W. E., Littlefield, J. L. and Nunamaker, R. A. (1989) Hydrocarbons of the cuticle and hemolymph of the adult honey bee. *Ann. Entomol. Soc. Am.* 82: 486–494.

Franks, N. R., Blum, M., Smith, R.-K. and Allies, A. B. (1990) Behavior and chemical disguise of cuckoo ant Leptothorax kutteri in relation to its host *Leptothorax acervorum.*. *J. Chem. Ecol.* 16: 1431–1444.

Free, J. B. (1959) The effect of moving colonies of honeybees to new sites on their subsequent foraging behaviour. *J. Agr. Sci.* 53: 1–9.

Frumhoff, P. C. and Schneider, S. (1987) The social consequence of honeybee polyandry: kinship influences worker interactions within colonies. *Anim. Behav.* 35: 255–262.

Gamboa, G. J., Foster, R. L. and Richards, K. W. (1987) Intraspecific nest and brood recognition by queens of the bumble bee, *Bombus occidentalis*. *Can. J. Zool.* 65: 2893–2897.

Gamboa, G. J., Reeve, H. K. and Pfennig, D. W. (1986a) The evolution and ontogeny of nestmate recognition in social wasps. *Ann. Rev. Entomol.* 31: 431–454.

Gamboa, G. J., Reeve, H. K., Ferguson, I. and Wacker, T. L. (1986b) Nestmate recognition in social wasps: the origin and acquisition of recognition odours. *Anim. Behav.* 34: 685–695.

Getz, W. M. (1991) The honey bee as a model kin recognition system. In: Kin Recognition (Hepper, P. G., ed.). pp. 358–412. Cambridge University Press, Cambridge, U.K.

Getz, W. M. (1993) Odour detection in bees. *Nature* 362: 119–120.

Getz, W.M., and Page, R.E. (1991) Chemosensory kin-communication systems and kin recognition in honey bees. *Ethology* 87: 298–315.

Getz, W. M. and Smith, K. B. (1986) Honeybee kin recognition: learning self and nestmate phenotypes. *Anim. Behav.* 34: 1617–1626.

Getz, W. M. and Smith, K. B. (1987) Olfactory sensitivity and discrimination of mixtures in the honey bee, *Apis mellifera*. *J. Comp. Physiol.* A. 160:239–245.

Getz, W. M., Brückner, D. and Parisian, T. R. (1982) Kin structure and the swarming behavior of the honeybee, *Apis mellifera*. *Behav. Ecol. Sociobiol.* 10: 265–270.

Getz, W. M., Brückner, D. and Smith, K. B. (1989) The ontogeny of cuticular chemosensory cues in worker honey bees *Apis mellifera*. *Apidologie* 20: 105–113.

Getz, W. M. and Page, R. E. (1991) Chemosensory kin-communication systems and kin recognition in honey bees. *Ethology* 87: 298–315.

Greenberg, L. (1979) Genetic component of bee odor in kin recognition. *Science* 206: 1095–1097.

Hamilton, W. D. (1964) The genetical evolution of social behaviour. Parts I and II. *J. Theor Biol.* 7: 1–16.

Hefetz, A., Bergström, G., and Tengö, J. (1986) Species, individual, and kin specific blends in Dufour's gland secretions of halictine bees—Chemical evidence. *J. Chem. Ecol.* 12: 197–208.

Hefetz, A., Blum, M. S., Eickwort, G. C., and Wheeler, J. W. (1978) Chemistry of the Dufour's gland secretion of halictine bees. *Comp. Biochem. Physiol.* 61B: 129–132.

Hefetz, A., Fales, H. M., and Batra, S. W. T. (1979) Natural polyesters: Dufour's gland macrocyclic lactones form brood cell laminesters in *Colletes* bees. *Science* 204: 415–417.

Hepburn, H. R. (1986) *Honey Bees and Wax*. Springer-Verlag, Berlin, Germany.

Hepper, P. G. (1991) *Kin Recognition*. Cambridge University Press, Cambridge, U.K.

Hölldobler, B. and Michener, C. D. (1980) Mechanisms of identification and discrimination in social Hymenoptera. In: *Evolution of Social Behavior: Hypotheses and Empirical Tests*. (Markl, H., ed.). pp. 35–58. Verlag Chemie, Weinheim, Germany.

Isigrini, M., Lenoir, A. and Jaisson, P. (1985) Preimaginal learning as a basis of colony-brood recognition in the ant, *Cataglyphis cursor*. *Proc. Natl. Acad. Sci. USA*. 82: 8545–8547.

Kaitala, V., Smith, B.H., and Getz, W.M. (1990) Nesting strategies of primitively eusocial bees: A model of nest usurpation during the solitary state of the nesting cycle. *J. Theor. Biol.* 144: 445–471.

Kalmus, H. and Ribbands, C. R. (1952) The origin of odors by which honey bees distinguish their companions. *Proc. Royal Soc.* (B) 140: 50–59.

Knudsen, J. T., Tollsten, L. and Bergström, L. G. (1993) Floral scents—A checklist of volatile compounds isolated by head-space techniques. *Phytochem.* 33: 253–280.

Kukuk, P. F., Breed, M. D., Sobti, A., and Bell, W. J. (1977) The contributions of kinship and conditioning to nest recognition and colony member recognition in a primitively eusocial bee, *Lasioglossum zephyrum*. *Behav. Ecol. Sociobiol.* 2: 319–327.

Michener, C. D. (1974) *The Social Behavior of the Bees*. Harvard University Press, Cambridge, MA.

Michener, C. D. (1978) The parasitic groups of the Halictidae (Hymenoptera, Apoidea). *University Kansas Sci. Bull.* 51: 291–339.

Michener, C. D. and Smith, B. H. (1987) Kin recognition in primitively social insects. In: Kin recognition in animals (Fletcher, D.J.C. and Michener, C.D., eds.). pp. 209–242. John Wiley, Chichester, U.K.

Morel, L. and Vander Meer, R. K. (1987) Nestmate recognition in *Camponotus floridianus*: Behavioral and chemical evidence for the role of age and social experience. In: *Chemistry and Biology of Social Insects* (Eder, J. and Rembold, H., eds.). pp. 471–472. Verlag J. Peperny, Munich, Germany.

Moritz, R. F. A. and Crewe, R. M. (1988) Chemical signals of queens in kin recognition of honeybees, *Apis mellifera* L. *J. Comp. Physiol.* A 164: 83–89.

Moritz, R. F. A. and Southwick, E. E. (1987) Metabolic test of volatile odor labels as kin recognition cues in honey bees (*Apis mellifera*). *J. exp. Zool.* 243: 503–507.

Moure, J. S., C. M. F. and Hurd, P. D. (1987) *An Annotated Catalog of the Halictid Bees of the Western Hemisphere (Hymenoptera; Halictidae)*. Smithsonian Inst. Press, Washington, D.C.

Noonan, K. C. (1986) Recognition of queen larvae by worker honey bees (*Apis mellifera* L.). *Ethology* 73: 295–306.

Obin, M. S. (1986) Nestmate recognition cues in laboratory and field colonies of *Solenopsis invicta* Buren: effect of environment and the role of cuticular hydrocarbons. *J. Chem. Ecol.* 12: 1965–1975.

Packer, L. (1991) The evolution of social behavior and nest architecture in sweat bees of the subgenus *Evylaeus* (Hymenoptera: Halictidae): a phylogenetic approach. *Behav. Ecol. Sociobiol.* 29: 153–160.

Page, R. E. and Erickson, E. (1986) Kin recognition and virgin queen acceptance by worker honey bees (*Apis mellifera* L.). *Anim. Behav.* 34: 1061–1069.

Page, R. E. Jr., Metcalf, R. A., Metcalf, R. L., Erickson, E. H. Jr. and Lampman, R. L. (1991) Extractable hydrocarbons and kin recognition in the honey bee. *J. Chem. Ecol.* 17: 745–756.

Pfennig, D. W., Gamboa, G. J., Reeve, H. K., Shellman-Reeve, J. and Ferguson, I. D. (1983) The mechanism of nestmate discrimination in social wasps (*Polistes*, Hymenoptera, Vespidae). *Behav. Ecol. Sociobiol.* 13: 299–305.

Plateaux-Quénu, C. (1959) Un nouveau type de société d'insectes: *Halictus marginatus* Brullé (Hymenoptera, Apoidea). *Ann. Biol.* 35: 325–444.

Ratnieks, F.L.W. (1991) The evolution of genetic odorcue diversity in social Hymenoptera. *Am. Nat.* 137: 202–226.

Reeve, H. K. (1989) The evolution of conspecific acceptance thresholds. *Am. Nat.* 133: 407–435.

Singer, T. L. and Espelie, K. E. (1992) Social wasps use nest paper hydrocarbons for nestmate recognition. *Anim. Behav.* 44: 63–68.

Singer, T. L., Camann, M. A., and Espelie, K. E. (1992a) Discriminant analysis of cuticular hydrocarbons of social wasp *Polistes exclamans* Viereck and surface hydrocarbons of its nest paper and pedicel. *J. Chem. Ecol.* 18: 785–797.

Singer, T. L., Espelie, K. E., and Himmelsbach, D. S. (1992b) Ultrastructural and chemical examination of paper and pedicel from laboratory and field nests of the social wasp *Polistes metricus* Say. *J. Chem. Ecol.* 18: 77–86.

Smith, B. H. (1983) Recognition of female kin by male bees through olfactory signals. *Proc. Natl. Acad. Sci. USA* 80: 4551–4553.

Smith, B. H. (1987) Effects of genealogical relationship and colony age on the dominance hierarchy of the primitively eusocial bee *Lasioglossum zephyrum* (Hymenoptera: Halictidae). *Anim. Behav.* 35: 211–217.

Smith, B. H. (1992) Merging mechanism and adaptation: an ethological approach to learning and generalization. In: *Insect Learning: Ecological and Evolutionary Perspectives* (Papaj, D.R. and Lewis, A.C., eds.). pp. 126–158. Chapman and Hall, New York, N.Y.

Smith B. H., and Ayasse M. (1987) Kin-based male mating preferences in two species of halictine bees (Hymenoptera: Halictidae). *Behav. Ecol. Sociobiol.* 20: 313–318.

Smith, B. H., and Weller, C. (1989) Social competition among gynes in halictine bees: the influence of bee size and pheromones on behavior. *J. Insect Behav.* 2: 397–411.

Smith, B. H., and Wenzel, J. W. (1988) Pheromonal covariation and kinship in social bee *Lasioglossum zephyrum* (Hymenoptera: Halictidae). *J. Chem. Ecol.* 14: 87–94.

Smith, B. H., Carlson, R. G., and Frasier, J. (1985) Identification and bioassay of macrocyclic lactone sex pheromone of the halictine bee *Lasioglossum zephyrum*. *J. Chem. Ecol.* 11: 1447–1456.

Tulloch, A. P. (1980) Beeswax—composition and analysis. *Bee World* 61: 47–62.

Vander Meer, R. K., Saliwanchik, D. and Lavine, B. (1989) Temporal changes in colony cuticular hydrocarbon patterns of *Solenopsis invicta*: Implications for nestmate recognition. *J. Chem. Ecol.* 15: 2115–2125.

Wcislo, W. T. (1987) The role of learning in the mating biology of a sweat bee *Lasioglossum zephyrum* (Hymenoptera: Halictidae). *Behav. Ecol. Sociobiol.* 20: 179–185.

Wcislo, W. T. (1993) Attraction and learning in mate-finding by solitary bees, *Lasioglossum (Dialictus) figuersi* Wcislo and *Nomia triangulifera Vachal* (Hymenoptera: Halictidae). *Behav. Ecol. Sociobiol.* 31: 139–148.

Westrich, P. (1989) Die Wildbienen Baden-Württembergs; Spezieller Teil. Verlag Eugen Ulmer, Stuttgart, Germany.

Yanega, D. (1988) Social plasticity and early-diapausing females in a primitively eusocial bee. *Proc. Natl. Acad. Sci. USA.* 85: 4374–4377.

9

Chemical Communication in the True Bugs and Parasitoid Exploitation

Jeffrey R. Aldrich
Insect Chemical Ecology Laboratory, USDA-ARS, Agricultural Research Center

1. Introduction

Heteroptera suck vertebrate and invertebrate blood, or plant sap, and they do it above or below ground, under or on water, even on salt water. There are some 35,000 species of Heteroptera worldwide, including 2,500 predaceous species, yet sexual pheromones are fully known for only a few species. Nevertheless, scattered behavioral, morphological, and chemical evidence suggests that attractant pheromones are widespread in the advanced terrestrial infraorders of Heteroptera: Cimicomorpha and Pentatomomorpha (Aldrich, 1988a, 1988b) (Fig. 9.1).

The meagerness of heteropteran pheromone data is, in part, a self-fulfilling prophecy from lack of effort. The diverse nature of the group itself has hindered progress, and the copious defensive secretions from the metathoracic scent glands found in most true bugs can contaminate sex pheromone samples [e.g., Aldrich et al. (1987)]. In certain taxa, however, long-range attractant pheromones are probably not prevalent. Male water striders (Gerridae) call females, and ward off competing males, with coded surface waves of water (Wilcox, 1979; Wilcox and Stefano, 1991). In truly aquatic Heteroptera, there is a tendency for exocrine glands to be lost (Cobben, 1978), while sound-producing organs evolved many times (Aiken, 1985). The ease with which water conducts waves favors this mode of distant communication in aquatic and semiaquatic bugs. Still, some aquatic heteropterans employ pheromones during courtship; for example, *(E)*-2-hexenyl acetate may be an aphrodisiac in giant water bugs (Belostomatidae) (Butendandt and Tam, 1957).

This review is limited to the terrestrial infraorders of Heteroptera. The morphology and chemistry of heteropteran defensive glands is briefly summarized; readers can consult other reviews for more details (Aldrich, 1988a; Blum, 1981; Carayon, 1971; Cobben, 1978; Staddon, 1979, 1986). Taxa for which there is chemical data associated with adult aggregation and courtship will be emphasized; however,

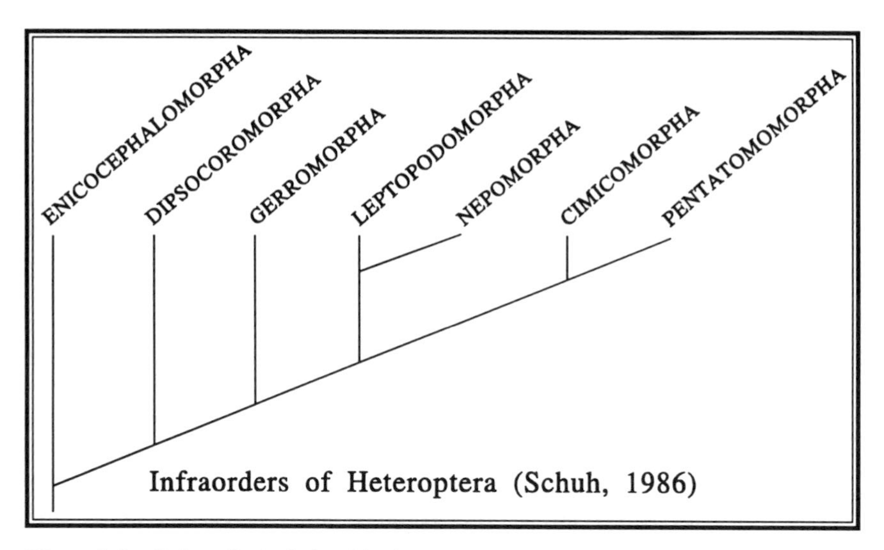

Figure 9.1. Infraordinal relationships in the Heteroptera. Enicocephalomorpha, Cimicomorpha, and Pentatomomorpha are terrestrial groups, while members of the other infraorders live in, on, or around water.

consideration of contact chemoreception will be excluded. A complex of parasitoids exploit heteropteran allomones and pheromones—the best-known cases of symbiont usurpation will be discussed.

2. Exocrine Glands

2.1. Overview of Defensive Glands and Chemistry

In general, Heteroptera produce allomones in dorsal abdominal scent glands as immatures, and in metathoracic scent glands as adults (Aldrich, 1988a; Staddon, 1986). The dorsal abdominal glands of nymphs are invaginated cuticle-lined sacs that open intersegmentally, with exocrine gland cells forming the wall. Commonly there are three or four dorsal abdominal glands, the exact number and position being characteristic of a taxon (Cobben, 1978). At metamorphosis these metameric glands are usually lost, but not always, and the metathoracic scent gland opening between the meso- and metathoracic legs, becomes functional. In the phylogenetically-advanced infraorder, Pentatomomorpha, cells of the metathoracic gland are segregated from the reservoir wall forming two large lateral glands, each emptying into a median reservoir. In addition, there are small secondary accessory glands attached to, or embedded in, the reservoir wall (Aldrich, 1988a; Carayon, 1971).

There are innumerable exceptions described to the "normal" heteropteran allomonal gland pattern (Staddon, 1979, 1986). The theme uniting many of these exocrinological deviants is that species specializing on poisonous plants frequently have reduced or modified metathoracic scent glands, and are aposematic (Aldrich, 1988a). Most, if not all, cotton stainer bugs (Pyrrhocoridae) selectively feed on toxic plants in the order Malvales, and have reduced metathoracic glands with the reservoir completely divided in some species [e.g., Everton et al. (1979)]. The scentless plant bugs, Rhopalidae, are so named because they supposedly lack metathoracic glands. This is only true for the Serinethinae (including the familiar boxelder bug), many of which feed on seeds of Sapindaceae containing poisonous cyanolipids (Aldrich et al., 1990b; Braekman et al., 1982). In seed bugs (Lygaeidae), virtually the entire assemblage of Lygaeinae sequester cardiac glycosides from Apocynales, many are aposematic, and in most species the metathoracic gland is reduced (Aldrich, 1988a). Thus, in the adults of various lineages, acquisition of host-plant toxins relieved the selective pressure for *de novo* synthesis of allomones. Nevertheless, this tendency is less pronounced in the corresponding nymphs, perhaps because the flightless immatures require longer-lasting protection (Aldrich and Yonke, 1975).

Assassin and ambush bugs (Reduviidae and Phymatidae) are sit-and-wait predators. Their sedentary mode of hunting, and use of the rostrum for stinging defense, probably contributed to reduction of the metathoracic and nymphal dorsal abdominal glands, and to evolution of new exocrine glands (Carayon, 1971; Cobben, 1978; Staddon, 1986). The new glands—Brindley's glands (situated laterally under the first abdominal tergite), and the ventral glands (at the junction of the thorax and abdomen)—are both paired glands that are identical in the sexes, and occur either by themselves or coexist with the metathoracic gland (Staddon, 1986). Secretions from Brindley's glands in hematophagous reduviids (Triatominae) often contain isobutyric acid (Blum, 1981), whereas in the lone chemically investigated triatomine having only metathoracic glands, *Dipetalogaster maximus*, 3-methyl-2-hexanone is the major secretory component (Rossiter and Staddon, 1983).

Notwithstanding peculiarities cited already and others [e.g., Borges and Aldrich, (1992)], plus various oddities due to secondarily evolved sexual functions (Section 2.2), both nymphal and adult allomones of many higher terrestrial bugs conform to patterns (Aldrich, 1988a). Dorsal abdominal gland secretions of nymphs typically include C6, C8, and C10 *(E)*-2-alkenals, C6 and C8 *(E)*-4-oxo-2-alkenals, and sometimes alkanes such as *n*-tridecane. In adults, allomones are most often mixtures of even-numbered C4–10 saturated or α, β-unsaturated aldehydes, alcohols, acetate or butyrate esters of these alcohols, C2, C4, or C6 acids, and sometimes *n*-alkanes (principally tridecane). At least in some coreids and pentatomids (Aldrich, 1988a), esters are secreted by the primary accessory glands into the reservoir where the corresponding alcohols, aldehydes, and acids are derived via the action of enzymes secreted by the secondary accessory glands.

Two features of the allomone system of pentatomids should also be mentioned because they constitute signals parasitoids use to recognize stink bug hosts (Section 5). First, when pentatomids ecdyse, the contents of their dorsal abdominal glands are shed with the exuviae (Borges and Aldrich, 1992). Second, molecular chain lengths of metathoracic gland irritants differ characteristically between some genera within the family (Table 9.1). Shedding defensive secretion with exuviae is universal for bugs [e.g., Aldrich et al. (1991c, 1993a)]; chain length and degree of saturation are probably distinctive at the generic to tribal levels or higher for most Heteroptera [e.g., Aldrich et al. (1979, 1990b)].

2.2. Sexually Dimorphic Exocrine Glands

Males of most species in the superfamily Pentatomoidea have unicellular glands in the abdominal integument (Carayon, 1981; Farine, 1987; Staddon, 1990; Staddon and Edmunds, 1991). These glands are either absent or less abundant and histologically different in females. Usually, the glandular cells are scattered among unmodified epidermal cells as in the southern green stink bug, *Nezara viridula* (Pentatominae) (Carayon, 1981). There has been confusion about this point (Aldrich, 1988b) because males of *Nezara*, and some other pentatomids, have areas on the anterior abdominal sternites that are visibly smoother than surrounding cuticle (Aldrich, 1988a), but these areas are not involved in pheromone production (Pavis, 1986; Staddon and Edmunds, 1991). In some Scutelleridae, Acanthosomatidae, lygaeids in the subfamily Oxycareninae, and in several genera of predaceous pentatomids (Asopinae: *Perillus, Stiretrus, Mineus,* and *Oplomus*), males have gland cells grouped into discrete, laterally paired areas on the abdominal sternites (Aldrich, 1988b; Carayon, 1981, 1984). These sternal glands (also called androconial glands) are obvious setal patches (Fig. 9.2) or areas of sculptured cuticle (Fig. 9.2) (Aldrich et al., 1986b; Carayon, 1984). An interesting variation occurs in males of the phytophagous pentatomid, *Aelia acuminata*, where the sternal glands are abundant beneath the abdominal spiracles, and have ducts opening into the tracheae (Staddon and Edmunds, 1991).

Two isolated instances of female-specific sternal glands are known: females of some of the Reduviinae have sternal glands (Carayon, 1987), as do some

Table 9.1. Pentatomid metathoracic scent gland secretions (%).

	alkanes		alk-2-enals			alk-2-enyl acetates			4-oxo-alk-2-enals	
Species	C_{12}	C_{13}	C_6	C_8	C_{10}	C_6	C_8	C_{10}	C_6	C_8
Nezara viridula[a]	2	54	1	0.5	26	3	0.5	4	7	—
Euschistus tristigmus[b]	5	49	4	6	—	—	20	1	4	—
Biprorulus bibax[c]	2	40	10	—	3	10	—	20	10	—

[a]Gilby and Waterhouse, 1967.

[b]Aldrich, unpublished.

[c]MacLeod *et al.*, 1975.

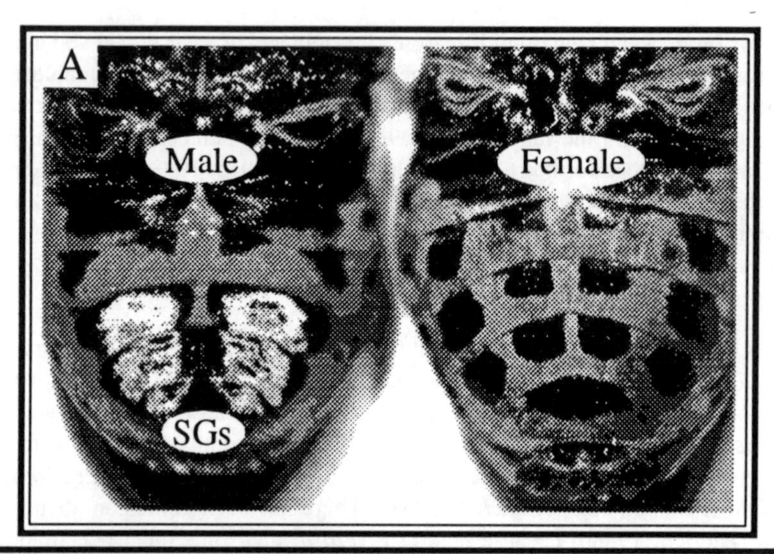

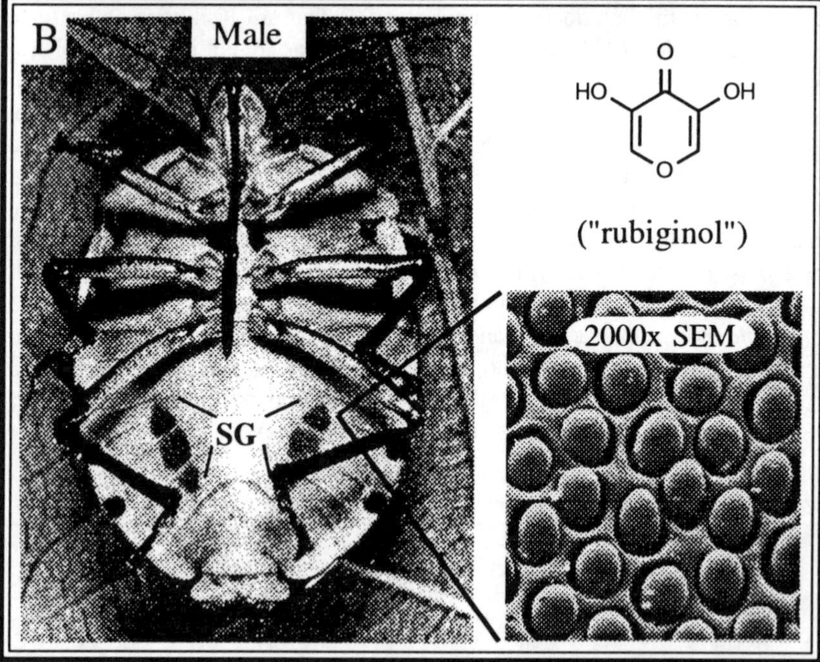

Figure 9.2. (A) Sternal glands (SGs) of a *Perillus bioculatus* male (left) having setae wet with secretion. Females (right) lack SGs. (B) Male-specific SGs of *Tectocoris diophthalmus* showing the sculptured cuticle overlying the glands (scanning electron micrograph; 2000X) [from Carayon (1984)], and the crystalline pyrone, rubiginol, secreted from the glands.

acanthosomatid females (Staddon, 1990). In the latter case, the glands are probably associated with egg guarding (Aldrich, 1988a) by these highly maternalistic bugs, but in assassin bugs the glands are presumed to produce sexual pheromones (Carayon, 1987).

More "sophisticated," multicellular exocrine glands, with storage reservoirs from which secretion can be selectively released, have evolved in males of various terrestrial bugs. In several genera of predaceous stink bugs, including *Podisus, Zicrona, Arma, Alcaeorrhynchus, Supputius, Tinacantha*, and *Oechalia*, males have a pair of enlarged glands opening underneath the wings (Fig. 9.3) (Aldrich, 1988a, 1988b, 1991, and unpublished; Thomas, 1992). Similar male-specific glands are known for two species of phytophagous bugs: *Sphaerocoris annulus* (Scutelleridae) (Fig. 9.3) (Gough et al., 1986), and *Biprorulus bibax* (Pentatomidae) (James and Warren, 1989). The large Australian assassin bug, *Pristhesancus plagipennis* (Redviidae), has three median dorsal abdominal glands in a row, with the anterior and posterior glands being much larger in males than in females or nymphs (Aldrich, 1991). In the Enicocephalomorpha (Fig. 9.1), the metathoracic gland has only been found in males, where it has an unusual tubular form (Azfal and Sahibzaba, 1989). Sexually dimorphic metathoracic glands are again encountered in the phylogenetically advanced Pentatomomorpha. Males of the large milkweed bug, *Oncopeltus fasciatus* (Lygaeinae), have a valve that prevents secretion from the primary accessory glands from entering the median reservoir, causing the bases of the accessory glands to swell (Everton and Staddon, 1979). An identical dimorphism has been discovered for an Australian broadheaded bug, *Mirperus scutellaris* (Alydidae) (Fig. 9.4), and in another Australian alydid, *Riptortus serripes*, the primary accessory glands are much larger in males than females (Fig. 9.4) (Aldrich et al., 1993a).

The function of glands opening near the genitalia remain a challenge for future researchers. Female wheel bugs, *Arilus cristatus* (Aldrich, unpublished) and the South American *A. carinatus* (Reduviidae) (Barth, 1961), have a bright orange subrectal gland that is everted like a papilionid caterpillar osmeterium when the bugs are disturbed, in the wheel bug releasing a nauseatingly sweet odor. Another kind of genitalic gland producing volatiles in many coreoid and lygaeoid males are ventral abdominal glands (Aldrich, 1988a). Ventral glands often coexist with other sexually dimorphic glands, and produce aromas wonderful to our sense of smell, but the function of ventral gland bouquets for bugs is mysterious (Section 3.2).

3. Sexual Pheromones in Terrestrial Bugs

Discovery of sexually dimorphic exocrine glands in Heteroptera, usually excessively developed in males, provided early circumstantial evidence for the existence of sex pheromones. This suspicion has now been confirmed, but the phero-

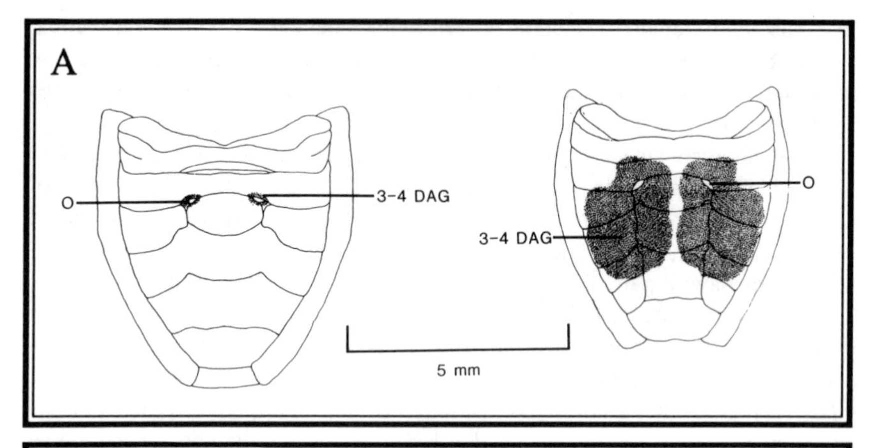

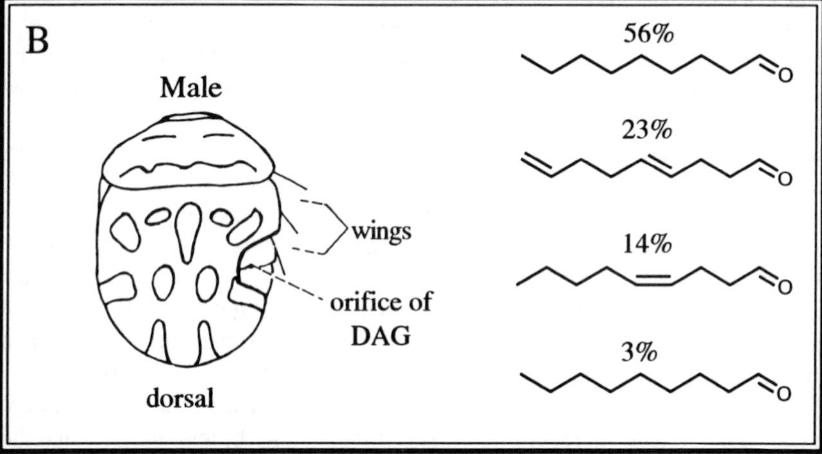

Figure 9.3. (A) Drawing of the abdominal terga of *Podisus maculiventris* showing the hypertrophied dorsal abdominal glands (DAGs) characteristic of a males compared to the small DAGs of a female (o = orifice). (B) Drawing showing the position of the opening for enlarged DAGs of male *Sphaerocoris annulus*, and the series of C9 aliphatic aldehydes comprising the glandular secretion [from Knight et al. (1985)].

mone situation for bugs is even more complicated because various species produce sex- and species-specific secretions from nonsexually dimorphic glands, or exhibit striking age-dependent shifts in the chemicals secreted by a gland. Although behavioral correlates for the latter types of exocrines are meager, at best, the chemistry as it is currently understood will be described as a challenge for future students of Heteroptera.

3.1. Pentatomoidea (Pentatomidae, Scutelleridae, and Acanthosomatidae)

Many predaceous pentatomids (Asopinae) produce enormous quantities of pheromones compared to phytophagous pentatomoids and most other insects (Aldrich,

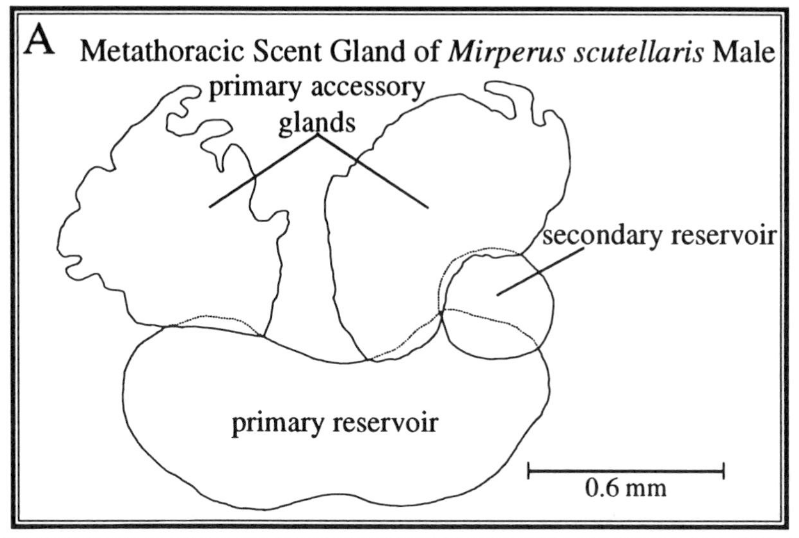

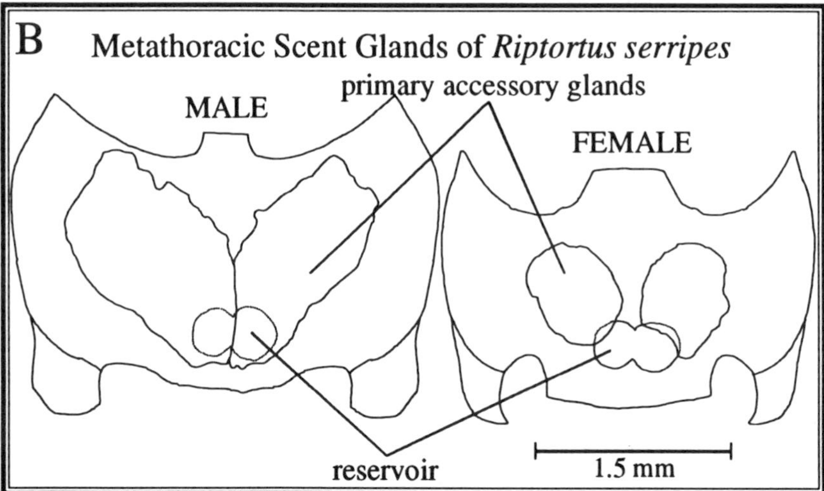

Figure 9.4. (A) Drawing of the metathoracic scent gland of a sexually mature *Mirperus scutellaris* male showing a secondary reservoir at the base of one accessory gland (secretion was removed from the other secondary reservoir); females lack secondary reservoirs. (B) The metathoracic scent glands of adult *Riptortus serripes* showing the enlarged primary accessory glands of a mature male (left) compared to those of a mature female (right).

1988a). Predators are inherently more dispersed than phytophagous species; therefore, the generalists species cannot rely on specific kairomones to find their menagerie of prey. Less than a tenth of the 2,500 species of Pentatomidae are predaceous, but the extreme reliance of asopines on attractant pheromones facilitated research on these species.

Males of many asopines, and some plant-feeding stink bugs and shield bugs, have a pair of greatly enlarged dorsal abdominal glands. Even in those species that do not, the small pair of homologous anterior dorsal glands found in nymphs remain active in adults (Evans et al., 1990).

3.1.1. Asopinae

The first attractant pheromone identified for a true bug was for the spined soldier bug, *Podisus maculiventris* (Aldrich et al., 1978, 1984a). This generalist predator of caterpillars, beetle larvae, and other foliage feeding insects maintains large populations throughout much of North America (Aldrich, 1985; Thomas, 1992). When they choose to, spined soldier bug males can call mates by releasing pheromone from massive dorsal abdominal glands (>1 mg of secretion/male) from openings underneath their wings (Fig. 9.3) (Aldrich et al., 1984a). Overwintered males emerge just prior to the bud burst of deciduous trees and begin pheromone calling without feeding; spring and summer generation males probably first search for prey, and then call mates (Table 9.2). The pheromone is a distinctive combination of low molecular weight, previously known compounds (Fig. 9.5). Minor components might be communicative during courtship, but are not essential for long-range attraction. Although males produce exclusively (+)-R-α-terpineol (V), the antipode is behaviorally invisible; a mixture of *(E)*-

Table 9.2. Podisus maculiventris *emergence data ($\overline{X}\pm SD$, degree-days >30°F; Beltsville, MD, 1982–1990), and phenologies of associated plants.*

Overwintered Generation		New Generation	
Start	Peak	Start	Peak
690 ± 120	1033 ± 146	2702 ± 636	3396 ± 440

Indicator Plants (from Orton and Green, 1989) Common Name (stage; degree-days > 30°F) / *Genus species*	
Red Maple (blooming; 400–650) *Acer rubrum*	Washington Hawthorn (bloom; 2000–2400) *Crataegus phaenopyrum*
Pussy Willow (catkins yellow; 400–650) *Salix discolor*	Amur Privet (blooming; 2500–2600) *Ligustrum amurense*
Silver Maple (buds breaking; 400–650) *Acer saccharinum*	Queen Anne's Lace (blooming; 3000–3300) *Daucus carota*
Saucer Magnolia (pink bud; 400–650) *Magnolia X soulangiana*	American Elder (blooming; 3000–3300) *Sambucus canadensis*

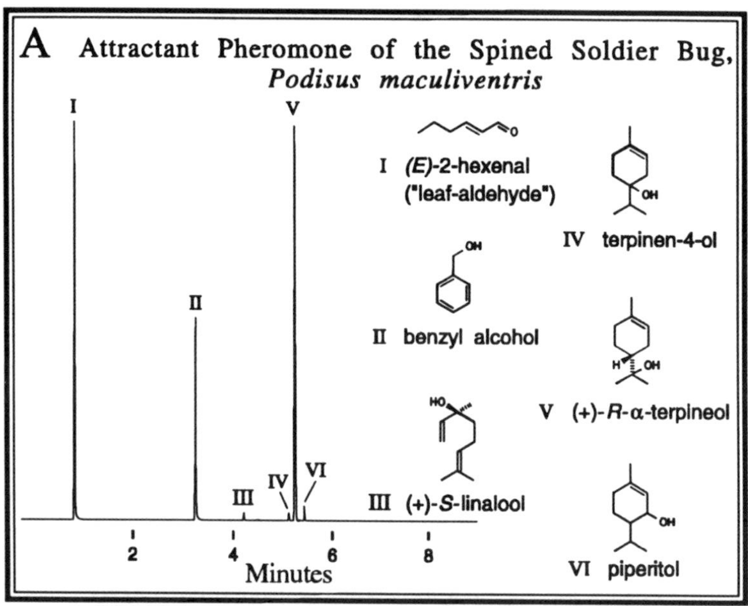

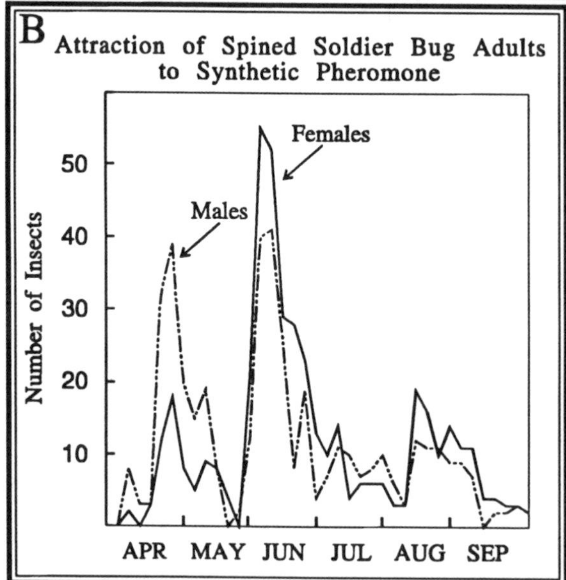

Figure 9.5. (A) Gas chromatogram and chemistry of the attractant pheromone of the spined soldier bug, *Podisus maculiventris*. (B) Attraction of male and female *P. maculiventris* to synthetic pheromone in the field.

2-hexenal (I) and racemic α-terpineol (1:2 by volume, respectively) is as attractive as the complete blend (Fig. 9.5) (Aldrich, 1988b). The small dorsal glands of *P. maculiventris* females produce their own unique blend: *(E)*-2-hexenal and the acid, with traces of benzaldehyde and nonanal (Aldrich et al., 1984b).

Female and male spined soldier bugs fly to pheromone-calling conspecific males, and immatures walk to calling males. As such, by conventional jargon the pheromone is considered an aggregation pheromone, but this terminology is ill-advised because it implies that a male emitting pheromone gains some advantage by attracting other conspecific males and nymphs. Immatures and, to some extent, males and females, may be drawn to the pheromone because it is associated with available prey. Parasitoids also powerfully influence this communication system (Section 4).

Another species, *P. neglectus* (=*fretus*) (Thomas, 1992), was encountered in a coniferous forest while field-testing artificial pheromones for *P. maculiventris* in Maryland (Aldrich et al., 1986a). The major monoterpenol in the pheromone of *P. neglectus* is (+)-S-linalool (III, Fig. 5), but otherwise the secretion of this sympatric species is much like that of the spined soldier bug (Aldrich et al., 1986a). *(E)*-2-Hexenal and racemic linalool (1:2) is an effective artificial pheromone for *P. neglectus* but, in contrast to *P. maculiventris*, significantly more females than males are captured in pheromone-baited traps. In the laboratory, male *P. neglectus* paired with female *P. maculiventris* produced a few viable hybrid males whose dorsal glands contained equivalent amounts of linalool and α-terpineol, plus *(E)*-2-hexenal, benzyl alcohol, and minor components of each parent species (Aldrich, 1988b). Pheromone traps baited with artificial hybrid pheromone caught both species simultaneously, indicating that neither predator is inhibited by the major monoterpenol of the other's pheromone (Section 4; Table 3). It is likely that minor components help maintain species isolation.

Male-specific secretions of two Neotropical species, *P. nigrispinus* (=*connexivus*) (Thomas, 1992) and *Supputius cincticeps* (=*Podisus* sp.), are much like the secretions of *P. maculiventris* and *P. neglectus*, respectively (Aldrich et al.,

Table 9.3. Attraction of Biprorulus bibax *adults to lemon groves in NSW, Australia, treated with dorsal abdominal gland extracts, caged live males, or untreated (James et al., 1991).*[a]

Grove Treatment	Sightings of Adults / 1 Hour Search (1990 & 1991)							
	18 Dec	31 Dec	7 Jan	14 Jan	22 Jan	31 Jan	13 Feb	Total
blank	0	0	0	0	0	1	2	3
blank	0	0	0	0	0	0	5	5
DAGs	0	6	2	0	8	8	10	44
DAGs	0	16	36	49	52	65	68	286
males	0	18	8	5	9	14	12	66

[a]insecticide pretreatment of five lemon groves (0.5–1.0 ha) on 15 December, 1990.

1991b). To the contrary, males of two other eastern North American species produce (along with *(E)*-2-hexenal and *(E)*-2-hexenol) characteristic aliphatic and aromatic volatiles instead of monoterpenols; in *P. placidus*, 9-hydroxy-2-nonanone is a major constituent, and in *P. macuronatus*, *(E)*-2-hexenyl tiglate is abundant.

An unexpected spin-off from research on *Podisus* pheromones was the discovery that yellowjackets in the *Vespula vulgaris* species group (Hymenoptera: Vespidae) are attracted to *(E)*-2-hexenal with either α-terpineol or linalool (Aldrich et al., 1985, 1986a). One explanation is that the combination of *(E)*-2-hexenal (known as leaf aldehyde) and these monoterpenols (among the most common in plants) constitutes a damaged-leaf odor [e.g., Turlings et al., (1991)] that yellowjackets use to find insects feeding on leaves. Furthermore, it is plausible that phytophagous ancestors of present-day *Podisus* spp. first responded to this damaged-leaf kairomone, and males subsequently evolved glands producing these semiochemicals as pheromones.

Some asopine males have sternal glands conspicuous as pubescent patches on the fourth to sixth abdominal sternites (Fig. 9.2) (Aldrich et al., 1986b; Aldrich and Lusby, 1986; Thomas, 1992). In the laboratory, starvation (simulating migration?) followed by engorgement on Colorado potato beetle larvae, *Leptinotarsa decemlineata* (Chrysomelidae), stimulated *Perillus bioculatus* males to secrete 6,10,13-trimethyltetradecyl isovalerate (Aldrich et al., 1986b). In *Stiretrus anchorago*, males secrete 6,10,13-trimethyltetradecanol, and adults and immatures are attracted to the synthetic racemic alcohol, especially on very hot days (>35°C) (Fig. 9.6) (Kochansky et al., 1989). The four stereoisomers of 6,10,13-trimethyl-

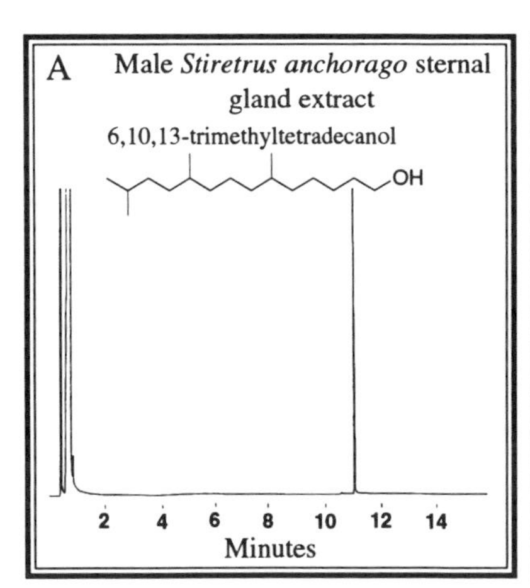

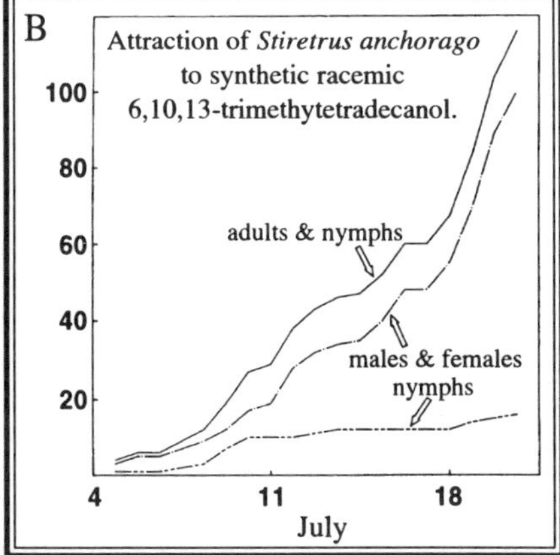

Figure 9.6. (A) Gas chromatogram of sternal gland extract from a *Stiretrus anchorago* male showing that the secretion is almost pure 6,10,13-trimethyltetradecanol. (B) Results of field test showing synthetic racemic alcohol attracts adults and nymphs of *S. anchorago* (July 1987, Beltsville Agricultural Research Center; Kochansky et al., 1989).

tetradecanol have been synthesized (Mori and Wu, 1991), but in preliminary field tests the bugs did not discriminate between stereoisomers (Aldrich, unpublished). At close range, sternal gland exudates might be aphrodisiac-like because males transfer secretion to their tibiae and stroke it on females during courtship (Aldrich and Lusby, 1986; Carayon, 1984). Alternatively, this behavior could serve to mark females as mated to deter recourting or courtship by other males (Aldrich, 1988a).

In summary, asopine males having enlarged dorsal abdominal glands are able to behaviorally control pheromone emission; whereas, in species with sternal glands, once the pheromone is excreted onto the sternum, these relatively nonvolatile compounds probably evaporate more or less continuously for days. The sternal gland pheromones, ignoring chirality, usually consist of only one novel molecule of low volatility. The known *Podisus* pheromones are highly volatile blends in which *(E)*-2-hexenal is omnipresent, with idiosyncratic compounds being produced by sympatric species. Some asopines have neither of these types of pheromone glands (e.g., *Picromerus, Dinorhynchus, Euthyrhynchus,* and *Cermatulus*) (Aldrich, 1988a, 1991; Thomas, 1992).

3.1.2. Phytophagous Pentatomidae

Male-specific volatiles have been isolated from several phytophagous stink bugs lacking consolidated pheromone glands (Aldrich, 1988a, 1988b). Foremost among these species is the southern green stink bug, *N. viridula*, a native of the Ethiopian region but now a cosmopolitan pest (Todd, 1989). Aeration extracts of mature males attracted females in an olfactometer (Borges et al., 1987), and adults and nymphs in the field (Aldrich et al., 1987). However, compounds from *N. viridula* males that are reportedly attractive to conspecific females (Pavis and Malosse, 1986) are actually artifacts formed by dimerization of *(E)*-4-oxo-2-hexenal (Aldrich et al., 1993b). The true pheromone typically includes *(Z)*-α-bisabolene, *trans*- and *cis*-1,2-epoxides of *(Z)*-α-bisabolene, *(E)*-nerolidol, and *n*-nonadecane (Fig. 9.7) (Aldrich et al., 1987; Baker et al., 1987). Females responded only to the synthetic (−)-(1S,2R,4S)-*trans*-epoxide in the laboratory (Baker et al., 1987). Syntheses for racemic mixtures of these *trans*-and *cis*-epoxides have been published (Marron and Nicolaou, 1989; Tomioka and Mori, 1992).

The ratio of *trans*- to *cis-(Z)*-α-bisabolene epoxide from Japanese *N. viridula* is essentially 1:1 (Fig. 9.7), whereas for populations from Mississippi (Fig. 9.7), California, Hawaii, Italy, France, Brazil, and Australia, this ratio is consistently much higher (2–4:1) (Aldrich et al., 1989, 1993b). *Nezara viridula* from Japan and the United States readily interbred, and F1 males produce an intermediate quotient of *trans-/cis-(Z)*-α-bisabolene epoxide (Aldrich et al., 1989). McLain (1992) demonstrated a rare-male mating advantage for *N. viridula*, which could be partially based on pheromone variation. On the other hand, males of the

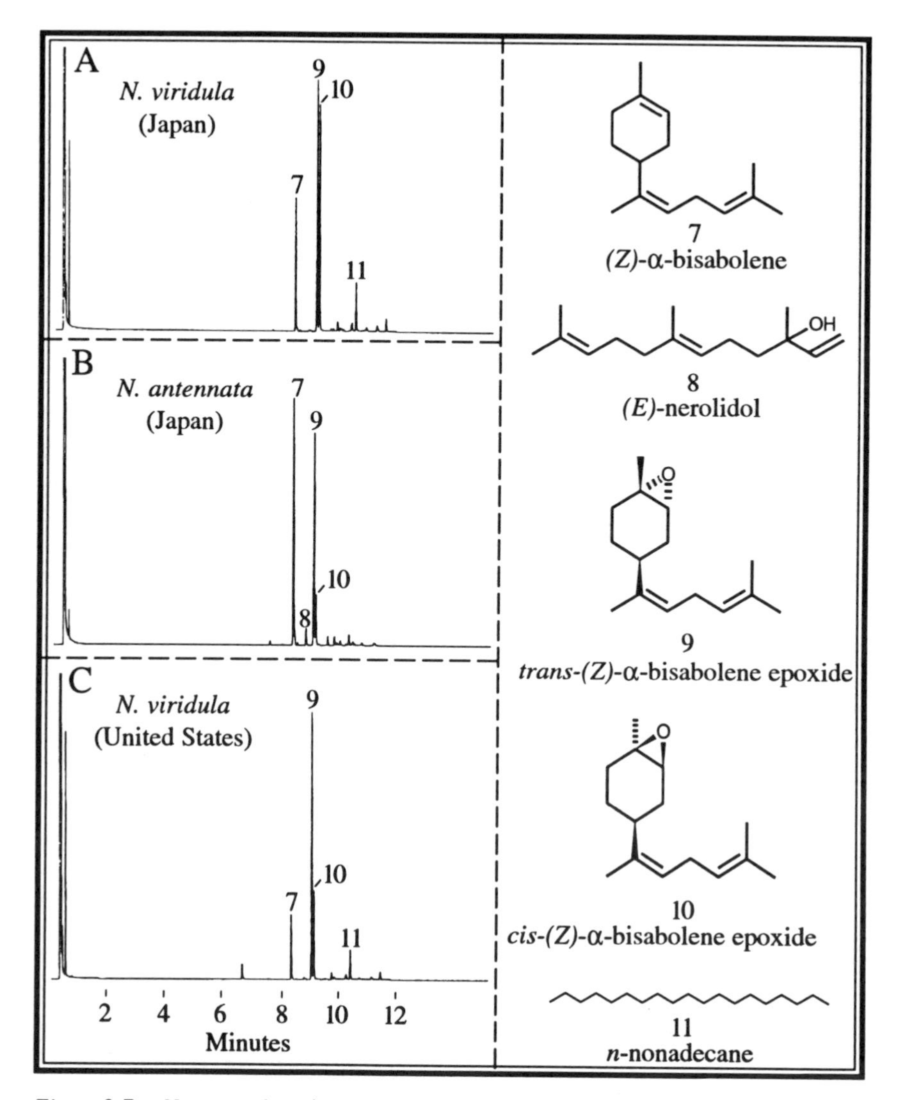

Figure 9.7. *Nezara* males release distinctive sesquiterpenoids from integumental glands. The ratios of *trans-* and *cis-(Z)*-α-bisabolene epoxides differ for *N. viridula* from Japan (A) versus populations of *N. viridula* from other geographically isolated regions such as the United States. (C). The epoxide ratios for the native Japanese species, *N. antennata* (B), are similar to that for the U. S. strain of *N. viridula*.

native Japanese species, *N. antennata*, produce a blend similar to those for most *N. viridula* populations (Fig. 9.7) (Aldrich et al., 1989, 1993b). Interspecific mating still occurs naturally between *N. antennata* and *N. viridula* in some areas, although sperm is not transferred (Kon et al., 1993, 1994).

Surprisingly, males of the North American green stink bug, *Acrosternum hilare*, emit the same compounds as *Nezara*, but with the relative abundances of the epoxide isomers reversed (Aldrich et al., 1989). A similar pattern is evident for the South American species, *A. aseadum*, and for the Central American species, *A. marginatum*, but in *A. pennsylvanicum* (host-specific on bracken fern) the blend of male-specific volatiles is nearly identical to that of Japanese *N. viridula* (Aldrich et al., 1989, 1993b). Discovery that suspected pheromones of members of the *Nezara/Acrosternum* complex consist of distinctive ratios of the same sesquiterpenes suggests that species in these sister genera evolved pheromones analogous to those of many Lepidoptera where canalized isomeric ratios are paramount for reproductive isolation. Indeed, Brezot et al. (1994) found that *N. viridula* females responded significantly greater to a 25% *cis-*/75% *trans*-epoxide ratio than to the *trans*-isomer alone in a laboratory bioassay. Pheromone release rate may also be an important factor as *(Z)*-α-bisabolene and the epoxide derivatives are released at a much higher rate by *A. pennsylvanicum* males (20 μg/male/day) than by *A. hilare* males (0.3 μg/male/day), with the rate for *N. viridula* males being intermediate (9 μg/male/day). Competitive exclusion (e.g., in Japan) and parasitization by tachinids (Section 4) may have provided selective pressure leading to contemporary *N. viridula* pheromone populations (Aldrich et al., 1989, 1993b). Further tests are needed to establish how many components in the male-specific *Nezara* and *Acrosternum* blends are required for pheromonal activity.

Euschistus stink bugs are among the most prolific of all pentatomids worldwide (>60 spp.) (Henry and Froeschner, 1988), and they manifest yet another pheromone theme. In the North American species, *E. conspersus*, *E. tristigmus*, *E. servus*, *E. politus*, and *E. ictericus,* males produce predominantly methyl *(2E,4Z)*-decadienoate, presumably from integumental glands (Aldrich et al., 1991a). Curiously, this ubiquitous *Euschistus* pheromone component is just one carbon short of the so-called pear ester (ethyl *(2E,4Z)*-decadienoate), and *E. conspersus* is a pest of pears [ref. in Aldrich et al. (1991a)]. In the South American soybean pest, *E. heros*, methyl *(2E,4Z)*-decadienoate is a minor volatile from males; the major volatile is methyl 2,6,10-trimethyltridecanoate (Fig. 9.8) (Aldrich et al., 1994). The Central American species, *E. obscurus*, is unique from its chemically known northern and southern congeneric neighbors in that both the *(2E,4Z)*-decadienoate and 2,6,10-trimethyltridecanoate methyl esters are abundantly liberated from males (Fig. 9.8) (Aldrich et al., 1994). Females, males, and nymphs of *E. conspersus, E. tristigmus, E. servus,* and *E. politus* are significantly attracted to methyl *(2E,4Z)*-decadienoate by itself in the field (Aldrich et al., 1991a); *E. obscurus* females respond to males and extracts from

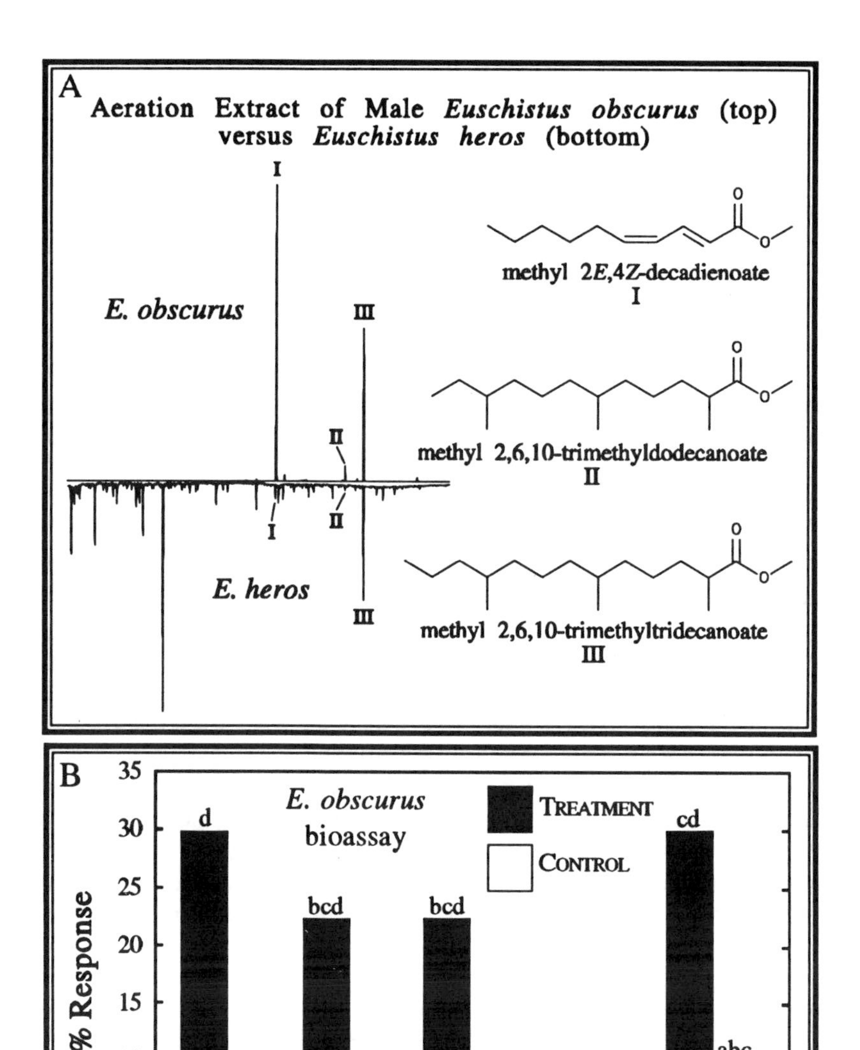

Figure 9.8. (A) Gas chromatograms and chemistry of aeration extracts from males of the Nearctic species, *Euschistus obscurus*, and the Neotropical species, *Euschistus heros*. (B) Bioassay responses of *E. obscurus* females to two live males, 1.3 individual equivalents of male extract, the total male extract trapped from the gas chromatograph (GC) effluent, GC-trapped fraction <2 min (region containing stink gland secretion), and GC-trapped fraction >2 min (region containing male-specific esters). Treatments followed by the same letters do not differ significantly (Borges and Aldrich, 1994).

males in an olfactometer (Fig. 9.8) (Borges and Aldrich, 1994). The role, if any, of minor male-specific *Euschistus* volatiles in reproductive isolation is unknown.

Adults of all the phytophagous pentatomids described above retain small, but active anterior dorsal abdominal glands which, at most, are only slightly sexually dimorphic. The secretions from these glands in *N. viridula* are not particularly remarkable for bugs (2-hexenal, hexanal, hexanol, and tridecane) (Aldrich et al., 1978). However, *Euschistus* spp. emit species-specific blends from these glands that may actually be epideictic pheromones promoting spacing in natural habitats (Aldrich, unpublished). This is a form of communication that may be widespread in terrestrial Heteroptera, judging by the mushrooming number of previously unreported exocrines identified from these tiny glands (later in this section and 3.3–5).

One phytophagous stink bug has finally been discovered whose males have gigantic dorsal abdominal glands homologous to the pheromone glands of *Podisus* (James and Warren, 1989). Gland extracts and live males of the Australian spined citrus bug, *Biprorulus bibax*, induced colonization of pesticide-sterilized citrus groves (Table 9.4) (James et al., 1991). The male-specific secretion is dominated by the novel hemiacetal, (1*E*)-3,4-*bis*(1′-butenyl)tetrahydro-2-furanol, and also includes lesser amounts of linalool, nerolidol, and farnesol isomers (Fig. 9.9) (Oliver et al., 1992). The insect produces only the (3S,4R)-stereoisomer (Mori et al., 1992). In a shadehouse bioassay permitting flight, synthetic (3S,4R,1′*E*)-3,4-*bis*(1′-butenyl)tetrahydro-2-furanol alone is biologically active, but a mixture mimicking the natural blend is significantly more attractive (James et al., 1994a). The unusual hemiacetal may be formed by enzymatic dimerization of 2-hexenal

Table 9.4. Podisus *spp. and symbionts caught in pheromone-baited traps (Aldrich, 1988b).*

H/L-T/B	*P. neglectus*	*P. maculiventris*	*E. flava*	*H. aurata*	*F. crinita*	*T. calvus*	chloropids[b] & milichiids
H/T/B	0	308	279	41	18	2	163
H/L/B	83	0	19	30	0	5	12
H/L-T/B	41	151	211	19	4	10	41
Totals:	124	459	509	90	22	17	216

[a]Compounds: H = (*E*)-2-hexenal, T = α-terpineol, L = linalool, B = benzyl alcohol; blend ratio = 1/2/0.17 by volume, 2 live- and 2 sticky-traps/treatment, March 27–May 29, 1985; one *P. maculiventris* and no symbionts in a control traps.

[b]Chloropidae [total # individuals]: *Olcella* new sp. near *tigramma* (Loew) [116 females, 32 males], *O. cinerea* (Loew) [3], Genus undetermined new sp. ("puzzling" intermediate between *Aphanotrigonum* and *Conioscinella*) [6], and *Rhopalopterum carbonarium* (Loew) [6].

[c]Milichiidae [total # individuals]: *Milichiella* sp. prob. *lacteipennis* Loew [15 females], and *Phyllomyza securicornis* (Fallen) [4 females].

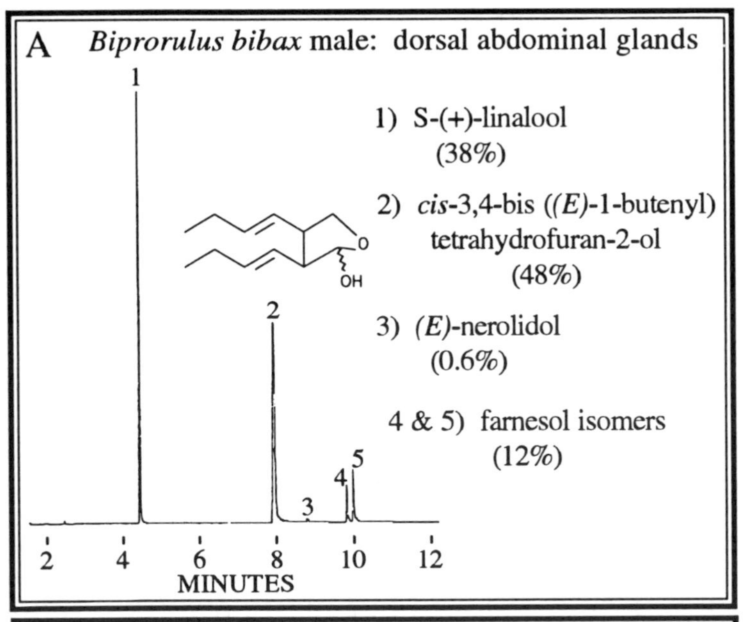

Figure 9.9. (A) Gas chromatogram and chemistry of the attractant pheromone of the spined citrus bug, *Biprorulus bibax*, produced by enlarged dorsal abdominal glands in males. (B) Hypothetical biosynthetic pathways leading from the ubiquitous heteropteran exocrine compound, *(E)*-2-hexenal, to the esoteric hemiacetal that is the key pheromone component.

(Fig. 9.8B) (Oliver, personal communication). *(E)*-2-Hexenal is a key synergist in *Podisus* pheromones, and occcurs in allomones of a multitude of Heteroptera (Aldrich, 1988a). Thus, spined citrus bug males appear to have created a distinctive chemical messenger by simply extending an ancient biosynthetic pathway.

3.1.3. Scutelleridae

Shield bugs are all phytophagous, but males of some species possess either hypertrophied dorsal abdominal glands or sternal glands, each of which are homologous to glands in asopines (Carayon, 1984; Knight et al., 1985). Secretions of two scutellerids (one for each gland type) have been chemically analyzed: the cottonwood harlequin bug, *T. diophthalmus* (sternal glands) (Fig. 9.2B), and *S. annulus* (dorsal glands) (Fig. 9.3B). Behavioral data as to the *raison d'être* for the secretions are wanting, but it is likely that one or both secretions constitute long-range attractants. The sternal gland secretion of *T. diophthalmus* males was originally postulated to be an aphrodisiac (Carayon, 1984) even though the crystalline glandular product, rubiginol (Fig. 9.2B), readily sublimes (Gough et al., 1986). Cottonwood harlequin bug males and females also retain small, identically appearing anterior dorsal abdominal glands which produce principally nonanal (Staddon et al., 1987; Aldrich, unpubl.). Liquid secretion is not actually emitted from these little glands, instead the adults rapidly open and close a flap exposing the inner glandular surface. Other Australian shield bugs exhibit similar behavior (Aldrich, 1991). A West African shield bug, *Hotea gambiae*, common on malvaceous weeds exhibits another exocrinological quirk involving its paired anterior dorsal glands. In nymphs the secretion consists of mono- and sesquiterpenes (Gough et al., 1985b); while in adults, only males retain the glands (without dramatic enlargement), which then secrete purely *(E)*-2-hexenol (Hamilton et al., 1985). Finally, males of the Middle Eastern wheat pest, *Eurygaster integriceps*, attract nearby females with vanillin, ethyl acrylate, and presumably a novel homosesquiterpenoid (Aldrich, 1988a; Staddon et al., 1994).

3.1.4. Acanthosomatidae

A hodgepodge of glands evolved in these semisocial bugs, producing trail and alarm pheromones in nymphs, substances involved in egg-guarding, and probably attractant pheromones from males (Aldrich, 1988a). Species-specific mixtures, composed entirely of monoterpenes (e.g., Fig. 10, compounds 25–28), are created in the small anterior dorsal abdominal glands of adults (Evans et al., 1990). In some species the dorsal glands are somewhat larger in males to compensate for reduced sternal glands (Staddon, 1990).

3.2. Coreoidea (Coreidae, Alydidae, and Rhopalidae)

From the wide array of atypical compounds identified from coreoids (Figure 9.10), it is apparent that some members of the superfamily make the same

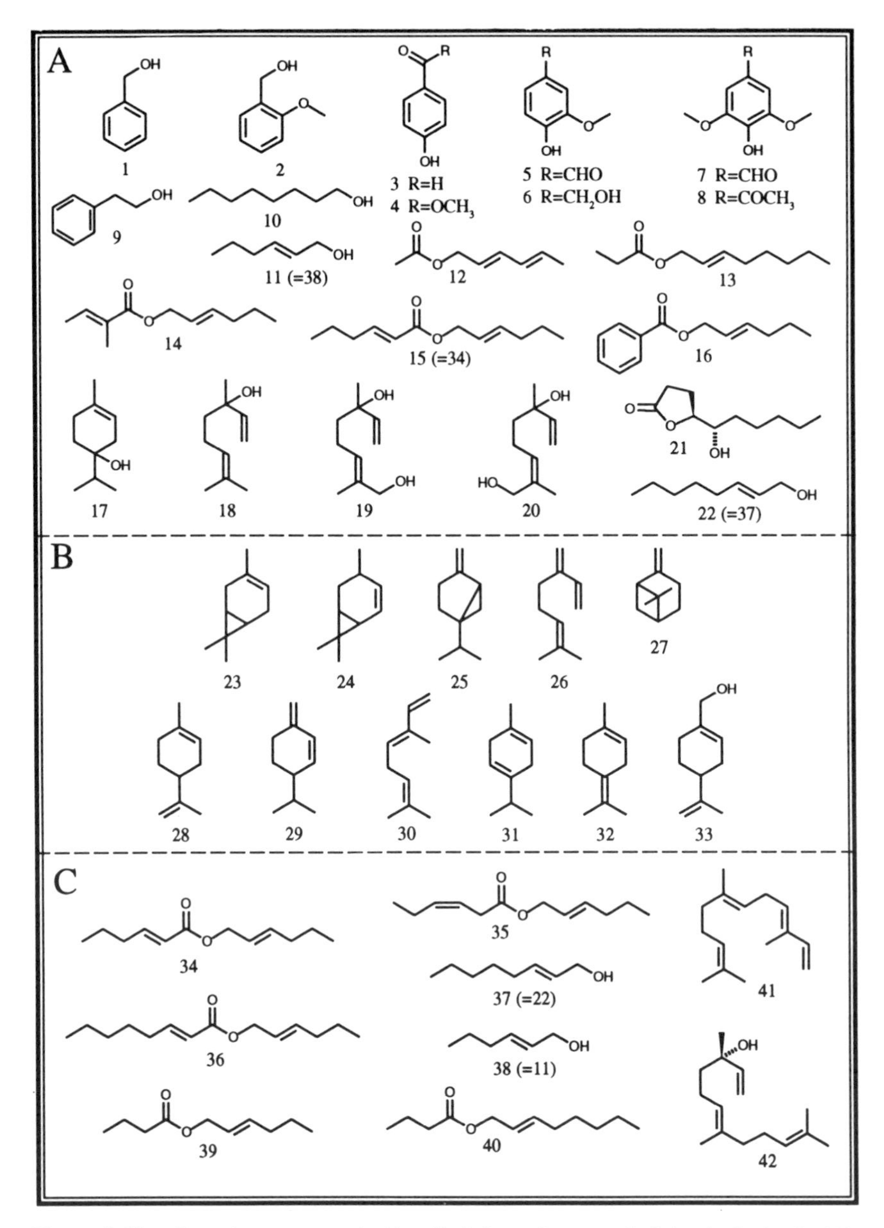

Figure 9.10. Exocrine compounds identified from the ventral abdominal glands (A), dorsal abdominal glands (B), and aeration extracts or metathoracic scent glands (C) of Coreoidea.

compounds in different glands, and many of these molecules are known pheromone components of other bugs. This wealth of chemical evidence suggests that coreoid bugs have sexual pheromones but, until recently, there has been virtually no behavioral proof.

3.2.1. Coreidae

Familiar-smelling compounds reminiscent of vanilla, cinnamon, cherries, and roses, for examples, are distinguishing notes in the ventral abdominal gland perfumes of several male leaffooted bugs, *Leptoglossus* spp. (Fig. 9.10, 1–13) (Aldrich et al., 1979; Gough et al., 1985a). Originally, an attractant role was proposed for these exudates, but blends mimicking secretions are unattractive to *Leptoglossus* in the field (Aldrich, 1988a). The fact that viscous liquid oozes from the glands during copulation, smearing females with very persistent odorants, suggests the alternate hypothesis that ventral abdominal gland secretions mark mated females. Still, the contingency that males call females with ventral gland secretions cannot be discounted, especially since compounds such as benzyl alcohol (1), *(E)*-2-hexenol (10), and *(E)*-2-hexenyl tiglate (11) are components of known or suspected *Podisus* attractant pheromones (Fig. 9.10) (Aldrich et al., 1991b).

In *Leptoglossus* species, males must also have integumental gland cells because they release sesquiterpenes into the air, although there are no obvious multicellular pheromone glands (Aldrich, 1988a; Aldrich et al., 1993b). Moreover, in another species, *L. australis*, males reportedly attract conspecific adults [ref. in Numata et al. (1990); Yasuda and Tsurumachi, 1994, in Japanese). In the Australian banana-spotting bug, *Amblypelta lutescens lutescens*, males lack ventral abdominal glands, yet liberate a species-specific odor composed primarily of (−)-(3R)-*(E)*-nerolidol (42), plus other sesquiterpenes (e.g., 41) (Fig. 9.10). Reexamination of earlier aeration extracts of *L. phyllopus* showed that males of this North American leaffooted bug release the same stereoisomer of nerolidol found in banana-spotting bugs (Aldrich et al., 1993b). Although (+)-(3S)-*(E)*-nerolidol is exceedingly common in plants, its antipode is not (Cane et al., 1990). Field tests are still needed to establish behavioral roles for these suspected pheromones.

3.2.2. Alydidae

Broadheaded bugs present another confusing picture of species with and without ventral glands, and sexually dimorphic metathoracic glands (Aldrich, 1988a; Aldrich et al., 1993b). In North American *Alydus* and *Megalotomus* spp. (remarkable ant and wasp mimics), males lack ventral glands and the metathoracic glands are not dimorphic, producing rancid secretions dominated by short-chain fatty acids in both sexes (Aldrich, 1988a). Interestingly, these foul-smelling alydids are among the few heteropterans attracted to carrion and feces (Payne et al., 1968; Schaefer, 1980).

The first solid chemical information, substantiated by behavioral data, that alydid males emit attractant pheromones is for *Riptortus* bean bug species. Caged live males of the Japanese bean bug, *R. clavatus*, draw both sexes of adults in the field (Numata et al., 1990). The attractive blend has been isolated by aeration, and active components (pinpointed by the gas chromatography-electroantennogram detector technique) have been identified, including (*E*)-2-hexenyl (*E*)-2-hexenoate (34), (*E*)-2-hexenyl (*Z*)-3-hexenoate (35) (Fig. 9.10), and myristyl isobutyrate. A synthetic blend of these compounds acts synergistically to attract adults and nymphs (Leal and Kadosawa, 1992; Leal et al., 1993a, 1993b; Leal, personal communication). The hypertrophied lateral glands of males are the source of *Riptortus* pheromones (Fig. 9.4) (Aldrich et al., 1993b). The metathoracic gland extracts and aeration samples from *R. serripes* males are dominated by (*E*)-2-hexenyl (*Z*)-3-hexenoate (35) and (*E*)-2-hexenyl (*E*)-2-octenoate (36), with a lesser amount of compound 34 (Fig. 9.10), all of which are absent from the females (Aldrich et al., 1993b).

Secretions from the primary and lateral scent gland reservoirs, and the ventral glands, were analyzed separately for the other alydid found to have sexually dimorphic metathoracic glands, *M. scutellaris* (Fig. 9.4) (Aldrich et al., 1993b). The chemistry of the primary reservoir is similar for males and females, except (*E*)-2-octenal is also a major component in males. In the male-specific lateral gland reservoir, alk-2-enals are replaced by alk-2-enols (37 and 38), and (*E*)-2-hexenyl butyrate (39) is the major component with a trace of compound 40 (Fig. 9.10). (*E*)-2-Octenol is simultaneously abundant in the secondary reservoir, and predominant in the ventral gland secretion of *M. scutellaris* males (Fig. 9.10, 22=37).

3.2.3. Rhopalidae

The common name, scentless plant bugs, could hardly be further from the truth. In reality, rhopalids express a dazzling exocrine vocabulary from dorsal abdominal glands in immatures and adults (Fig. 9.10, 23–33), and ventral abdominal glands in males (Fig. 9.10, 17–21). In addition, the Rhopalinae have fully formed metathoracic scent glands which, in the only species studied (*Niesthrea louisianica*), produce monoterpenes 27–29 (Fig. 9.10) and thymol (Aldrich et al., 1990b). Comparative analyses showed that ventral and dorsal abdominal gland secretions are species-specific except for boxelder bugs, *Boisea* (=*Leptocoris*) spp., which casts some doubt on the validity of separate *Boisea* species in the eastern and western continental United States. Serinthines specializing on Sapindaceae containing certain cyanolipids even excrete a metabolite (4-methyl-2(5H)-furanone) attractive to conspecifics (Aldrich et al., 1990a).

Rhopalids are generally "predators" of poisonous seeds, so it is possible scent gland odors serve as warning signals associated with blood-borne toxins, in addition to the intrinsic irritancy of their monoterpene and aliphatic vapors.

However, the species specificity of dorsal gland secretions begs the question of whether these exocrine blends are involved in recognition and aggregation. Huge, sometimes multispecies, aggregations of *Jadera* (Serinthinae) occur on and under host plants, often with more males than females (Carroll, 1988; Tanaka and Wolda, 1988). In this context, ventral gland secretions may be involved in courtship and female marking, as well as in male-male encounters. For example, in aggregations of *J. haematoloma*, males are mounted by other males but quickly rejected after the courting male antennates the genitalia of his partner (Carroll, 1988, 1991).

3.3. Pyrrhocoridae

The Central American cotton stainer, *Dysdercus bimaculatus*, is representative of the fascinating ecology of cotton stainers (Carroll and Loye, 1990). The bugs form huge, multispecies aggregations on toxic seed crops from widely spaced, asynchronously maturing malvaceous trees. Females histolyze their wing muscles to maximize egg production. Sex ratios are heavily male-biased, and males are the principal colonizers of new seed crops. Consequently, males are forced to intensely compete for females during both local and regional phases of their life cycle.

Pheromones help coordinate the contrasting life histories of male and female cotton stainers, but behavioral roles for identified pyrrhocorid exocrines remain sketchy. *Dysdercus* adults retain the anterior two dorsal abdominal glands of the three found in nymphs. Secretions of the anterior glands are thought to be involved in maintaining aggregations [e.g., Calam and Scott (1969)]. Adult *D. cingulatus* manufacture odd combinations of monoterpenoid and straight-chain compounds, such as 1-decyne and *m*-cymene, in the anterior glands (Farine et al., 1992a). During the first few days of adulthood in *D. intermedius*, the primary accessory glands of the reduced, divided metathoracic gland complex secrete hexenyl and octenyl acetates but, by the seventh day, linalool replaces the aliphatic esters (Everton et al., 1979; Staddon, 1986). The shift to linalool synthesis may be characteristic of the genus, as the compound is present in at least three other *Dysdercus* spp. (Aldrich, 1988a), but *Pyrrhocoris apterus* adults do not produce linalool (Daroogheh and Olagbemiro, 1982; Farine et al., 1992b). In *D. koenigii* females, juvenile hormone initiates synthesis of an olfactory signal from the metathoracic glands that stimulates courtship (Hebbalkar and Sharma, 1982); however, in *D. intermedius,* linalool is equally present in the sexes (Everton et al., 1979). Finally, males of *D. cingulatus* secrete methyl *(4Z)*-decenoate, β-caryophyllene, α-humulene, and minor compounds, from epithelial cells of the abdominal sternum (Farine et al., 1992a). It is noteworthy that among the Heteroptera, methyl *(4Z)*-decenoate is most similar to the main attractant pheromone component of Nearctic *Euschistus* spp.: methyl *(2E,4Z)*-decadienoate (Aldrich et al., 1991a).

3.4. Lygaeidae

Most Rhyparochrominae are dull brown bugs that feed on seeds in ground litter, a microhabitat where semiochemicals are likely to be very important (Sweet, 1964). About all that is known in this regard, however, is males of some *Ligyrocoris* spp. have pleasant odors (possibly from ventral abdominal glands) that may be attractive to females (Harrington, 1991).

Milkweed bugs (Lygaeinae) are the butterflies of the bug world—black on red associated with poison may deter predators, but it attracts scientists! Indeed, the large milkweed bug was long ago declared a "research animal" (Feir, 1974). In the Lygaeinae, poisons from host plants have evidently relaxed selection pressure on the metathoracic gland for defense, allowing sexual functions to evolve. Secretions of sexually dimorphic metathoracic glands have been analyzed in *O. fasciatus* (the large milkweed bug) and *Spilostethus rivularis* with the outstanding feature of both being an excess of alk-2-enyl and/or alka-2,4-dienyl acetates in secretions of males (Games and Staddon, 1973; Staddon et al., 1985). In another seed bug protected by sequestered toxins, the cotton seed bug (Oxycareninae: *Oxycarenus hyalinipennis*), synthesis in the metathoracic accessory glands switches from *(E)*-2-octenyl acetate to mainly *(Z,E)*-α-farnesene as bugs mature, similar to the situation in *Dysdercus intermedius* (Section 3.3) (Olagbemiro and Staddon, 1983; Knight et al., 1984a; Staddon, 1986).

Despite early recognition that *O. fasciatus* males apparently release a pheromone from their modified metathoracic scent gland (Lener, 1969; Games and Staddon, 1973), and field observations that the search for mates and host plants is olfactory (Ralph, 1977), insect migratory theory based heavily on *Oncolpeltus* research almost totally ignored the role of males in the migration/colonization process (Dingle, 1985). In the Pentatomomorpha, males have clearly been favored as the emitters of attractant pheromones. The finding that live males and male-specific gland extracts of the spined citrus bug, *B. bibax* (see Section 3.1.2), are capable of inducing bouts of colonization suggests that male-produced pheromones are an important component of the immigration process. The discovery that, at least in the Pentatomomorpha, males may initiate colonization, underscores the inappropriateness of the original descriptor of specialized migratory behaviors as the "oogenesis-flight syndrome" (Dingle, 1985). In a more recent treatment of insect migration, in fact, the implicit sexual bias has been eliminated from the terminology by omitting the word oogenesis (Dingle, 1989). Testing the hypothesis that semiochemicals mediate colonization would be particularly appropriate for milkweed bugs given the wealth of knowledge on the migratory syndrome of these insects.

3.5. Reduviidae

A breakthrough is unfolding in this, the largest family of predaceous land bugs (Schaefer, 1988)—the first attractant pheromone has been identified for an assassin bug. Males of *Pristhesancus plagipennis* (=*P. papuensis*) arch the wings and deflex the abdomen while releasing a pleasant-smelling odor from sexually dimorphic dorsal abdominal glands (Section 2.2) (Aldrich, 1991). Secretion from the anterior gland of this Australian species is reminiscent of coreid ventral gland exudates, with >75% 3-methylbutanol, plus 2-phenylethanol (9) (Fig. 9.10) and a hexenoic acid (Moore et al., 1993). The posterior dorsal gland extract is completely different; the minor components (decanal and *(Z)*-3-hexenol) were easily identified, but the major component (>90%) and a homolog are novel molecules (Moore et al., 1993). Identification of the principal unknown natural product has been accomplished, and verified by synthesis, as *(Z)*-3-hexenyl (R)-2-hydroxy-3-methylbutanoate (James et al., 1994b). In olfactometer bioassays, extracts of male posterior abdominal glands are attractive to females, as is the (2R)-enantiomer, but neither the antipode nor the racemate are active (James et al., 1991, 1994, 1994b). In Queensland, *P. plagipennis* is part of the natural enemy complex that keeps the spined citrus bug, *B. bibax* (Section 3.1), below the economic threshold level. However, where biocontrol agents do not occur, the citrus bug has become a serious pest (James, 1992). A synthetic lure for *P. plagipennis* may prove useful in an unusual classical biocontrol program within the country to mass-trap predators in Queensland for establishment in southern regions.

Blood-sucking kissing bugs (Triatominae) also reportedly produce sex pheromones, but the evidence for their existence is conflicting and none have been chemically identified (Antich, 1968; Ondarza et al., 1986; Rojas et al., 1991).

3.6. Miridae

Female plant bugs produce attractant pheromones which represents a reversal of what is currently known for other Heteroptera (Aldrich et al., 1988; Graham, 1987; King, 1973; Strong et al., 1970; Thistlewood et al., 1989). At least in cocoa capsids, *Distantiella theobroma*, and mullein bugs, *Campyloma verbasci*, females raise their abdomen while releasing pheromone (Fig. 9.11). Only males are attracted, usually at a particular time of day (*Lygus* spp. and *D. theobroma*) or night (*Helopeltis clavifer* and *C. verbasci*). Mated females become unattractive for a few days in most species, but mated cocoa capsid females never resume pheromone calling. Removal of the two terminal antennal segments of *L. hesperus* males has little effect on their ability to locate females, suggesting that pheromone receptors on the first and second antennal segments are sufficient for orientation (Graham, 1988). This is a general characteristic of Miridae as the third and fourth segments are diminutive (Slater and Baranowski, 1978).

A long-sought breakthrough in mirid sex pheromone chemistry has been achieved for the mullein bug, *C. verbasci*. Airborne and thoracic (but not abdominal) extracts of females attract males in the field, suggesting an attractive pheromone comes from the metathoracic scent gland (Thistlewood et al., 1989). Hexyl butyrate is abundant in males, whereas females contain both butyl and hexyl butyrates as

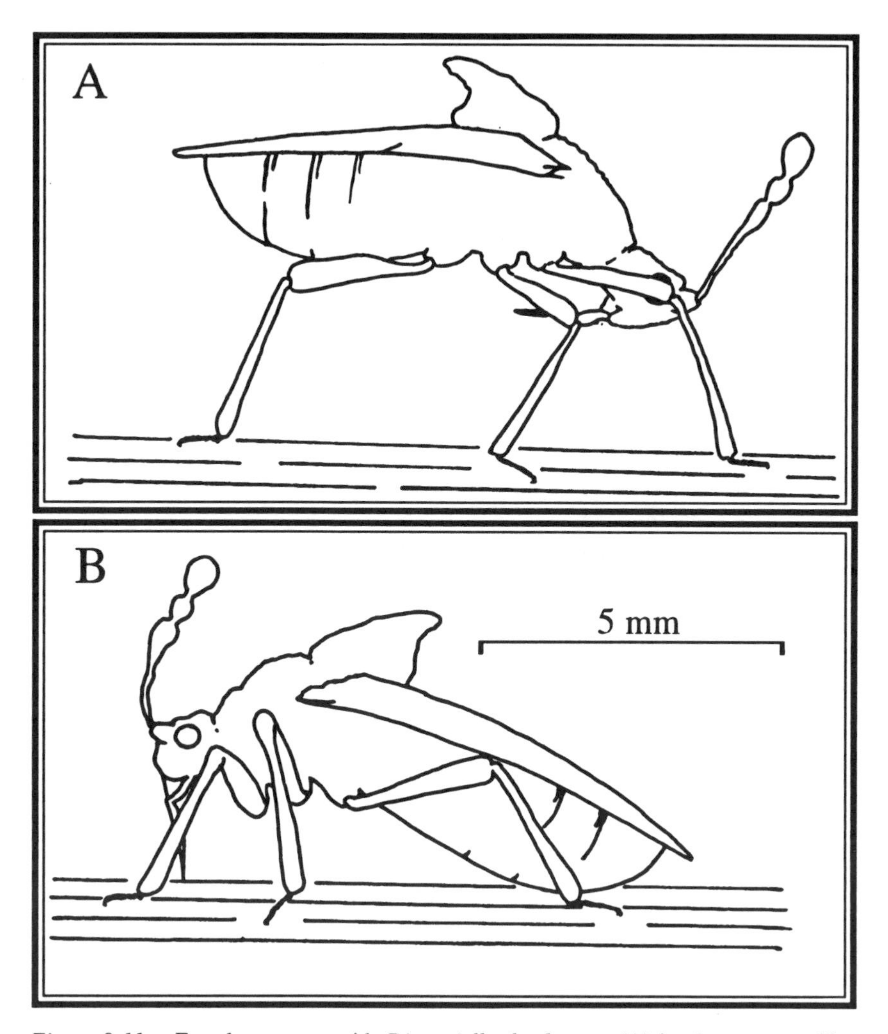

Figure 9.11. Female cocoa capsid, *Distantiella theobroma*, (A) in pheromone calling position, and (B) in feeding position [from King (1973)].

major constituents, and butyl butyrate is about three times more concentrated in airborne than thoracic extracts of females. Still, neither butyl butyrate alone or combined with hexyl butyrate are attractive. *(E)*-Crotyl butyrate [*(E)*-2-butenyl butyrate] has now been identified as a trace synergist (inactive alone); a 16:1 ratio of butyl butyrate with *(E)*-2-butenyl butyrate rivals the attractiveness of five live females (Smith et al., 1991; McBrien et al. 1994a, 1994b).

Esters are key components for several plant bugs, but pheromones for other mirids have yet to be fully deciphered [e.g., Hedin et al. (1985)]. In the predaceous mirid, *Pilophorus perplexus*, butyl and hexyl butyrates dominate the secretions of both sexes, with octyl acetate and an isomer of *(E)*-2-hexenyl butyrate being female-specific minor constituents (Knight et al., 1984b). Female *Lygus lineolaris* and *L. elisus* produce large and equivalent amounts of *(E)*-2-hexenyl and hexyl butyrates, whereas males produce much less of the unsaturated ester (Aldrich et al., 1988; Gueldner and Parrott, 1978), but no such dimorphism occurs in *L. hesperus* (Aldrich et al., 1988). Terpenyl acetates occur in one mirid; headspace analyses of agitated *Harpocera thoracica* adults showed that female-derived volatiles are enriched with geranyl and neryl acetates (Hanssen and Jacob, 1982, 1983). Male *H. thoracica* may be heavily reliant on a pheromone to find mates because they only live for a week, whereas females are long-lived and disperse aerially before ovipositing (Southwood and Leston, 1959).

From the foregoing account, it seems that mirids are another taxon in which the metathoracic scent gland has acquired a sexual function, with esters being the key components (Knight et al., 1984b; Aldrich, 1988a). However, Graham (1988) presented convincing evidence that pheromone in *L. hesperus* originates near the ovipositor. Judging from the sex pheromone system of the mullein bug, females of other mirids may well attract males with synergistic blends involving trace components. Still, in view of the richness of mirid species (>10,000 species), it is perhaps unrealistic to expect that there is but one type of pheromone system for the group.

3.7. Enicocephalidae

Unique-headed bugs are rare predators, and are the only bugs to form aerial leks (Hickman and Hickman, 1981; Stys, 1981). Males of this group have hypertrophied exocrine glands (Section 2.2). In addition, the anterior dorsal abdominal gland remains active in both sexes of adults (Aldrich, 1988a). Individual genera form male-dominated, female-dominated, or mixed swarms, with male-dominated leks possibly most common (Stys, 1981). Vision is important for mate location as males often have larger eyes than females (Kritsky, 1977). Pheromones of enicocephalids are unknown.

4. Exploitation of Pheromones as Kairomones

Orientation to plant volatiles is often an important step in the host-selection process (Tumlinson et al., 1992; Vinson, 1984), and for some Heteroptera, such as lygaeine specialists on Asclepiadaceae [e.g., Anderson (1991)], this phase is dominant. However, research on Pentatomidae shows that pheromones are exploited as kairomones to a surprising extent. For example, in three seasons, over 17,800 parasitoids were attracted to pheromone-baited traps for *P. maculiventris*—more than five times the number of spined soldier bugs caught (Aldrich, 1985).

4.1. Diptera

Tachinidae, Ceratopogonidae, Milichiidae, and Chloropidae flies are masters of chemical espionage when it comes to pentatomid pheromones (Table 9.3) (Aldrich, 1985, 1988a, 1988b).

4.1.1. Tachinidae

In the eastern United States, two tachinid species, *Euclytia flava* and *Hemyda aurata*, are highly attracted to *P. maculiventris* and *P. neglectus* pheromones (Fig. 9.12; Table 9.3). Both species lay macroeggs, principally on adults. *Hemyda aurata* is a specialist on *Podisus*, whereas *E. flava* also parasitizes several phytophagous stink bugs (Aldrich, 1985). Feral male spined soldier bugs usually have about four times more tachinid eggs on them when collected than do females, and some 10% of overwintered *P. maculiventris* are parasitized by *H. aurata* (Aldrich et al., 1984a). Surprisingly, in eight years of pheromone-trapping, not a single *E. flava* has been reared from spring-collected soldier bugs (Aldrich et al., 1984a, unpublished).

Despite the total absence of *E. flava* from overwintered soldier bugs, this species is invariably the first to appear at pheromone traps in the spring (Fig. 9.12). *Euclytia flava* males frequently perch on or nearby pheromone-baited live traps, but both sexes of *E. flava* are susceptible to sticky traps. Spined soldier bugs apparently modify their behavior to avoid parasitization. For example, *P. maculiventris* captures in live traps abruptly decline, and counts remain low for the rest of the season, once *E. flava* appears, in marked contrast to captures in sticky traps (Fig. 9.12). Spring emergence of *P. maculiventris* is past its peak by the time *E. flava* is present and, likewise, these flies arrive late for the emergence of new-generation *P. maculiventris* adults (Fig. 9.12). In addition, *E. flava* seems to not waste it's time on *P. neglectus* (Table 9.3) which, as a univoltine species confined to coniferous forests, is rare compared to *P. maculivetris*. In contrast, the specialist, *H. aurata*, is perfectly synchronized with the upsurge of new-generation adults at the end of May (Fig. 9.12), and attacks *P. neglectus* as well as *P. maculiventris* (Table 9.3).

Curiously, *H. aurata* will not enter live traps, and females greatly outnumber males in sticky traps, even though individuals captured from surrounding foliage are always male. On the other hand, the sexes of *E. flava* are equally susceptible to both types of trap when baited with pheromone containing racemic or (+)-α-terpineol. Nevertheless, these two tachinids behave similarly under natural conditions because in traps baited with live *P. maculiventris* males, the preponderance of *E. flava* are female (Aldrich et al., 1984a). High concentrations of synthetic pheromone appear to superstimulate *E. flava* males causing more of them to be trapped. *Euclytia flava* and *H. aurata* males are territorial around pheromone-calling spined soldier bugs—this interpretation is substantiated by

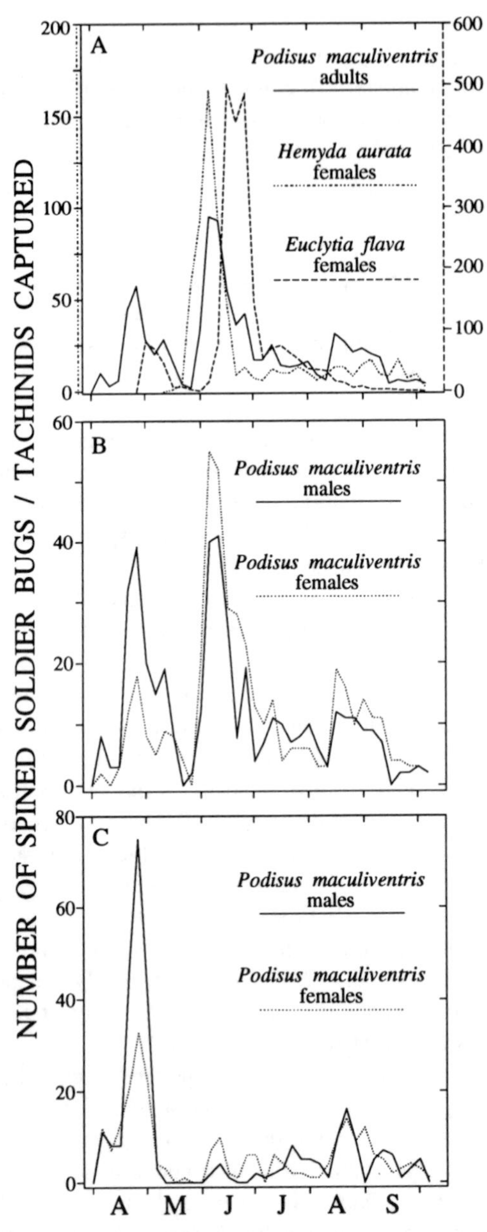

Figure 9.12. (A) Field attraction of *Podisus maculiventris* adults, and females of the tachinid parasitoids, *Euclytia flava* (scale on right) and *Hemyda aurata*, to seven sticky traps baited with synthetic *P. maculiventris* pheromone in 1983. *P. maculiventris* adults caught on or near seven sticky traps (B), or inside five live traps (C) baited with synthetic pheromone in 1983. Each point is the sum of catches for five consecutive days. All tests conducted at the Beltsville Agricultural Research Center (Aldrich et al., 1984a).

the fact that males of each tachinid are significantly larger than females (the opposite is true for *Podisus* spp.) (Aldrich, 1985).

By August, almost no *E. flava* are attracted to traps baited with *P. maculiventris* pheromone (Fig. 9.12); this generalist evidently shifts its attack to plant-feeding pentatomids toward the end of the growing season. For example, more than 90% of adult *Elasmostethus cruciatus* (Acanthosomatidae) collected during July and August were parasitized by *E. flava* (Jones and McPherson, 1980), and *E. flava* is one of a complex of tachinids that use the pheromone of *Euschistus* spp. (Section 3.1) as a host-finding kairomone (Aldrich et al., 1991a). Reproduction, as evidenced by pheromone responsiveness (Fig. 9.13), is skewed toward the early season in *P. maculiventris* to take advantage of the onslaught of foliage-feeding larvae, while populations of phytophagous stink bugs build up later in the season as host plants mature. Whether parasitoids switch from responding to one kairomone to another with time, or simply choose the extant kairomone at a given point in time, is an open question.

Other tachinids "home-in" on attractant pheromones of phytophagous stink bugs [e.g., Moriya and Masakazu (1984)], with the best-known case being attraction of *Trichopoda* spp. to pheromone-releasing *N. viridula* males (Aldrich

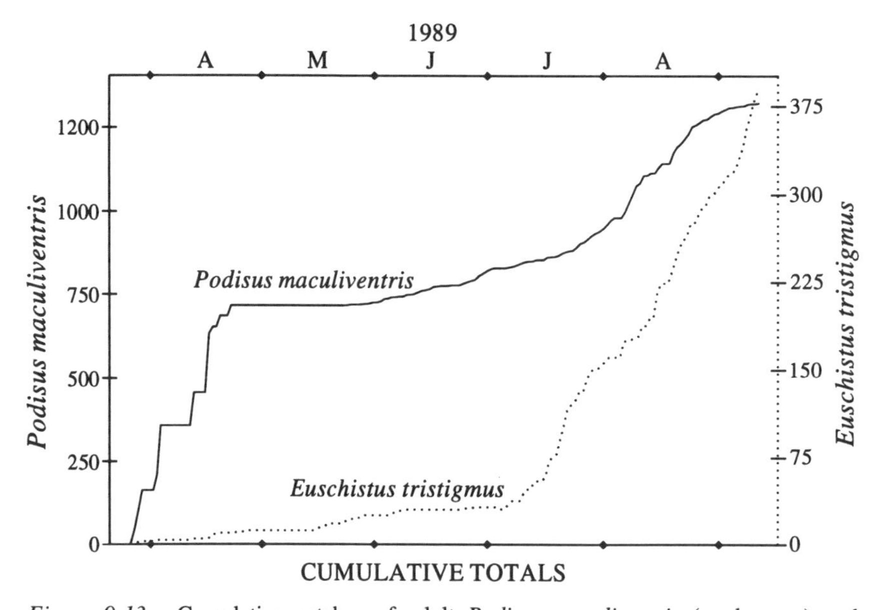

Figure 9.13. Cumulative catches of adult *Podisus maculiventris* (predaceous) and *Euschistus tristigmus* (phytophagous) in live traps baited with synthetic pheromones. Field test conducted at the Beltsville Agricultural Research Center; eight traps/each set; traps for *P. maculiventris* baited every three days with synthetic formulation (Table 3) (Aldrich et al., 1984a); traps for *E. tristigmus* baited every other day with 10 μl of a 0.3 mg/μl solution of methyl (2*E*,4*Z*)-decadienoate in heptane.

et al., 1987, 1989). In fact, the preference of *T. pennipes* females for male bugs, and the specificity of *Trichopoda* spp. for certain geographically isolated *N. viridula* populations, were early indicators that males produce a pheromone and that pheromone strains of *N. viridula* exist (Aldrich et al., 1993a; Todd, 1989). In the Americas, at least six tachinid species have adapted to *N. viridula* and are now more abundant on this exotic host than native hosts (Jones, 1988; Hokkanen, 1986), including *T. giacomellii* in Argentina which parasitizes *N. viridula* in a density dependent fashion (Liljesthrom and Bernstein, 1990). *Trichopoda pennipes* is the only Nearctic tachinid to accomplish this host shift. The chemical similarity of native *Acrosternum* spp. and *Nezara* pheromones (Section 3.1) almost certainly facilitated adoption of *N. viridula* by New World tachinids (Aldrich et al., 1989). Moreover, it is possible that pheromone strains of *N. viridula* evolved due to tachinid selection pressure.

Both predaceous and phytophagous pentatomid males risk tachinid attack when they advertise their presence with pheromones. Indeed, the risk of parasitization for *N. viridula* females (and probably females of other bugs) derives primarily from their association with males during courtship and copulation (McLain et al., 1990). *Podisus* males have pheromone glands capable of opening and closing (perhaps itself a response to parasite pressure), and a cheater male counterstrategy has apparently evolved whereby "silent" males respond to the pheromone of conspecifics in order to intercept females without incurring parasitization (Aldrich et al., 1984a; Aldrich, 1985). Parasite pressure is abated on the relatively uncommon *P. neglectus,* which may be the reason the noncalling strategy is depressed (Table 9.3) (Aldrich, 1988b). *Nezara viridula* males cannot precisely control pheromone emission, but in the 200 years since arriving in the United States, parasitism by *T. pennipes* seems to have shortened the preoviposition period and lengthened the fifth nymphal stage, thereby minimizing exposure to tachinids (Hokkanen, 1983).

4.1.2. Ceratopogonidae

Females of the ectoparasitic biting midge, *Forcipomyia crinita*, use the pheromone of *P. maculiventris* as a beacon to find potential hosts (Table 9.3) (Aldrich et al., 1984a). The flies swell with the blood like a mosquito while feeding through thick pronotal cuticle, apparently with little adverse affect on hosts. Bugs seem unaware of the presence of biting midges.

These ectoparasites are supreme generalists and eavesdroppers. A given ceratopogonid species often feeds on hosts from different orders (Wirth and Messersmith, 1971). Some ceratopogonids are attracted to stridulating wood-boring beetles (Wirth and Messersmith, 1971), while others go to cantharidin from blister beetles [ref. in Wirth (1980)]. Biting midges probably respond to different heteropteran pheromones such as that of *A. pennsylvanicum* (Section 3.1), a species also fed upon by *F. crinita* (Aldrich, unpublished).

4.1.3. Milichiidae and Chloropidae

Several species of small milichiid and chloropid flies routinely show up in traps baited with synthetic *P. maculiventris* pheromone (Table 9.4) (Aldrich, unpublished). By far the most frequently encountered is a new species of chloropid in the genus *Olcella* (Table 9.3).

Milichiids and chloropids, including an *Olcella* sp., fly to heteropterans trapped in spider webs when the bugs emit defensive secretion (Section 5.1). As such, there is some question whether attraction of these symbionts to synthetic *Podisus* pheromones is a nonadaptive by-product of the fact that the widespread allomonal component, *(E)*-2-hexenal, is also a *Podisus* pheromone component (Eisner et al., 1991). However, attraction to the soldier bug pheromone appears to be specific and adaptive because significantly more chloropids and milichiids are caught in traps baited with synthetic pheromone for *P. maculiventris* than for *P. neglectus* (Table 9.3; $c^2 = 30.7$, $p<0.005$). The association of some of these symbionts with asopines probably qualifies as kleptoparasitism (Vollrath, 1984) since individuals of *Olcella* n. sp. feed on prey captured by *P. maculiventris* inside live traps (Aldrich, unpublished). *Phyllomyza securicornis* (Milichiidae) is another consistent visitor (Table 9.3), and members of this genus are sometimes phoretic (Robinson and Robinson, 1977). Other milichiids are associated with assassin bugs (Richards, 1953; Robinson and Robinson, 1977).

4.2. Hymenoptera

Females of the egg parasitoid, *Telonomus calvus* (Hymenoptera: Scelionidae), come to the vicinity of *P. maculiventris* or *P. neglectus* males releasing pheromone and wait for a female bug to arrive, whereupon one or more wasps become phoretic on the inseminated bug (Aldrich, 1985; Orr et al., 1986a) (Table 9.3). Host-finding by *T. calvus* is unusual because it entails first orienting to a male-produced pheromone and then recognizing female hosts, possibly by the sex-specific odor from the small dorsal abdominal glands of females (Aldrich et al., 1984b; Aldrich, 1985). *Telonomus calvus* requires *Podisus* eggs <12 hr old to successfully develop (Orr et al., 1986a). Phoresy ensures that fresh eggs are found, and enables *T. calvus* to compete with larger, nonphoretic species having less stringent host-egg age requirements, such as *T. podisi* (Orr et al., 1986b).

Telonomus podisi parasitizes eggs of several phytophagous pentatomids and, in agroecosystems, is the dominant egg parasitoid of the spined soldier bug (Orr et al., 1986b; Orr, 1988). In contrast to tachinids, attraction of *T. podisi*, *Trissolcus basalis*, *T. euschistii*, and other nonphoretic oligophagous scelionids, to sexual pheromones (Section 3.1) must be weak, if it occurs at all (Borges and Aldrich, unpublished; Mattiacci et al., 1993). Nonetheless, a few *T. podisi* females were caught in pheromone-baited traps for *P. maculiventris* (Aldrich, 1985). Even though all previously identified long-range kairomones attractive to parasitoids are host sex pheromones (Tumlinson et al., 1992), other signals are more important for distant orientation by scelionid parasitoids of Heteroptera (Section 5.2).

5. Exploitation of Allomones as Kairomones

Data demonstrating exploitation of allomones by parasites or predators of Heteroptera are sparse or anecdotal, yet the phenomenon is undoubtedly real for certain Tachinidae, Milichiidae, and Chloropidae (Diptera); and probable for some Braconidae, Sphecidae, and Chrysididae (Hymenoptera). It is difficult to believe that heteropteran allomones are not exploited as kairomones by several species of tarsonemid mites (Prostigmata) which live in the metathoracic scent gland of Coreidae (Fain, 1979).

5.1. Diptera

Tachinids occasionally parasitize *Podisus* nymphs, suggesting that these flies use signals other than sex pheromones to find immature bugs. Since the contents of nymphal scent glands are shed with the exuviae (Section 2.1), an experiment was designed to determine if tachinids respond to the allomone of *P. maculiventris* nymphs (Aldrich, 1988b). Artificial nymphal stink gland secretion attracts both *E. flava* and *H. aurata*, albeit less so than artificial adult pheromone (Table 9.5). It is easy to imagine that responding to secretion leaching from exuviae is highly adaptive for tachinids because, in so doing, the flies find vulnerable hosts that will not ecdyse before their externally laid macroeggs have a chance to hatch.

Milichiid and chloropid flies, including species congeneric with those attracted to *Podisus* pheromones (Table 9.3), also profit by the emission of heteropteran allomones (Eisner et al., 1991; Sivinski and Stowe, 1980; Sivinski, 1984). Bugs

Table 9.5. Attraction of tachinid fly parasitoids to artificial Podisus maculiventris *adult pheromone*[a] *or nymphal allomone*[b] *(Aldrich, 1988b).*[c]

	Tachinid Species			
	Euclytia flava		*Hemyda aurata*	
Lure[d]	female	male	female	male
(*E*)-2-hexenal + benzyl alcohol + α-terpineol[a]	249	171	42	35
(*E*)-4-oxo-2-hexenal + linalool + tridecane[b]	88	8	11	18
linalool	0	0	0	2
unbaited	0	0	0	0

[c]Six live-traps and 6 sticky-traps/treatment monitored daily, April 25–June 17, 1986

[d]Blend ratios: (*E*)-2-hexenal/benzyl alcohol/α-terpineol as in Table 8.4, equal volumes of (*E*)-4-oxo-2-hexenal/linalool/tridecane; 10 μl/trap/day.

snared in webs of orb-weaving spiders, *Nephila clavipes*, spray scent gland secretion to little effect against the attacking spider, but flies zigzag their way from downwind toward the spider to share in the meal (Eisner et al., 1991). This scenario is repeated with *N. viridula* and other pentatomids and coreids producing *(E)*-2-hexenal; in fact, the synthetic compound alone is significantly attractive (Eisner et al., 1991). The nutritional cost of these symbionts to *Nephila* varies from being inconsequential to substantial, depending on age of the spider and the number of flies stealing food (Eisner et al., 1991). Thus, the relationship blurs the distinction between commensalism and kleptoparasitism (Vollrath, 1984).

Miltogrammine flies that attack sphecid wasps carrying heteropteran nymphs may prove to be an extremely convoluted example of message stealing (Section 5.2).

5.2. Hymenoptera

Peristenus and *Leiophron* spp. (Braconidae) parasitize the nymphs of many mirid species (Day 1987, personal communication), with certain hosts (Day and Saunders, 1990) and habitats (Snodgrass and Fayad, 1991) preferred over others. Circumstantial evidence suggests *Peristenus* spp. use the allomones of nymphs as kairomones. Females of both *P. pseudopallipes* (Lim and Stewart, 1976) and *P. stygicus* (Condit and Cate, 1982) sometimes seize lygus bug exuviae and attempt to oviposit in them. Furthermore, *P. stygicus* females attack Mirinae, Orthotylinae, and Phylinae species, but totally ignore species of Bryocorinae (Condit and Cate, 1982), a subfamily whose nymphs curiously lack external openings for their dorsal abdominal glands (Aryeetey and Kumar, 1973).

Another situation where allomones might undermine a bug's safety involves sand wasps (Sphecidae: Bembicini), some of which provision their nests exclusively with immature Heteroptera (Evans, 1966). Double espionage may occur here too by distantly related culprits: cuckoo wasps (Chrysididae), and miltogrammine flies (Sarcophagidae). Cuckoo wasps oviposit in heteropteran nymphs, before or after they are captured by stink bug hunters, as a means of ultimately parasitizing the sphecids (Carrillo S. and Caltagirone, 1970). Similarly, swarms of miltogrammine flies commonly larviposit on heteropteran prey as they are flown in the grasp of sphecids returning to their nests (Evans, 1966). The extent to which these wasps and flies rely on allomones to find prey is unknown.

Finally, a concrete case of chemical attraction by a parasitoid to a heteropteran allomone has been discovered for *Trissolcus basalis* (Scelionidae), the most important egg parasitoid of the southern green stink bug, *N. viridula*. Although *T. basalis* females are not appreciably drawn to the attractant pheromone of *N. viridula*, they are attracted to and stimulated by *(E)*-2-decenal, the main scent gland aldehyde of adult southern green stink bugs (Mattiacci et al., 1993). A second kairomone emanating from the glue used to fasten down egg masses comes into play at short range for final recognition (Bin et al., 1993). Pentatomids can be grouped according to the chain length of scent gland alkenals (Table 9.1). Therefore, differential long-range attraction to stink bug allomones may explain why some scelionids exhibit clear host preferences in the field [e.g., Orr et al. (1986b); Foerster and de Queiroz (1990)], but are indiscriminate in the laboratory (Orr, 1988; Jones, 1988).

The discovery of long-range parasitoid attraction to the allomone of *Nezara* is puzzling because ovipositing hosts do not normally emit scent gland secretion nor is *(E)*-2-decenal detectable from eggs (Borges and Aldrich, 1992; Mattiacci et al., 1993). It is possible that low levels of scent gland compounds give bugs a body odor sensed by parasitoids. Alternatively, wasps could be attracted to stink bug allomones because they are indicative of dense populations of potential hosts. The distinction between individual versus population-level explanations for parasitoid response to heteropteran allomones is more than an esoteric argument. Intensive efforts are under way to mass produce and to manage wild *T. basalis* for augmentation and conservation. Synthetic oviposition stimulants are essential to artificially rearing *T. basalis* (and other parasitoids) (Volkoff et al., 1992), and attractant kairomones are potentially invaluable tools for husbandry of wild parasitoids (Aldrich, 1991; Tumlinson et al., 1992).

6. Conclusions

Challenges abound for chemical ecologists of the Heteroptera. There is a long list of identified exocrines from a smattering of species in the order, and a short list of assigned functions for known molecules. Behavioral research on heteropteran semiochemicals needs to be emphasized, especially in the field. Pheromones of predaceous heteropterans, and kairomones for parasitoids of noxious heteropterans, are potentially powerful semiochemicals for agriculture in the 21st century.

Many true bugs are relatively large, with long-lived adults and, as such, are choice targets for parasites and predators. Protection provided by potent defensive secretions sometimes backfires by exposing Heteroptera to adapted enemies. Except for the Miridae, heteropteran pheromones identified to date come from males in species with large adults. Evolution of males as the attractive sex may, in part, be due to selection pressure from parasitoids. While, predaceous Heteroptera appear to be extremely reliant on pheromones to congregate, they suffer from a tremendous barrage of parasites homing in on their pheromones. Mechanisms allowing control over pheromone release evolved in some heteropteran species, enabling them to restrict exposure to parasitoids. This evolutionary step created the opportunity for males to cheat by retaining their pheromone and intercepting females responding to calling males.

Discovery of the first pheromones for moths renewed hope for biorational pest control. It is increasingly clear that both predators and parasites frequently rely on semiochemicals to find prey or hosts, and each other. In the Heteroptera, unique opportunities exist for managing beneficials with innocuous chemicals, particularly for key predators conspicuously absent from this discussion: pirate bugs (Anthocoridae), damsel bugs (Nabidae), and bigeyed bugs (Geocorinae).

Acknowledgments

I dedicate this chapter to Thomas R. Yonke who sparked my interest in Heteroptera. I will always be grateful to Murray S. Blum and William S. Bowers for allowing me the freedom to pursue my fascination with bugs. I also thank those who have listened to me talk about bugs over the years, especially my wife Barbara.

References

Aiken, R. B. (1985) Sound production by aquatic insects. *Biol. Rev.*, 60: 163–211.

Aldrich, J. R. (1985) Pheromone of a true bug (Hemiptera-Heteroptera): Attractant for the predator, *Podisus maculiventris*, and kairomonal effects. In: *Semiochemsitry: Flavors and Pheromones* (Acree, T. E. and Soderlund, D. M., eds.) pp. 95–119. de Gruyter, Berlin.

Aldrich, J. R. (1988a) Chemical ecology of the Heteroptera. *Annu. Rev. Entomol.*, 33: 211–38.

Aldrich, J. R. (1988b) Chemistry and biological activity of pentatomoid sex pheromones. In: *Biologically Active Natural Products: Potential Use in Agriculture* (Cutler, H. G., ed.) pp. 417–31. ACS Symposium Series No. 380, Washington, D.C.

Aldrich, J. R. (1991) Pheromones of good and bad bugs. *News Bull. Entomol. Soc. Queensland*, 19: 19–27.

Aldrich, J. R. and Lusby, W. R. (1986) Exocrine chemistry of beneficial insects: Male-specific secretions from predatory stink bugs (Hemiptera: Pentatomidae). *Comp. Biochem. Physiol.*, 85B: 639–42.

Aldrich, J. R. and Yonke, T. R. (1975) Natural products of abdominal and metathoracic scent glands of coreoid bugs. *Ann. Entomol. Soc. Am.*, 68: 955–60.

Aldrich, J. R., Blum, M. S. and Fales, H. M. (1979) Species specific male scents of leaf-footed bugs, *Leptoglossus* spp. (Heteroptera: Coreidae). *J. Chem. Ecol.*, 5: 53–62.

Aldrich, J. R., Kochansky, J. R. and Abrams, C. B. (1984a) Attractant for a beneficial insect and its parasitoids: Pheromone of the predatory spined soldier bug, *Podisus maculiventris* (Hemiptera: Pentatomidae). *Environ. Entomol.*, 13: 1031–6.

Aldrich, J. R., Kochansky, J. P. and Sexton, J. D. (1985) Chemical attraction of the eastern yellowjacket, *Vespula maculifrons* (Hymenoptera: Vespidae). *Experientia*, 41: 420–2.

Aldrich, J. R., Lusby, W. R. and Kochansky, J. P. (1986a) Identification of a new predaceous stink bug pheromone and its attractiveness to the eastern yellowjacket. *Experientia,* 42: 583–5.

Aldrich, J. R., Blum, M. S., Lloyd, H. A. and Fales, H. M. (1978) Pentatomid natural products: Chemistry and morphology of the III-IV dorsal abdominal glands of adults. *J. Chem. Ecol.,* 4: 161–72.

Aldrich, J. R., Kochansky, J. P., Lusby, W. R. and Borges, M. (1991b) Pheromone blends of predaceous bugs (Heteroptera: Pentatomidae: *Podisus* spp.). *Z. Naturforsch.,* 46C: 264–9.

Aldrich, J. R., Lusby, W. R., Kochansky, J. P. and Abrams, C. R. (1984b) Volatile compounds from the predatory insect *Podisus maculiventris* (Hemiptera: Pentatomidae): Male and female metathoracic scent gland and female dorsal abdominal gland secretions. *J. Chem. Ecol.,* 10: 561–8.

Aldrich, J. R., Neal, J. W., Jr., Oliver, J. E. and Lusby, W. R. (1991c) Chemical and maternalistic trade-offs in lace bugs (Heteroptera: Tingidae): Alarm pheromones and exudate defense in *Corythucha* and *Gargaphia* spp. *J. Chem. Ecol.,* 17: 2307–22.

Aldrich, J. R., Oliver, J. E., Lusby, W. R. and Kochansky, J. P. (1986b.) Identification of male-specific exocrine secretions from predatory stink bugs (Hemiptera: Pentatomidae). *Arch. Insect Biochem. Physiol.,* 3: 1–12.

Aldrich, J. R., Oliver, J. E., Lusby, W. R., Kochansky, J. P. and Borges, M. (1994) Identification of male-specific volatiles from Nearctic and Neotropical stink bugs (Heteroptera: Pentatomidae). *J. Chem. Ecol.*, 20: 1103–1111.

Aldrich, J. R., Oliver, J. E., Lusby, W. R., Kochansky, J. P. and Lockwood, J. A. (1987) Pheromone strains of the cosmopolitan pest, *Nezara viridula* (Heteroptera: Pentatomidae). *J. Exp. Zool.*, 244: 171–5.

Aldrich, J. R., Carroll, S. P., Lusby, W. R., Thompson, M. J., Kochansky, J. P. and Waters, R. M. (1990a) Sapindaceae, cyanolipids, and bugs. *J. Chem. Ecol.,* 16: 199–210.

Aldrich, J. R., Carroll, S. P., Oliver, J. E., Lusby, W. R., Rudmann, A. A. and Waters, R. M. (1990b) Exocrine secretions of scentless plant bugs: *Jadera, Boisea* and *Niesthrea* species (Hemiptera: Heteroptera: Rhopalidae). *Biochem. Syst. Ecol.,* 18: 369–76.

Aldrich, J. R., Hoffmann, M. P., Kochansky, J. P., Lusby, W. R., Eger, J. E. and Payne, J. A. (1991a) Identification and attractiveness of a major pheromone component for Nearctic *Euschistus* spp. stink bugs (Heteroptera: Pentatomidae). *Environ. Entomol.,* 20: 477–83.

Aldrich, J. R., Lusby, W. R., Kochansky, J. P., Hoffmann, M. P., Wilson, L. T. and Zalom, F. G. (1988) *Lygus* bug pheromones *vis-a-vis* stink bugs. *Proc. Beltwide Cotton Production Conferences,* 1988: 213–6.

Aldrich, J. R., Lusby, W. R., Marron, B. E., Nicolaou, K. C., Hoffmann, M. P. and Wilson, L. T. (1989) Pheromone blends of green stink bugs and possible parasitoid selection. *Naturwissenschaften,* 76: 173–5.

Aldrich, J. R., Numata, H., Borges, M., Bin, F., Waite, G. K. and Lusby, W. R. (1993b) Artifacts and pheromone blends from *Nezara* spp. and other stink bugs (Heteroptera: Pentatomidae). *Z. Naturforsch.,* 48C: 73–79.

Aldrich, J. R., Waite, G. K., Moore, C., Payne, J. A., Lusby, W. R. and Kochansky, J. P. (1993a) Male-specific volatiles from Nearctic and Australasian true bugs (Heteroptera: Coreidae and Alydidae). *J. Chem. Ecol.,* 19: 2767–81.

Anderson, D. B. (1991) Host choice and its consequences in the parasitic fly *Leucostoma crassa. Vaxtskyddsrap. Avhandlingar,* 22: 1–22.

Antich, A. V. (1968) Attraccion por alor en ninfas y adultos de *Rhodnius prolixus. Rev. Inst. Med. Trop. Sao Paulo,* 10: 242–6.

Aryeetey, E. A. and Kumar, R. (1973) Structure and function of the dorsal abdominal gland and defence mechanism in coca-capsids (Miridae: Heteroptera). *J. Entomol. Ser.,* 47A: 181–9.

Azfal, M. and Sahibzaba, S. S. (1989) Scent apparatus of superfamily Enicocephalioidea. *Biologia,* 35: 65–7.

Baker, R., Borges, M., Cooke, N. G. and Herbert, R. N. (1987) Identification and synthesis of *(Z)*-(1′S,3′R,4′S)(−)-2-(3′,4′-epoxy-4′-methylcyclohexyl)-6-methylhepta-2,5-diene, the sex pheromone of the southern green stinkbug, *Nezara viridula* (L.). *Chem. Commun.,* 1987: 414–6.

Barth, R. (1961) Sobre o orgao abdominal glandular de *Arilus carinatus* (Forster, 1771) (Heteroptera, Reduviidae). *Mem. Inst. Oswaldo Cruz,* 59: 37–43.

Bin, F., Vinson, S. B., Strand, M. R., Colazza, S. and Jones, W. A., Jr. (1993) Source of an egg kairomone for *Trissolcus basalis* a parasitoid of *Nezara viridula. Physiol. Entomol.,* 18: 7–15.

Blum, M. S. (1981) *Chemical Defenses of Arthropods.* Academic Press, New York.

Borges, M. and Aldrich, J. R. (1992) Instar-specific defensive secretions of stink bugs (Heteroptera: Pentatomidae). *Experientia,* 48: 893–6.

Borges, M. and Aldrich, J. R. (1994) Attractant pheromone for the Nearctic stink bug, *Euschistus obscurus* (Heteroptera: Pentatomidae): Insight into a Neotropical relative. *J. Chem. Ecol.,* 20: 1095–1102.

Borges, M., Jepson, P. C. and Howse, P. E. (1987) Long-range mate location and close-range courtship behaviour of the green stink bug, *Nezara viridula* and its mediation by sex pheromones. *Entomol. exp. appl.,* 44: 205–12.

Braekman, J. C., Daloze, A. and Pasteels, J. M. (1982) Cyanogenic and other glucosides in a neo-Guinean bug *Leptocoris isolata*: Possible precursors in its host-plant. *Biochem. Syst. Ecol.,* 10: 355–64.

Brézot, P., Malosse, C., Mori, K. and Renou, M. (1994) Bisabolene epoxides in sex pheromone in *Nezara viridula* (Heteroptera: Pentatomidae): Role of *cis* isomer and relation to specificity of pheromone. *J. Chem. Ecol.* 20: 3133–3147.

Butenandt, A. and Tam, N.-D. (1957) Uber einen geschlechtsspezifischen Duftstoff der Wasserwanze *Belostoma indica* Vitalis (*Lethocerus indicus* Lep.). *Hoppe-Seyler Z. Physiol. Chem.,* 308: 277–83.

Calam, D. H. and Scott, G. C. (1969) The scent gland complex of the adult cotton stainer bug, *Dysdercus intermedius. J. Insect Physiol.,* 15: 1695–1702.

Cane, D. E., Ha, H.-J., McIlwaine, D. B. and Pascoe, K. O. (1990) The synthesis of (3R)-nerolidol. *Tetrahedron Lett.*, 31: 7553–4.

Carayon, J. (1971) Notes et documents sur l'appareil odorant metathoracique des Hemipteres. *Ann. Soc. Entomol. Fran.*, 7: 737–70.

Carayon, J. (1981) Dimorphisme sexuel des glandes tegumentaires et production de pheromones chez les Hemipteres Pentatomoidea. *C. R. Acad. Sci. Ser. III*, 292: 867–70.

Carayon, J. (1984) Les androconies de certains Hemipteres Scutelleridae. *Ann. Soc. Entomol. Fran.*, 20: 113–134.

Carayon, J. (1987) *Reduvius (Pseudoreduvius) dicki* n. sp. et ses allies. Premiere mention de glandes a pheromones chez des hemipteres reduviides. *Rev. Fran. Entomol.*, 9: 190–6.

Carrillo S., J. L. and Caltagirone, L. E. (1970) Observations on the biology of *Solierella peckhami, S. blaisdelli* (Sphecidae), and two species of Chrysididae (Hymenoptera). *Ann. Entomol. Soc. Am.*, 63: 672–81.

Carroll, S. P. (1988) Contrasts in the reproductive ecology of tropical and temperate populations of *Jadera haematoloma* (Hemiptera: Rhopalidae), a mate-guarding insect. *Ann. Entomol. Soc. Am.*, 81: 54–63.

Carroll, S. P. (1991) The adaptive significance of mate guarding in the soapberry bug, *Jadera haematoloma* (Hemiptera: Rhopalidae). *J. Insect Behav.*, 4: 509–30.

Carroll, S. P. and Loye, J. E. (1990) Male-biased sex ratios, female promiscuity, and copulatory mate guarding in an aggregating tropical bug, *Dysdercus bimaculatus*. *J. Insect Behav.*, 3: 33–48.

Cobben, R. H. (1978) *Evolutionary Trends in Heteroptera. Part II. Mouthpart-structures and Feeding Strategies*. Veenman and Zonen, Wageningen, the Netherlands.

Condit, B. P. and Cate, J. R. (1982) Determination of host range in relation to systematics for *Peristenus stygicus* (Hymenoptera: Braconidae), a parasitoid of Miridae. *Entomophaga*, 27: 203–10.

Day, W. H. (1987) Biological control efforts against *Lygus* and *Adelphocoris* spp. infesting alfalfa in the United States, with notes on other associated mirid species. In: *Economic Importance and Biological Control of* Lygus *and* Adelphocoris *in North America* (Hedlund, R. C. and Graham, H. M., eds.) pp. 20–39. USDA ARS-64, Washington, D.C.

Day, W. H. and Saunders, L. B. (1990) Abundance of the garden fleahopper (Hemiptera: Miridae) on alfalfa and parasitism by *Leiophron uniformis* (Gahan) (Hymenoptera: Braconidae). *J. Econ. Entomol.*, 83: 101–6.

Daroogheh, H. and Olagbemiro, T. O. (1982) Linalool from the cotton stainer *Dysdercus superstitiosus* (F.) (Heteroptera: Pyrrhocoridae). *Experientia*, 38: 421–2.

Dingle, H. (1985) Migration. In: *Comprehensive Insect Physiology, Biochemistry and Pharmacology*, Vol. 9 (G. A. Kerkut and L. I. Gilbert, eds.) pp. 193–224. Pergamon, Oxford.

Dingle, H. (1989) The evolution and significance of migratory flight. In: *Insect Flight* (G. J. Goldsworthy and C. H. Wheeler, eds.) pp. 99–114. CRC Press, Boca Raton, Fla.

Eisner, T., Eisner, M. and Deyrup, M. (1991) Chemical attraction of kleptoparasitic flies to heteropteran insects caught by orb-weaving spiders. *Proc. Natl. Acad. Sci. USA*, 88: 8194–7.

Evans, C. E., Staddon, B. W. and Games, D. E. (1990) Analysis of gland secretions of Pentatomoidea (Heteroptera) by gas chromatography-mass spectrometry techniques. In: *Chromatography and Isolation of Insect Hormones and Pheromones* (A. R. McCaffery and I. D. Wilson, eds.) pp. 321–8. Plenum Press, New York.

Evans, H. E. (1966) *Bicyrtes*, a genus of stinkbug-hunters. In: *The Comparative Ethology and Evolution of the Sand Wasps* (Evans, H. E.) pp.144–75. Harvard University Press, Cambridge, Mass.

Everton, I. J. and Staddon, B. W. (1979) The accessory gland and metathoracic scent gland function in *Oncopeltus fasciatus*. *J. Insect Physiol.*, 25: 133–41.

Everton, I. J., Knight, D. W. and Staddon, B. W. (1979) Linalool from the metathoracic scent gland of the cotton stainer *Dysdercus intermedius* Distant (Heteroptera: Pyrrhocoridae). *Comp. Biochem. Physiol.*, 63B: 157–61.

Fain, A. (1979) Les Coreitarsoneminae (Prostigmata, Tarsonemidae) parasites de la glands odoriferante d'Hemipteres de la famille Coreidae. *Acarologia*, 21: 279–90.

Farine, J. P. (1987) The exocrine glands of *Dysdercus cingulatus* (Heteroptera, Pyrrhocoridae): Morphology and function of nymphal glands. *J. Morphol.*, 194: 195–207.

Farine, J. P., Bonnard, O., Brossut, R. and Le Quere, J. L. (1992a) Chemistry of pheromonal and defensive secretions in the nymphs and the adults of *Dysdercus cingulatus* Fabr. (Heteroptera, Pyrrhocoridae). *J. Chem. Ecol.*, 18: 65–76.

Farine, J. P., Bonnard, O., Brossut, R. and Le Quere, J. L. (1992b) Chemistry of defensive secretions in nymphs and adults of fire bug, *Pyrrhocoris apterus* L. (Heteroptera, Pyrrhocoridae). *J. Chem. Ecol.*, 18: 1673–82.

Feir, D. (1974) *Oncopeltus fasciatus*: A research animal. *Annu. Rev. Entomol.*, 19: 81–96.

Foerster, L. A. and de Queiroz, J. M. (1990) Incidencia natural de parasitismo em ovos de pentatomideos da soja no centro-sul do Parana. *An. Soc. Entomol. Brasil*, 19: 221–31.

Gilby, A. R. and Waterhouse, D. F. (1965) The composition of the scent of the green vegetable bug, *Nezara viridula*. *Proc. R. Soc.*, 162B: 105–120.

Games, D. E. and Staddon, B. W. (1973) Chemical expression of a sexual dimorphism in the tubular scent glands of the milkweed bug, *Oncopeltus fasciatus* (Dallas) (Heteroptera; Lygaeidae). *Experientia*, 29: 532–533.

Gough, A. J. E., Games, D. E., Staddon, B. W. and Olagbemiro, T. O. (1985a) Male produced volatiles from coreid bug *Leptoglossus australis* (Heteroptera). *Z. Naturforsch.*, 40C: 142–4.

Gough, A. J. E., Hamilton, J. G. C., Games, D. E. and Staddon, B. W. (1985b) Multichemical defense of plant bug *Hotea gambiae* (Westwood) (Heteroptera: Scutelleridae). Sesquiterpenoids from abdominal gland in larvae. *J. Chem. Ecol.*, 11: 343–52.

Gough, A. J. E., Games, D. E., Staddon, B. W., Knight, D. W. and Olagbemiro, T. O. (1986) C9 aliphatic aldehydes: Possible sex pheromone from male tropical West African shield bug, *Sphaerocoris annulus*. *Z. Naturforsch.*, 41C: 93–6.

Graham, H. M. (1987) Attraction of *Lygus* spp. males by conspecific and congeneric females. *Southwest. Entomol.,* 12: 147–55.

Graham, H. M. (1988) Sexual attraction of *Lygus hesperus* Knight. *Southwest. Entomol.,* 13: 31–6.

Gueldner, R. C. and Parrott, W. L. (1978) Volatile constituents of the tarnished plant bug. *Insect Biochem.,* 8: 389–91.

Hamilton, J. G. C., Gough, A. J. E., Staddon, B. W. and Games, D. E. (1985) Multichemical defense of plant bug *Hotea gambiae* (Westwood) (Heteroptera: Scutelleridae): *(E)*-2-hexenol from abdominal gland in adults. *J. Chem. Ecol.,* 11: 1399–1409.

Hanssen, H.-P. and Jacob, J. (1982) Monoterpenes from the true bug *Harpocera thoracica* (Hemiptera). *Z. Naturforsch.,* 37C: 1281–2.

Hanssen, H.-P. and Jacob, J. (1983) Sex-dependent monoterpene formation in *Harpocera thoracica* (Hemiptera). *Naturwissenschaften,* 70: 622–3.

Harrington, B. J. (1991) *Ligyrocoris* (Heteroptera: Lygaeidae: Rhyparochrominae) male-produced scents suggest a biochemical character system for systematic analysis. *J. New York Entomol. Soc.,* 99: 478–86.

Hebbalkar, D. S. and Sharma, R. N. (1982) An experimental analysis of mating behavior of the bug *Dysdercus koenigii* F. (Hemiptera: Pyrrhocoridae). *Indian J. Exp. Biol.,* 20: 399–405.

Hedin, P. A., W. L. Parrott, W. L., Teders, W. L. and Reed, D. K. (1985) Field responses of the tarnished plant bug to its own volatile constituents. *J. Miss. Acad. Sci.,* 30: 63–6.

Henry, T. J. and Froeschner, R. C. (1988) *Catalog of the Heteroptera, or True Bugs, of Canada and the Continental United States.* Brill, New York.

Hickman, V. V. and Hickman, J. L. (1981) Observations on the biology of *Oncylocotis tasmanicus* (Westwood) with descriptions of the immature stages (Hemiptera, Enicocephalidae). *J. Nat. Hist.,* 15: 703–15.

Hokkanen, H. (1983) The effect of 200 years of coevolution between the green stink bug, *Nezara viridula* (Hemiptera, Pentatomidae), and the parasite *Trichopoda pennipes* (Diptera, Tachinidae) on the host-parasite relationship. In: *Interspecific Homeostasis, Pest Problems, and the Principles of Classical Biological Pest Control* (Hokkanen, Ph.D. Thesis), pp. 85–132. Cornell University, Ithaca, New York.

Hokkanen, H. (1986) Polymorphism, parasites, and the native area of *Nezara viridula* (Hemiptera, Pentatomidae). *Ann. Entomol. Fenn.,* 52: 28–31.

James, D. G. (1992) Effect of temperature on development and survival of *Pristhesancus plagipennis* (Hem.: Reduviidae). *Entomophaga,* 37: 259–64.

James, D. G. and Warren, G. N. (1989) Sexual dimorphism of dorsal abdominal glands in *Biprorulus bibax* Breddin (Hemiptera: Pentatomiade). *J. Aust. Entomol. Soc.,* 28: 75–6.

James, D. G., Aldrich, J. R., Oliver, J. E. and Moore, C. (1991) Attractant pheromones in the spined citrus bug, and an assassin bug predator: Prospects for spined citrus bug management. *Proc. 1st Natl. Conf. Aust. Soc. Hort. Sci. (Sydney),* pp. 369–73.

James. D. G., Moore, C. J. and Aldrich, J. R. (1994b) Identificiation, synthesis, and bioactivity of a male-produced aggregation pheromone in assassin bug, *Pristhesancus plagipennis* (Hemiptera: Reduviidae). *J. Chem. Ecol.* 20: 3281–3295.

James, D. G., Mori, K., Aldrich, J. R. and Oliver, J. E. (1994a) Flight-mediated attraction of *Biprorulus bibax* (Breddin) (Hemiptera: Pentatomidae) to natural and synthetic aggregation pheromone. *J. Chem. Ecol.,* 20: 71–80.

Jones, W. A., Jr. (1988) World review of the parasitoids of the southern green stink bug, *Nezara viridula* (L.) (Heteroptera: Pentatomidae). *Ann. Entomol. Soc. Am.,* 81: 262–73.

Jones, W. A., Jr. and McPherson, J. E. (1980) The first report of the occurrence of acanthosomatids in South Carolina. *J. Georgia Entomol. Soc.,* 15: 286–9.

King, A. B. S. (1973) Studies of sex attraction in the cocoa capsid, *Distantiella theobroma* (Heteroptera: Miridae). *Entomol. Exp. Appl.,* 16: 243–54.

Knight, D. W., Rossiter, M. and Staddon, B. W. (1984a) *(Z,E)*-α-Farnesene: Major component of secretion from metathoracic scent gland of cotton seed bug, *Oxycarenus hyalinipennis* (Costa) (Heteroptera; Lygaeidae). *J. Chem. Ecol.,* 10: 641–8.

Knight, D. W., Rossiter, M. and Staddon, B. W. (1984b) Esters from the metathoracic scent gland of two capsid bugs, *Pilophorus perplexus* Douglas and Scott and *Blepharidopterus angulatus* (Fallen) (Heteroptera: Miridae). *Comp. Biochem. Physiol.,* 78B: 237–9.

Knight, D. W., Staddon, B. W. and Thorne, M. J. (1985) Presumed sex pheromone from androconial glands of male cotton harlequin bug *Tectocoris diophthalmus* (Heteroptera; Scutelleridae) identified as 3,5-dihydroxy-4-pyrone. *Z. Naturforsch.,* 40C: 851–3.

Kochansky, J. P., Aldrich, J. R. and Lusby, W. R. (1989) Synthesis and pheromonal activity of 6,10,13-trimethyltetradecanol for the predatory stink bug, *Stiretrus anchorago* (Heteroptera, Pentatomidae). *J. Chem. Ecol.,* 15: 1717–28.

Kon, M., Oe, A. and Numata, H. (1993) Intra- and interspecific copulations in the two congeneric green stink bugs, *Nezara antennata* and *N. viridula* (Heteroptera, Pentatomidae), with reference to postcopulatory changes in the spermatheca. *Japan Ethol. Soc.,* 11: 83–9.

Kon, M., Oe, A. and Numata, H. (1994) Ethological isolation between the two congeneric green stink bugs, *Nezara antennata* and *N. viridula* (Heteroptera, Pentatomidae). *Japan Ethol. Soc.,* 12: 67–71.

Kritsky, G. (1977) Observations on the morphology and behavior of the Enicocephalidae (Hemiptera). *Entomol. News,* 88: 105–10.

Leal, W. S. and Kadosawa, T. (1992) *(E)*-2-Hexenyl hexanoate, the alarm pheromone of the bean bug *Riptortus clavatus* (Heteroptera: Alydidae). *Biosci. Biotech. Biochem.,* 56: 1004–5.

Leal, W. S., Kadosawa, T. and Nakamori, H. (1993a) Aggregation pheromone of the bean bug, *Riptortus clavatus*. *Proc. 37th Ann. Meeting Japan Soc. Biosci., Biotech. & Agrochem.,* p. 212.

Leal, W. S., Nakamori, H. and Kadosawa, T. (1993b) Ecological significance of *Riptortus clavatus* aggregation pheromone to adults and nymphs. *Proc. 37th Ann. Meeting Soc. Appl. Entomol. Zool.,* p. 9.

Lener, W. (1969) Pheromone secretion in the large milkweed bug, *Oncolpeltus fasciatus*. *Am. Zool.*, 9: 1143.

Lim, K. P. and Stewart, R. K. (1976) Laboratory studies on *Peristenus pallipes* and *P. pseudopallipes* (Hymenoptera: Braconidae), parasitoids of the tarnished plant bug, *Lygus lineolaris* (Hemiptera: Miridae). *Can. Entomol.*, 108: 815–21.

Liljesthrom, G. and Bernstein, C. (1990) Density dependence and regulation in the system *Nezara viridula* (L.) (Hemiptera: Pentatomidae), host and *Trichopoda giacomellii* (Blanchard) (Diptera: Tachinidae), parasitoid. *Oecologia*, 84: 45–52.

MacLeod, J. K., Howe, I., Cable, J., Blake, J. D., Baker, J. T. and Smith, D. (1975) Volatile scent gland components of some tropical Hemiptera. *J. Insect Physiol.*, 21: 1219–24.

Marron, B. E. and Nicolaou, K. C. (1989) Stereocontrolled synthesis of the sex pheromone of the green stink bug. *Synthesis Commun.*, 1989: 537–9.

Mattiacci, L., Vinson, S. B., Williams, H. J., Aldrich, J. R. and Bin, F. (1993) A long range attractant kairomone for the egg parasitoid *Trissolcus basalis*, isolated from the defensive secretion of its host, *Nezara viridula*. *J. Chem. Ecol.*, 19: 1167–81.

McBrien, H. L., Judd, G. J. R. and Borden, J. H. (1994a) *Campylomma verbasci* (Heteroptera: Miridae): Pheromone-based seasonal flight patterns and prediction of nymphal densities in apple orchards. *J. Econ. Entomol.* 87: 1224–1229.

McBrien, H. L., Judd, G. J. R., Borden, J. H. and Smith, R. F. (1994b) Development of sex pheromone-baited traps for monitoring *Campylomma verbasci* (Heteroptera: Miridae) *Environ. Entomol.* 23: 442–446.

McLain, D. K. (1992) Preference for polyandry in female stink bugs, *Nezara viridula* (Hemiptera: Pentatomidae). *J. Insect Behav.*, 5: 403–10.

McLain, D. K., March, N. B., Lopez, J. R. and Drawdy, J. A. (1990) Intravernal changes in the level of parasitization of the southern green stink bug (Hemiptera: Pentatomidae), by the feather-legged fly (Diptera: Tachinidae): Host sex, mating status, and body size as correlated factors. *J. Entomol. Sci.*, 25: 501–9.

Moore, C. J., Aldrich, J. R. and James, D. G. (1993) Some aspects of the scent gland chemistry of the assassin bug *Pristhesancus plagipennis* Walker (Hemiptera: Reduviidae). In: *Pest control and Sustainable Agriculture* (Corey, S. A., Dall, D. J. and Milne, W. M., eds.) pp. 300–2. CSIRO Publications, Australia.

Mori, K. and Wu, J. (1991) Synthesis of the four stereoisomers of 6,10,13-trimethyl-1-tetradecanol, the aggregation pheromone of predatory stink bug, *Stiretrus anchorago*. *Liebigs Ann. Chem.*, 1991: 783–8.

Mori, K., Amaike, M. and Oliver, J. E. (1992) Synthesis and absolute configuration of the hemiacetal pheromone of the spined citrus bug *Biprorulus bibax*. *Liebigs Ann. Chem.*, 1992: 1185–90.

Moriya, S. and Masakazu, S. (1984) Attraction of the male brown-winged green bug, *Plautia stali* Scott (Heteroptera: Pentatomidae) for males and females of the same species. *Appl. Entomol. Zool.*, 19: 317–22.

Numata, H., Kon, M. and Hidaka, T. (1990) Male adults attract conspecific adults in the bean bug, *Riptortus clavatus* Thunberg (Heteroptera: Alydidae). *Appl. Ent. Zool.*, 25: 144–5.

Olagbemiro, T. O. and Staddon, B. W. (1983) Isoprenoids from the metathoracic scent gland of cotton seed bug, *Oxycarenus hyalinipennis* (Costa) (Heteroptera: Lygaeidae). *J. Chem. Ecol.*, 9: 1397–1412.

Oliver, J. E., Aldrich, J. R., Lusby, W. R., Waters, R. M. and James, D. G. (1992) A male-produced pheromone of the spined citrus bug. *Tetrahedron Lett.*, 33: 891–4.

Ondarza, R. N., Gutierrez-Martinez, A. and Malo, E. A. (1986) Evidence for the presence of sex and aggregation pheromones from *Triatoma mazzottii* (Hemiptera: Reduviidae). *J. Econ. Entomol.*, 79: 688–92.

Orr, D. B. (1988) Scelionid wasps as biological control agents: A review. *Fla. Entomol.*, 71: 506–28.

Orr, D. B., Russin, J. S. and Boethel, D. J. (1986a) Reproductive biology and behavior of *Telenomus calvus* (Hymenoptera: Scelionidae), a phoretic egg parasitoid of *Podisus maculiventris* (Hemiptera: Pentatomidae). *Can. Entomol.*, 118: 1063–72.

Orr, D. B., Russin, J. S., Boethel, D. J. and Jones, W. A. (1986b) Stink bug (Hemiptera: Pentatomidae) egg parasitism in Louisiana soybeans. *Environ. Entomol.*, 15: 1250–4.

Orton, D. A. and Green, T. L. (1989) *Coincide*. Plantsmen's Publications, Flossmoor, Ill.

Pavis, C. (1986) *Aspects de la Communication Pheromonale et de la Sensibilite Olfactive chez un Heteroptere Pentatomidae:* Nezara viridula *(L.)*. Ph.D. Thesis, Universite de Paris-Sud Centre d'Orsay, Paris.

Pavis, C. and C. Malosse, C. (1986) Mise en evidence d'un attractif sexuel produit par les males de *Nezara viridula* (L.) (Heteroptera: Pentatomidae). *C. R. Acad. Sci. Ser. III*, 303: 273–6.

Payne, J. A., Mead, F. W. and King, E. W. (1968) Hemiptera associated with pig carrion. *Ann. Entomol. Soc. Am.*, 61: 565–567.

Ralph, C. P. (1977) Search behavior of the large milkweed bug, *Oncopeltus fasciatus* (Hemiptera: Lygaeidae). *Ann. Entomol. Soc. Am.*, 70: 337–42.

Richards, O. W. (1953) A communication on commensalism of *Desmometopa* (Diptera: Milichiidae) with predacious insects and spiders. *Proc. Roy. Entomol. Soc. Lond.*, 18C: 55–6.

Robinson, M. H. and Robinson, B. (1977) Associations between flies and spiders: Bibio-commensalism and dipsoparasitims?. *Psyche*, 84: 150–7.

Rojas, J. C., Ramirez-Rovelo, A. and Cruz-Lopez, L. (1991) Search for a sex pheromone in *Triatoma mazzottii* (Hemiptera: Reduviidae). *J. Med. Entomol.*, 28: 469–70.

Rossiter, M. and Staddon, B. W. (1983) 3-Methyl-2-hexanone from the triatomine bug *Dipetalogaster maximus* (Uhler) (Heteroptera; Reduviidae). *Experientia*, 39: 380–1.

Rudmann, A. A. and Aldrich, J. R. (1988) Chirality determinations for a tertiary alcohol: Ratios of linalool enantiomers in insects and plants. *J. Chromatogr.*, 407: 324–9.

Schaefer, C. W. (1980) The host plants of the Alydinae, with a note on heterotypic feeding aggregations (Hemiptera: Coreoidea: Alydidae). *J. Kans. Entomol. Soc.*, 53: 115–122.

Schaefer, C. W. (1988) Reduviidae (Hemiptera: Heteroptera) as agents of biological control. *BICOVAS Proc.*, 1: 27–33.

Schuh, R. T. (1986) The influence of cladistics on heteropteran classification. *Annu. Rev. Entomol.*, 31: 67–93.

Sivinski, J. (1984) Mating by kleptoparasitic flies (Diptera: Chloropidae) on a spider host. *Fla. Entomol.*, 68: 216–22.

Sivinski, J. and Stowe, M. (1980) A kleptoparasitic cecidomyiid and other flies associated with spiders. *Psyche,* 87, 337–48.

Slater, J. A. and Baranowski, R. M. (1978) *How to Know the True Bugs*. Brown, Dubuque, Iowa.

Smith, R. F., Pierce, Jr., H. D. and Borden, J. H. (1991) Sex pheromone of the mullein bug, *Campyloma verbasci* (Meyer) (Heteroptera: Miridae). *J. Chem. Ecol.,* 17: 1437–47.

Snodgrass, G. L. and Fayad, Y. H. (1991) Euphorine (Hymenoptera: Braconidae) parasitism of the tarnished plant bug (Heteroptera: Miridae) in areas of Washington County, Mississippi disturbed and undisturbed by agricultural production. *J. Entomol. Sci.*, 26: 350–6.

Southwood, T. R. E. and Leston, D. (1959) *Land and Water Bugs of the British Isles*. Warne, London.

Staddon, B. W. (1979) The scent glands of Heteroptera. *Adv. Insect Physiol.*, 14: 351–418.

Staddon, B. W. (1986) Biology of scent glands in the Hemiptera-Heteroptera. *Ann. Soc. Entomol. Fr.*, 22: 183–90.

Staddon, B. W. (1990) Male sternal pheromone glands in acanthosomatid shield bugs from Britain. *J. Chem. Ecol.*, 16: 2195–2201.

Staddon, B. W. and Edmunds, M. G. (1991) Gland regional and spacial patterns in the abdominal sternites of some pentatomoid Heteroptera. *Ann. Soc. Entomol. Fran.*, 27: 189–203.

Staddon, B. W., Abdollahi, A., Parry, G., Rissiter, M. and Knight, D. W. (1994) Major component in male sex pheromone of cereal pest *Eurygaster integriceps* Puton (Heteroptera: Sculelleridae) identified as a homosesquiterpenoid. *J. Chem. Ecol.*, 20:2721–2731.

Staddon, B. W., Thorne, M. J. and Knight, D. W. (1987) The scent glands and their chemicals in the aposematic cotton harlequin bug, *Tectocoris diophthalmus* (Heteroptera: Scutelleidae). *Aust. J. Zool.*, 35: 227–34.

Staddon, B. W., Gough, A. J. E., Olagbemiro, T. O. and Games, D. E. (1985) Sex dimorphism for ester production in the metathoracic scent gland of the lygaeid bug *Spilostethus rivularis* (Germar) (Heteroptera). *Comp. Biochem. Physiol.*, 80B: 235–239.

Strong, F. E., Sheldahl, J. A., Hughes, P. R. and Hussein, E. M. K. (1970) Reproductive biology of *Lygus hesperus* Knight. *Hilgardia,* 40: 105–47.

Stys, P. (1981) Unusual sex ratios in swarming and light-attracted Enicocephalidae (Heteroptera). *Acta Entomol. Bohemoslov.*, 78: 430–2.

Sweet, M. H. (1964) The biology and ecology of the Rhyparochrominae of New England (Heteroptera: Lygaeidae), Part II. *Entomol. Am.*, 44: 1–201.

Tanaka, S. and Wolda, H. (1988) Oviposition behavior and diel rhythms of flight and reproduction in two species of tropical seed bugs. *Proc. Kon. Nederl. Akad. Wetensch.*, 91C: 165–74.

Thistlewood, H. M. A., Borden, J. H., Smith, R. F., Pierce, Jr., H. D. and McMullen, R. D. (1989) Evidence for a sex pheromone in the mullein bug, *Campyloma verbasci* (Meyer) (Heteroptera: Miridae). *Can. Entomol.*, 121: 737–44.

Thomas, D. B. (1992) *Taxonomic Synopsis of the Asopine Pentatomidae (Heteroptera) of the Western Hemisphere*. The Thomas Say Foundation, Volume XVI, Entomol. Soc. Am., Lanham, MD.

Todd, J. W. (1989) Ecology and behavior of *Nezara viridula*. *Annu. Rev. Entomol.*, 34: 273–92.

Tomioka, H. and Mori, K. (1992) Pheromone synthesis. 138. Simple synthesis of a diastereomeric mixture of 2-(3′,4′-epoxy-4′-methylcyclohexyl)-6-methyl-2,5-heptadiene, the pheromone of the green stink bug. *Biosci. Biotech. Biochem.*, 56: 1001.

Tumlinson, J. H., Turlings, T. C. J. and Lewis, W. J. (1992) The semiochemical complexes that mediate insect parasitoid foraging. *Agricult. Zool. Rev.*, 5: 221–52.

Turlings, T. C. J., Tumlinson, J. H., Heath, R. R., Proveaux, A. T. and Doolittle, R. E. (1990) Isolation and identification of allelochemicals that attract the larval parasitoid, *Cotesia marginiventris* (Cresson), to the microhabitat of one of its hosts. *J. Chem. Ecol.*, 17: 2235–2251.

Vinson, S. B. (1984) Parasitoid-host relationship. In: *Chemical Ecology of Insects* (Bell, W. J. and Carde, R. T., eds.) pp. 205–33. Chapman and Hall, New York.

Volkoff, N., Vinson, S. B., Wu, Z. X. and Nettles, W. C., Jr. (1992) *In vitro* rearing of *Trissolcus basalis* [Hym., Scelionidae] an egg parasitoid of *Nezara viridula* [Hem., Pentatomidae]. *Entomophaga,* 37: 141–8.

Vollrath, F. (1984) Kleptobiotic interactions in invertebrates. In: *Producers and Scroungers: Strategies of Exploitation and Parasitism* (Barnard, C., ed.) pp. 61–94. Croom Helm, London.

Wilcox, R. S. (1979) Sex discrimination in *Gerris remigis*: Role of a surface wave signal. *Science,* 206: 1325–1327.

Wilcox, R. S. and Stefano, J. D. (1991) Vibratory signals enhance mate-guarding in a water strider (Hemiptera: Gerridae). *J. Insect Behav.*, 4: 43–50.

Wirth, W. W. (1980) A new species and corrections in the *Atrichopogon* midges of the subgenus *Meloehelea* attacking blister beetles (Diptera: Ceratopogonidae). *Proc. Entomol. Soc. Wash.*, 82: 124–39.

Wirth, W. W. and Messersmith, D. H. (1971) Studies on the genus *Forcipomyia*. 1. The North American parasitic midges of the subgenus *Trichohelea* (Diptera: Ceratopogonidae). *Ann. Entomol. Soc. Am.*, 64: 15–26.

Yasuda, K. and Tsurumachi, M. (1994) Immigration of leaf-footed plant bug, *Leptoglossus australis* (Fabricius) (Heteroptera: Coreidae), into cucurbit fields on Ishigaki Island, Japan, and role of male pheromone. *Jpn. J. Appl. Entomol. Zool.* 38: 161–67.

10

Propaganda, Crypsis, and Slave-making

Ralph W. Howard, USDA-ARS
U.S. Grain Marketing Research Laboratory

Roger D. Akre[1]
Department of Entomology, Washington State University

1. Introduction and Definitions

In complex ecological communities, the ability to survive and reproduce is often tenuous at best. Many organisms do so by simply "bullying" their way through life, but for many others a more furtive approach is necessary. These latter animals resort to either blending into their environment, or to mimicking specific species-characteristic cues of other, usually dominant, members of their community. Sometimes, rather than taking such a defensive approach, they resort instead to an offensive approach, using what have been called propaganda substances. They may even combine both approaches as has been found for some slave-making ants. Although a variety of cues may be used to produce these various strategies, by far the most commonly used cues are chemical, and they will be the only ones considered in this chapter.

A major concept that has evolved in the last few years is that many, if not most insects and other arthropods, possess species-specific recognition cues, usually on the outside of their bodies, and that these cues are used by both conspecifics and heterospecifics in making behavioral response choices when two organisms meet one another (Fletcher and Michener, 1987; Hölldobler and Carlin, 1987; Howard, 1993). Such recognition cues are sometimes volatile substances perceived at a moderate distance, but for most animals it appears as if the cues are cuticular-borne and are not perceived except at close range, often after mutual antennal tapping. These recognition cues serve to convey a diversity of messages. For solitary arthropods, they may convey only species, gender, and lifestage cues. In social organisms, these cues may convey additional information such as kinship, nestmate status, and caste status. Although controversy has

[1]Deceased

arisen over the ability of animals to make some of these precise distinctions (Moritz, 1991), there is general agreement that cuticular-borne semiochemical cues are used by many arthropods on a daily basis in their interactions with one another (Fletcher and Michener, 1987; Howard, 1993; Lockey, 1988; Stowe, 1988).

Given that such cues exist, this chapter explores the ways that different arthropods have exploited these species-recognition cues to their own advantage. In some cases our concept of the terms used to describe these phenomena is somewhat broader than those espoused by the original authors (Stowe, 1988). We will present examples of *propaganda* in which one organism releases chemicals that present "false advertising" to another organism, fooling it into believing that the releasing organism is something other than it really is, for example, perhaps bigger, or fiercer . Many of these propaganda semiochemicals are used in interspecific interactions, although they are also known in intraspecific situations. We will present other examples of organisms that use semiochemicals to *camouflage* themselves. In this case they attempt to blend into the environment by covert acquisition from the environment of the species-specific recognition cues of their competitors or hosts. Although frequently effective, such covert acquistion of environmental semiochemicals is not without risk. Finally, we will present examples of organisms that have managed to evolve a suite of chemical cues that are exact or nearly exact replicates of the cues born by their competitors or hosts, a phenomenon we describe as *chemical mimicry*. Unlike the organisms that practice camouflage, the organisms that practice chemical mimicry biosynthesize their own cuticular-borne cues and are not dependent upon their competitors or hosts to provide them. Although the lines demarcating each of these three categories of chemical cues are sometimes vague, we hope that the examples provided will be relatively clear in the context in which they are presented. We will not explore the voluminous literature dealing with organisms that acquire dietary chemicals, sequester them, and then use them in various mimicry complexes involving learned responses from predators.

2. Propaganda Substances

A common problem among all animals is to convince a superior adversary that one is bigger or stronger or more dangerous than one actually is. At the simplest level, if an animal can produce a personal "repellent" that gives it a brief period to escape, then the costs involved in making such a chemical would seem to be minimal. The review by Blum (1981) indicates that many insects and other arthropods use exocrine gland products in just such a manner against a multitude of predators or competitors. Many chemicals adopted for this simple repellency are noxious elements such as various quinones, acids, reactive aldehydes, and proteins, and all seem to function primarily by causing immediate discomfiture or toxicity to the receiving organism.

2.1. Phalangids and Ant Alarm Pheromones

For many organisms ants are their most common and important predator. Ants use a variety of semiochemical cues to regulate almost all facets of their daily lives (Hölldobler and Wilson, 1990). One such cue is known as alarm pheromones (Blum, 1969; 1984), and nearly all ants produce and use them. As their name suggests, these semiochemicals are released to tell other colony members of the need for assistance in dealing with some perceived threat to the colony (Blum, 1984; Buschinger and Maschwitz, 1984). They are also useful, however, in conveying the message that the danger to the colony is so great that no defense should be mounted, and that the colony should disperse instead. The difference in the message "come help me attack a foe" and "man the lifeboats" can be one of intensity of signal rather than differences in chemical constituents. It makes sense therefore, that if an arthropod could produce one or more of the common ant alarm pheromones and produce it in very high concentrations, then when confronted by attacking ants such an organism could get the ants to let it escape. A classic example of such propaganda semiochemical communication is used by at least nine species of phalangids ("daddy longlegs") in the genus *Leiobunum* or *Hadrobonus* to defend themselves when attacked by ants: they emit copious quantities of volatile chemicals such as 4-methyl-3-heptanone (Blum and Edgar, 1971; Jones et al., 1976), (*E*)-4,6-dimethyl-6-octen-3-one and (*E*)-4,6-dimethyl-6-nonen-3-one (Meinwald et al., 1971; Jones et al., 1976; 1977), (*E*)-4-methyl-4-hexen-3-one and 4-methyl-3-hexanone (Jones et al., 1977), or (*E*)-4-methyl-4-hepten-3-one (Jones et al., 1976) from specialized scent glands located on the flanks of their prosoma just behind the first pair of coxae. These chemicals are all identical to known ant alarm pheromones or are closely related analogues (Hölldobler and Wilson, 1990). The openings for the glands are narrow slits displaced from the center of the double-walled depression which surrounds them. The floor of a depression is textured to provide a greater surface area or possibly a more retentive surface for the emitted secretion. Furthermore, the double wall ensures that a temporary external reservoir exists for the released secretions, thus providing for a relatively steady emission of the volatile secretions for a short time (Blum and Edgar, 1971). When the leg of a *Leiobunum* is held by a pair of forceps or ant mandibles, the glandular secretion is forcibly ejected, uniformly wetting the dorsal surface of the prosoma and opisthosoma. The odor of the secretion is readily detected (at least by humans) for several minutes after the initial stimulus has been applied. The adaptive significance of this propaganda message is easily understood, since ants are major predators of phalangids, and the volatiles released by the phalangids are alarm pheromones for ants (Blum, 1984) in a multitude of genera in several subfamilies. At low levels ant alarm pheromones elicit attraction to the emitting source, at somewhat higher levels the ants begin to show aggressive behavior, and at even higher levels they disperse (Hölldobler and Wilson, 1990). The levels of these compounds released by phalangids are thought to be well above that required to elicit dispersal behavior by the ants. Blum and Edgar (1971) reported that when a phalangid was introduced onto the foraging platform of a fire ant colony (*Solenopsis spp.*), the ant workers quickly grabbed a leg of the phalangid, which then immediately released its secretions, often contaminating the attacking ant. Ants so contaminated immediately released the phalangid and began grooming themselves. Ants near the phalangid appeared to be highly irritated and exhibited abnormal ambulatory movements. The phalangid immediately moved away from the ants, and a zone free of ants was maintained around the phalangid for several minutes (Blum and Edgar, 1971). The only caveat in the above experiment is that *Solenopsis* is not known to use these ketones as an alarm pheromone, but many other myrmecines do. Although no field experiments have been conducted with these organisms, it is reasonable to presume that use of these propaganda substances by phalangids would result in escaping predation by the ants in most situations.

2.2. *Other Arthropods and Ant Alarm Pheromones*

Many other arthropods are also known to produce substantial quantities of one or more common ant alarm pheromones in their defensive secretions (Blum, 1981;1984; Fales et al., 1980; Hölldobler and Wilson, 1990), and it is widely believed that these chemicals were evolved in response to ant predation. As with the experiments with the phalangids, much of the literature is circumstantial and does not involve field experiments where the attacked organism (mutillid wasp, beetle, cockroach, etc.) is actually shown to release the putative alarm pheromones in the appropriate concentrations, or that the ants respond as predicted. It is particularly important to conduct such experiments with prey and ants that have evolved with one another. For example, Zalkow et al. (1981) reported that soldiers of the termite *Reticulitermes flavipes* produce massive quantities of two "defensive" sesquiterpenes in their frontal glands, but that when confronted with what seemed to be an appropriate ant predator (*Solenopsis invicta*), they failed to release any of these sesquiterpenes, relying instead solely on their formidable mandibles. Zalkow et al. (1981) therefore concluded that these compounds were not "defensive" in nature. A few months later, the first writer received a personal communication from M. S. Blum (University of Georgia) saying that he had found that when soldiers of *R. flavipes* were confronted with workers of a native sympatric thief ant (*Solenopsis sp.*), then the termite soldiers readily doused the ants with the sesquiterpenes, and that they were very effective defensive chemicals. *S. invicta*, of course, is in the United States known as the **imported** red fire ant, and *R. flavipes* did not evolve with this ant species as one of its predators. It did evolve, however, with various diminutive thief ants, which are highly effective predators of subterranean termite workers.

2.3. Ant Venoms and Resource Competition

Propaganda semiochemicals are not restricted to defensive uses. Many ant species use such chemicals for offensive purposes during food procurement. An example of this strategy was reported by Hölldobler (1973) and Blum et al. (1980) in their studies on the raiding strategy of the European thief ant, *Solenopsis (Diplorhoptrum) fugax*. This little ant (in common with the other known members of this subgenus) lives near colonies of other ant species. Its scouting workers build tunnel systems that connect to the brood chamber of target prey species, and once these tunnels are complete scouts returning to their nest lay a chemical trail from their Dufour's gland. They then recruit masses of their colony mates back through the trail-pheromone-laden tunnels to the host brood chamber and there proceed to steal the brood and carry it back to their own nest. They are able to do so with almost no interference from the host ant brood tenders because when the *S. fugax* raiders enter the brood chamber, they raise their gasters and release from their poison glands a potent repellent that causes the host ants to abandon the brood chambers and to not return for at least an hour. In *S. fugax* this repellent was shown to be 2-butyl-5-heptylpyrrolidine dissolved in a small amount of n-alkanes (C_{17}-C_{24}) (Blum et al., 1980) and was effective against workers from 18 different ant species (Hölldobler, 1973). In their laboratory bioassays Blum et al. (1980) showed that the dialkylpyrollidine by itself at concentrations equivalent to that contained by a single *S. fugax* worker produced all of the behavioral responses seen when the live thief ants were used. Thus, when a small glass rod contaminated with the dialkylpyrollidine was introduced into the brood chamber of any of six species of ants (*Camponotus ligniperda, C. planatus, Lasius flavus, Formica fusca, F. peripilosa,* or *Myrmica rubra*), the brood tending workers strongly avoided the rod and frequently rushed to their larvae and picked them up and attempted to flee the brood chamber. Blum et al. (1980) also showed that when host larvae were treated directly with the dialkylpyrollidine and offered to host workers in a foraging chamber, they were almost never picked up by the workers, but instead were avoided. In contrast, brood which was untreated was always immediately picked up and returned to the nest brood chamber. A variety of unsymmetrical 2,5-dialkylpyrollidines are now known from many *Solenopsis (Diplorhoptrum)* species (Jones et al., 1982a; Blum, 1985; Escoubas and Blum, 1990), and it is likely that these species also use their venom in a similar manner, although relevant behavioral studies have yet to be conducted.

Studies have been conducted with several ant species which also use their venom chemicals for interference competition (see Chapter 9, Traniello and Robson), although in many cases the actual semiochemicals involved have not been identified. Examples include *Solenopsis invicta* (Baroni Urbani and Kannowski, 1974), *Monomorium spp.* (Hölldobler, 1973; Adams and Traniello, 1981; Jones et al., 1982a;1982b; 1988; 1989; Andersen et al., 1991); *Meranoplus spp.* (Hölldobler and Wilson, 1990); and *Forelius (=Iridomyrmex)* (Hölldobler, 1982). These ants are known to apply their venoms topically to food resources they encounter, and the studies above show that, at least in the laboratory, such marked food items are not acceptable to other ant species for at least a moderate period. By placing such a repellent onto their newly found food resource, these ants gain enough time to recruit other members of their own colony to the food resource and retrieve it for themselves, even in the presence of larger and presumably more effective competitors. A convincing field test of this hypothesis was conducted by Andersen et al. (1991) with *Monomorium "rothsteini"* in Australia. This species is a slow-moving, nonaggressive ant that is a very successful member of several Australian ant communities but possesses no obvious morphological specializations to account for its success in competing with the highly aggressive *Iridomyrmex spp.* that dominate these communities (Greenslade, 1976; Greenslade and Halliday, 1983; Greenslade and Greenslade, 1989). When Andersen et al. (1991) placed honey baits on microscope slides out in the field, the *M. "rothsteini"* found them and if other ants such as the *Iridomyrmex spp.* approached them, they raised and waved their gasters at the other ants. This invariably caused the other ants to abandon the honey bait. Such "gaster-flagging" behavior has also been observed in other species of *Monomorium* (Adams and Traniello, 1981). Andersen et al. (1991) hypothesized that *M. "rothsteini"* was releasing dialkylpyrollidines during this gaster-flagging behavior. They indeed found that this species possesses two dialkylpyrollidines in a roughly 1:11.5 ratio; the minor component was *trans*-2-ethyl-5-undecylpyrollidine and the major component was *trans*-2-ethyl-5-tridecylpyrollidine. When these pure compounds were field-tested on the microscope slides at levels equal to those found in single *M. "rothsteini"* (about two μg), they were equally effective as the live ants in preventing the access of other ants to the honey baits. It thus seems that this species is indeed using its dialkylpyrollidines as very effective propaganda substances to allow it to control access to food resources, even in the face of what seem to be superior competitors. One of the species field-tested against these alkaloids was the sympatric and congeneric *M. pharaonis*, and although this species possesses its own (different) dialkylpyrollidines, it was still completely repelled by the *M. "rothsteini"* compounds.

2.4. Stingless bees and Alarm Pheromones

A similar usage of propaganda chemicals by robber bees in the genus *Lestrimellita* has been known since the mid-1960's (Blum, 1966), with the vast majority of the studies being conducted with the neotropical species, *L. limao*. These are eusocial bees, with queens and a worker caste, but they lack pollen-collecting or pollen-carrying leg structures and are never seen on flowers. Rather, they rob the nests of various species of *Trigona* (to which they are closely related) or less commonly, they rob various species of *Melipona* or even *Apis* (Michener, 1974; Sakagami et al., 1993), carrying the mixed honey and pollen stores of their victims back in their extraordinarily enlarged crop. The raids are initiated by one or more *Lestrimellita* workers laying an aerial trail of citral (Stejskal, 1962) from the mandibular gland (Blum, 1966; Blum et al., 1970) to the *Trigona* nest that is to be robbed. The initially small number of robber bees enter the *Trigona* nest where they are immediately attacked and may even be killed. However, in the course of their struggles they release massive amounts of their alarm pheromone citral (a mixture of the monoterpenes geranial and neral) (but see below), which causes the resident *Trigona* workers to retreat into the recesses of the nest or to flee the nest (Blum et al., 1970; Nogueira-Neto, 1970; Sakagami et al., 1993). Meanwhile, the aerial plume trail and the citral emanating from the mouth of the nest will have attracted dozens or even hundreds of *L. limao* that can now plunder the nest at will. The robbers carry off not only the brood cell provisions (wax, resins, honey, and pollen), but also the cerumen that they use for repairing their own nest. The brood cells containing small larvae or eggs on top of the comb stack were opened and emptied by pillaging bees who completely remove much of the comb wax and cerumen used to construct the brood cells (Michener, 1974; Sakagami et al., 1993). Contrary to earlier beliefs, the release of citral by the bees appears to be limited to the early and final stages of the raiding process (Sakagami et al., 1993)

Not all *Trigona* or other stingless bee species are so readily plundered. Many species, especially those that have wide nest entrances, respond to aliens in the nest environment by evoking rapid physical attacks by guard bees often followed by prompt recruitment of additional defending bees (Michener, 1974). Wittman (1985) and Wittman et al. (1990) have extended the knowledge of the actual semiochemicals present in the heads of *L. limao* and have shown how one host species, *Trigona (Tetragonisca) angustula* resists raids of *L. limao*. Head capsules of *L. limao* contain four hydrocarbons (heneicosane, (Z)-9-heneicosane, tricosane, and (Z)-9-tricosene); three alcohols (hexadecanol, (Z)-7-hexadecenol and (Z)-9-octadecenol); two acetates ((Z)-7-hexadecenyl acetate and (Z)-9-octadecenyl acetate); four ethyl esters (ethyl tetradecanoate, ethyl hexadecanoate, ethyl octadecanoate and ethyl (Z,Z)-9,12-octadecadienoate); one ketone (6-methyl-5-hepten-2-one) and the major component, citral (75%). Except for (Z)-9-tricosene that makes up 10% of the mixture, all other components make up 2.5% or less. Although far less is known of the semiochemistry of *T. angustula* (Schröder, 1985), it notably lacks either citral or 6-methyl-5-hepten-2-one, the two key compounds in this story. In a series of behavioral studies Wittman et al. (1990) found that *T. angustula* always has several guard bees hovering downwind of the entrance to its nest, and as approaching insects of the same general size and profile as stingless bees approach the nest, they rise to attack if they are *L. limao*. They further showed that this discrimination process involved more than just visual cues, since when dried, extracted *L. limao* dummies were presented to them they did not attack, whereas the presentation of live tethered *L. limao* triggered an attack. When Wittman et al. (1990) tested individual *L. limao* head chemicals or mixtures of head chemicals by presenting them on a small wax dummy outside the entrance to the nest, and counting for five minutes the number of attacks on the wax dummy, they found that the hydrocarbons, esters, and alcohols did not elicit a defensive response from the *T. angustula* guards. Presentation of citral or the ketone, however, caused large numbers of the guards to fly out of the nest and to take up hovering positions. When all 16 compounds were presented in the same ratios and concentrations as that found in the head of one *L. limao*, they elicited the strongest reaction of all, reaching a response of 60% of that provoked by a live robber bee.

Although it was clear to Wittmann et al. (1990) that the guard bees were responding directly to the semiochemicals produced by the invading robber bees, it was not immediately clear whether the additional *T. augustula* recruited from inside the hive were also responding to the robber bee kairomones, or whether the guard *T. angustula* were releasing an alarm pheromone of their own that initiated the recruitment response. Subsequent studies (Schröder and Francke, unpublished, cited in Wittman et al., 1990) have shown that benzaldehyde serves as an alarm pheromone for *T. angustula* , and that this compound elicits the massive recruitment response observed. Although a mechanical stimulus to the nest such as banging or shaking it may cause the *T. angustula* workers to release benzaldehyde and take up hovering positions outside the nest, such mechanical stimuli do not elicit the massive recruitment response observed when *L. limao* chemicals are present. This is clearly a case where use by a species of a propaganda substance has backfired. Perhaps, however, there are enough other species of stingless bees in the environment of *L. limao* that do use citral as their own alarm pheromone (such as *T. fulviventris*) (Johnson, 1987) that the payoff is worth the risks involved.

Other stingless bees are also known to use propaganda chemicals in their raids on other bees. One such species is the fire bee *Trigona (Oxytrigona) mellicolor* (Moure, 1946; Bian et al., 1984) which frequently raids hives of the honey bee, *Apis mellifera,* causing the honey bees to either hang motionless on the hive, to flee the hive and form a cluster of bees outside the hive, or to just wander in a seemingly disoriented manner over the surface of the comb. The cephalic glands of this species contain two hydrocarbons (n-tetradecane and n-pentadecane), three simple ketones (2-heptanone, 2-decanone and 3-hepten-2-one), three esters (dodecyl acetate, tetradecyl acetate and hexadecyl acetate) and two diketones ((*E*)-3-nonene-2,5-dione and (*E*)-3-heptene-2,5-dione) (Bian et al., 1984). Rinderer et al. (1988) gathered bioassay data on avoidance behavior and propensity by the honey bee to sting a rubber septum treated with each of these chemicals singly and as mixtures . They found that only the two diketones elicited similar reactions to those observed from live *T. mellicolor,* and the presence of the other eight chemicals did not have any synergistic effects. The primary behavior observed from the exposed honey bees was a negative chemotaxis. The authors note that although *T. mellicolor* readily and successfully raids nests of *A. mellifera* (an introduced species), they are not thought to be able to successfully raid the nests of other sympatric, native neotropical bees (Roubik et al., 1987). The reasons why this is so remain to be explained.

2.5. Bolas Spiders and Moth Sex Pheromones

In a somewhat different vein, various arthropods use propaganda semiochemicals to falsely portray themselves as sexually receptive individuals. Perhaps the most fascinating example of this ploy is found in the manner in which large immature and mature female bolas spiders (Araneae: Araneidae, commonly called orb weavers) in five genera attract male moths (Eberhard, 1977; Stowe et al., 1987; Stowe, 1988; Yeargan, 1988), and capture them by striking them with a sticky silk "bola" (Fig. 10.1). The biology of bolas spiders was recently reviewed by Yeargan (1994). Several other araneid genera, including *Celaenia*, *Taczanowskia*, and *Kaira*, do not use a bola, but hang upside down from a leaf or silk strand and capture the lured male moths with their outstretched legs. A similar chemically mediated attraction and prey capture method may also be used by *Phoroncidia studo* (Araneae: Theridiidae) in its predation of nematocerous Diptera (Eberhard, 1981). The most extensive studies have been conducted with the bolas spiders; with most of the work being on various members of the genus *Mastophora* (Stowe, 1988; Yeargan, 1994).

The spider initiates hunting activities by placing a more or less horizontal web line between two attachment points. The spider then drops a second thread and draws it out to a length of a few centimeters, and hanging from it by the spinnerets, the spider then uses its hind legs in alternating strokes to comb viscous material (both silk and liquid material) from the spinneret to produce the nascent bola (Eberhard, 1980). At this point it is assumed that the spider begins emission of the chemicals that attracts its prey, although no direct evidence is available on this point. At any rate, the exact behavior that follows is different for each bolas spider group in the three major regions of distribution (North and South America, Australia, and Africa). In the Americas, *Mastophora* species hold the bola essentially stationary with a front leg until the moth is close, then she cocks the leg

---

Figure 10.1. Female of *Mastophora hutchinsoni* with bola poised for capturing moth prey.

Figure 10.2. Nymph of *Redivius personatus* (masked hunter) (Reduviidae) with patches of lint attached to its sticky dorsal surface to provide camouflage (ecological homologue of "wolf-in-sheep's-clothing" discussed in text.)

Figure 10.3. Primary queen of *Reticulitermes flavipes* with phoretic *Trichopsensis frosti* on her thorax and abdomen.

Figures 10.4–10.6. Termitophiles associated with *Reticulitermes virginicus*: (a) *Trichopsenius depressus;* (b) *Xenistusa hexagonalis;* and (c) *Philotermes howardi*. Note unidentified mite (arrow) on abdomen of *X. hexagonalis* in Figure 10.4, and the extended versus upcurved abdomen of *P. howardi* in Figs. 10.5 and 10.6, respectively.

Plate 1.

and swings the bola toward the prey with very rapid, pendulum-like strokes (Hutchison, 1903; Yeargan, 1994). In Australia, however, *Ordgarius* spp. rapidly whirl the bola while the moth approaches (Coleman, 1976; McKeown, 1952; Mascord, 1970). Finally, an African *Cladomelea* spp. has been reported to start whirling her bola as soon as it is completed, whirling it continuously for up to fifteen minutes whether or not a moth is present (Akerman, 1923). Such a process would not seem to be very effective, and the behavior of this species would bear reexamination. *Mastophora* does not always successfully hit the approaching moth, and the same bola may be repeatedly swung at the moth as it makes additional approaches to the spider. However, if the moth is hit with the bola, it rarely escapes (Eberhard, 1980). The spider quickly descends to the prey (or may draw the prey up by the line) and bites it. A few seconds later the spider wraps the prey in silk and feeds upon it immediately, or attaches it to the horizontal line and commences to form a new bola to capture more prey (up to eight moths) before beginning to feed (Eberhard, 1980; Longman, 1922; Stowe, 1986; Yeargan, 1988).

Since the early part of this century it has been suggested that bolas spiders possibly enticed their prey into range by emitting an attractive odor (Hutchison, 1903). It was not until the late 1970's, however, that Eberhard proved this was so (Eberhard, 1977). He showed that only male moths were attracted and captured, that they always approached the spider from downwind, that even if the spider was visually hidden by a baffle that partially deflected the airflow the moth always approached by way of the air pattern around the trailing edge of the baffle; and that even though dozens of species of moths and other insects were active in the general area, only a limited number of moth species were captured. Since then, Stowe (1986; 1988) and Yeargan (1988) have shown that other *Mastophora* species elicit the same behavior from various moth species. It remained for Stowe et al. (1987) to show that the chemical cues emitted by *M. cornigera* females were components of the female sex pheromones of the moth species being captured. They identified three compounds, (Z)-9-tetradecenyl acetate, (Z)-9-tetradecenal, and (Z)-11-hexadecenal, all of which are well-known pheromone components of several prey species of *M. cornigera*. The isolation and identification of these compounds were major achievements since the components are apparently produced only while the female spider is hunting, and such hunting behavior is difficult to elicit in the laboratory. Their experiments required them to collect from eight females over a period of two months to obtain sufficient material for identification (Stowe et al., 1987). The exact source of the chemicals is not known, but apparently comes from the spider rather than from the bola (Eberhard, 1980). Stowe et al. (1987) extracted webs and bolas from these spiders and were unable to find any of the pheromones on the silk. Furthermore, Eberhard (1977) had shown that glue droplets alone were not attractive to male moths, whereas hunting spiders without droplets were attractive.

More than 40 species of male moths in seven families have been reported as prey of bolas spiders (Yeargan, 1994). Five species of *Mastophora* are known to capture 28 species of noctuids and one or more species of gelechiids, geometrids, pyralids, plutellids, tineids, and tortricids. Many of these prey species have extensive geographic distributions and are often economic pests, suggesting that the prey of these spiders are abundant and ubiquitous on both a geographic and local scale.

The biology and chemistry of moth sex pheromones are better known than almost any other aspect of insect chemical ecology (Arn et al., 1992; McNeil, 1991). Almost all species use mixtures of two or more components to make up their semiochemical signal, and the chemical nature of known blends seems to fall into two major groups. In one, the mixtures consist of unsaturated (one to three double bonds in either the Z or E configuration) straight-chain aliphatic aldehydes, alcohols, or acetates with chain lengths of 10–18 carbons (Baker, 1989). In the second pheromone blend group, the components are straight-chain or branched hydrocarbons that are usually multiply unsaturated at the 1-, 3-, 6-, 9-, 12-, or 15-position (with Z configuration), and may possess an epoxide at one of these double bond positions (Baker, 1989). Most moths appear to use compounds from one or the other of these two groups, but not from both. Exceptions are known, however, such as the arctiids *Estigmene acrea* and *Hyphantria cunea,* which use mixtures of aldehydes and epoxides (Arn et al., 1992).

The volatiles emitted by *M. cornigera* are all representative of the acetate/aldehyde/alcohol group of pheromones, and all known prey of this bolas spider are moths that have pheromones in this group. Other bolas spiders, however, may produce volatiles from the hydrocarbon/epoxide pheromone blend group, or possibly even volatiles from both moth pheromone blend groups. Thus, *M. bisaccata* has been reported to capture at least two species of moths that use the acetate/aldehyde/alcohol blends, and at least three moth species that are believed to use hydrocarbons/epoxides for pheromones (Yeargan, 1994). Similarly, *M. phrynosoma* captures moths from both pheromone blend groups (Yeargan, 1994), as does *M. hutchinsoni* (Yeargan, 1988). Since it is well known that moths also have distinct, species-specific diel periods of sex pheromone emission (and corresponding temporally linked patterns of male responsiveness), it is possible that bolas spiders either alter the blend of chemicals they emit during the night to track the different populations of moths, or more likely that they emit a single compromise blend. Indeed, it is known that *Mastophora* capture particular prey species at different times of night (Stowe, 1986; Yeargan, 1988). Further understanding of this semiochemical propaganda will require analysis of the attractant emissions of many more species of bolas spiders.

3. Mimicry and Camouflage in Nonsocial Arthropods

3.1. "Wolves-in-Sheep's-Clothing"

Several predatory arthropods have developed what Eisner et al. (1978) have called the "wolf-in-sheep's-clothing" strategy. Larvae of the lacewing *Chrysopa slossonae* (Neuroptera: Chrysopidae) carefully approach the wax-bloom covered nymphs of the aphid *Prociphilus tesselatus* and remove packets of the wax from the aphids which they then place onto their own body. When the lacewing larvae have covered themselves with enough wax, they are then able to move among the aphids without being recognized as an alien individual. This ploy serves two purposes: it allows the lacewing to get close enough to its prey to grab it without causing any disturbance to the aggregated aphids, and it prevents the ants that tend the aphids (and which normally serve as effective defenders) from realizing that their charges are being taken from right under their own antennae.

Several species of assassin bugs (Hemiptera: Reduviidae) which are specialist predators of termites also use a "wolf-in-sheep's-clothing" strategy (Fig. 10.2). Thus Odhiambo (1958a, 1958b) reported that nymphs of *Acanthaspis petax* that occur in the ventilation tunnels of mounds of the tropical termites *Macrotermes spp*. cover themselves with nest debris and hide in crevices awaiting the opportunity to attack the termite nymphs. If discovered by the termite workers, the covering on the cuticle of the reduviid served in most cases to prevent its being detected as foreign. McMahan (1982, 1983) discovered that nymphs of a neotropical reduviid, *Salyavata variegata*, commonly associated with the carton nests of various *Nasutitermes spp*. in Costa Rica also coat their body with bits of the nest material as a means of disguising their presence. In this case, the bugs do not live inside the nest but rather are found on the outside. Termite nests are usually thought of as providing substantial protection against invading predators, and this is mostly true. But the nests of *Nasutitermes* cannot be expanded to allow for colony growth except by the termites regularly breaching the nest walls, and so an opportunity exists for predation. The *S. variegata* nymphs scrape small pieces of carton with their pretarsal claws and pat it onto their body surfaces, which are covered with sticky setae (McMahan, 1982), completely covering their dorsal surfaces. This camouflage conceals the nymphs from visual predators and from tactile or olfactory discovery by the blind termite workers and soldiers engaged in repairing the breach in the carton. The *S. variegata* nymphs, however, have taken their strategy one step further than the lacewing larvae studied by Eisner et al. (1978). They slowly approach a termite nymph engaged in nest repair, stab it with their beak, and then withdraw away from the nest opening to suck the prey contents out. Leaving the prey carcass on the end of its beak, the reduviid nymph then slowly approaches the termite workers again and waves the termite carcass at them. Because termites eat their dead nestmates to recover their valuable nitrogen content (Moore, 1969), a termite worker grabs for the carcass, but while doing so is itself grabbed by the reduviid nymph and quickly hauled away and impaled. Although in laboratory studies the bugs would take both small and large termite workers (and occasionally soldiers), field studies suggested that the bugs mostly capture large termite workers. This probably is a function of these being the caste that is most readily available and which respond to the proffered bait (McMahan, 1983).

Intermediate between these two ploys is the foraging tactic taken by two species of spiders [*Strophius nigricans* (Thomisidae) and *Aphantochilus rogersi* (Aphantochilidae)] that prey on ants (Oliveira and Sazima, 1985). These spiders catch a foraging ant and then after imbibing its juices, hold the carcass aloft as a mobile shield and again approach foraging ants to entice them to hold still long enough for the spider to grab them. Since the ant carcass would be covered with species- and nest-mate-specific cuticular lipids, the investigating ant would have no reason to initially suspect the presence of a predator.

3.2. Sexual mimicry

Solitary insects are also known to engage in a chemically based form of sexual mimicry. One of the best examples of this is found in the carrion-inhabiting staphylinid beetle, *Aleochara curtula* (Peschke, 1983; 1985; 1987a; 1987b; Peschke and Metzler, 1987). The cuticle of this species contains over 127 cuticular hydrocarbons, of which 64 could be identified, representing 91% of the hydrocarbon mass. Normal alkanes comprised 19%, monoenes 4%, monomethyl alkanes 57%, and dimethyl alkanes 11%. Both sexes contain predominantly the same cuticular hydrocarbons, but females were shown to contain a series of C_{21} and C_{23} monoenes in μg quantities that were absent from the males. Using a bioassay where crude models dosed with test chemicals released genitalic grasping behavior from males, Peschke and Metzler (1987) showed that (*Z*)-7-heneicosene and (*Z*)-7-tricosene were sex pheromones for this species; indeed, this is the first identified sex pheromone from the family Staphylinidae. This pheromone acts over only a short distance, because the compounds are of intrinsic low volatility, and are dissolved in the mass of cuticular waxes that retard their volatility. Females of this species release other components from their tergal gland, including (*Z*)-4-tridecene, n-dodecanal, and (*Z*)-5-tetradecenal. These compounds appear to act as long-range attractants, and to act as synergists for the cuticular hydrocarbon pheromone by lowering the threshold for the genitalic grasping behavior. In addition, at low concentrations they may act as additional aphrodisiacs for the males of this species.

As with other carrion beetles, *A. curtula* males must actively compete with other males for the females, and this includes male-male combat in which large males typically win (Peschke, 1985; 1987a). This combat involves males pushing against each other with their heads and mandibles and drumming the ends of their upcurved abdomens onto the body of the other beetle. In early studies Peschke found that cuticular hydrocarbon extracts from young, starved, or multiply mated males elicited genitalic grasping responses from other males. In contrast, hydrocarbons from sexually isolated, well-fed males did not trigger such responses, nor did small males. Peschke (1987b) found that the young, starved, or multiply mated males all had substantial quantities of the female sex pheromones [the (*Z*)-7-monenes] on their cuticle, and he concluded that these males were mimicking the females. No such female pheromone was found on small males. He notes that by mimicking females the males can gain access to a carcass and feed on blow fly maggots, thus replenishing their energy reserves for forthcoming reproductive cycles. If these males do not feed, they transfer small spermatophores and thus fertilize fewer eggs. The amount of intermale aggression and homosexual interactions was modulated by the quantities of these female pheromones on the male cuticle; when the levels were high, male-male aggression was low. Peschke (1987b) also found, however, that the females would actively repulse those males that contained the sex pheromones on their cuticle. Peschke (1985; 1987a; 1987b) argued that the females were thus choosing competent, nonmimicking males that were capable of withstanding the aggressive interactions at the mating site, and hence of transferring a large spermatophore. A complicating factor in these analyses is that relatively small changes in environmental rearing temperatures of the females (22 °C vs. 27 °C) drastically reduced male sexual responsiveness to the females. This variance was shown to be correlated with the virtual absence of the female sex pheromones on the females reared at the higher temperature. Peschke (1987b), unfortunately, was unable to distinguish whether this temperature effect was caused by differential evaporation of the pheromones (doubtful in our opinion), or whether the beetles actively alter the pattern of hydrocarbon biosynthesis at different temperatures (Hadley, 1977; Toolson and Hadley, 1977). It is strongly suggested by the wording in Peschke's papers that the males are *de novo* producing the female pheromones, but no direct evidence is presented. Although the ultimate behavioral consequences may be the same whether the males manufacture these chemicals themselves or somehow acquire them from contact with females, the evolutionary implications are different.

Other examples of males that use pheromones to elicit long-range attraction or courtship behavior from conspecific males are known: *Drosophila spp*. [references in Peschke, 1985 and Hall, 1986]; many solitary bees (Cane and Tengö, 1981; and references therein); and staphylinids in the genus *Leistotrophus* (Alcock and Forsyth, 1988). The detailed chemistry in most of these examples is not well known, complicating the interpretation of the observations. Such attraction need not be restricted to males only, of course. Many examples are known in which females of parthenogenetic species actively solicit copulations from males of closely related species (Lloyd, 1979), and in these cases chemical signals almost certainly play a role (Woodroffe, 1958; Lloyd, 1984; Mitter and Klun, 1987).

4. Mimicry and Camouflage in Social Insects

4.1. General Concepts

The world of social insects is a fascinating one, and in many respects an individual colony can be viewed as a self-contained community, with the social insects being the dominant species. Such colonies are, however, members of the larger insect community and are often viewed as either being of an open or closed nature with respect to conspecific colonies (Clément, 1986; Provost, 1985). Open colonies are those which admit with fairly high frequencies members of other conspecific colonies into their territory and nest, whereas closed colonies rarely admit conspecific individuals, but rather vigorously exclude them from their territory and nest. Both open and closed colonies tend to defend their territories with varying degrees of intensity from heterospecific insects. Such defenses are not absolute, however, and an amazing diversity of insects have managed to insert themselves into the social fabric of a colony to a greater or lesser extent. Such insertions range from individuals that use the social insect nest only as a hiding place from which to launch predatory attacks on the nest members (such individuals are subject to immediate attack if discovered) to individuals that are for all intents and purposes considered full-fledged members of the colony and are never attacked (even if they are obligatory predators), and to every degree of integration in between (Hölldobler and Wilson, 1990; Kistner, 1969; 1979; 1982). The behavior of guests in terms of their interactions with their hosts can also span a range of traits, from being obligatorily predatory to exclusively mutualistic (although no single individual or species shows this sort of a range). Other groups of arthropods such as Collembola and mites use the nest only as a habitat, avoiding any interactions with the social insects, and feeding on fungi or small organisms in the nest.

To some extent the degree of interactions between these aliens and the social insects is dictated by environmental constraints such as the complexity and size of the host nest (number of niches), whether the nest is an open or closed type of colony, and the absolute population size of the colony. Young colonies or colonies that just inherently are small rarely have any guests (the exceptions to this are slave-making species; see later). In larger colonies, however, more opportunities are available for invasion and almost every social insect colony of any size has some aliens ensconced somewhere within its boundaries. To the extent that a nest is diffuse and crenellated, it may be possible for many of the nonintegrated species to avoid most interactions with the host insects. But many guests will at some time or another be discovered by the social insects and will at that point have to either fool their hosts into accepting them as members of the colony, or they must have some defenses to allow them to escape being killed forthright or to be excluded forcibly from the nest or walled off into some abandoned section of the nest. The various strategies (primarily chemical) that insects and other arthropods have evolved to meet this confrontation with their hostile social insect hosts are the subject of the remainder of this section.

The social insects themselves have a multitude of semiochemical cues that they use in their day-to-day interactions. Perhaps arguably the most important of these in terms of species recognition are their cuticular hydrocarbons, followed in close order by various glandular secretions and lastly by various environmental odors ("nest odors") (Fletcher and Michener, 1987; Howard, 1993; Kistner, 1990). Alien insects can thus achieve integration by mimicking one or more of these semiochemical cues of their hosts. A moderate amount of knowledge is now available on the mimicking of cuticular hydrocarbons, less so on glandular exudates by various inquilines, and knowledge of the chemistry of "nest odors" remains virtually unknown. One would predict that as an inquiline more closely matched any of these host semiochemical cues, the more readily it would be accepted into the social fabric of the colony, and this seems to be the case. Individuals that possess only a limited component of the species' odor signature may be accepted for a brief time, but to be tolerated for any significant period they require a much more extensive suite of complementary chemicals.

4.2. Cuticular Hydrocarbons and Chemical Mimicry

Our greatest understanding of chemical mimicry is that of the use of cuticular hydrocarbons. A substantial body of evidence has now accumulated to suggest that cuticular hydrocarbons are very important (and maybe the most important) component of nestmate recognition (or species recognition) in ants, termites, bees, and wasps [see references in (Howard, 1993) and Chapter 8 by Smith and Breed in this volume]. These hydrocarbons exist as highly complex, species-specific, multicomponent mixtures that are accordingly thought to possess a high information content. Although each species has a characteristic overall profile, it is now known that the profiles of individual colony members can differ slightly from the average species profile. It is also known that both individual and species profiles slowly change with time and through environmental interactions. Such changes, however, appear to be small enough to be within the informational processing and learning capabilities of the insects.

Because there is some variability in the species-specific or nestmate-specific cuticular hydrocarbon profiles within any given social insect colony, it is reasonable to expect that potential inquilines need not face an insurmountable obstacle while evolving cuticular hydrocarbon profiles which at least partially mimic that of the host colony. One would predict that as an inquiline became more integrated into the life of a social insect colony, its hydrocarbon profile should more closely approximate that of its host (Howard et al., 1980; 1982b). Evidence is slowly accumulating to support this hypothesis (Howard, 1993).

4.3. Chemical Mimicry and Termitophiles

Subterranean termites in the genus *Reticulitermes* have been successfully invaded by several staphylinid beetles, most of which have evolved to become highly integrated host-specific members of the termite societies, and which are totally dependent for their existence upon their hosts (Kistner, 1969; 1979; Howard, 1991). In southern Mississippi two species of *Reticulitermes* are sympatric and five known staphylinid inquilines occur with them, with only one species (*Anacyptus testaceus*) (Staphylinidae, Hypocyphtinae) being common to the two species. *R. flavipes* has one major termitophile in that area, *Trichopsenius frosti* (Staphylinidae, Trichopseniinae) (Fig. 10.3), whereas *R. virginicus* hosts three major termitophiles, *T. depressus* and *Xenistusa hexagonalis* (both Staphylinidae, Trichopseniinae) (Fig. 10.4) and *Philotermes howardi* (Staphylinidae, Aleocharinae) (Figs. 10.5 and 10.6). All these inquilines except *A. testaceus* are highly integrated; *A. testaceus* is only moderately integrated.

'As measures of integration, the highly integrated species are never attacked by the termite hosts (Howard, 1976; Howard et al., 1980; 1982b), they receive all of their food from their hosts in the same manner that the dependent termite caste forms do (direct solicitation of stomatodeal and proctodeal fluids) (Howard, 1978), they participate in allogrooming with their hosts (and receive in turn allogrooming from their hosts and from other termitophiles) (Howard, 1976), they exhibit the same "head-banging" behavior that the termites exhibit when the nest is disturbed (Howard, 1991), they attack parasitic mites found on their termite hosts (Howard, 1991), one species courts and mates on the abdomens of the physogastric termite queens (Howard, 1979); and they follow the termites trail pheromones even better than the termites (Howard, 1980). All these characteristics are exhibited by the adult termitophiles. Data on immature stages of these termitophiles are much more scarce (Kistner and Howard, 1980), but they do not appear to interact extensively with the termites, perhaps because of the vast disparity in size between them and the termites. Adults of the moderately integrated species *A. testaceus* are occasionally attacked by a termite, but they are rarely injured because of their limuloid body shape (the beetle simply "pops" out of the termites mandibles with no structural damage to its integument) (Howard, 1991). This beetle is much smaller than the other termitophiles in these colonies and has not been observed obtaining food from its termite hosts. It has been observed, however, allogrooming the tarsi of the termites (the only body parts which it can reach). No data are available on the ability of *A. testaceus* to follow the trail pheromones of *Reticulitermes spp*. Immatures of *A. testaceus* also do not appear to interact with the termites in the nest.

A comparison of the cuticular hydrocarbons obtained from the highly integrated termitophiles associated with *R. flavipes* and *R. virginicus* showed that they were the same as those of their host termites, with no components present on the beetles that were not also found on the cuticle of the termites (Howard et al., 1978; 1980; 1982b). Earlier studies had shown that the cuticular hydrocarbons of both termites were species-specific (Howard et al., 1978; 1982c), and that they appeared to show caste-specific proportions of the individual components. Clearly, if the hydrocarbon profiles of the termites show caste-specific proportions, the termitophiles cannot mimic all of the possibilities of each different caste, nor do they do so. The cuticular hydrocarbons of the termitophiles are most like the caste forms they predominantly interact with [queens for *T. frosti*

(Howard et al., 1980) and workers for *T. depressus*, *X. hexagonalis*, and *P. howardi* (Howard et al. 1982b)]. Although such similarities might arise because the termitophiles somehow acquired by mechanical transfer the cuticular hydrocarbons of their hosts, radiolabeling experiments with ^{14}C-acetate showed that *T. frosti* and *X. hexagonalis* biosynthesized all of their cuticular hydrocarbons, and that the label was incorporated in each hydrocarbon in the same proportion as that hydrocarbon was found on the cuticle of the termitophiles (Howard et al., 1980; 1982b). Such findings strongly argue for the termitophiles acquiring little if any cuticular hydrocarbons directly from their hosts.

The other termitophile commonly found with *R. flavipes* and *R. virginicus* in Mississippi, *A. testaceus*, is not highly integrated [indeed, it has been reported from other termite species and even from ant nests (Kistner, 1982)] and therefore would not be expected to show a complete congruence of hydrocarbons with either *R. flavipes* or *R. virginicus*, nor does it do so. Preliminary examinations of the cuticular hydrocarbons of this termitophile showed that the unusual conjugated diene found on the cuticle of the termites was also found on the beetle, as were some hydrocarbons found only with *R. flavipes* and with *R. virginicus* (Howard and McDaniel, unpublished). It is not known yet whether these host-specific components were obtained during the allogrooming behavior of this beetle, or whether the beetle actually makes the compounds itself. These beetles are extremely tiny (hence there is a low absolute amount of available hydrocarbon) and infrequently collected, making it difficult to conduct definitive experiments on their cuticular hydrocarbon composition. Although the beetles examined were collected with a colony of *R. flavipes*, they could have interacted in the recent past with a colony of *R. virginicus*, because it is well known that termites are mobile and frequently move into termite galleries previously occupied by the other sympatric species (Howard et al., 1982a).

4.4. *Chemical Mimicry and Myrmecophiles*

Since most known termitophiles are host-specific (Kistner, 1969; 1979), it is not too difficult to imagine the evolution of chemical mimicry of host cuticular hydrocarbons to the refined state observed in the above species. More inquilines are known from ant nests, and these inquilines are sometimes not host specific, even at the generic level. It is also known that ants, like termites, possess species-specific cuticular hydrocarbons, and that they use these hydrocarbons as prime nestmate recognition cues [see references in Howard, 1993]. The ant workers and soldiers (when present) are also well known to vigorously exclude non-nestmates from the colony, either dragging them outside the nest or killing them outright (Hölldobler and Wilson, 1990). Given these facts, it was difficult to see how myrmecophiles that parasitize multiple species of ants could be using chemical mimicry of cuticular hydrocarbons as an integrating mechanism. A partial answer to this dilemma has been discovered in studies with two species of syrphid flies in the genus *Microdon* (Garnett et al., 1985; Howard et al., 1990a; 1990b). These flies have unusual modified larvae which are obligate predators of the brood of various ants. These fly larvae are highly integrated myrmecophiles and are almost never attacked by the ants. If the colony is disturbed, exposed second and early third instars (Figs. 10.7 and 10.8) assume the shape of an ant cocoon and are carried by the host ants to safety (Akre et al., 1988). Larval *Microdon* feeding on prepupae inside cocoons are also carried to safety when the nest is disturbed. Mature third instars overwinter deep in the nest with the ants. Early the following spring they migrate to the surface of the nest where they form a hardened puparium (Fig. 10.9). They emerge as adults in the early morning hours before the ants began to actively forage. If they are discovered by the foraging ants, however, the newly emerged adult flies are immediately attacked and dismembered (Akre et al., 1973; 1988). The fly larvae also have a number of unusual morphological features whose functions are presently unknown (Figs. 10.10–10.12).

The larval *Microdon* have been shown to possess the same cuticular hydrocarbons as those found on their hosts, and to biosynthesize their own hydrocarbons rather than getting them from the ants (Howard et al., 1990a;1990b). However, since both species of *Microdon* studied are known to occur with multiple hosts, how do they mimic the cuticular hydrocarbons of each of their different hosts, each of which is likely to have rather different cuticular hydrocarbon profiles? The key to this puzzle is that although the adult ants from different species have very different total cuticular hydrocarbon profiles, the larval ants from several different species in two subfamilies have cuticular hydrocarbon profiles that are very similar to one another, being dominated by n-alkanes, with much smaller proportions of the worker-specific types of hydrocarbons (alkenes and methyl-branched hydrocarbons) (Howard et al., 1990a;1990b; unpublished data; Kaib et al., 1993; Yamaoka, 1990). It is probable that the known tolerance of many ant species for alien brood (Hölldobler and Wilson, 1990) is a reflection of this apparent commonality of brood hydrocarbon profiles. If true, then all the *Microdon* larvae have to mimic is the cuticular hydrocarbon profile of generalized ant brood rather than the much more diverse cuticular hydrocarbon profiles of the worker ants. And such shared larval hydrocarbon profiles seems to be the case for at least two common species of *Microdon* in the Pacific Northwest of the United States. *M. piperi* is predominantly associated with *Camponotus modoc*, but also occurs with several other species of *Camponotus* (Formicinae). *M. albicomatus* is usually associated with various *Formica spp.* (Formicinae), but has also been collected with *Myrmica incompleta* (Myrminacae) (Akre et al., 1988). Examination of the cuticular hydrocarbons from *M. piperi* collected with *C. modoc* and of the cuticular hydrocarbons from *M. albicomatus* collected with *M. incompleta* indicated in both cases an exact congruence of cuticular hydrocarbons between fly larvae and host ant larvae (Howard et al., 1990a;1990b).

Although the fly larvae are highly integrated myrmecophiles, as noted above, the adult flies are not so integrated and are readily attacked if they do not manage to escape the ant nest upon eclosion. Examination of the cuticular hydrocarbons of adult *M. piperi* revealed that although they possessed many cuticular hydrocarbons of the ant-mimicking larval flies, they also possessed a large (30% of the total) proportion of n-alkenes that were absent on either the immature flies or any form of the ants (Howard et al., 1990a). Although no behavioral tests have been conducted, it is likely that this high proportion of alien hydrocarbons on the adult fly cuticle is the primary cue to the investigating ant worker that triggers the aggressive response.

4.5. Chemical Camouflage and Myrmecophiles

Studies to date suggest that many species of *Microdon* are highly integrated myrmecophiles that have managed to achieve true chemical mimicry with their various hosts. Many other myrmecophiles are known, however, which are not highly integrated into the ant society, and yet commonly occur there. Many of these inquilines probably also use cuticular hydrocarbons as a means of fooling their hosts, but they do so by a process better characterized as chemical camouflage (acquiring the masking cues from the host rather than producing them by *de novo* biosynthesis) rather than chemical mimicry as defined earlier (but see Dettner and Liepert (1994) for an alternative viewpoint). In the first chemically characterized example of this phenomenon by a myrmecophile, Vander Meer and Wojcik (1982) reported that the scarab beetle *Myrmecaphodius excavaticollis* (now *Martinezia dutertrei*), associated with fire ants (*Solenopsis* spp.) in the southern United States, had a cuticular hydrocarbon profile that was dominated by the same hydrocarbons as the host ant, and that this ant-like hydrocarbon profile allowed them to move about the colony with minimal antagonism from the ants. This beetle is a predator that feeds on the ant larvae as well as being a general scavenger that feeds on dead ants and other insects. Each species of

---

Figure 10.7. Second larval instar of *Microdon piperi* on a cocoon of its ant host, *Camponotus modoc*.

Figure 10.8. Second and early third larval instars of *M. cothurnatus* with cocoons of their ant host, *Formica obscuripes*.

Figure 10.9. Puparia of *M. cothurnatus*.

Figure 10.10. Ventral surface of third larval instar of *M. piperi* showing morphological structures of unknown function (arrows). The large number of stout setae present are thought to provide a surface for releasing viscous fluids which help the larvae to move on a substrate.

Figure 10.11. Enlarged structure of unknown function from Figure 10.10.

Figure 10.12. Dorsal surface of *M. piperi* showing extensive cuticular structures which resemble flowers.

Plate 2.

Solenopsis that has been examined has a characteristic, species-specific hydrocarbon profile (Lok et al., 1975; Nelson et al., 1980; Vander Meer and Lofgren, 1990), and these hydrocarbons are considered important nestmate recognition cues (Vander Meer et al., 1990). The scarab beetle also possesses, however, higher molecular weight hydrocarbons that are not found on the host ants. Wojcik (1975; 1990) reported that under stress situations the ants will attack the beetles. Vander Meer and Wojcik (1982) also noted that upon first entering a fire ant colony, the beetles are immediately attacked, and many are killed. Those that manage to survive, however, are relatively immune to further attack. Upon leaving one mound of ants and entering another mound (often of another species of *Solenopsis*) the beetles must go through the same initial attack/survival response, but if they are successful they now have acquired the cuticular hydrocarbons of their new host. Although the authors did not directly show how the beetles acquired their hydrocarbon camouflage, it is likely that as the ants initially attacked them, they were smearing the contents of their postpharyngeal glands onto the cuticle of the beetles. It is well known that this gland is a concentrated source of hydrocarbons that are the same as the ants' cuticular hydrocarbons (Attygalle et al., 1985; Bagnères and Morgan, 1991). These scarab beetles are clearly not highly integrated inquilines in these ant colonies, but for those beetles that manage to survive, their hydrocarbon coat provides enough protection to fool most of the ants, they have access to a very concentrated, high-energy, protein-rich food supply; and they are not subject to other predators while feeding.

Many other inquilines are known from fire ant nests, and some evidence exists that at least three of them may use cuticular hydrocarbons as integrating mechanisms (Wojcik, 1990). These include two beetles, a staphylinid, *Myrmecosaurus spp.*, and the pselaphid *Fustiger elegans* (Fig. 10.13) and a parasitic wasp, the eucharatid, *Orasema* spp. (Wojcik, 1990). Although at least 22 species of staphylinids are known in the genus *Myrmecosaurus*, behavioral data are available for only *M. ferrugineus* (Wojcik, 1980), which is a nonpredaceous symphile that is highly integrated into colonies of *Solenopsis invicta*. Wojcik

---

Figure 10.13. *Adranes taylori* (Coleptera:Clavergidae) walking on larvae of the host ant, *Lasius pallitarsus* (ecological homologue of pselaphids discussed in text).

Figure 10.14. Worker of *Camponotus modoc* feeding a staphylinid beetle, *Xenodusa reflexa* (ecological homologue of *Lomechusa* discussed in text).

Figure 10.15. *Tetradonia marginalis* running in a raiding column of *Eciton hamatus* (ecological homologue of *Pella* discussed in text).

Figure 10.16. *Hetaerius tristriatus* (Coleoptera:Histeridae) soliciting food from a worker of the host ant, *Formica podzolica*.

Figure 10.17. Female of *Dolichovespula adulterina* being pursued by a worker of the host, *D. arenaria* (arrow) during invasion of an early nest.

Figure 10.18. Female of *Vespula austriaca* in trophallaxis with a worker of the host, *Vespula acadica*.

Plate 3.

(1990) analyzed the cuticular hydrocarbons of a *Myrmecosaurus spp*. that was collected with *Solenopsis richteri* and other *Solenopsis spp*. nests from Argentina, and found that the beetle hydrocarbons were similar to those of the ant hosts. Somewhat more data are available for the pselaphid *F. elegans*, which is also a symphile in nests of various *Solenopsis spp*. in Argentina and Uruguay. These beetles are well integrated into the fire ant society and may build up to substantial numbers in the mounds (Bruch, 1930; San Martin, 1968; Wojcik, 1990). The beetles carry ant eggs in their mandibles and are usually found close to the brood pile in the nest, suggesting that they are predators on the ant eggs (Silveira Guido et al., 1963; San Martin, 1968). Like many pselaphids, these beetles have well-developed trichomes that are highly attractive to the ants (Wojcik, 1990), and presumably contain chemical secretions that influence the behavior of the ants. Wojcik (1990) also found that this beetle contains some cuticular hydrocarbons that are the same as those found on the host ants, but did not report data on whether the beetle has other hydrocarbons that are not present on the ants. Somewhat more evidence is available for the situation with the eucharatid *Orasema spp*. and its host ants. All known members of this wasp family are parasites of ants (Kistner, 1982). A summary of the known biology of these parasites is presented by Wojcik (1990). Salient points include the females ovipositing into plant tissues with their scimitar-shaped ovipositor (Heraty, 1985; Clausen, 1940), followed by the newly emerged first larval instar (known as planidia) climbing onto ant hosts and riding phoretically back to the ant nest where they decamp and transfer to the ant brood. The planidia then burrow under the cuticle of the ant larvae (Wheeler and Wheeler, 1937; Wojcik, 1990) and go into arrested development until the ant larvae reach the late prepupal stage (Wheeler, 1907; Clausen, 1941). The parasite then emerges from the host and feeds as an ectoparasite on the host prepupae or pupae (Wojcik, 1990), or sometimes it also feeds at least partially as an endoparasite (Wheeler and Wheeler, 1937). After the mature wasp larva stops feeding, it drops off the ant pupal carcass and is kept with the ant brood, where it is cared for by the host ants. The parasite is assisted in pupation by the worker ants, and the resulting wasp pupae are mixed in with the ant pupae, where they are tended as if they were ants (Silveira Guido et al., 1963; Wojcik, 1990). The adult wasps are fed and groomed by the host ants and appear to be completely integrated into the ant society (Silveira Guido et al., 1963; Williams, 1980; Wojcik, 1990). At some unknown age, the adult parasites leave the ant nest and mate in the surrounding vegetation (Williams, 1980; Wojcik, 1990). Given the degree of integration by the parasites in the ant colony, it is not surprising that when Jouvenaz et al. (1988) and Vander Meer et al. (1989a) examined the cuticular hydrocarbons of an *Orasema sp*. associated with *Solenopsis spp*. in Florida they found the parasites to have cuticular hydrocarbons nearly identical to those of their hosts. The only wasp stages they examined, however, were the pupae and adults, and not the actively feeding ectoparasitic larvae. The wasp pupae have cuticular hydrocarbon profiles that are dominated by ant-like patterns, whereas the newly emerged adult wasps possess hydrocarbon profiles that are only about 75% identical to those of the ants. Once the wasps leave the ant nest, the proportion of ant-like cuticular hydrocarbons rapidly diminish to about 14% (Vander Meer et al., 1989a). Although these authors argue against the parasite larvae biosynthesizing their own hydrocarbons, it is possible, however, that the wasp does. The differences observed between the cuticular hydrocarbons of the parasitoid's pupal form, and the hydrocarbons of the adult parasitoid are very similar to the situation found by Howard et al. (1990a;1990b) for the two species of *Microdon* where larval biosynthesis was unequivocally shown.

4.6. Appeasement Chemicals as Integrating Mechanisms

Long before cuticular hydrocarbons were recognized as being of semiochemical significance, various workers had noted that myrmecophiles and termitophiles interacted with their hosts in such a manner as to suggest a semiochemical interaction, and terms such as appeasement substances were suggested. These observations, and the inferences drawn from them, were strengthened by numerous morphological examinations of the inquilines [reviewed in Hill et al. (1976); Kistner (1979); Hölldobler and Wilson (1990); Kistner and Jacobson (1975; 1990); Kistner (1977; 1984); Shower and Kistner (1977); Spicer and Kistner (1980); and Jacobson et al. (1986)] that showed that many of these inquilines had well-developed glandular systems in exactly those regions of the body of the inquiline where the host ant or termite directed most of their attention. Furthermore, several researchers were able to show that free-living relatives of these inquilines did not have these glandular systems (Kistner, 1979; Table 10.1). Therefore with good reason a number of researchers concluded that these glandular exudates were the major integrating mechanisms by which these inquilines insert themselves into the society of their hosts (Kistner, 1969; 1979; 1990). Despite this reasonable and attractive hypothesis, the supporting chemical documentation has proven to be extremely elusive. A series of studies testing this glandular exudate hypothesis have been conducted by Hölldobler and his associates with European myrmecophiles in the genera *Lomechusa*, (Fig. 10.14) *Atemeles*, and *Pella* (Fig. 10.15) (Hölldobler, 1967; 1970; 1971; 1972; Hölldobler et al., 1981; extensively summarized in Hölldobler and Wilson, 1990). The first two genera have members which are highly integrated myrmecophiles, whereas members of the genus *Pella* are not highly integrated.

The life histories of these beetles are complex, especially those of *Atemeles* spp. which live with different ant hosts at different times of the year. The first challenge for the adult beetles is to gain entrance into an ant colony, and Hölldobler elucidated how this is achieved by *A. pubicollis*. The adult beetle approaches an ant worker (*Formica sp.* or *Myrmica sp.*) and lightly taps the ant with its antennae while simultaneously raising the tip of its abdomen over its prothorax toward the frons of the ant. This region of the beetle contains a complex set of glands that constitute what Hölldobler called the appeasement gland complex. As the tip of the abdomen comes close to the face of the ant a secretion is released which the ant immediately licks up. This secretion seems to suppress any aggressive behavior by the ant. The ant then seems to be attracted by a second set of glands (termed "adoption glands") located along the lateral margins of the abdomen. As the ant approaches this part of the beetle, the beetle lowers its abdomen to allow the ant access to these glands. The openings of these glands are surrounded by bristles that the ant grasps and uses to pick up the beetle. *A. pubicollis* then assumes a characteristic tightly curled posture (death feigning) while the ant carries it into the nest and deposits it into the brood chambers. Hölldobler has postulated that the adoption glands produce a semiochemical mimic of host ant brood pheromones, but as yet no chemicals have been isolated or identified. Once in the nest the adult beetles appear to be accepted by the ant workers, and the workers provide the beetles with regurgitated food in the same manner that they do for other nestmates. The beetles then lay their eggs in the ant nest, where the resulting larvae are also highly integrated myrmecophiles that are fed and protected by the ants. These beetle larvae and adults are, however, predators on the ant brood and on each other.

The behavioral and chemical means by which *A. pubicollis* larvae are accepted by the ants has also been carefully studied by Hölldobler. The beetle larvae show a characteristic begging behavior, involving rearing up as soon as they are touched by an ant worker to make contact with the labium of the ant with their own mouthparts, thereby eliciting a regurgitation of food from the ant. This behavior is apparently a "superstimulus" or "overoptimal" stimulus compared to the weaker begging behavior of the ant larvae, thereby gaining the beetle larvae preferential treatment. The ant brood tenders also frequently and intensively groom the beetle larvae, and Hölldobler (1970) showed that this was facilitated by a series of glandular cells located beneath the dorsolateral area of each abdominal segment of the beetle larva. To demonstrate that a chemical exchange was occurring, Hölldobler carried out an experiment in which he lacquered some beetle larvae and placed them outside of the ant nest along with some freshly killed, but otherwise untreated control beetle larvae. Ants quickly recovered the control beetle larvae and returned them to the brood pile, whereas the shellacked larvae were either ignored or carried to the kitchen middens. He also showed that acetone extracts of beetle larvae, when placed on filter paper dummies or on thoroughly extracted *A. pubicollis* larvae, resulted in the treated objects being carried into the brood area of the nest. Unfortunately, nothing is known of the chemical nature of these extracts.

Hölldobler has also conducted extensive studies with *Lomechusa strumosa*, a staphylinid that occurs with *Formica sanguinea*. Although it has only one ant host, the adult appears to migrate in the spring from the nest where it overwintered, to another nest of the same species, where mating takes place. The adoption procedure for this beetle in the nest of the new host ant nest is complicated. Even though *L. strumosa* has well-developed tergal glands that are filled with benzoquinone, methyl benzoquinone, ethyl benzoquinone, and n-tridecane [typical insect defensive components, Blum (1981)], it almost never uses these compounds against its hosts. Rather, it first presents the ants the trichomes on its legs, slowly circling in one spot with its legs widely spread so that the ants can readily reach the trichomes. At the same time the beetle antennates the ants. Then it slowly points the tip of its abdomen at the ants, and releases a whitish, viscous fluid presumably from the battery of exocrine glands at the tip of the abdomen (Hölldobler and Wilson, 1990). This secretion, which is proteinaceous and has no detectable amount of carbohydrate, is readily consumed, causing the activity level of the ants to moderate. The *Lomechusa* then orients so that the ants have access to additional trichomes along the lateral margins of its abdomen which project above the adoption glands. The ants lick and grasp the trichome bristles, and then drag or carry the beetle into the brood chamber of their nests. Nothing is known of the chemistry of these glands. Inside the nest the adult *Lomechusa* are fed by the ants in a manner very much like that used by the ants to feed their larvae, and the beetle continues to appease the ants with secretions from its abdominal appeasement glands.

In addition to these well-integrated myrmecophiles, Hölldobler and his associates have also studied a number of species in the genus *Pella*. These beetles are important because they are representative of taxa thought to be close to the ancestral forms from which more specialized staphylinid myrmecophiles were derived. These little staphylinids are associated primarily with the formicine ant *Lasius fuliginosus* and do not live within the ant society, but instead occupy the foraging trails and garbage dump of the nest. The adults are scavengers of dead ants, but are also occasionally predators of live ants. The *Pella* larvae are strictly scavengers, feeding on ant carcasses in the kitchen midden.

Like other aleocharines, the adult *Pella* have well-developed tergal defense glands, and if severely attacked by the ants they release a pungent-smelling, brownish acidic secretion from the glands. When the attacked beetle releases the secretion, the ant immediately releases its grip on the beetle and commences to wipe its mouthparts and antennae on the substrate. The beetle then immediately runs away, thus avoiding the other ants that rapidly approach the ant cleaning itself, presumably because that ant has released alarm pheromone. Hölldobler et al. (1981) never observed the beetles to release this tergal gland secretion when they were predators on the ants. The beetles that had been released by the ant, however, sometimes had a smell similar to that of the ant mandibular gland secretions, but Hölldobler et al. suggested that this was because of contamination rather than the beetles having similar compounds in their tergal glands. *L. fulginosus* mandibular gland components are known to be predominately farnesal, 6-methyl-5-hepten-2-one, perillene, and dendrolasin (Quilico et al., 1957; Bernardi et al., 1967), whereas the tergal glands of *Pella* contain p-benzoquinone, ethyl benzoquinone, n-undecane, n-tridecane, and a series of short-chain branched fatty acids (Kolbe and Proske, 1973; Hölldobler et al., 1981). Of these compounds, only the undecane is also known from *L. fuliginosus*, where it occurs in the Dufour's gland and functions as an alarm pheromone (Dumpert, 1972). It is doubtful, however, that *Pella* could be releasing only the undecane and none of the other components in its tergal gland secretions as a propaganda substance. Indeed, Hölldobler et al. (1981) found that isolated tergal gland secretions elicited only a repellent reaction from ants rather than stereotypical alarm behavior, and concluded that the secretions of this gland are for defensive use only.

Kistner and Blum (1971) have studied the interactions of the Japanese myrmecophiles *Pella japonicus* (and possibly *P. comes*) with their host *Lasius spathepus* when they are on foraging trails. When no ants are nearby the beetle walks in the column with its abdomen extended, but when an ant approaches, the beetle recurves its abdomen and presents it to the ant in such a manner that the ant can palpate the beetle's abdomen with its antenna. This usually results in the ant moving on and ignoring the beetle. If for some reason either the ant or the beetle is hyperactive, the ant will sometimes make more aggressive movements toward the beetle, and when it does so the beetle pumps its recurved abdomen back and forth to an extended position two to three times, causing the ant to pull back and reverse the direction that it had been traveling. This reversal behavior was considered to be identical to that produced by the ants when they were exposed to alarm pheromones from their mandibular glands. Kistner and Blum (1971) analyzed the *L. spathepus* alarm pheromone and found it to be mainly citronellal. They also noted that when they picked up live specimens of the *Pella* that the beetles smelled to them like citronellal. In simple field bioassays, using ethanol extracts of the beetles applied to filter paper or ant extracts applied to filter paper, they showed that the foraging ants responded to these treated papers as if they were alarm pheromone-emitting objects. Although Kistner and Blum (1971) postulated that the tergal glands of the *Pella* contained citronellal, and that the beetles were using this compound as an allomone to deter the ants from attacking them, it is possible that the odors noted on the beetle surface were contamination from the ants rather than their own tergal gland products.

The *Pella sp.* studied by Hölldobler et al. (1981) feigned death when attacked mildly by ants, especially in early spring. This appeasement behavior consists of falling on their side, legs and antennae folded tightly against the body and abdomen curved upward. The ants would then either ignore the beetles or pick them up and carry them around for awhile until they were placed on the kitchen midden. Rarely were the beetles harmed by such activities. Later in the year, when the temperatures were higher and the ants were more active, the beetles resorted to a different sort of appeasement behavior that was similar to that used by *Lomechusa* and *Atemeles*. Like these beetles, *Pella* has an appeasement gland complex at its abdominal tip, and like them it appears that at least one of these glands produces an appeasement substance that "calms" the ants. *Pella* also produces a white, viscous droplet from its abdominal tip that is readily consumed by the ants. Again, nothing is known of the chemical nature of these secretions.

This story would not be complete without at least a brief mention of the inquilines associated with army ants in Central and South America (Akre, 1968; Akre and Rettenmeyer, 1966; Akre and Torgerson, 1968, 1969; Torgerson and Akre, 1969) and Africa (Kistner, 1968, 1993). Of particular interest, again, are staphylinids that run the gamut from generalized predators (*Tetradonia*, Fig. 10.15) to extremely myrmecoid species that are even the same color as their hosts. All follow the chemical trails of the ants and are highly integrated into the ant societies. In addition, there are several species of histerids (Fig. 10.16) that are also well integrated into the ant colonies. All groom the ants, and species of *Euxenister* have tibial brushes that aid in acquiring the colony odor (Akre, 1968). Although there are no data on chemicals produced or used by these inquilines, they contain numerous exocrine glands that possibly produce secretions that aid in their integration into the ant colonies. An excellent summary of these and other studies of guests found with army ants are found in Kistner (1982).

5. Social Parasitism and Slavery

5.1. General Concepts and Definitions

Social parasitism occurs among all groups of social Hymenoptera, with approximately 270 species of socially parasitic ants (Heinze, 1991), and 3,500 species of socially parasitic bees (Bohart, 1970; Field, 1992) but only a small number of socially parasitic vespid species among the approximately 860 social species (Akre, 1982; Field, 1992). Social insect colonies are incredibly rich reservoirs of stored food and slave labor, and many species of social and solitary insects have evolved to take advantage of this bounty. In nearly all cases the social parasites are closely related sympatric species that have probably evolved from conspecifics that availed themselves of the nest or brood of their sisters (Bourke and Franks, 1991; Buschinger, 1986; 1990; Heinze, 1991; LeMoli and Mori, 1987). Presumably this close relatedness has allowed the parasites ready access to the host colonies as they evolved toward workerless caste systems. Some of these social parasites have also evolved slightly different morphology, which when coupled with their behavior makes them efficient social predators.

The complexity of social parasitism in insects is diverse and can conveniently be divided into four basic types: (1) Xenobiosis or guests; (2) temporary parasitism; (3) permanent parasitism with slavery; and (4) permanent parasitism without slavery or inquilinism (Buschinger, 1986). The first and last categories are at present known only from ants, but the other two categories are known for ants, bees, and wasps. Although the term slavery has been preempted by myrmecologists, there is no reason that it cannot also be used for the bees and wasps. As Wilson (1975) notes, however, insect slavery is not comparable to human slavery, and it is only one among many genetic adaptations that insects have evolved to meet the demands of life. Knowledge of the extent and mechanistic structure of social parasitism is greatest for ants (Hölldobler and Wilson, 1990) and will be discussed in some detail. Although much less is known about social parasitism in bees and wasps, it appears as if they may use many of the same strategies as ants (Michener, 1974; Akre, 1982). A summary of the major findings with bees and wasps will be presented first, although additional information on these insects will also be presented later in sections dealing with specific mechanistic aspects of social parasitism.

Approximately 15% of the 4,000–5,000 species of bees in North America are parasitic, with a total of 3,500 species of parasitic bees worldwide (Bohart, 1970). Host bees rarely attack these parasites or try to expel them from the nest. Obligate permanent social parasites (*Psithyrus* spp.) of bumble bees display an array of types of behavior in invading and becoming established in the nest (Fisher, 1983a;1983b; 1983c; 1984a; 1984b; 1985; 1987a; b; Fisher and Sampson, 1992; Fisher and Weary, 1988). Some species invade only very small, young colonies with few workers, coexist with the foundress queen, and seem unable to control the ovarian development of workers (Fisher, 1983a; 1985; 1987b). Other species invade larger colonies, displace or kill the queen, and suppress worker ovarian development. The latter observation strongly suggests that the parasite queen is using pheromonal control in preventing ovarian development [the evidence suggests queens of some species of *Bombus*, including temporary parasites, also have queen pheromones (summarized in Free, 1987)]. However, if some species of *Psithyrus* can control the ovaries of workers by physical dominance such as mauling, this may allow these species to have more host species. Those species proposed as having a chemical that helps in appeasement or ovarial suppression may be relegated to fewer and more specific hosts. At this time these possibilities are somewhat speculative.

North American yellowjackets have two permanent social parasites. *Dolichovespula adulterina* (Fig. 10.17) occurs with its hosts *Dolichovespula arenaria*, *D. alpicola*, and *D. norwegica*; and *Vespula austriaca* (Fig. 10.18) takes over colonies of *Vespula acadica*. Neither of these parasites has workers. *D. adulterina* females passively invade colonies of the aerial yellowjacket, *D. arenaria*, but ultimately dominate the foundress queen (Greene et al., 1978). Initially there is a period of coexistence with both the parasite and the host queen present in the colony. The parasite then gradually starts to take control of the colony by dominating the queen and the workers and by laying eggs into cells of the nest. Conflict with the queen usually results, and the host queen is frequently killed by the female parasite. Workers are definitely attracted to the parasite, but the parasite is unable to control ovarian development of the workers after the queen dies or is killed, and she ultimately kills the most aggressive workers (those with the greatest ovarian development). While it is unclear what role chemicals might play in the interactions among these species, behavioral evidence suggests that the inquiline produces a compound or compounds that not only pacified its hosts but attracted host workers.

Species of *Polistes* are also attacked by social parasites, all of which are species of *Sulcopolistes* [=*Polistes*, see (Carpenter, 1991)] [summarized by (Akre, 1982)]. *P. sulcifer, P. semenowi*, and *P. atrimandibularis* are permanent social parasites of other species of *Polistes* (Turillazzi 1992). The usurpation of *Polistes biglumis bimaculatus* nests by *Polistes atrimandibularis* (Lorenzi *et al.* 1992) is similar to usurpation behavior previously reported for *D. arctica* (=*adulterina)* by Greene *et al.* (1978). The parasite invades the nest when only the foundress is present, and initially the invader avoids conflict with the queen. After a few hours she becomes active and dominant. This sequence strongly suggests mediation and control by semiochemicals (Cervo et al., 1990). Unfortunately, while many other species of presumed social parasites are known, information on these parasites is nil.

As mentioned, some of the best examples to illustrate all four types of social parasitism occur among the ants. Guests (also known as xenobiotic individuals) are insects that live within the nest material or nest walls of their hosts. They build their own nests that are well separated from the brood chambers of their hosts, and they care for their own brood. They are dependent upon the host in that they steal food from their hosts while they are engaged in trophallaxis, thus being trophic parasites. This trophic parasitism is thought to have arisen from plesiobiosis, the habitual nesting of two species in very close proximity. A well-documented example of guest species are ants in the genus *Formicoxenus* (Leptothracini) that have small colonies that live in the nests of larger species of *Formica* or *Myrmica* (Lenoir et al., 1992a; 1992b).

Temporary parasitism involves species that depend on their host species only during the initial founding of new colonies. In a common scenario, the parasitic queen conducts her nuptial flight, mates, and then seeks out a host colony which she enters and displaces the resident queen, either by killing her or by chasing her away. The parasitic queen then takes over the egg laying, thus producing future workers. The existing workers of the original host queen continue to work for the colony until they die, thus allowing the parasite queen to focus exclusively on egg production, and the ultimate generation of a colony consisting solely of her species. Representative examples of this level of social parasitism include the ants *Lasius umbratus* (the parasite) and *L. niger* (the host) (Gösswald, 1938), and *Bothriomyrmex spp.* (the parasite) and *Tapinoma spp.* (the host) (Forel, 1929).

Permanent parasitic species are those in which the invading queen kills or drives out most of the host workers and the host queen, and then takes over the remaining host brood. When the host brood emerge as new adult workers, they become slaves of the parasite queen as they take over from her the normal colony duties. When the brood of the parasite queen emerges, they have one functional role: conducting raids on other colonies of the host species to gather additional brood to become workers within the parasitized colony. Examples of this level of parasitism are the ants *Polyergus spp.* (the parasite) and *Formica spp.* (the host) (Topoff, 1990).

Finally, examples are known of permanently parasitic species that do not enslave any host workers, but rather live with and depend on their hosts for their entire lives. In this case, the parasite usually lacks a worker caste altogether, and puts all of her reproductive output into producing future reproductive forms. The parasite queen lives alongside of (or on top of) the host queen, and the host workers raise the parasite queen's sexual brood in the same manner that they do their own brood. The classic example of this stage of parasitism is that of the ant *Teleutomyrmex schneideri* and its host *Tetramorium caespitum* (Kutter, 1969). Additional examples of ants that exemplify each of these four stages of social parasitism are described in Hölldobler and Wilson (1990).

The proximate and ultimate factors that have resulted in these fascinating relationships have been the subject of numerous investigations for well over a hundred years. Much of the effort has gone into trying to understand the behavioral, biochemical, and physiological mechanisms whereby the invading parasite queen is able to first locate and then enter a host colony and survive long enough to either take over or be accepted, and in the similar mechanistic considerations that control the actions of her subsequent brood. Although we are far from having a solid understanding of these processes, much progress has been made in recent years, and our understanding is rapidly accelerating as more scientists delve into the biology of these host-parasite relationships.

5.2. Host Location

No matter what stage of social parasitism is involved, the initial problem of the dispersing parasite queen or foraging slave scout is to locate a suitable host colony. Although vision is certainly an element in this searching behavior, most investigators have assumed that the major cue used is some sort of chemical emanating from the host nest. As one might expect, a demonstration of this in a field situation is extremely difficult, and accordingly the best evidence available on nest location from odor cues comes from laboratory studies. Bees in the genus *Psithyrus* are obligate brood parasites of bumble bees. Female *Psithyrus* emerge in the spring after their host bumble bee queens have emerged and initiated their nest and brood rearing. The parasitic females search for the host nest by flying low over the vegetation or by walking on the ground looking for the host nests, which are often located in abandoned rodent burrows (Alford, 1975). Odors coming from the bumble bee nest were suspected by early students of these bees to be important host-finding cues for the parasite (Wheeler, 1928; Free and

Butler, 1959). Cederberg (1977) showed a trail-following behavior by *P. rupestris* females to worker *Bombus* extracts streaked on a paper substrate, and Cederberg (1979, 1983) suggested that *P. rupestris* followed trails that host workers of *Bombus lapidarious* deposited on the ground as the *Bombus* workers wandered in and out of the nest. Fisher (1983c) used laboratory bioassays to show that the parasite *P. ashtoni* was able to discriminate between volatile odors of nesting material from two bumble bee hosts (*B. terricola* and *B. affinis*), a non-bumble bee host (*B. bimaculatus*), and nest material alone. Direct contact with these materials was not necessary, strongly suggesting the parasite is able to use this sort of airborne cue in a field situation without having to actually track a substrate borne trail cue. Fisher (1983c) did not determine how far away the parasite could detect such odors. Plath (1934) noted that while excavating nests of *B. bimaculatus* from the soil that *P. citrinus* females flying nearby were immediately attracted and landed on the opened nest. Normally the host bumble bee nests are intact and odor emission from the nest ought to be somewhat lower. Fisher (1983c) notes that *Psithyrus* females characteristically fly in a slow searching pattern close to the ground, often alighting and searching extensively in the area under leaves and twigs, investigating all dark areas that might contain rodent burrows, although he was unable to link this observed behavior with known sources of nest odor. He notes that the underground nests of the two major host species of *P. ashtoni* often have extensive tunnels that lead underground to the nest proper, and the *Psithyrus* female may at first be guided primarily by sight to locate appropriate looking nest entrances, and then only at close distance use odor cues emanating from the nest opening. Fisher et al. (1993) have recently identified several volatile secretions from two host *Bombus* species that are possible odor cues used by two species of *Psithyrus*, including a series of methyl ketones, esters, and hydrocarbons. They showed that these mixtures were distinctive for each host species and that the parasites preferentially responded to host odor sources over nonhost odor sources.

A different approach to locating the host nest is used by the slave-making ant queens in the genus *Polyergus*: they either follow (Cool-Kwait and Topoff, 1984; Marlin, 1968; Talbot, 1968) or participate (Topoff, 1990) in raiding swarms of their mother nest to a new *Formica* nest. In either case, the queen's penetration of the new host colony is likely facilitated by the disorganization produced by the raid.

The problem of parasite queen dissemination by guest parasites has also been examined. Stäger (1925) found that *Formicoxenus nitidulus* follows the natural above-ground foraging trails of its host, the red wood ant *Formica pratensis*, and confirmation of this observation was made in the laboratory by Elgert and Rosengren (1977). Lenoir et al. (1992a; 1992b) have shown that *Formicoxenus provancheri* also readily follows the trail pheromone of its host *Myrmica incompleta*, whose foraging trails are mainly subterranean. Inasmuch as *M. incompleta* forms very large unicolonial populations in dispersed habitats with extensive and numerous abandoned galleries (Lenoir et al., 1992a; 1992b), the ability of *F. provancheri* to follow the trail pheromone of its host is clearly adaptive.

5.3. Queen Admittance into the Host Colony

Once the parasite queen has found the host nest, the next major challenge facing her is how to gain admittance into the colony without being killed by either the host workers or the resident host queen, since it has been shown that slave species are able to recognize slave-makers as enemies (Alloway, 1990). In keeping with the diversity of types of social parasites, there are a number of strategies which have been taken by these insects, including sometimes more than one approach by the same parasite. Thus Stuart (1984) found with laboratory studies that the parasite queen of *Harpagoxenus canadensis* when entering a *Leptothorax spp.* colony would engage in one of two tactics: in one tactic she would move quickly and erratically about the nest interior so as to elicit attacks from the *Leptothorax* workers, raising sufficient attention to herself (causing her to be perceived as a much larger raiding party?) so as to cause the host workers to release enough alarm pheromone to cause the colony to abandon the nest. This tactic, while leaving the nest in control of the parasite queen, also results in a nest with severely depleted host brood to serve as future slaves. An alternative approach used by the *H. canadensis* queens was described by Stuart (1984) as a passive one, in which the parasite queen enters the nest slowly and reaches out with her antennae to gently touch any host workers that she encounters. In contrast to the frantic technique, this furtive technique does not result in alarm behavior by the host workers, and allows the parasite to take over the nest while keeping all of the host brood and to even be accepted by a large proportion of the resident adult workers. What little aggression there was toward the parasite was limited to one-on-one encounters and appeared to dissipate quickly. The parasite in this case seized the aggressive individual in its mandibles and dragged or carried it outside of the nest, where she released the offending worker unharmed. This behavior was in every respect different from the aggressive behavior observed during the frantic approach. This dichotomous behavior by the invading parasite queens is certainly unusual and is worthy of further study. As Stuart (1984) notes, however, the taxonomic status of the host he used is in some confusion (Heinze et al., 1992), and although he believes that the different invasion strategies he observed were in response to a single host, it is possible that what he observed were responses to two different hosts. Even if this is true, it is still intriguing that one species of parasite could modify its behavior to this degree.

Some tantalizing information on a different tactic was obtained by Allies et al. (1986) in laboratory studies with the cuckoo ant *Leptothorax kutteri* and the slave-making ant, *Harpagoxenus sublaevis*, both of which share a common host, *Leptothorax acervorum*. They found that if they held single *L. kutteri* queens with single *L. acervorum* workers for a period of one hour, that the latter always attacked the *L. kutteri* queen, but apparently without immediately fatal results for either participant, since the *L. acervorum* workers were then utilized for tests in which they were reintroduced into their colony. Allies et al. (1986) observed that when the *L. kutteri* queen was attacked, she consistently smeared the *L. acervorum* worker with a clear, viscous fluid that emanated from the tip of the queen's gaster. This behavior was not observed in interactions between *L. acervorum* workers and conspecific queens or workers from another nest. *L. acervorum* workers so marked when returned to their home colonies were attacked by their nestmates. To try to find out where the clear, viscid fluid originated, Allies et al. (1986) prepared crude extracts of the Dufour's gland or poison vesicles of queen *L. kutteri*, solvent alone (mineral oil), and extracts of queen *L. acervorum* Dufour's gland and applied these extracts to *L. acervorum* workers and then released the marked workers back into their home colony. Only the workers marked with Dufour's gland extracts from the *L. kutteri* queen were attacked. Allies et al. (1986) therefore concluded that the *L. kutteri* queens were using a propaganda substance derived from their Dufour's gland to cause workers of their host species to attack each other and cause enough confusion to allow the intruding parasite queen to gain entrance into the nest where she could hide long enough to acquire some of the nest odor. The strength of their argument is diluted somewhat by their observation that 15 separate attempts to introduce a single *L. kutteri* queen into an established *L. acervorum* laboratory colony all resulted in the death of the introduced parasite queen. Clearly, additional studies are needed to elucidate the functional significance of the use of Dufour's gland secretion by *L. kutteri* queens in attempts at colony usurpation.

Allies et al. (1986) also examined the mechanism by which founding queens of the slave-making *H. sublaevis* gained entry into colonies of *L. acervorum*. Unlike *L. kutteri*, the queens of this species possess large, scissor-like mandibles with which they dismember opponents. Allies et al. (1986) found that *H. sublaevis* queens also possess hypertrophied Dufour's glands, and that they too marked *L. acervorum* workers with clear, viscous secretions from that gland, and that the marked workers were attacked by their nestmates when they returned to the home nest. This phenomenon was first reported for this parasite-host complex by Buschinger (1974), and similar reports exist for slave raids by *H. americanus* on its North American *Leptothorax* hosts (Alloway, 1979). Allies et al. (1986) unfortunately did not report whether the introduction of single *H. sublaevis* queens into *L. acervorum* laboratory colonies resulted in their successful establishment. Assuming that the *H. sublaevis* queens are more effective than *L. kutteri* queens were in their marking of host workers to cause internecine battles, then it is not difficult to see how this would be adaptive for the intruding parasite, since her goal is to drive all adult host workers away from the brood.

The Dufour's gland contents of invading parasite queens do not, however, always lead to the host workers attacking one another. The western slave-making species *Polyergus breviceps* is associated with several *Formica spp.*, including *Formica gnava*. The *P. breviceps* queens are incapable of rearing brood, and accordingly must penetrate the *Formica* colony, kill the resident *Formica* queen and gain the acceptance of the host workers. Topoff et al. (1988) have found that the parasite queen accomplishes this in the following manner. When she first enters the *Formica* colony she is immediately attacked by the *Formica* workers, who pin the intruding parasite queen down and spray her with formic acid. The *Polyergus* queen responds by vigorously flexing her pinned appendages and by piercing the heads of her attackers with her formidable mandibles until she is able to break free. She then immediately flings herself onto the *Formica* queen, using her mandibles to bite the host queen all over her body, usually killing her in less than an hour. During this fight to the death, the gaster of the *Polyergus* queen was consistently oriented toward the *Formica* queen. Workers who attempted to intervene usually quickly withdrew, wiping their mandibles on the substrate (Topoff et al., 1988). Once the host queen was killed, the parasite queen then groomed herself while standing on top of the killed queen, or she rolled on the substrate while holding on to the dead queen, or laid still with the dead queen on top of her. Topoff (1990) suggested that this rubbing and rolling behavior is perhaps for the purpose of transferring host queen odors to the *Polyergus* queen, thus hiding her alienness. Within as few as 15 minutes after such activities, host workers began to groom the parasite queen, with only occasional attacks on her. Within 12–48 hours the parasite queen was completely accepted into the colony and was no longer attacked. The successful invasion of a *Formica* colony in the laboratory required the initial presence of the *Formica* queen.

This sequence of actions led Topoff et al. (1988) to postulate that the *Polyergus* queen was often using secretions from her Dufour's gland as part of her means of overcoming the colony's initial resistance to her, and Topoff (1990) suggests that she uses these same chemicals to continue to quell any antagonism toward her in future days. To test this hypothesis, they used a modification of the experimental protocol of Allies et al. (1986), treating a *Pogonomyrmex occidentalis* worker with aqueous extracts from the Dufour's, pygidial, and poison glands of queen *P. breviceps* [Topoff et al. (1988) argued that they could not fairly use *P. breviceps* queens for tests because they could not keep the queens from releasing additional chemicals during the test, thus confounding their interpretations; we would note also that Dufour gland contents, at least, have very limited water solubility]. When these treated *Pogonomyrmex* workers were introduced into a laboratory nest of *Formica*, those treated with the Dufour's gland extract caused the workers to withdraw and to show a compete lack of aggression, whereas the other treatments resulted in the ant being immediately attacked by the *Formica* workers. Topoff et al. (1988) thus concluded that the secretions from the Dufour's gland of the *P. breviceps* queen were acting as an appeasement substance rather than as a propaganda substance. Although Topoff et al. (1988)

conducted no chemical analyses on the contents of the Dufour's gland of this species, they noted that Bergström and Löfquist (1968) had reported that the Dufour's gland of a related European species, *Polyergus rufescens*, contained many of the same volatile components as its two slave species, *Formica fusca* and *F. rufibarbis*, and that therefore, by analogy, perhaps *P. breviceps* queens were mimicking the volatiles of their host species. Perusal of Bergström and Löfquist (1968) reveals that the chemicals shared by *P. rufescens* and its hosts included n-nonane, n-decane, n-undecane (all somewhat minor), and a farnesene isomer (in moderate abundance). Notably, the major components of the Dufour's gland of the two *Formica spp.* (an extensive series of normal and branched alkanes and alkenes in *F. fusca* and medium-chain-length acetates, corresponding alcohols and 2-tridecanone, as well as the hydrocarbons in *F. rufibarbis*) were not present in the *P. rufescens* Dufour's gland volatiles, despite an assertion to the contrary by Topoff et al. (1988). We suggest, therefore, that Topoff et al.'s (1988) conclusion that the Dufour's gland secretion of the *Polyergus* queen is mimicking that of the host *Formica* queen is possibly premature. Similarly, the hypothesis that the invading parasite queen acquires persistent host odors from the dead queen is also in need of additional testing. Even if the parasite acquired some cuticular-borne chemicals from the dead queen, it is rather unlikely that they would continue to persist for very long as the cuticular lipids of the parasite queen are undoubtedly being continuously replenished by biosynthesis throughout her life span. It is possible that the parasite queen has either cuticular hydrocarbons or other cuticular lipids that mimic those of the host ants, as is known for various non-ant inquilines [see references in (Howard, 1993)], but until the chemicals of the two species that Topoff et al. (1988) worked with are examined directly, all of these conclusions must be considered as working hypotheses.

In each of the slave-maker:slave systems discussed above, chemicals from the Dufour's gland were directly implicated in successful parasitization of the host colony. Ollett et al. (1987) have examined the chemistry of the Dufour's glands from the slave-making species *Harpagoxenus sublaevis* and Ali et al. (1987) have examined the correponding Dufour's gland products from two enslaved species, *Leptothorax acervorum* and *L. nylanderi*. Both researcher groups found that the Dufour's gland contents were complex mixtures of hydrocarbons and sesquiterpenes. The mixture of hydrocarbons in *H. sublaevis* is close in composition to that of *L. acervorum*, but rather different from that of *L. nylanderi*. In both *H. sublaevis* and *L. acervorum* a heptadecene and a heptadecadiene are major components, but the slave-maker also has substantial quantities of (E)—β-farnesene which is absent from the enslaved *Leptothorax* species (Ali et al., 1987). Similarly, the *Leptothorax* species have a different and, as yet only partially identified, sesquiterpene called Tetramorene-2 which is absent from *H. sublaevis*. No essential differences were found in the Dufour's glands composition of the mated queens or workers of *L. acervorum* (Ali et al., 1987). These authors conclude that the sesquiterpenes in the slavemaker may indeed be responsible for the disruptive effects caused by the raiding *H. sublaevis* workers in the *L. acervorum* nest. Ollett et al. (1987) had earlier noted that these sesquiterpenes would have limited stability in air, but that dissolved in the much greater quantities of alkanes and alkenes, they would probably have an adequate lifetime to meet the disruptive behavioral needs of the raid, while dissipating sufficiently to not cause alarm by their *Leptothorax* slaves back in their home nest.

5.4. Queen Usurpation and Acceptance

Despite the above uncertainties, it is clear that successful parasitic queens manage to find a way to exist in the host colonies and to be accepted as full-fledged nestmates of their slaves or hosts. Little is known of the detailed means by which this occurs, but Franks et al. (1990) have discovered one possible set of options that is used by the cuckoo ant *Leptothorax kutteri* in its interactions with its host *Leptothorax acervorum*. In detailed behavioral studies with these ants, Franks et al. (1990) found that *L. kutteri* queens groomed both the *L. acervorum* queens and workers at extremely high rates, but that they themselves were not groomed by the host workers at rates any greater than other colony workers. Since the *L. kutteri* queen never does any work for the host colony, or even care for its own brood, Franks et al. (1990) hypothesized that the parasite queen engaged in such intensive grooming behavior as a means of acquiring host odors. As a preliminary test of that hypothesis, these scientists showed that if host workers, queens, and parasite were isolated from each other and maintained on identical diets for ten days and then returned to the main colony, the isolated workers and host queen were quickly reaccepted with only a couple of seconds of grooming, whereas the *L. kutteri* queen was thoroughly examined by a number of workers and intensively groomed for several minutes. Franks et al. (1990) concluded from this that the parasite which had been unable to intensively groom host ants for ten days had lost or had diminished quantities of an important colony recognition signal. To further test their hypothesis, Franks et al. (1990) made cuticular extracts of both the hosts and the parasite and examined them by gas chromatography-mass spectrometry (GC-MS). Although a large number of sometimes poorly resolved components were found, these authors focused on the major components which included 18 hydrocarbons and 6 fatty acids. The cuticular hydrocarbons chosen were a series of n-alkanes, 3-methyl and 11-methylalkanes, 9-alkenes, and dienes of unknown double-bond location and stereochemistry. The carbon number distribution ranged from C27 to C34. The fatty acids were (Z)-9-14:1, 14:0, (Z)-9-16:1, 16:0, (Z)-9-18:1, and 18:0. Franks et al. (1990) then constructed a similarity index [$SI=S(x-y)*(x-y)/E$, where x and y are the normalized fractions of the same compound in the individuals being compared, and E is the average amount of that compound for all four types of individuals examined (*L. kutteri* queens, *L. acervorum* workers from a nest containing *L. kutteri* queens, *L. acervorum* workers from nests without *L. kutteri* queens, and *L. acervorum* queens)]. The smaller the index, the greater the similarity, and the results were then summed by class (hydrocarbons versus fatty acids). They found that the *L. kutteri* queens and *L. acervorum* workers that had been exposed to the parasite were most similar to each other in terms of fatty acid profiles, and only slightly less so in terms of cuticular hydrocarbons. In both cases, they showed a high degree of correspondence, and were more like one another than were *L. acervorum* workers and queens that had not been exposed to the parasite. Franks et al. (1990) therefore concluded that the *L. kutteri* queens were to a large extent passing themselves off as colony members by possessing cuticular lipid profiles that were very similar to host colony members, and that this congruence was a result of the intense grooming activity exhibited by the *L. kutteri* queens.

The means by which one temporary social parasite invades its host species has also been worked out. *Bothriomyrmex syrius* is a temporary parasite of *Tapinoma spp*. in Israel and presumably behaves very similarly during host colony invasion to various *Bothriomyrmex* species studied by Santschi (1920) in Italy. The parasite female sheds her wings after the nuptial flight, and then searches for a *Tapinoma* nest where she is accosted by the *Tapinoma* workers and dragged into the nest. The parasite female offers no resistance, and once inside the nest she climbs on the back of the host queen and slowly decapitates her. The *Tapinoma* workers do not seem to know that their queen has been eliminated, and they proceed to feed the parasite and her brood until all of the *Tapinoma* workers have died off. By that time the *Bothriomyrmex* workers have eclosed, and they assume normal colony maintenance functions. Santschi (1920) concluded that the parasite queen succeeded for two reasons: she is the same general size and color as the workers of the *Tapinoma* colony that she is attempting to invade, and more importantly, the *Bothriomyrmex* female possesses the typical *Tapinoma* smell. In 1956, Trave and Pavan chemically identified the compound which accounts for the *Tapinoma* smell as 6-methyl-5-heptene-2-one and showed that it originated from the anal gland of the ants. Lloyd et al. (1986) thus hypothesized that the Israeli *Bothriomyrmex* queen would also possess this ketone in her anal gland, and this was how she was able to pass inspection while invading the host colony. As expected, the parasite queen indeed did have substantial quantities of this ketone in her anal gland, as well as trace quantities of the terpene α-terpineol. This ketone is also commonly found in the anal glands of many other dolichoderine ants (Trave and Pavan, 1956; McGurk et al., 1968; Hefetz and Lloyd, 1983). The *B. syrius* workers produced after the host workers are dead do not have any of this ketone in their anal glands, but rather have (*Z*)-3-dodecenoic acid, (*E*)-2-dodecenoic acid, and oleic acid as their major components. Inasmuch as they do not need to interact with any of the *Tapinoma* workers, it is perhaps not surprising that these workers do not possess any of the *Tapinoma* ketone.

5.5. Propaganda Substances in Social Parasitism and Slavery

In addition to studies on the usurpation and acceptance behavior of social parasite queens, a fair amount of attention has also been given to the chemical ecology of the worker forces of these parasitized colonies. One of the best documented of these situations is the use of chemical allomones or propaganda substances by raiding slave-making ants in the genera *Formica* and *Polyergus*. As detailed above, Bergström and Löfquist (1968) reported that the slave-making ant *Formica sanguinea* had Dufour's gland constituents that were very similar to those of its two primary slave species, *Formica fusca* and *F. rufibarbis*. Although these authors indicated that they knew that Dufour's gland chemistry was frequently involved with alarm pheromone activity, they did not ascribe any particular behavioral significance to their observation that the slave-maker and its slaves had similar Dufour's gland chemistry. In the discussion section of their paper on stingless bees, Blum et al. (1970) noted the results of these scientists and hypothesized that the *F. sanguinea* raiding workers were using the contents of their Dufour's gland in an allomonal manner to trigger alarm in the raided *Formica spp.* colonies, and that "the pheromones secreted by raiding ants may be primarily responsible for the breakdown in resistance which characterizes the behaviour of the host workers in raided nests." They further stated, "The dissolution of organized social behaviour which occurs when certain species of ants are exposed to high concentrations of alarm pheromones almost guarantees that they will be incapable of maintaining sustained resistance to the raiding ants" (Blum et al., 1970). In 1971, Regnier and Wilson published their study of two slave-making species in the *Formica sanguinea* group (*F. pergandei* and *F. subintegra*) and showed that these two slave-makers did indeed use their Dufour's gland contents in exactly the manner predicted by Blum et al.(1970). They found that the hypertrophied Dufour's glands of both *F. pergandei* and *F. subintegra* contained approximately 10% of their body weight as three esters: decyl acetate, dodecyl acetate, and tetradecyl acetate, along with smaller quantities of tridecane. They further demonstrated that these esters were alarm pheromones for these ants by crushing a single Dufour's gland onto an applicator stick, and presenting the treated stick to workers in a laboratory nest. The exposed workers were at first highly excited and attracted to the odor source, and several workers attacked the stick. The general alarm behavior did not die out for over 30 minutes. Similar results were obtained when pure acetates at a concentration equivalent to that found in a single ant were introduced on a paper square into the colony.

The major host species of these two slave-makers is the little ant *Formica subsericea*, and Regnier and Wilson (1971) found that its Dufour's gland contained tridecane and 2-tridecanone but no acetates. When they exposed these ants to the Dufour's gland contents of either of the slave-makers, or to single gland equivalents of the pure acetates, they were not attracted to the odor source, but rather tended to scatter quickly if exposed to it for more than a few seconds.

This observation led Regnier and Wilson to test the hypothesis that during their raids the slave-makers sprayed their victims with the contents of their Dufour's gland, and this resulted in an abandonment of the nest. Indeed, they did observe the raiding slave-makers to spray their victims, and gas chromatographic analysis of the sprayed victims showed that they had been coated with an amount of secretions of the slave-maker Dufour's gland nearly equal to the full equivalent of the victims own Dufour's gland contents [we estimate based on Fig. 3 of Regnier and Wilson (1971) that each slave-maker coats its victim with perhaps one-fifth of its total Dufour's gland contents]. Victims so sprayed did indeed run away and abandon the nest. As Regnier and Wilson (1971) noted, the acetates are highly effective propaganda substances because they are released in such large quantities, and because being somewhat less volatile than many ant alarm pheromones, they remain in the nest environment for long periods of time. Indeed, it has long been known (Huber, 1810) that after slave-making raids by these *Formica spp.* (and related congeners) that the raided workers that have dispersed frequently never return to their old nest . Although Regnier and Wilson (1971) base their argument on the semiochemical properties of the acetates, the Dufour's gland of the slave-maker ant also contains moderate quantities of tridecane, a probable alarm pheromone for *F. subsericea*, and this too may play a role in the observed behavior of the raided ants.

5.6. Cuticular Hydrocarbons, Chemical Mimicry, and Slavery

In addition to these various exocrine gland products, cuticular hydrocarbons have frequently been postulated to serve as species recognition cues in ants (Howard, 1993), and indeed heterospecific ant colonies do not usually tolerate one another, nor do they have similar cuticular hydrocarbon profiles. It is possible, however, under some circumstances to mix species that would ordinarily fight each other and produce amicable mixed colonies (Fielde, 1903; 1904; 1905; Hölldobler and Markl, 1989)). Errard et al. (1988) and Bagnères et al. (1991) have conducted experiments with two such species, *Formica selysi* and *Manica rubida*. They have shown that the two species when held separately have their own distinctive species specific cuticular hydrocarbon profiles [only saturated alkanes for *M. rubida* but saturated and unsaturated (monoenes and dienes) for *F. selysi*]. When mixed in artificial laboratory colonies, however, the two species each acquired some hydrocarbons of the other. These changes were of both a qualitative and a quantitative nature, and allowed the ants to inhabit the nest without displaying interspecific aggression. The implications of these findings for social parasitism are obvious. The authors contend that the observed changes are the result of active biosynthetic processes rather than a passive transfer mechanism.

Not only have cuticular hydrocarbons been strongly implicated in species and nestmate recognition processes of ants, they are increasingly also being shown to be important in brood and caste recognition (Bonavita-Cougourdan et al.,

1989; 1990). These scientists have conducted many experiments in which they tested brood recognition both within a colony and between colonies of *Camponotus vagus*. They showed that workers of a given colony of *C. vagus* would transport their own larvae in preference to non-nestmate larvae, and that a primary cue governing this choice was the cuticular hydrocarbons present on the larvae. As in studies with a different *Camponotus* species (Howard et al., 1990a), the *C. vagus* workers and larvae have the same cuticular hydrocarbons, but in substantially different proportions. Also, as in their interspecific adoption experiments (Bagnères et al., 1991), they found that when *C. vagus* larvae from one colony are placed in another colony, their cuticular hydrocarbon profile slowly changes during their adoption period to approach that of their new host. Bonavita-Cougourdan et al. (1989; 1990) have also clearly shown that all caste forms (larvae, workers, alate males and females, and queens) have distinctive cuticular hydrocarbon profiles.

An underlying assumption of all of these studies is that the cuticular hydrocarbons are actually used by the insects themselves for the postulated recognition process. Although most of the earlier studies were based on circumstantial evidence [reviewed in (Howard, 1993)], more recent work has resulted in bioassay data that directly show that at least ants (and probably all social insects) do indeed use cuticular hydrocarbons as at least one cue for recognition of species, colony, or nestmate status (Howard et al., 1982c; Haverty et al., 1988; Bagnères et al., 1991). These data were primarily obtained by applying hydrocarbons to surrogate dummies of one sort or another and observing the response of the social insects to the surrogates. A more direct demonstration of the role of hydrocarbons in species recognition was provided by Bonavita-Cougourdan et al. (1987) where they placed the contents of *Camponotus vagus* nestmate or non-nestmate postpharyngeal glands directly onto live nestmates. When these individuals were treated with nestmate glandular chemicals no agonism was seen. When they were treated with non-nestmate glandular material they were immediately attacked. As noted earlier, the contents of ant postpharyngeal glands are identical to the cuticular hydrocarbon profiles.

Hydrocarbon profiles are not necessarily a straightforward phenotypic parameter, however. Obin (1986) and Vander Meer et al. (1989b) reported the fire ant *Solenopsis invicta* has cuticular hydrocarbon profiles that are distinctive for any given colony and appear to change with time. As they note, recognition cues on an insect's cuticle can be heritable characters or from an environmental source, and accordingly it is to be expected that the cuticular hydrocarbon profile of an individual will be dynamic rather than static. Although we agree that an individual's hydrocarbon profile probably varies with time, the experiments in Vander Meer et al. (1989b) are slightly confounded by changes in both colony size, caste composition, and age structure during the four to eight month observational period. As the authors note, no one as yet has been able to separate the relative importance of genetic variation from environmental variation in cuticular hydrocarbons. Numerous studies have shown that nestmate recognition processes are influenced by experience (Errard, 1986; Morel et al., 1988; Carlin and Hölldobler, 1983; Morel and Blum, 1988), although a component part of their recognition abilities may not be learned (Isingrini et al., 1985). Vander Meer et al. (1989b) make the point that because of the dynamic change in colony odor (the environmental component) workers in all likelihood must be able to update continually what constitutes their species and colony recognition template, including any changes in cuticular hydrocarbon profiles.

Even given these caveats, it is not unreasonable to predict that cuticular hydrocarbons (or other cuticular-borne lipids) will prove to be of some importance in the establishment and maintenance of social parasitism. Currently, only limited experimental efforts have been directed toward testing this hypothesis. Franks et al. (1990), as discussed above, have approached the problem from the context of the parasite acquiring the host label by grooming, but did not discuss the possibility that the parasite might be capable of biosynthesizing the requisite cuticular hydrocarbons (or cuticular fatty acids) without having to obtain them from the hosts. Vander Meer and Wojcik [cited in Wojcik (1989)] have reported that the social parasite *Solenopsis (Labauchena) daguerrei* has a cuticular hydrocarbon profile that is essentially identical to that of the parasitized host queen, *Solenopsis richteri*. Although they present no chemical data per se (or any gas chromatography traces), their wording implies that the parasite queen has some cuticular hydrocarbon components that do not match those of the *S. richteri* queen.

Direct evidence on the role of cuticular hydrocarbons in social parasitism has been provided by three research groups. Yamoaka (1990) has compared both chemically and behaviorally the hydrocarbons from the slave-maker ant *Polyergus samurai* and two of its hosts, *Formica japonica* and *Formica sp*. 5. He found that *P. samurai* always had the same cuticular hydrocarbon profile as whichever host it was associated with (and each host had distinctive cuticular hydrocarbon profiles). He further found that if *P. samurai* which had been cultured with *Formica sp*. 5 was then transferred to a colony of *F. japonica*, that within 10 days the cuticular hydrocarbon profile of *P. samurai* now matched that of its new host. When Yamoaka (1990) analyzed *P. samurai* workers either two or 14 days old which had been held isolated from their hosts, he found that their hydrocarbon profile was very larval-like (almost exclusively n-alkanes and then in low abundance). He concluded from this that the slave-maker acquired its adult hydrocarbons primarily by allogrooming of its hosts. It is also possible that the slave-maker biosynthesizes adult-type cuticular hydrocarbons and that this biosynthesis is triggered by interactions with its slaves. Such a hypothesis requires that the slave-maker have a set of genetic instructions that allows for multiple host phenotypes, and that host-specific cues then trigger specific genetic instructions to turn on one gene set and not another. A test of this hypothesis could possibly be done by placing newly eclosed slave-makers into a host colony where they could receive host odors but not host allogrooming. The alternative hypothesis (allogrooming) could best be tested by labeling host workers in such a way that they had radioactive cuticular hydrocarbons and showing that the slave-makers acquired the host label in a time-dependent manner.

Kaib et al. (1993) have conducted similar studies with the slave-maker *Harpagoxenus sublaevis* and two of its hosts, *Leptothorax acervorum* and *L. muscorum*, and found essentially the same results as Yamoaka (1990). They found that *L. acervorum* had 11% n-alkanes, 39% branched hydrocarbons, and 59% alkenes; *L. muscorum*, in contrast, had 28% n-alkanes, 72% branched alkanes, and only traces of alkenes. They further found that colonies of either *Leptothorax* sp. without the slave-maker had the same species-specific profiles. *H. sublaevis* had the same profile as whichever host it was with, and when it had both hosts in a common nest, then all three species ended up with a blended profile (Kaib et al., 1993). These authors also favored the hypothesis of allogrooming to explain their findings. In contrast to both Yamoaka (1990) and Kaib et al. (1993), Habersetzer and Bonavita-Cougourdan (1993) and Habersetzer (1993) have reported that the slave-maker *Polyergus rufescens*, and its slave *Formica rufibarbis* do not have identical cuticular profiles, nor did the free-living populations of *F. rufibarbis* have identical profiles to the enslaved populations. These researchers, however, were not analyzing only cuticular hydrocarbons, but rather total hexane extracts of the ants with no subsequent purification. In subsequent studies these authors have shown by GC-MS that the slave and slave-maker share a great number of hydrocarbon components, but that each species also possesses some components that are not present in the other (personal communication, A. Bonavita-Cougourdan).

In preliminary studies with a single colony, we have found that raiding workers of the slave-maker *Polyergus breviceps* have nearly identical cuticular hydrocarbon profiles to the slave workers with which they were collected (*Formica podzolica*). A total analysis of these cuticular hydrocarbons has not been completed, but the major components are a series of monoenes (C23:1 to C27:1) and a prominent C25 diene, with the next most abundant components being the corresponding n-alkanes. Much smaller quantities of monomethylalkanes are also present in both species. We have no experimental information whatsoever regarding the biosynthetic source of the hydrocarbons on the slave-maker and slave, but we would not be surprised to find that each species produced its own hydrocarbons. In our opinion, the effort to analyze the cuticular hydrocarbons from a set of species representing each of the social parasitism types would be well worth the trouble and would, we believe, lead to a much greater understanding of the semiochemical mechanisms operating to maintain these host-parasite relationships.

6. Conclusions and Future Research Needs

Many, if not most, arthropods use semiochemicals in one or more of the manners that have been discussed. Our understanding of the semiochemicals involved, and the ways in which they are used, has greatly increased in the last quarter of a century, but we have still just barely scratched the surface of this subdiscipline of chemical ecology. Indeed, much of the early literature needs to be reexamined in the light of current theory and technology, and new studies need to be conducted with these classic systems as well as with many new systems. Many of the concepts set forth were predicated solely on laboratory studies, and the bioassays used frequently did not include ecologically relevant organisms. In many instances a single population was used, and no effort was made to investigate the effects of age, gender, life stage, or habitat on the postulated semiochemical function.

Cuticular semiochemistry is at the center of many, if not most, of the topics examined in this chapter. Cuticular hydrocarbons have been the chemicals most studied in this regard, but they are only one of many possible cuticular lipid classes that are found on the surface of arthropods (Lockey, 1988; Buckner, 1993; Howard, 1993). A need exists to isolate individual lipid classes and to bioassay them for semiochemical function, both singly and as mixtures. We realize the enormity of the task involved, but in the absence of such knowledge it is likely to be impossible to understand the true role of the various cuticular-borne semiochemicals in crypsis, camouflage, or social parasitism.

In a similar manner, the relative importance of glandular-derived semiochemicals versus cuticular chemicals will remain unknown unless comparative studies are made. Such studies should focus not only on the larger, more chemically rich glands (such as Dufour's or mandibular glands), but also on the many specialized exocrine glands found in various myrmecophiles and termitophiles. Although the products of these specialized glands of inquilines have been most frequently postulated to be involved with social integration with their hosts, it is possible that they are also used for semiochemical communication either with conspecifics (sex pheromones for example) or for communication with nonhost heterospecifics (defensive allomones against predatory mites or other predators). The development of relevant bioassays to identify the semiochemical functions of these various glandular products may well be as challenging as the chemical identifications.

As challenging as these studies will be, they are relatively easy compared to the studies that will be required to finally chemically characterize "nest odors." Dynamic vapor sampling methods (headspace analysis), coupled with sensitive GC-MS offers real hope of establishing whether such odors really exist. Such headspace samples must, of course, be compared to chemicals present on the insect cuticle and must be shown to elicit appropriate behavioral responses.

The above considerations have primarily dealt with methodological issues. Of equal, if not greater need, is the issue of comparative studies with a diversity of taxa. As we noted earlier, the concept of slavery in social insects has almost exclusively been considered the provenance of ant researchers, but bees and wasps have evolved very similar lifestyles and deserve attention. Indeed, the similarities and differences among these various social parasites should allow for definitive clarifications of many difficult issues. Likewise, although the literature on inquilines of social insects is abundant, much of it is taxonomic in nature, with only limited behavioral or ecological information. Substantial gains in our understanding of how such inquilines have evolved could be gained from detailed comparative analyses of selected taxa using a combination of behavioral, biochemical, ecological, and physiological approaches. The ways in which animals use propaganda-type semiochemicals is undoubtedly much more complex than the examples we have presented. Further gains in this area will require an open mind and a willingness by researchers to look for these types of phenomena in many ecological settings. Such examinations are most likely to be fruitful if conducted in the field, rather than in controlled laboratory settings. Indeed, we believe that a major need of all of chemical ecology is to refocus on the ecological side of the problem so as to put the chemical findings into a biologically meaningful context.

Acknowledgments

We thank M. Blum, B. Hölldobler, J. Heinze, D. Kistner, H. Reed, W. Tschinkel, and K. Yeargan for reviewing an earlier draft of this chapter; K. Yeargan for providing a prepublication copy of his review of bolas spiders; K. F. Haynes for providing a photograph of a bolas spider; and J. Krchma for the SEM of the *Microdon* larvae. We thank E. Myhre for assistance in obtaining references and in preparing figures. All programs and services of the U. S. Department of Agriculture are offered on a nondiscriminatory basis without regard to race, color, national origin, religion, sex, age, marital status, or handicap.

References

Adams, E. A. and Traniello, J. F. A. (1981). Chemical interference competition by *Monomorium minimum* (Hymenoptera: Formicidae). *Oecologia* 51: 265–270.

Akerman, C. (1923). A comparison of the habits of a South African spider, *Cladomelea*, with those of an Australian *Dicrostichus*. *Ann. Natal Mus.* 5: 83–88.

Akre, R. D. (1968). The behavior of *Euxenister* and *Pulivinister*, histerid beetles associated with army ants (Coleoptera: Histeridae: Hymenoptera: Formicidae: Doryline). *Pan-Pacific Entomol.* 44: 87–101.

Akre, R. D. (1982). The social wasps, In *Social Insects*, vol. 4, (Herman, H. R., ed.), pp. 1–105 Academic Press, New York.

Akre, R. D. and Rettenmeyer, C. W. (1966). Behavior of Staphylinidae associated with army ants (Formicidae: Ecitonini). *J. Kansas Entomol. Soc.* 39: 745–782.

Akre, R. D. and Torgerson, R. L.. (1968). The behavior of *Diploeciton nevermanni*, a staphylinid beetle associated with army ants. *Psyche* 75: 211–215.

Akre, R. D. and Torgerson, R. L. (1969). Behavior of *Vatesus* beetles associated with army ants (Coleoptera: Staphylinidae). *Pan-Pacific Entomol.* 45: 269–281.

Akre, R. D., Alpert, G., and Alpert, T. (1973). Life cycle and behavior of *Microdon cothurnatus* in Washington (Diptera: Syrphidae). *J. Kans. Entomol. Soc.* 46: 327–338.

Akre, R. D., Garnett, W. B., and Zack, R. S. (1988). Biology and behavior of *Microdon piperi* in the Pacific Northwest (Diptera: Syrphidae). *J. Kans. Entomol. Soc.* 61: 441–452.

Alcock, J. and Forsyth, A. (1988). Post-copulatory aggression towards their mates by males of the rove beetle *Leistotrophus versicolor* (Coleoptera: Staphylinidae). *Behav. Ecol. Sociobiol.* 22: 303–308.

Alford, D. V. (1975). *Bumblebees*. Davis-Poynter, London.

Ali, M. F., Morgan, E. D., Attygalle, A. B., and Billen, J. P. J. (1987). Comparison of Dufour gland secretions of two species of *Leptothorax* ants (Hymenoptera: Formicidae). *Z. Naturforsch.* 42c: 955–960.

Allies, A. B., Bourke, A. F. G., and Franks, N. R. (1986). Propaganda substances in the cuckoo ant *Leptothorax kutteri* and the slave-maker *Harpagoxenus sublaevis*. *J. Chem. Ecol.* 12: 1285–1293.

Alloway, T. (1979). Raiding behavior of two species of slave-making ants, *Harpagoxenus americanus* (Emery) and *Leptothorax duloticus* (Wesson) (Hymenoptera: Formicidae). *Anim. Behav.* 27: 202–210.

Alloway, T. M. (1990). Slave-species ant colonies recognize slavemakers as enemies. *Anim. Behav.* 39: 1218–1220.

Andersen, A. N., Blum, M. S., and Jones, T. H. (1991). Venom alkaloids in *Monomorium "rothsteini"* Forel repel other ants: Is this the secret to success by *Monomorium* in Australian ant communities? *Oecologia* 88: 157–160.

Arn, H., Toth, M., and Priesner, E. (1992). *List of the Sex Pheromones of Lepidoptera and Related Attractants*, 2nd Ed., International Organization for Biological Control, Montfavet, France.

Attygalle, A. B., Billen, J. P. J., and Morgan, E. D. (1985). The postpharyngeal glands of workers of *Solenopsis geminata* (Hym. Formicidae). *Actes Coll. Insectes Soc.* 2: 79–86.

Bagnères, A.-G. and Morgan, E. D. (1991). The postpharyngeal glands and cuticle of Formicidae contain the same characteristic hydrocarbons. *Experientia* 47: 106–111.

Bagnères, A.-G., Errard, C., Mulheim, C., Joulie, C., and Lange, C. (1991). Induced mimicry of colony odor in ants. *J. Chem. Ecol.* 17: 1641–1664.

Baker, T. C. (1989). Sex pheromone communication in the Lepidoptera: new research progress. *Experientia* 45: 248–262.

Baroni Urbani, C. and Kannowski, P. B. (1974). Patterns in the red imported fire ant settlement of Louisiana pasture: Some demographic parameters, interspecific competition and food sharing. *Environ. Entomol.* 3: 755–760.

Bergström, G. and Löfquist, J. (1968). Odour similarities between the slave-keeping ants

Formica sanguinea and *Polyergus rufescens* and their slaves *Formica fusca* and *Formica rufibarbis*. *J. Insect Physiol.* 14: 995–1011.

Bernardi, R., Cardani, C., Ghiringhella, D., Selva, A., Baggini, A., and Pavan, M. (1967). On the components of secretion of mandibular glands of the ant *Lasius (Dendrolasius) fuliginosus*. *Tetrahedron Lett.* 40: 3893–3896.

Bian, Z., Fales, H. M., Blum, M. S., Jones, T. H., Rinderer, T. E., and Howard, D. F. (1984). Chemistry of cephalic secretion of fire bee *Trigona (Oxytrigona) tataira*. *J. Chem. Ecol.* 10: 451–460.

Blum, M. S. (1966). Chemical releasers of social behavior. VIII. Citral in the mandibular gland secretion of *Lestrimelitta limao* (Hymenoptera: Apoidea: Melittidae). *Ann. Entomol. Soc. Am.* 59: 962–964.

Blum, M. S. (1969). Alarm pheromones. *Annu. Rev. Entomol.* 14: 57–80.

Blum, M. S. (1981). *Chemical Defenses of Arthropods*. Academic Press, New York.

Blum, M. S. (1984). Alarm pheromones, In *Comprehensive Insect Physiology, Biochemistry and Pharmacology*, Vol. 9, (Kerkut, G. A. and Gilbert, L. I., eds.), pp. 193–224. Pergamon, Oxford, U.K.

Blum, M. S. (1985). Alkaloidal ant venoms: Chemistry and biological activities, In *Bioregulators for Pest Control,* (Hedin, P. A., ed.), pp. 393–408, American Chemical Society, Washington, D.C.

Blum, M. S., and Edgar, A. L. (1971). 4-methyl-3-heptanone: identification and role in opilionid exocrine secretions. *Insect Biochem.* 1: 181–188.

Blum, M. S., Jones, T. H., Hölldobler, B., Fales, H. M., and Jaouni, T. (1980). Alkaloidal venom mace: Offensive use by a thief ant. *Naturwissenschaften* 67: 144–145.

Blum, M. S., Crewe, R. M., Kerr, W. E., Keith, L. H., Garrison, A. W., and Walker, M. M. (1970). Citral in stingless bees: isolation and function in trail laying and robbing. *J. Insect Physiol.* 16: 1637–1648.

Bohart, G. S. (1970). The evolution of parasitism among bees. Forty-first Honor Lecture, The Faculty Association, Utah State Univ., Ogden, Utah.

Bonavita-Cougourdan, A., Clément, J.-L., and Lange, C. (1987). Nestmate recognition: the role of cuticular hydrocarbons in the ant *Camponotus vagus* Scop. *J. Entomol. Sci.* 22: 1–10.

Bonavita-Cougourdan, A., Clément, J.-L., and Lange, C. (1989). The role of cuticular hydrocarbons in recognition of larvae by workers of the ant *Camponotus vagus*: changes in the chemical signature in response to social environment (Hymenoptera: Formicidae). *Sociobiology* 16: 49–74.

Bonavita-Cougourdan, A., Clément, J.-L., and Poveda, A. (1990). Les hydrocarbures cuticulaires el les processus de reconnaissance chez les Fourmis: Le code d'information complexe de *Camponotus vagus*. *Actes Coll. Insectes Soc.* 6: 273–280.

Bourke, A. F. G. and Franks, N. (1991). Alternative adaptations, sympatric speciation and the evolution of parasitic, inquiline ants. *Biol. J. Linn. Soc.* 43: 157–178.

Bruch, C. (1930). Notas preliminares acera de *Labauchena daguerrei* Santschi. *Rev. Soc. Entomol. Argent.* 3: 73–80.

Buckner, J. S. (1993). Cuticular polar lipids of insects, In *Insect Lipids: Chemistry, Biochemistry and Biology*, (Stanley-Samuelson, D. W. and Nelson, D. R., eds.), pp. 227–270, Univ. Nebraska, Lincoln, Neb.

Buschinger, A. (1974). Experimente und Beobachtung zur Gründung und Entwicklung neuer Sozietäten der sklavenhaltenden Ameise *Harpagoxenus sublaevis* (Nyl.). *Insectes Soc.* 21: 381–406.

Buschinger, A. (1986). Evolution of social parasitism in ants. *Tree* 1: 155–160.

Buschinger, A. (1990). Sympatric speciation and radiative evolution of socially parasitic ants-Heretic hypotheses and their factual background. *Z. Zool. Syst. Evol.* 28: 241–260.

Buschinger, A. and Maschwitz, U. (1984). Defensive behavior and defensive mechanisms in ants, In *Defensive Mechanisms in Social Insects*, (Herman, H. R., ed.), pp. 95–150, Praeger, New York.

Cane, J. H. and Tengö, J. (1981). Pheromonal cues direct mate-seeking behavior of male *Colletes cunicularis* (Hymenoptera: Colletidae). *J. Chem. Ecol.* 7: 427–436.

Carlin, N. F. and Hölldobler, B. (1983). Nestmate and kin recognition in interspecific mixed colonies of ants. *Science* 222: 1027–1029.

Carpenter, J. M. (1991). Phylogenetic relationships and the origin of social behavior in the Vespidae, In *The Social Behavior of Wasps* (Ross, K. G. and Matthews, R. W., eds.), pp. 7–32, Comstock/Cornell, Ithaca, New York.

Cederberg, B. (1977). Evidence for trail-marking in *Bombus terrestris* workers (Hymenoptera, Apidae). *Zoon* 5: 143–146.

Cederberg, B. (1979). Odour guided host selection in *Psithyrus* (Hym., Apidae). *Ent. Tidskr.* 100: 128–129.

Cederberg, B. (1983). The role of trail pheromones in host selection by *Psithyrus rupestris* (Hymenoptera: Apidae). *Ann. Entomol. Fenn.* 49: 11–16.

Cervo, R., Lorenzi, M. C., and Turillazzi, S. (1990). Nonaggressive usurpation of the nest of *Polistes biglumis bimaculatus* by the social parasite *Sulcopolistes atrimandibularis* (Hymenoptera: Vespidae). *Insectes Soc.* 37: 333–347.

Clausen, C. P. (1940). The oviposition habits of the Eucharidae (Hymenoptera). *J. Washington Acad. Sci.* 30: 504–516.

Clausen, C. P. (1941). The habits of the Eucharidae. *Psyche* 48: 57–69.

Clément, J.-L. (1986). Open and closed societies in *Reticulitermes* termites (Isoptera, Rhinotermitidae): Geographic and seasonal variations. *Sociobiology* 11: 311–323.

Coleman, C. (1976). Notes on a local fishing or bolas spider *Ordgarius monstrosus*. *North Queensland Nat.* 44: 2–4.

Cool-Kwait, E. and Topoff, H. (1984). Raid organization and behavioral development in the slave-making ant *Polyergus lucidus* Mayr. *Insectes Soc.* 31: 361–374.

Dettner, K. and Liepert, C. (1994). Chemical mimicry and camouflage. *Annu. Rev. Entomol.* 39: 129–154.

Dumpert, K. (1972). Alarmstoffrezeptoren auf der Antenne von *lasius fuliginosus* (Latr.) (Hymenoptera, Formicidae). *Z. vergl. Physiol.* 76: 403–425.

Eberhard, W. G. (1977). Aggressive chemical mimicry by a bolas spider. *Science* 198: 1173–1175.

Eberhard, W. G. (1980). The natural history and behavior of the bolas spider *Mastophora dizzydeani* sp. n. (Araneidae). *Psyche* 87: 143–169

Eberhard, W. G. (1981). The single line web of *Phoroncidia studo* Levi (Araneae: Theridiidae): A prey attractant? *J. Arachnol.* 9: 229–232.

Eisner, T., Hicks, K., Eisner, M., and Robson, D. S. (1978). "Wolf-in-sheep's-clothing" strategy of a predaceous insect larva. *Science* 199: 790–794.

Elgert, B. and Rosengren, R. (1977). The guest ant *Formicoxenus nitidulus* follows the scent trail of its wood ant host (Hymenoptera, Formicidae). *Mem. Soc. Fauna Flora Fenn.* 53: 35–38.

Errard, C. (1986). Role of early experience in mixed-colony recognition in the ants *Manica rubida* and *Formica selysi*. *Ethology* 72: 243–249.

Errard, C., Bagnères, A.-G., and Clément, J.-L. (1988). Les signaux chimiques de la reconnaissance interspecifique chez les fourmis. *Actes Coll. Insectes Soc.* 5: 285–292.

Escoubas, P. and Blum, M. S. (1990). The biological activities of ant-derived alkaloids, In *Applied Myrmecology: A World Perspective*, (Vander Meer, R. K., Jaffe, K., and Cedeno, A., eds.), pp. 482–489, Westview, Boulder, Colorado.

Fales, H. M., Jaouni, T. M., Schmidt, J. O., and Blum, M. S. (1980). Mandibular gland allomones of *Dasymutilla occidentalis* and other mutillid wasps. *J. Chem. Ecol.* 6: 895–903.

Field, J. (1992). Intraspecific parasitism as an alternative reproductive tactic in nest-building wasps and bees. *Biol. Rev.* 67: 79–126.

Fielde, A. M. (1903). Artificial mixed nests of ants. *Biol. Bull. Mar. Biol. Lab. Woods Hole* 5: 320–325.

Fielde, A. M. (1904). Power of recognition among ants. *Biol. Bull. Mar. Biol. Lab. Woods Hole* 7: 227–250.

Fielde, A. M. (1905). The progressive odor of ants. *Biol. Bull. Mar. Biol. Lab. Woods Hole* 10: 1–16.

Fisher, R. M. (1983a). Behavioral interactions between a social parasite, *Psithyrus citrinus* (Hymenoptera: Apidae), and its bumble bee hosts. *Proc. Entomol. Soc. Ontario* 114: 55–60.

Fisher, R. M. (1983b). Inability of the social parasite *Psithyrus ashtoni* to suppress ovarian development in workers of *Bombus affinis* (Hymenoptera: Apidae). *J. Kansas Entomol. Soc.* 56: 69–73.

Fisher, R. M. (1983c). Recognition of host nest odour by the bumble bee social parasite *Psithyrus ashtoni* (Hymenoptera: Apidae). *N. Y. Entomol. Soc.* 91: 503–507.

Fisher, R. M. (1984a). Dominance by a bumble bee social parasite (*Psithyrus citrinus*) over workers of its host (*Bombus impatiens*). *Anim. Behav.* 32: 303–304.

Fisher, R. M. (1984b). Evolution and host specificity: A study of the invasion success of a specialized bumble bee social parasite. *Can. J. Zool.* 62: 1641–1644.

Fisher, R. M. (1985). Evolution and host specificity: Dichotomous invasion success of *Psithyrus citrinus* (Hymenoptera: Apidae), a bumble bee social parasite in colonies of its two hosts. *Can. J. Zool.* 63: 977–981.

Fisher, R. M. (1987a). Temporal dynamics of facultative social parasitism in bumble bees (Hymenoptera: Apidae). *Anim. Behav.* 35: 1628–1636.

Fisher, R. M. (1987b). Queen-worker conflict and social parasitism in bumble bees (Hymenoptera: Apidae). *Anim. Behav.* 35: 1026–1036.

Fisher, R. M. and Sampson, B. J. (1992). Morphological specializations of the bumble bee social parasite *Psithyrus ashtoni* (Cresson) (Hymenoptera: Apidae). *Can. Entomol.* 124: 69–77.

Fisher, R. M. and Weary, D. M. (1988). Buzzing bees: Communication between bumble bee social parasites (Hymenoptera: Apidae) and their hosts. *Bioacoustics* 1: 3–12.

Fisher, R. M., Greenwood, D. R. and Shaw, G. J. (1993). Host recognition and the study of a chemical basis for attraction by cuckoo bumble bees (Hymenoptera: Apidae). *J. Chem. Ecol.* 19: 771–786.

Fletcher, D. J. C. and Michener, C. D. eds. (1987). *Kin Recognition in Animals*. Wiley and Sons, New York.

Forel, A. (1929). *The Social World of the Ants Compared with that of Man*, Vol. 1 and 2. Albert and Charles Boni, New York.

Franks, N., Blum, M. S., Smith, R.-K., and Allies, A. B. (1990). Behavior and chemical disguise of cuckoo ant *Leptothorax kutteri* in relation to its host *Leptothorax acervorum*. *J. Chem. Ecol.* 16: 1431–1444.

Free, J. B. (1987). *Pheromones of Social Bees*. Comstock, New York

Free, J. B., and Butler, C. G. (1959). *Bumblebees*. Macmillan , New York.

Garnett, W. B., Akre, R. D., and Sehlke, G. (1985). Chemical mimicry and predation by myrmecophilous Diptera (Diptera: Syrphidae). *Fla. Entomol.* 68: 615–621.

Gösswald, K. (1938). Grundsätzliches über parasitische Ameisen unter besonderer Berücksichtigung der Abhängigen Koloniegründung von *Lasius umbratus mixtus* Nyl. *Z. Wiss. Zool.* 151: 101–148.

Greene, A., Akre, R. D., and Landolt, P. J. (1978). Behavior of the yellowjacket social parasite, *Dolichovespula arctica* (Rohwer) (Hymenoptera: Vespidae). *Melanderia* 29: 1–28.

Greenslade, P. J. M. (1976). *A Guide to the Ants of South Australia*, South Australia Museum, Adelaide, Australia.

Greenslade, P. J. M. and Greenslade, P. (1989). Ground layer invertebrate fauna, In *Mediterrean Landscapes in Australia: Mallee Ecosystems and Their Management*, (Nobel, J. C. and Bradstock, R. A., eds.), pp. 266–284, CSIRO and ANU, Canberra, Australia.

Greenslade, P. J. M. and Halliday, R. B. (1983). Colony dispersion and relationships of meat ants *Iridomyrmex purpureus* (Hymenoptera: Formicidae) and allies in an arid locality in South Australia. *Insectes Soc.* 30: 82–99.

Habersetzer, C. (1993). Cuticular spectra and inter-individual recognition in the slave-making ant *Polyergus rufescens* and the slave species *Formica rufibarbis*. *Physiol. Entomol.* 18: 167–175.

Habersetzer, C. and Bonavita-Cougourdan, A. (1993). Cuticular spectra in the slave-making ant *Polyergus furicens* and the slave species *Formica rufibarbis*. *Physiol. Entomol.* 18: 160–166.

Hadley, N. F. (1977). Epicuticular lipids of the desert tenebrionid beetle, *Eleodes armata*: Seasonal and acclimatory effects on composition. *Insect Biochem.* 7: 277–283.

Hall, J. C. (1986). Mutants of biological rhythms and conditioned behavior in *Drosophila* courtship, In *Evolutionary Genetics of Invertebrate Behavior: Progress and Prospects*, (Huettel, M. D., ed.), pp. 237–246, Plenum, New York.

Haverty, M. I., Page, M., Nelson, L. J., and Blomquist, G. J. (1988). Cuticular hydrocarbons of dampwood termites, *Zootermopsis*: Intra- and intercolony variation and potential as taxonomic characters. *J. Chem. Ecol.* 14: 1035–1058.

Hefetz, A. and Lloyd, H. A. (1983). Identification of new components from anal glands of *Tapinoma simrothi* Pheonicium. *J. Chem. Ecol.* 9: 607–613.

Heinze, J. (1991). Biochemical studies on the relationship between socially parasitic ants and their hosts. *Biochem. Sys. Ecol.* 19: 195–206.

Heinze, J., Stuart, R. J., Alloway, T. M. and Buschinger, A. (1992). Host specificity in the slave-making ant *Harpagoxenus canadensis* M. R. Smith. *Can. J. Zool.* 70: 167–170.

Heraty, J. M. (1985). A revision of the nearctic Eucharitinae (Hymenoptera: Chalcidoidea: Eucharitidae). *Proc. Entomol. Soc. Ontario* 116: 61–103.

Hill, W. B., Akre, R. D., and Huber, J. D. (1976). Structure of some epidermal glands in the myrmecophilous beetle *Adranes taylori* (Coleoptera: Pselaphidae). *J. Kansas Entomol.* Soc. 49: 367–384.

Hölldobler, B. (1967). Zur Physiologie der Gast-Wirt-Beziehungen (Myrmecophilie) bei Ameisen. I. Das Gastverhältnis der *Atemeles-* und *Lomechusa* Larven (Col. Staphylinidae) zu *Formica* (Hym. Formicidae). *Z. vergl. Physiol.* 56: 1–21.

Hölldobler, B. (1970). Zur Physiologie der Gast-Wirt-Beziehungen (Myrmecophilie) bei Ameisen. II. Das Gastverhältnis des imaginalen *Atemeles pubicollis* Bris. (Col. Staphylinidae) zu *Myrmica* und *Formica* (Hym. Formicidae). *Z. vergl. Physiol.* 66: 176–189.

Hölldobler, B. (1971). Communication between ants and their guests. *Sci. Am.* 224: 86–91.

Hölldobler, B. (1972). Verhaltensphysiologische Adaptationen an ökologische Nischen in ameisennestern. *Verh. Dtsch. Zool. Ges.* 65: 137–143.

Hölldobler, B. (1973). Chemische strategie beim nahrungserwerb der Diebsameise (*Soleonopsis fugax* Latr.) und der Pharaoameise (*Monomorium pharaonis* L.). *Oecologia* 11: 371–380.

Hölldobler, B. (1982). Interference strategy of *Iridomyrmex pruinosum* (Hymenoptera: Formicidae) during foraging. *Oecologia* 52: 208–213.

Hölldobler, B. and Carlin, N. F. (1987). Anonymity and specificity in the chemical communication signals of social insects. *J. Comp. Physiol.* A 161: 567–581.

Hölldobler, B. and Markl, H. (1989). Notes on interspecific, mixed colonies in the harvester ant genus *Pogonomyrmex*. *Psyche* 96: 237–238.

Hölldobler, B. and Wilson, E. O. (1990). *The Ants*. Belknap/Harvard, Cambridge, Mass.

Hölldobler, B., Möglich, M., and Maschwitz, U. (1981). Myrmecophilic relationship of *Pella* (Coleoptera: Staphylinidae) to *Lasius fulgininosus* (Hymenoptera: Formicidae). *Psyche* 88: 347–374.

Howard, R. W. (1976). Observations on behavioral interactions between *Trichopsenius frosti* Seevers (Coleoptera: Staphylinidae) and *Reticulitermes flavipes* (Kollar) (Isoptera: Rhinotermitidae). *Sociobiology* 2: 189–192.

Howard, R. W. (1978). Proctodeal feeding by termitophilous Staphylinidae associated with *Reticulitermes virginicus* (Banks). *Science* 201: 541–543.

Howard, R. W. (1979). Mating behavior of *Trichopsenius frosti*: utilization of physogastric *Reticulitermes flavipes* queens as sexual aggregation centers. *Ann. Entomol. Soc. Am.* 72: 127–129.

Howard, R. (1980). Trail-following by termitophiles. *Ann. Entomol. Soc. Am.* 73: 36–38.

Howard, R. W. (1991). Guess who's coming to dinner? The behavioral and chemical ecology of termites and termitophiles. *Melandria* 47: 59–67.

Howard, R. W. (1993). Cuticular hydrocarbons and chemical communication, In *Insect Lipids: Chemistry, Biochemistry and Biology*, (Stanley-Samuelson, D. W. and Nelson, D. R., eds.), pp. 179–226, Univ. Nebraska, Lincoln, Neb .

Howard, R. W., Akre, R. D., and Garnett, W. B. (1990a). Chemical mimicry in an obligate predator of carpenter ants (Hymenoptera: Formicidae). *Ann. Entomol. Soc. Am.* 83: 607–616.

Howard, R. W., McDaniel, C. A., and Blomquist, G. J. (1978). Cuticular hydrocarbons of the eastern subterranean termite, *Reticulitermes flavipes* (Kollar) (Isoptera: Rhinotermitidae). *J. Chem. Ecol.* 4: 233–245.

Howard, R. W., McDaniel, C. A., and Blomquist, G. J. (1980). Chemical mimicry as an integrating mechanism: cuticular hydrocarbons of a termitophile and its host. *Science* 210: 431–433.

Howard, R. W., McDaniel, C. A., and Blomquist, G. J. (1982b). Chemical mimicry as an integrating mechanism for three termitophiles associated with *Reticulitermes virginicus* (Banks). *Psyche* 89: 157–167.

Howard, R. W., Stanley-Samuelson, D. W., and Akre, R. D. (1990b). Biosynthesis and chemical mimicry of cuticular hydrocarbons from the obligate predator, *Microdon albicomatus* Novak (Diptera: Syrphidae) and its ant prey, *Myrmica incompleta* Provancher (Hymenoptera: Formicidae). *J. Kansas Entomol. Soc.* 63: 437–443

Howard, R. W., Jones, S. C., Mauldin, J. K., and Beal, R. H. (1982a). Abundance, distribution, and colony size estimates for *Reticulitermes spp*. (Isoptera: Rhinotermitidae) in southern Mississippi. *Environ. Entomol.* 11: 1290–1293.

Howard, R. W., McDaniel, C. A., Nelson, D. R., Blomquist, G. J., Gelbaum, L. T., and Zalkow, L. H. (1982c). Cuticular hydrocarbons of *Reticulitermes virginicus* (Banks) and their role as potential species- and caste-recognition cues. *J. Chem. Ecol.* 8: 1227–1239.

Huber, P. (1810). *Recherches sur les Moeurs des Fourmis Indigènes*. Paschoud, Paris.

Hutchison, C. E. (1903). A bolas throwing spider. *Sci. Am.* 89: 172.

Isingrini, M., Lenoir, A., and Jaisson, P. (1985). Preimaginal learning as a basis of colony-brood recognition in the ant *Cataglyphis cursor*. *Proc. Natl. Acad. Sci.* USA 82: 8545–8547.

Jacobson, H. R., Kistner, D. H., and Pasteels, J. M. (1986). Generic revision, phylogenetic classification, and phylogeny of the termitophilous tribe Corotocini (Coleoptera: Staphylinidae). *Sociobiology* 12: 1–245.

Johnson, L. K. (1987). The pyrrhic victory of nest robbing bees: did they use the wrong pheromone? *Biotropica* 19: 188–189.

Jones, T. H., Blum, M. S., and Fales, H. M. (1982a). Ant venom alkaloids from *Solenopsis* and *Monomorium* species, recent developments. *Tetrahedron* 38: 1949–1958.

Jones, T. H., Blum, M. S., Escoubas, P., and Musthak, A. T. M. (1989). Novel pyrrolidines in the venom of the ant *Monomorium indicum*. *J. Nat. Prod.* 52: 779–784.

Jones, T. H., Meinwald, J., Hicks, K., and Eisner, T. (1977). Characterization and synthesis of volatile compounds from the defensive secretions of some "daddy longlegs" (Arachnida: Opiliones: *Leiobunum spp*.). *Proc. Natl. Acad. Sci.* USA 74: 419–422.

Jones, T. H., Blum, M. S., Andersen, A. N., Fales, H. M., and Escoubas, P. (1988). Novel 2-ethyl-5-alkylpyrrolidines in the venom of an Australian ant of the genus *Monomorium*. *J. Chem. Ecol.* 14: 35–45.

Jones, T. H., Conner, W. E., Kluge, A. F., Eisner, T., and Meinwald, J. (1976). Defensive substances of Opilionids. *Experientia* 32: 1234–1235.

Jones, T. H., Blum, M. S., Howard, R. W., McDaniel, C. A., Fales, H. M., DuBois, M. B., and Torres, J. (1982b). Venom chemistry of ants in the genus *Monomorium*. *J. Chem. Ecol.* 8: 285–300.

Jouvenaz, D. P., Vander Meer, R. K. and Wojcik, D. P. (1988). Chemical mimicry in a parasitoid, *Orasema sp*. (Hymenoptera: Eucharitidae) of fire ants. *Proc. 18th Intern. Congr. Entomol.*, Abstr., p. 243, International Congress of Entomology, Vancouver, British Columbia.

Kaib, M., Heinze, J. and Ortius, D. (1993). Cuticular hydrocarbon profiles of adults and pupae in the slave-making ant *Harpagoxenus sublaevis* and its hosts. *Naturwissenschaften* 80: 281–285.

Kistner, D. H. (1968). Revision of the myrmecophilous species of the tribe Myrmedonini.

part II. The genera *Aenictonia* and *Anommatochara*-Their relationships and behavior. *Ann. Entomol. Soc. Am.* 61: 971–986.

Kistner, D. H. (1969). The biology of termitophiles, In *Biology of Termites*, Vol. 1, (Krishna, K. and Weesner, F. M., eds.), pp. 525–557 Academic, New York.

Kistner, D. H. (1977). New species and records of the termitophilous genus *Schizelythron* from Malaya with a brief note on their behavior (Coleoptera: Staphylinidae, Trichopseniinae). *Sociobiology* 2: 297–304.

Kistner, D. H. (1979). Social and evolutionary significance of social insect symbionts, In *Social Insects*, Vol. 1, (Herman, H., ed.), pp. 340–413, Academic, New York.

Kistner, D. H. (1982). The social insects' bestiary, *In Social Insects*, Vol. 3, (Herman, H., ed.), pp. 1–244, Academic, New York.

Kistner, D. H. (1984). A review of the termitophilous genus *Termitodiscus* with an analysis of the relationships of the species and a review of their behavior and relationship to their termite hosts (Coleoptera: Staphylinidae; Isoptera: Termitidae). *Sociobiology* 8: 225–285.

Kistner, D. H. (1990). The integration of foreign insects into termite societies or why do termites tolerate foreign insects in their societies? *Sociobiology* 17: 191–215.

Kistner, D. H. (1993). Cladistic analysis, taxonomic restructuring and revision of the Old World genera formerly classified as Dorylomimini with comments on their evolution and behavior (Coleptera: Staphylinidae). *Sociobiology* 22: 151–374.

Kistner, D. H. and Blum, M. S. (1971). Alarm pheromone of *Lasius (Dendrolasius) spathepus* (Hymenoptera: Formicidae) and its possible mimicry by two species of *Pella* (Coleoptera: Staphylinidae). *Ann. Entomol. Soc. Am.* 64: 589–594.

Kistner, D. and Howard, R. W. (1980). First report of larval and pupal Trichopseniinae: external morphology, taxonomy and behavior. *Sociobiology* 5: 3–20.

Kistner, D. H. and Jacobson, H. R. (1975). A review of the myrmecophilous Staphylinidae associated with *Aenictus* in Africa and the Orient (Coleoptera; Hymenoptera, Formicidae) with notes on their behavior and glands. *Sociobiology* 1: 21–73.

Kistner, D. H. and Jacobson, H. R. (1990). Cladistic analysis and taxonomic revision of the ecitophilous tribe Ecitocharini with studies of their behavior and evolution (Coleoptera, Staphylinidae, Aleocharinae). *Sociobiology* 17: 333–465.

Kolbe, W. and Proske, M. G. (1973). Iso-Valeriansäure im Abwehrsekret von *Zyras humeralis* (Col. Staphylinidae) an Waldameisen. *Entomol. Blätter* 69: 57–60.

Kutter, H. (1969). Die sozialparasitischen Ameisen der Schweiz, *Viert. Naturf.Gesell. Zurich*, 113: 1–62.

LeMoli, F. and Mori, A. (1987). The problem of enslaved ant species: origin and behavior, In *From Individual to Collective Behavior in Social Insects*, (Pasteels, J. M. and Deneubourg, J.-L., eds.), pp. 333–363, Birkhäuser Verlag, Basel, Switzerland.

Lenoir, A., Detrain, C., and Barbazanges, N. (1992a). Host trail following by the guest ant *Formicoxenus provancheri*. *Experientia* 48: 94–97.

Lenoir, A., Errard, C., Francoeur, A., and Loiselle, R. (1992b). Relations entre la fourmi parasite *Formicoxenus provancheri* et son hôte *Myrmica incompleta*. Données biologiques et éthogiques (Hym. Formicidae). *Insectes Soc.* 39: 81–97.

Lloyd, J. E. (1979). Mating behavior and natural selection. *Fla. Entomol.* 62: 17–34.

Lloyd, J. E. (1984). On deception, a way of all flesh, and firefly signalling and systematics, In *Oxford Survey of Evolutionary Biology*, (Dawkins, R. and Ridley, M., eds.), pp. 48–84, Oxford Univ., Oxford, U.K..

Lloyd, H. A., Schmuff, N. R., and Hefetz, A. (1986). Chemistry of the anal glands of *Bothriomyrmex syrius* Forel: Olfactory mimetism and temporary social parasitism. *Comp. Biochem. Physiol.* 83B: 71–73.

Lockey, K. H. (1988). Lipids of the insect cuticle: origin, composition and function. *Comp. Biochem. Physiol.* 89B: 595–645

Lok, J. B., Cupp, E. W., and Blomquist, G. J. (1975). Cuticular lipids of the imported fire ants, *Solenopsis invicta* and *richteri*. *Insect Biochem.* 5: 821–829.

Longman, H. A. (1922). The magnificent spider: *Dicrostichus magnificus* Rainbow. Notes on cocoon spinning and method of catching prey. *Proc. Roy. Soc. Queensland* 33: 91–98.

Lorenzi, M. C., Cervo, R., and Turillazzi. (1992). Effects of social parasitism of *Polistes atrimandibularis* on the colony cycle and brood production of *Polistes biglumis bimaculatus* (Hymenoptera, Vespidae) *Boll. Zool.* 59: 267–271.

Marlin, J. C. (1968). Notes on a new method of colony formation employed by *Polyergus lucidus* Mayr. *Illinois State Acad Sci.* 61: 207–209.

Mascord, R. (1970). *Australian Spiders in Colour.* Reed Books, Sydney, Australia.

McGurk, D. J., Frost, J., Waller, G. R., Eisenbraun, E. J., Vick, K., Drew, W. A., and Young, J. (1968). Iridodial isomer variation in dolichoderines. *J. Insect Physiol.* 14: 841–845.

McKeown, K. C. (1952). *Australian Spiders, Sirius Books*, (1963 rev. by N. L. Roberts), Angus and Robertson, Sydney, Australia.

McMahan, E. A. (1982). Bait-and-capture strategy of a termite-eating assassin bug. *Insectes Soc.* 29: 346–351.

McMahan, E. A. (1983). Adaptations, feeding preferences, and biometrics of a termite-baiting assassin bug (Hemiptera: Reduviidae). *Ann. Entomol. Soc. Am.* 76: 483–486.

McNeil, J. N. (1991). Behavioral ecology of pheromone-mediated communication in moths and its importance in the use of pheromone traps. *Annu. Rev. Entomol.* 36: 407–430.

Meinwald, J., Kluge, A. F., Carrel, J. E., and Eisner, T. (1971). Acyclid ketones in the defensive secretions of a daddy longlegs (*Leiobunum vittatum*) (Arachnida/Opiliones). *Proc. Natl. Acad. Sci. USA* 68: 1467–1468.

Michener, C. D. (1974). *The Social Behavior of the Bees.* Belknap/Harvard, Cambridge, Mass.

Mitter, C. and Klun, J. A. (1987). Evidence of pheromonal constancy among sexual and asexual females in a population of the fall cankerworm *Alsophila pometaria* (Lepidoptera: Geometridae). *J. Chem. Ecol.* 13: 1823–1830.

Moore, B. P. (1969). Biochemical studies in termites, In *Biology of Termites*, Vol. 1, (Krishna, K. and Weesner, F. M., eds.), pp. 407–432, Academic, New York.

Morel, L. and Blum, M. S. (1988). Nestmate recognition in *Camponotus floridanus* worker ants: are sisters or nestmates recognized? *Anim. Behav.* 36: 718–725.

Morel, L., Vander Meer, R. K., and Lavine, B. K. (1988). Ontogeny of nestmate recognition cues in the red carpenter ant (*Camponotus floridanus*): Behavioral and chemical evidence for the role of age and social experience. *Behav. Ecol. Sociobiol.* 22: 175–183.

Moritz, R. F. A. (1991). Kin recognition in honeybees: experimental artefact or biological reality?, In *The Behaviour and Physiology of Bees*, (Goodman, L. J. and Fisher, R. C., eds.), pp. 48–59, C. A. B. International, London, U.K.

Moure, J. S. (1946). Sobre a abelha tatairo ou cagafogo. *Chacaras Quintais* 73: 612–613.

Nelson, D. R., Fatland, C. L., Howard, R. W., McDaniel, C. A., and Blomquist, G. J. (1980). Re-analysis of the cuticular methylalkanes of *Solenopsis invicta* and *S. richteri*. *Insect Biochem.* 10: 409–418.

Nogueiro-Neto, P. (1970). Behavior problems related to the pillages made by some parasitic stingless bees (Meliponinae, Apidae), In *Essays in Memory of T. C. Schneirla*, (Aronson,R., Tobach, E., Lehrman, D. S. and Rosenblatt, J. S., eds.), pp. 416–464, Freeman, San Francisco.

Obin, M. S. (1986). Nestmate recognition cues in laboratory and field colonies of *Solenopsis invicta* Buren (Hymenoptera: Formicidae). Effect of environment and role of cuticular hydrocarbons. *J. Chem. Ecol.* 12: 1965–1975.

Odhiambo, T. R. (1958a). The camouflaging habits of *Acanthaspis petax* Stål (Hem., Reduviidae) in Uganda. *Entomol. Mon. Mag.* 94: 47.

Odhiambo, T. R. (1958b). Some observations on the natural history of *Acanthaspis petax* Stål (Hemi., Reduviidae) living in termite mounds in Uganda. *Proc. R. Entomol. Soc. London (A)* 33: 167–175.

Oliveira, P. S. and Sazima, I. (1985). Ant-hunting behaviour in spiders with emphasis on *Strophius nigricans* (Thomisidae). *Bull. Br. Arachnol. Soc.* 6: 309–312.

Ollett, D. G., Morgan, E. D., Attygalle, A. B., and Billen, J. P. J. (1987). The contents of the Dufour gland of the ant *Harpagoxenus sublaevis* Nyl. (Hymenoptera: Formicidae). *Z. Naturforsch.* 42c: 141–146.

Peschke, K. (1983). Defensive and pheromonal secretion of the tergal gland of *Aleochara curtula* II. Release and inhibition of male copulatory behavior. *J. Chem. Ecol.* 9: 13–31.

Peschke, K. (1985). Immature males of *Aleochara curtula* avoid intrasexual aggression by producing the female sex pheromone. *Naturwissenschaften* 72: 274–275.

Peschke, K. (1987a). Male aggression, female mimicry and female choice in the rove beetle, *Aleochara curtula* (Coleoptera: Staphylinidae). *Ethology* 75: 265–284.

Peschke, K. (1987b). Cuticular hydrocarbons regulate mate recognition, male aggression, and female choice of the rove beetle, *Aleochara curtula*. *J. Chem. Ecol.* 13: 1993–2008.

Peschke, K. and Metzler, M. (1987). Cuticular hydrocarbons and female sex pheromones of the rove beetle, *Aleochara curtula* (Goeze) (Coleoptera: Staphylinidae). *Insect Biochem.* 17: 167–178.

Plath, O. E. (1934). *Bumblebees and Their Ways*. Macmillan, New York.

Provost, E. (1985). The closure of ant societies. I. Analysis of interactions involved in experimental encounters between worker ants of the same species (genus *Leptothorax* or *Camponotus lateralis*) but from different societies. *Insectes Soc.* 32: 445–462.

Quilico, A., Piozzi, F. and Pavan, M. (1957). The structure of dendrolasin. *Tetrahedron* 1: 177–185.

Regnier, F. E. and Wilson, E. O. (1971). Chemical communication and "propaganda" in slave-maker ants. *Science* 172: 267–269.

Rinderer, T. E., Blum, M. S., Fales, H. M., Bian, Z., Jones, T. H., Buco, S. M., Lancaster, V. A., Danka, R. S., and Howard, D. F. (1988). Nest plundering allomones of the fire bee *Trigona (Oxytrigona) mellicolor*. *J. Chem. Ecol.* 14: 495–501.

Roubik, D. W., Smith, B. H., and Carlson, R. G. (1987). Formic acid in caustic cephalic secretions of stingless bee, *Oxytrigona* (Hymenoptera: Apidae). *J. Chem. Ecol.* 13: 1079–1086.

Sakagami, S. F., Roubik, D. W., and Zucchi, R. (1993). Ethology of the robber stingless bee, *Lestrimelitta limao* (Hymenoptera: Apidae). *Sociobiology* 21: 237–277.

San Martin, P. R. (1968). Notas sobre *Fustiger elegans* Raffray (Coleoptera, Pselaphidae) en el Uruguay y la Argentina. *Physis* 28: 59–64.

Santschi, F. (1920). Fourmis du genre "*Bothriomyrmex*" Emery (Systematique et moeurs). *Rev. Zool. Africaine*. 7: 201–224.

Schröder, W. (1985). Vergleichende Untersuchungen flüchtiger Inhaltsstoffe von solitären und sozialen Bienen. Ph. D. Dissertation. Fachbereich Chemie, Univ. Hamburg, Germany.

Shower, J. A. and Kistner, D. H. (1977). The natural history of the myrmecophilous tribe Pygostenini (Coleoptera: Staphylinidae). Section 4. Glandular anatomy of the Pygostenini. *Sociobiology* 2: 305–321.

Silveira Guido, A., San Martin, P., Crisci-Pisano, C., and Carbonell-Bruhn, J. (1963). Investigations on the biology and biological control of the fire ant, *Solenopsis saevissima richteri* Forel in Uruguay. Second and third reports. Dept. Sanidad Veg., Fac. Agron., Univ. Republic, Montevideo, Uruguay.

Spicer, R. M. and Kistner, D. H. (1980). The behavior and exocrine gland system of *Termitoides marginatus* (Coleoptera, Staphylinidae) with a redescription of the genus and species. *Sociobiology* 5: 215–247.

Stäger, R. (1925). Das Leben der Gastameise (*Formicoxenus nitidulus* Nyl.) in neuer Beleuchtung. *Z. Morph. Ökol. Tiere* 3: 452–476.

Stejskal, M. (1962). Duft als 'Sprache' der tropischen Bienen. *Südwestdeut. Imker* 49: 271.

Stowe, M. K. (1986). Prey specialization in the Araneidae, In *Spiders: Webs, Behavior and Evolution*, (Shear, W. A., ed.), pp. 101–131, Stanford Univ., Stanford, Calif.

Stowe, M. K. (1988). Chemical mimicry, In *Chemical Mediation of Co-evolution*, (Spencer, K. C., ed.), pp. 513–580, Academic, New York.

Stowe, M. K., Tumlinson, J. H. and Heath, R. R. (1987). Chemical mimicry: Bolas spiders emit components of moth prey species sex pheromones. *Science* 236: 964–967.

Stuart, R. J. (1984). Experiments on colony foundation in the slave-making ant *Harpagoxenus canadensis* M. R. Smith (Hymenoptera: Formicidae). *Can. J. Zool.* 62: 1995–2001.

Talbot, M. (1968). Flights of the ant *Polyergus lucidus* Mayr. *Psyche* 75: 46–52.

Toolson, E. C. and Hadley, N. F. (1977). Cuticular permeability and epicuticular lipid composition in two Arizona vejovid scorpions. *Physiol. Zool.* 50: 323–330.

Topoff, H. (1990). Slave-making ants. *Am. Sci.* 78: 520–528.

Topoff, H., Cover, S., Greenberg, L., Goodloe, L., and Sherman, P. (1988). Colony founding by queens of the obligatory slave-making ant. *Polyergus breviceps*: the role of Dufour's gland. *Ethology* 78: 209–218.

Torgerson, R. L. and Akre, R. D. (1969). Reproductive morphology and behavior of a Thysanuran, *Trichatelura manni,* associated with army ants. *Ann. Entomol. Soc. Am.* 62: 1367–1374.

Trave, R. and Pavan, M. (1956). Veleni degli insetti. Principi estratti dalla formica *Tapinoma nigerrimum* Nyl. *Chem. Ind. Milano* 38: 1015–1019.

Turillazzi, S. (1992). Nest usurpation and social parasitism in *Polistes* wasps: New acquisitions and current problems, In *Biology and Evolution of Social Insects*, (Billen, J. ed.), pp. 263–272, Leuven Univ., Leuven, Belgium.

Vander Meer, R. K. and Wojcik, D. P. (1982). Chemical mimicry in a myrmecophilous beetle *Myrmecaphodius excaviticollis*. *Science* 218: 806–808.

Vander Meer, R. K. and Lofgren, C. S. (1990). Chemotaxonomy applied to fire ant systematics in the United States and South America, In *Applied Myrmecology, a world perspective*, (Vander Meer, R. K., Jaffe, K. and Cedeno, A., eds.), pp. 75–84, Westview, Boulder, Colo.

Vander Meer, R. K., Jouvenaz, D. P., and Wojcik, D. P. (1989a). Chemical mimicry in a parasitoid (Hymenoptera: Eucharitidae) of fire ants (Hymenoptera: Formicidae). *J. Chem. Ecol.* 15: 2247–2261.

Vander Meer, R. K., Obin, M. S., and Morel, L. (1990). Nestmate recognition in fire ants: monogyne and polygyne populations, In *Applied Myrmecology, a world perspective*, (Vander Meer, R. K., Jaffe, K., and Cedeno, A., eds.), pp. 322–328, Westview, Boulder, Colo.

Vander Meer, R. K., Saliwanchik, D., and Lavine, B. (1989b). Temporal changes in colony cuticular hydrocarbon patterns of *Solenopsis invicta*. Implications for nestmate recognition. *J. Chem. Ecol.* 15: 2115–2125.

Wheeler, W. M. (1907). The polymorphism of ants, with an account of some singular abnormalities due to parasitism. *Bull. Amer. Mus. Nat. Hist.* 23: 1–93.

Wheeler, W. M. (1928). *The Social Insects: Their Origin and Function*. Harcourt, Brace, New York.

Wheeler, G. C. and Wheeler, E. H. (1937). New hymenopterous parasites of ants (Chalcidoidea: Eucharidae). *Ann. Entomol. Soc. Am.* 30: 163–173.

Williams, R. N. (1980). Insect natural enemies of fire ants in South America with several new records. *Proc. Tall Timbers Conf.* 7: 123–134.

Wilson, E. O. (1975). Slavery in ants. *Sci. Am.* 232: 32–36.

Wittmann, D. (1985). Aerial defense of the nest by workers of the stingless bee *Trigona (Tetragonisca) angustula* (Latreille) (Hymenoptera: Apidae). *Behav. Ecol. Sociobiol.* 16: 111–114.

Wittmann, D., Radtke, R., Zeil, J., Lubke, G., and Francke, W. (1990). Robber bees (*Lestrimelitta limao*) and their host: Chemical and visual cues in nest defense by *Trigona (Tetragonisca) angustula* (Apidae: Meliponinae). *J. Chem. Ecol.* 16: 631–641.

Wojcik, D. P. (1975). Biology of *Myrmecaphodius excavaticollis* (Blanchard) and *Euparia castanea* Serville (Coleoptera: Scarabaeidae) and their relationships to *Solenopsis* spp. (Hymenoptera: Formicidae), Ph. D. Dissertation, Univ. of Florida, Gainesville.

Wojcik, D. P. (1980). Fire ant myrmecophiles: Behavior of *Myrmecosaurus ferrugineus* Bruch (Coleoptera: Staphylinidae) with comments on its abundance. *Sociobiology* 5: 63–68.

Wojcik, D. P. (1989). Behavioral interactions between ants and their parasites. *Fla. Entomol.* 72: 43–51.

Wojcik, D. P. (1990). Behavioral interactions of fire ants and their parasites, predators and inquilines, In *Applied Myrmecology, a world perspective*, (Vander Meer, R. K., Jaffe, K., and Cedeno, A., eds.), pp. 329–344, Westview, Boulder, Colo.

Woodroffe, G. E. (1958). The mode of reproduction of *Ptinus clavipes panzer* from *mobilis* Moore (=*P. latro* auct.) (Coleoptera: Ptinidae). *Proc. R. Entomol. Soc. London Ser. A* 33: 25–30.

Yamaoka, R. (1990). Chemical approach to understanding interactions among organisms. *Physiol. Ecol. Japan* 27 (special no.): 31–52.

Yeargan, K. V. (1988). Ecology of a bolas spider, *Mastophora hutchinsoni*: phenology, hunting tactics and evidence for aggressive chemical mimicry. *Oecologia* 74: 265–284.

Yeargan, K. V. (1994). Biology of bolas spiders, Annu. Rev. Entomol. 39: 81–99.

Zalkow, L. H., Howard, R. W., Gelbaum, L. T., Gordon, M. M., Deutsch, H. M., and Blum, M. S. (1981). Chemical ecology of *Reticulitermes flavipes* (Kollar) and *R. virginicus* (Banks) (Rhinotermitidae): Chemistry of the soldier cephalic secretions. *J. Chem. Ecol.* 7: 717–731.

Subject Index

alarm pheromone 366–372, 398–402
allogrooming 406–408
allomones 187–188, 320–321, 350–352, 364–408
alpha conditioning 72–73, 76
anemotaxis (attraction) 23–5, 31–32, 71, 112–120, 166–174, 191, 297–299, 372–375, 378
appeasement chemicals 389–393
associative learning 49–55, 68–69,70–73, 79–81, 90
attraction (amenotaxis) 23–5, 31–32, 71, 112–120, 166–174, 191, 297–299, 372–375, 378
aversion learning 52–55

beta conditioning 72–73, 76
biosynthesis 184–187, 334–336, 378, 382,384, 389, 408
Brindley's glands 320

camouflage (olfactory) 375–393
chemotaxis 105, 120–122, 171, 191, 247, 371
circadian (diel) rhythm 24, 26, 28, 31, 32, 124–125, 375
coevolution 163–164, 188–191
commensalism 351
conditioning 72–73, 76

defensive behaviors 262–263, 318–321, 364–410
diel (circadian) rhythm 24, 26, 28, 31, 32, 124–125, 375
dorsal gland 319–321, 328,334, 336, 339, 342
Dufour's gland 243–244, 247–250, 252, 254, 256, 294, 296, 299–301, 310, 399–401

egg load 91–92
experience 22, 66, 70–71, 74, 76–78, 81, 407

feeding 27–29, 47–61, 155–157, 176–178, 216–218, 224–225, 388
frontal gland 367

glands
  Brindley's glands 320
  dorsal gland 319–321, 328,334, 336, 339, 342
  Dufour's gland 243–244, 247–250, 252, 254, 256, 294, 296, 299–301, 310, 399–401
  frontal gland 367
  metaplural gland 243
  metathoracic gland 318–323, 344, 338–340, 342, 344
  Pavan's gland 248
  poison gland 243, 248, 250–252, 254, 400
  postpharyngeal gland 406
  pygidial gland 243–244, 249, 260, 400
  sternal gland 243–244, 249–250, 257, 262–263, 321–323, 336
  venom gland 248
genetics (heritability) 128–138, 163–165, 296

habituation 47–48, 68, 80, 106
heritability (genetics) 128–138, 163–165, 296
host finding 29–35, 165–176, 396–398
host races 164–165

induction 55–58
instrumental learning 68, 81

kairomone 75, 80, 82, 88–91, 344–352, 371
kleptoparasitism 351
klinotaxis 247–248

landing 7, 18, 23–25, 30, 122–123, 161, 167, 170
learning 49–61, 66–93, 268–269, 291, 294, 296–297, 302, 407
  alpha conditioning 72–73, 76
  associative 49–55, 68–69,70–73, 79–81, 90
  aversion 52–55
  beta conditioning 72–73, 76
  experience 22, 66, 70–71, 74, 76–78, 81, 407
  habituation 47–48, 68, 80, 106
  induction 55–58
  instrumental 68, 81
  sensitization 48–49, 68, 79
leks 344

metaplural gland 243
metathoracic gland 318–323, 344, 338–340, 342, 344
mimicry 375–393, 405–408

olfactory reception 136, 160–163, 216
oviposition 18, 218–224, 388

parasitoids 34–35, 65–93, 344–352
Pavan's gland 248
perching 106–197
pheromone
  alarm 366–372, 398–402
  queen 306, 400
  sex 24, 82, 84–85, 87, 92, 106–140, 154–193, 251, 318, 323–352, 372, 374–375, 377
  territorial 242–274
  trail 242–274, 397–398
physiological state 18, 73, 90–91
poison gland 243, 248, 250–252, 254, 400
polarized light 9–10
postpharyngeal gland 406
pygidial gland 243–244, 249, 260, 400

queen pheromone 306, 400

ranging 107–108
recruitment 246–274, 368–369

sensitization 48–49, 68, 79
sex pheromone 24, 82, 84–85, 87, 92, 106–140, 154–193, 251, 318, 323–352, 372, 374–375, 377
slavery 393–408
sternal gland 243–244, 249–250, 257, 262–263, 321–323, 336
synomone 70, 79, 190

tandem running 257, 259–260
territorial pheromone 242–274
territory 176
trail pheromone 242–274, 397–398
trophic levels 67, 87–88
trophallaxis 386–387, 395
tropotaxis 248

venom gland 248
visual responses 9–10, 18, 30–32, 50–51, 82, 84, 109–110, 109–110, 158–159, 219–224, 255, 344, 370, 396

Taxonomic Index

Arthropods

Acanthaspis petax 376
Acromyrmex 252
Acrosternum 332, 348
Acrosternum aseadum 332
Acrosternum hilare 332
Acrosternum marginatum 332
Acrosternum pennsylvanicum 332, 348
Adranes taylori 386–387
Aedes aegyptii 30–32, 118
Aelia acuminata 321
Agrotis segetum 11, 134
Alcaeorrhynchus 323
Aleochara curtula 377
Amblypelta lutescens 338
Amblyseius potentillae 109
Ambylopone australis 259
Amitermes 251
Amitermes unidentatus 251–252
Amorbia cuneana 134
Anacyptus testaceus 381–382
Anax imperator 125
Aneuretus 260
Aneuretus simoni 259
Anisandrus dispar 175
Aphaenogaster 252, 261
Aphantochilus rogersi 377
Apis 292, 310, 369
Apis mellifera 11, 287–288, 290, 304, 306–309, 371
Argyrotaenia velutinana 127, 131–132
Arilus critatus 323
Arma 323
Artogeia 215
Atemeles 389, 392
Atemeles pubicollis 389–390
Atta 251–252, 256, 260
Atta cephalotes 252, 255
Atta laevigata 256
Atta texana 252
Azteca 262
Azteca chartifex 257
Azteca trigona 256–257

Biprorulus bibax 321, 323, 328. 334, 341–342
Blaberus craniifer 108
Blastophagus 155
Blatella germanica 110, 124
Boisea 339
Bombus 292, 394, 397
Bombus affinis 397
Bombus bimaculatus 397
Bombus lapidarious 397
Bombus occidentalis 256
Bombus terricola 397
Bombyx mori 119, 251
Bothriomyrmex 395, 403
Bothriomyrmex syrius 402
Brachymeria intermedia 79, 81, 87
Bracon mellitor 76

Cadra cautella 119
Callosamia promethea 55
Campoletis sonorensis 35–36
Camponotus 260, 262, 383, 406
Camponotus beebi 257
Camponotus femoratus 257
Camponotus ligniperda 368

Camponotus modoc 383, 386–387
Camponotus pennsylvanicus 249, 255
Camponotus planatus 368
Camponotus socius 246
Camponotus vagus 406
Campyloma verbasci 342
Cataglyphis 258, 259
Celaenia 372
Cermatulus 330
Christoneura fumierana 110, 125–128
Christoneura occidentalis 110
Chrysopa slossonae 375–376
Cidaria albulata 111
Cladomelea 374
Coleophora laricella 122
Colias 217, 228
Colias eurytheme 138, 216
Colias philodice 138, 216
Coptotermes formosanus 251
Cotesia rubecula 22, 78
Cotinus nitida 110
Crematogaster 257
Cyclocephala lurida 137

Daceton armigerum 249–250, 252
Dacnusa sibirica 35, 37
Dacus 21
Dacus dorsalis 3
Delia antiqua 32–34
Delia brassicae 118
Dendroctonus 154, 157, 162–163, 170–171, 180, 183, 187, 189, 191
Dendroctonus brevicomis 155–156, 158, 162, 166–167, 175, 179–182, 186–191
Dendroctonus frontalis 154, 162–164, 175, 178–179, 181–183, 190
Dendroctonus jeffreyi 180
Dendroctonus micans 8, 169
Dendroctonus ponderosae 154, 159, 162, 164, 166, 168, 175, 177–178, 180–182, 189
Dendroctonus pseudotsugae 155, 162, 165–166, 175, 179, 181, 190
Dendroctonus rufipennis 162
Dendroctonus simplex 162
Dendroctonus terebrans 164
Dermestes ater 113
Dermestes maculatus 111
Diacrisia virginica 54
Dialictus 294
Dinorhynchus 330
Dipetalogaster maximus 320
Diplorhoptrum 368
Diplorhoptrum fugax 368
Distantiella theobroma 342–343
Dolichovespula 292
Dolichovespula adulterina 386–387, 394–395
Dolichovespula alpicola 394
Dolichovespula arctica 395
Dolichovespula arenaria 394
Dolichovespula norwegica 394
Drosophila 57, 75–76, 85, 89, 108, 118, 378
Drosophila melanogaster 26–27, 85
Drosophila simulans 85
Dryocoetes 162, 169
Dysdercus 340
Dysdercus bimaculatus 340
Dysdercus cingulatus 340
Dysdercus intermedius 340
Dysdercus intermedius 341
Dysdercus koenigii 340

Eciton 262, 268–269
Eciton burchelli 264, 268–269
Eciton hamatum 264, 268
Eciton hamatus 386–387
Eciton rapax 264, 268
Ectatomma ruidum 250
Elasmostethus cruciatus 347
Enoclerus lecontei 190
Epiphyas postvittana 23, 122
Estigmene acrea 375
Estigmene congrua 54
Eucera palestinae 256
Euclytia flava 334, 345–346, 350
Euphydryas editha 21
Eurygaster intergriceps 336
Euschistus 334, 340, 347
Euschistus conspersus 332
Euschistus heros 332, 333
Euschistus ictericus 332
Euschistus politus 332
Euschistus servus 332
Euschistus tristigmus 321–332, 347
Euthyrhynchus 330
Euxenister 393
Evylaeus 294

Fannia canicularis 81
Forcipomyia crinita 334, 348
Forelius 368
Formica 260, 383, 389, 395–397, 400, 404–405, 407
Formica fusca 246, 368, 401, 404
Formica gnava 400

Formica japonica 407
Formica obscuripes 384–385
Formica pergandei 404
Formica peripilosa 368
Formica podzolica 386–387, 408
Formica pratensis 397
Formica rufibarbis 401, 404, 408
Formica sanguinea 404
Formica schaufussi 244
Formica selysi 405
Formica strumosa 390
Formica subintegra 404
Formica subsericea 404–405
Formicoxenus 395
Formicoxenus nitidulus 397
Formicoxenus provancheri 397–398
Fustiger elegans 386–388

Geotrupes 108
Glossina 29–32, 119
Gnathotrichus 155
Gnathotrichus retusus 162
Gnathotrichus sulcatus 162
Grapholita molesta 25, 114–115, 121, 127, 130

Hadrobonus 366
Harpagoxenus americanus 399
Harpagoxenus canadensis 398
Harpagoxenus sublaevis 398–399, 401–402, 408
Harpocera thoracica 344
Helicoverpa zea 75, 79–80
Heliothis virescens 36
Helopeltis clavifer 342
Hemyda aurata 334, 345–346, 350
Hesperotettix viridis 54
Hetaerius tristriatus 386–387
Hodotermes 243
Hodotermes mossambicus 244, 247, 251, 255
Holomelina lamae 81
Hotea gambiae 336
Hyalophora cecropia 106
Hylastes 155
Hylemya antiqua 118
Hylurgops 171
Hylurgops palliatus 174, 175
Hymenolepis diminuta 125
Hyphantria cunea 375
Hypothenemus 155

Ips 155, 157–158, 163, 170, 183–184, 187, 189, 191–192
Ips amitinus 162, 184
Ips avulsus 162–163, 175
Ips calligraphus 162–164
Ips cembrae 162
Ips confusus 155
Ips duplicatus 162, 187–188
Ips grandicollis 155, 162–163, 168, 175
Ips paraconfusus 120, 155, 158, 162, 166–167, 170–171, 175–176, 178–179, 181, 184–190, 192
Ips pini 134, 137,155, 160–162, 165, 175, 188
Ips sexdentatus 165, 169
Ips typographus 155–159, 162, 166, 168–169, 175, 184, 187–188, 190
Iridomyrmex 368–369
Iridomyrmex humilis 248

Jadera 340
Jadera haematoloma 340

Kaira 372

Labauchena daguerrei 407
Lasioglossum 292, 294, 310
Lasioglossum malachurum 256, 298, 300–301
Lasioglossum zephyrum 287–288, 290, 293, 295, 299, 310
Lasius 260
Lasius flavus 217, 368
Lasius fuliginosus 246, 249, 391–392
Lasius neoniger 244, 249–250, 254, 256, 260–261, 271
Lasius niger 395
Lasius spathepus 392
Lasius umbratus 395
Laspeyresia pomonella 122
Leiobunum 366
Leiophron 351
Leistotrophus 378
Leptinotarsa decemlineata 329
Leptocoris 339
Leptogenys 259–260
Leptogenys chinensis 249
Leptogenys diminuta 249
Leptoglossus 338
Leptoglossus australis 338
Leptoglosus phyllopus 338
Leptopilina boulardi 89
Leptopilina fimbriata 89
Leptopilina heterotoma 75–78, 85, 89
Leptothorax 401–402

Leptothorax acervorum 398–399, 401–402, 408
Leptothorax affinis 255
Leptothorax kutteri 398–399, 402
Leptothorax muscorum 408
Leptothorax nylanderi
Leptothorax unifasciatus 255
Lestrimellita 370–371
Lestrimellita limao 370–371
Ligurotettix coquilletti 54
Ligyrocoris 341
Linepithema humile 248, 271
Liometopum apiculatum 265, 271
Liriomyza bryoniae 37
Locusta migratora 29–30, 48
Lomechusa 389, 391–392
Lomechusa strumosa 390–391
Lygus 342
Lygus elisus 344
Lygus hesperus 344
Lygus lineolaris 344
Lymantria dispar 81, 87, 110, 114, 116–118, 122–125, 126–127

Macrotermes 376
Macrotermes michaelseni 254
Mamestra brassicae 48
Manduca sexta 11, 54–56
Manica 251
Manica rubida 405
Martinezia dutertrei 384
Mastophora 372, 374–375
Mastophora bisaccata 375
Mastophora cornigera 374–376
Mastophora hutchinsoni 372–373, 375
Mastophora phrynosoma 375
Mastotermes 243–251
Mayetiola destructor 7, 19–20, 106
Medetera aldrichii 190
Medetera bistriata 190
Megaponera foetens 259
Melanoplus sanguinipes 50–51
Melipona 369
Meranoplus 368
Messor 251
Messor pergandei 268
Microdon 383–384, 389
Microdon albicomatus 383
Microdon cothurnatus 384–385
Microdon piperi 383–385
Microplitis croceipes 74–76, 79–80, 89–91
Mineus 321
Mirperus scutellaris 323, 325, 339
Monacis debilis 257
Moniliformis moniliformis 125
Monomorium 368
Monomorium minimum 261–262
Monomorium pharaonis 249
Monomorium rothsteini 369
Monomorium viridum 265–266, 271
Musca domestica 81
Muscidifurax zaraptor 81
Myrmecaphodius excavaticollis 384
Myrmecosaurus 386, 388
Myrmecosaurus ferrugineus 386
Myrmica 248, 251, 389, 395
Myrmica incompleta 383, 397
Myrmica ruba 250, 368
Myrmica sabuleti 244, 271
Myrmica scabrinodis 250

Nasutitermes 244, 262, 376
Nasutitermes costalis 249–250, 262–263
Nasutitermes exitiosus 250–251
Nasutitermes gravelosus 251
Nasutitermes kempai 252
Nasutitermes walkeri 251
Neivamyrmex 262
Neoponera apicalis 259
Neostruma 259
Nephila clavipes 351
Nezara 332, 348, 352
Nezara antennata 332
Nezara viridula 321, 330–334, 347–348, 351
Niesthrea louisanica 339
Novomessor 252, 261

Ocymyrmex 259
Odontotermes 251, 254
Oechalia 323
Oecophylla longinoda 256
Oecophylla smaragdina 256
Olcella 349
Oncopeltus 341
Oncopeltus fasciatus 323, 341
Onychomyrmex 259
Oplomus 321
Orasema 386, 388
Ordgarius 374
Orectognathus versicolor 259
Ostrinia nubilalis 135–136
Oxycarenus hyalinipennis 341

Pachycondyla apicalis 259
Pachycondyla laevigata 244, 259–260

Pachycondyla obscuricornis 259–260
Pachycondyla tesserinoda 255
Paltothyreus tarsatus 259
Papilio 230
Papilio polyxenes 217
Paraponera clavata 255, 259
Pectinophora gossypiella 127, 130, 132–133, 138–139
Pella 386, 389, 391–393
Pella comes 392
Pella japonicus 392
Perillus 321
Perillus bioculatus 322
Perillus bioculatus 329
Periplaneta americana 107, 111, 115, 119–121, 124–125
Peristenus 351
Peristenus pseudopallipes 351
Peristenus stygicus 351
Pheidole 251, 253, 260
Pheidole dentata 262
Pheidole hyatti 265, 271
Pheidole pallidula 248, 271
Philotermes howardi 372–373, 382
Phormia regina 107
Phoroncidia studo 372
Phyllomyza securicornis 349
Picromerus 330
Pieris 214, 216, 218–220, 225, 227, 230–231
Pieris brassicae 55, 87, 214–219,221–225, 227–230
Pieris napi 215, 217, 222, 224–225, 227–229
Pieris rapae 19, 50, 109, 214, 216–225, 227–230
Pieris virginiensis 227, 229
Pilophorus perplexus 344
Pimpla instigator 87
Pityogenes 155, 157–158, 170, 187
Pityokteines curvidens 162
Pityophthorus 155, 162
Pityophthorus chalcographus 158, 162, 166, 172, 175
Plodia interpunctella 114
Podisus 323, 329–330, 334, 338, 345, 347–350
Podisus connexivus 328
Podisus flava 334
Podisus fretus 328, 334–345, 348–349
Podisus maculiventris 324, 326–328, 334, 344–347, 349–350
Podisus macuronatus 329
Podisus neglectus 328
Podisus nigrispinus 328
Podisus placidus 329
Pogonomyrmex 249, 252, 254, 256, 260
Pogonomyrmex babius 250, 254–255
Pogonomyrmex barbatus 250, 259
Pogonomyrmex maricopa 259
Pogonomyrmex occidentalis 400
Pogonomyrmex rugosus 250, 254, 259
Polistes 288, 292, 301–302, 308, 395
Polistes atrimandibularis 395
Polistes biglumis 395
Polistes dominulus 243
Polistes exclamans 302
Polistes metricus 256, 302
Polistes semenowi 395
Polistes sulcifer 395
Polybia sericea 243
Polyergus 396–397
Polyergus breviceps 400–401, 408
Polyergus rufescens 401, 408
Polyergus samurai 407
Prionopelta amabilis 259
Pristhesancus papuensis 342
Pristhesancus plagipennis 323, 342
Pristomyrmex pugens 249
Prociphilus tesselatus 376
Pseudacanthotermes 251
Pseudaletia unipuncta 48, 127
Pseudomyrmex 243
Psithyrus 394, 396
Psithyrus ashtoni 397
Psithyrus citrinus 397
Psithyrus rupestris 397
Pyrrhocoris apterus 340

Redivius personatus 372–373
Reticulitermes 244, 251–252,380–381
Reticulitermes flavipes 250, 367, 381–382
Reticulitermes lucifugus 252
Reticulitermes speratus 251
Reticulitermes virginicus 251, 372–373, 381–382
Rhagoletis pomonella 50
Rhizophagus grandis 8
Riptortus 339
Riptortus clavatus 339
Riptortus serripes 323, 325, 339

Salyavata variegata 376
Schedorhinotermes 251
Schedorhinotermes lamanianus 247, 251–252, 262

Schistocerca americana 52–54, 59
Schistocerca gregaria 28–29, 48, 51, 59
Scolytus 155, 158, 171
Scolytus multistriatus 155, 162, 166, 177
Scolytus quadrispinosus 167, 169
Scolytus rugulosus 177
Scolytus ventralis 168, 180, 182
Smithistruma 259
Solenopsis 244, 248, 266, 367, 384, 386, 388
Solenopsis daguerrei 407
Solenopsis fugax 368
Solenopsis germinata 243
Solenopsis invicta 252, 367, 386, 406
Solenopsis richteri 388, 407
Sphaerocoris annulus 323, 324, 336
Spilostethus rivularis 341
Spodoptera littoralis 56
Stiretrus
Stiretrus anchorango 329
Strophius nigricans 377
Strumigenys 259
Sulcopolistes 395
Supella longipalpa 124–125
Supputius 323
Supputius cincticeps 328

Taczanowskia 372
Taeniopoda eques 59
Tapinoma 395, 403
Tatochila distincta 215
Tectocoris diophthalmus 321, 336
Teleutomyrmex schneideri 396
Telonomus clavus 349
Telonomus podisi 349
Temnochila virescens 190–191
Tenebrio molitor 125
Termitopone laevigata 259
Tetradonia 393
Tetradonia marginalis 386–387
Tetramorium 251
Tetramorium caespitum 248, 252, 265, 271, 396
Tetramorium guineense 252
Tetramorium impurum 249, 252
Thanasimus formicarius 190–191
Tinacantha 323
Tomicus 155, 170–171, 183, 187
Tomicus minor 157, 162
Tomicus piniperda 155–159, 163, 166, 168–175, 179, 190
Trachymyrmex 252
Trachymyrmex septentrionalis 252
Tribolium castaneum 134
Trichogramma minutum 34
Trichogramma spp. 70, 84
Trichoplusia ni 56, 122, 134, 136
Trichopoda 347
Trichopoda giacomellii 348
Trichopoda pennipes 348
Trichopsenius depressus 372–373, 381–382
Trichopsenius frosti 381–382
Trigona 243, 369–370
Trigona angustula 370–371
Trigona mellicolor 370
Trinervitermes 250–251
Trinervitermes bettonianus 250–252, 254
Trinervitermes germinatus 252
Trinervitermes gratiosus 251–252
Trinervitermes occidentalis 252
Trinervitermes togoensis 252
Trissolcus basalis 349, 351–352
Trissolcus euschistii 349
Trogoderma granarium 137
Trogoderma variabile 110–111, 115, 120
Trypodendron 171, 191
Trypodendron domesticum 175–176
Trypodendron lineatim 162, 165–166, 172

Vespula acadica 394
Vespula austriaca 386–387, 394
Vespula vulgaris 329

Xenistusa hexagonalis 372–373, 381–382
Xenodusa reflexa 386–387
Xyleborus 155
Xylocopa 256
Xylocopa pubescens 256

Yponomeuta padellus 129
Yponomeuta rorellus 130

Zacryptocerus maculatus 257
Zicrona 323
Zootermopsis 244, 246
Zygaena filipendulae 123

Plants, Fungi, Bacteria

Abies concolor 176
Abies grandis 182
Abies magnifica 176
Acer rubrum 326
Acer saccharinum 326
Aesculus octandra 177

Alliaria petiolata 227, 229
Alyssum saxatile 224
Astragalus 215

Bacillus cereus 175, 187
Barbarea vulgaris 220, 229
Bellis perennis 29
Betula 175
Betula pendula 174
Brassica 55, 230, 231
Brassica juncea 219
Brassica oleracea 217
Brassuca juncea 224
Bunias orientalis 214

Capparis spinosa 230
Capsella bursa-pastoris 220
Carica papaya 220
Carya ovata 167, 177
Ceratocystis 156, 182
Ceratocystis ips 182
Ceratocystis minor 182–183
Ceratocystis pilifera 182
Ceratocystis schrenkiana 182
Cheiranthus allionii 220, 223
Cleome spinosa 224
Conringia orientalis 224
Crataegus phaenopyrum 326

Daucus carota 326

Erysimum asperum 225, 227
Erysimum cheiranthoides 220–226

Fagus sylvatica 175
Fomes annosus 182

Gleophyllum trabeum 251

Iberis amara 220–226
Isatis tinctoria 220–221, 224, 230, 231

Larix occidentalis 179
Ligustrum amurense 326
Lunaria annua 224

Magnolia X soulangiana 326
Moricandia arvensis 230

Ophiostoma 156

Pedicularis 215
Phomopsis oblonga 177
Picea abies 165
Pimpinella anisum 306
Pinus contorta 168, 177, 181
Pinus echinata 179
Pinus elliotii 179
Pinus jeffreyi 180
Pinus lambertiana 176
Pinus palustris 179
Pinus ponderosa 180
Pinus sabiniana 182, 185–186
Pinus sylvestris 168, 172, 179
Pinus taeda 177, 179, 181–183
Populus deltoides 177
Populus tremula 174
Pseudotsuga menziesii 176, 181

Quercus 175
Quercus alba 167

Salix discolor 326
Sambucus canadensis 326
Senecio vulgaris 30
Sinapis arvensis 217

Trichosporium 156
Tropaeolum 55
Tropaeolum majus 219, 221, 224

Ulmus americana 177